Digital Video
Image Quality
and
Perceptual
Coding

Signal Processing and Communications

Digital Video Image Quality and Perceptual Coding

edited by
H.R. Wu and K.R. Rao

Taylor & Francis
Taylor & Francis Group
Boca Raton   London   New York

A CRC title, part of the Taylor & Francis imprint, a member of the
Taylor & Francis Group, the academic division of T&F Informa plc.

Published in 2006 by
CRC Press
Taylor & Francis Group
6000 Broken Sound Parkway NW, Suite 300
Boca Raton, FL 33487-2742

© 2006 by Taylor & Francis Group, LLC
CRC Press is an imprint of Taylor & Francis Group
No claim to original U.S. Government works
Printed in the United States of America on acid-free paper
10 9 8 7 6 5 4 3 2

International Standard Book Number-10: 0-8247-2777-0 (Hardcover)
International Standard Book Number-13: 978-0-8247-2777-2 (Hardcover)
Library of Congress Card Number 2005051404

Cover image of Claude E. Shannon reprinted with permission of Lucent Technologies Inc/Bell Labs.

Library of Congress Cataloging-in-Publication Data

Digital video image quality and perceptual coding / edited by Henry R. Wu, K.R. Rao.
    p. cm. -- (Signal processing and communications)
   Includes bibliographical references and index.
   ISBN 0-8247-2777-0
   1. Digital video. 2. Imaging systems--image quality. 3. Perception. 4. Coding theory. 5. Computer vision. I. Wu, Henry R. II. Rao, K. Ramamohan (Kamisetty Ramamohan) III. Series.
TK6680.5.D55 2006
006.6'96--dc22
                                           2005051404

Visit the Taylor & Francis Web site at
http://www.taylorandfrancis.com

and the CRC Press Web site at
http://www.crcpress.com

To those who have pioneered, inspired and persevered.

Copyrights release for ISO/IEC:

About Shannon image on the front cover:

The original image of Claude E. Shannon, the father of information theory, was provided by Bell Laboratories of Lucent Technologies. It was compressed using the JPEG coder that is information lossy. Its resolution was 4181×5685 pixels with white margins around all sides. This digital image had "scanning dust."

The original image was cropped down to 4176×5680 for compression and the "scanning dust" was removed by copying the surrounding pixels over the dust. The resultant image was used as the "original" image to produce the compressed Shannon image using the perceptual lossless image coder (PLIC) as described in Chapter 13 with an implementation intended for medical image compression. The PLIC was benchmarked against the JPEG-LS and the JPEG-NLS (d=2).

The coding error images were produced between the cropped original Shannon image and the compressed images. In the error images, the red color represents a positive error, the blue a negative error and the white a zero error. The black color is not actually black, it is small valued red or blue color. Comparing the two error images, it can be appreciated how the PLIC uses the human vision model to achieve perceptual lossless coding with a higher compression ratio than the benchmarks, whilst maintaining the visual fidelity of the original picture.

The top image is the "original" image that can be compressed to 3.179 bpp using the JPEG-LS. The mid-right image is the JPEG-NLS (i.e., JPEG near lossless) compressed at d=2 with Bitrate = 1.424 bpp, Compression Ratio = 5.6180:1, MSE = 1.9689 and PSNR = 45.1885 dB. The mid-left image is the difference image between the "original" and that compressed by the JPEG-NLS (d=2). The bottom-right is the PLIC compressed with Bitrate = 1.370 bpp, Compression Ratio = 5.8394:1, MSE = 2.0420 and PSNR = 45.0303 dB. The bottom-left image is a difference image between the "original" and that compressed by the PLIC.

Contributors

Alan C. Bovik, University of Texas at Austin, Austin, Texas, U.S.A.

Jorge E. Caviedes, Intel Corporation, Chandler, Arizona, U.S.A.

Tao Chen, Panasonic Hollywood Laboratory, Universal City, California, U.S.A.

François-Xavier Coudoux, Université de Valenciennes, Valenciennes, Cedex, France.

Philip J. Corriveau, Intel Media and Acoustics Perception Lab, Hillboro, Oregon, U.S.A.

Mark D. Fairchild, Rochester Institute of Technology, Rochester, New York, U.S.A.

Marc G. Gazalet, Université de Valenciennes, Valenciennes, Cedex, France.

Jae Jeong Hwang, Kunsan National University, Republic of Korea.

Michael Isnardi, Sarnoff Corporation Inc., Princeton, New Jersey, U.S.A.

Ryoichi Kawada, KDDI R&D Laboratories Inc., Japan.

Weisi Lin, Institute for Infocomm Research, Singapore.

Jeffrey Lubin, Sarnoff Corporation Inc., Princeton, New Jersey, U.S.A.

Makoto Miyahara, Japan Advanced Institute of Science and Technology, Japan.

Ethan D. Montag, Rochester Institute of Technology, Rochester, New York, U.S.A.

Franco Oberti, Philips Research, The Netherlands.

Albert Pica, Sarnoff Corporation Inc., Princeton, New Jersey, U.S.A.

K. R. Rao, University of Texas at Arlington, Arlington, Texas, U.S.A.

Hamid Sheikh, Texas Instruments, Inc., Dallas, Texas, U.S.A.

Damian Marcellinus Tan, Royal Melbourne Institute of Technology, Melbourne, Victoria, Australia.

Zhou Wang, University of Texas at Arlington, Arlington, Texas, U.S.A.

Stefan Winkler, Genista Corporation, Montreux, Switzerland.

Hong Ren Wu, Royal Melbourne Institute of Technology, Melbourne,
       Victoria, Australia.

Zhenghua Yu, National Information Communication Technology Australia (NICTA).

Michael Yuen, ESS Technology, Inc., Beijing, China

Jian Zhang, National Information Communication Technology Australia (NICTA).

Acknowledgments

The editors, H. R. Wu and K. R. Rao, would like to thank all authors of this handbook for their contributions, efforts and dedication without which this book would not have been possible.

The editors and the contributors have received assistance and support from many of our colleagues that has made this handbook, *Digital Video Image and Perceptual Coding*, possible. The generous assistance and support includes valuable information and materials used in and related to the book, discussions, feedback, comments on and proof reading of various parts of the book, recommendations and suggestions that shaped the book as it is. Special thanks are due to the following persons:

M. Akgun	Communications Research Center, Canada
J. F. Arnold	Australian Defence Force Academy
B. Baxter	Intel Corporation
J. Cai	Nanyang Technological University
N. Corriveau	Spouse of P. Corriveau
S. Daly	Sharp Laboratories of America
M. Frater	Australian Defence Force Academy
N. G. Kingsbury	University of Cambridge
L. Lu	IBM T. J. Watson Research Center
Z. Man	Nanyang Technological University
S. K. Mitra	University of California, Santa Barbara
K. N. Ngan	The Chinese University of Hong Kong
E. P. Simoncelli	New York University
C.-S. Tan	Royal Melbourne Institute of Technology
A. Vincent	Communications Research Center, Canada
M. Wada	KDDI R&D Laboratories
B. A. Wandell	Stanford University
S. Wolf	Institute for Telecommunication Sciences
D. Wu	Royal Melbourne Institute of Technology
C. Zhang	Nanyang Technological University
Z. Zhe	Monash University
All members	VQEG

A. C. Bovik acknowledges the support by the National Science Foundation under grant CCR-0310973.

H. R. Wu and Z. Yu acknowledge the support by Australian Research Council under grant A49927209.

Assistance and support to this book project which H. R. Wu received from Monash University where he lectured from 1990 to 2005, and from Nanyang Technological University where he spent his sabbatical from 2002 to 2003, are gratefully acknowledged.

Special thanks go to David Wu of Royal Melbourne Institute of Technology for his assistance in producing the final LATEX version of this handbook and the compressed Shannon images shown on the front cover of the book.

H. R. Wu and K. R. Rao would like to express their sincere gratitude to B. J. Clark, publishing consultant at CRC Press LLC, who initiated this book project and whose professional advice and appreciation of efforts which were involved in this undertaking made the completion of this book possible. Sincere thanks also go to Nora Konopka, our Publisher, and Jessica Vakili, our Project Coordinator, at CRC Press LLC for their patience, understanding and unfailing support that helped see this project through. We are most grateful to our Project Editor, Susan Horwitz, whose professional assistance has made significant improvement to the book's presentation. The work by Nicholas Yancer on the back cover of the book and the brilliant cover design by Jonathan Pennell are greatly appreciated.

Last but not least, without the patience and forbearance of our families, the preparation of this book would have been impossible. We greatly appreciate their constant and continuing support and understanding.

Preface

The outset of digital video image coding research is commonly acknowledged [Cla95] to be around 1950, marked by Goodall's paper on television by pulse code modulation (or PCM) [Goo51, Hua65], Cutler's patent on differential quantization of communication signals (commonly known as differential pulse code modulation or DPCM for short) [Cut52], Harrison's paper on experiments with linear prediction in television [Har52], and Huffman's paper on a method for the construction of minimum redundancy codes (commonly known as Huffman coding) [Huf52]; notwithstanding that some of the pioneering work on fundamental theories, techniques and concepts in digital image and video coding for visual communications can be traced back to Shannon's monumental work on the mathematical theory of communication in 1948 [Sha48], Gabor's 1946 paper on theory of communication [Gab46] and even as early as the late 1920s when Kell proposed the principle of frame difference signal transmission in a British patent [Kel29, SB65, Sey63, Gir03]. While international standardization of digital image and video coding [RH96] might be considered by many as the end of an era or, simply, research in the area, for others it presents new challenges and signals the beginning of a new era, or more precisely, it is high time that we addressed and, perhaps, solved a number of long standing open problems in the field.

A brief review of the history and the state-of-the-art of research in the field will reveal the fundamental concepts, principles and techniques used in image data compression for storage and visual communications. An important goal that was set fairly early by forerunners in image data compression is to minimize statistical (including source coding, spatio-temporal and inter-scale) and psychovisual (or perceptual) redundancies of the image data to either comply with a certain storage or communications bandwidth restrictions or limitations with the best possible picture quality, or to provide a certain picture quality service with the lowest possible amount of data or bit rate [Sey62]. It helped to set the course and to raise a series of issues widely researched, which have inspired and, in many ways, frustrated generations of researchers in the field. Some of these issues and associated problems are better researched, understood and solved than others.

Using information theory and optimization techniques, we understand reasonably well the definition of statistical redundancy and what is the theoretical lower bound set by Shannon's entropy in lossless image and video coding [Sha48, JN84]. We have statistically modelled natural image data fairly well, which has led to various optimal or sub-optimal compression techniques in the least mean square sense [Sey62, Cla85, NH95]. We routinely apply the rate-distortion theory with the mean squared error (MSE) as a distortion measure in design of constant bit rate coders. We have pushed the performance of a number of traditional compression techniques, such as predictive and transform coding, close to their limit in terms of decorrelation and energy packing efficiencies. Motion compensated prediction has been thoroughly investigated for inter-frame coding of video and image sequences, leading to a number of effective and efficient algorithms used in practical systems.

In model-, object- or segmentation-based coding, we have been trying to balance bit allocations between coding of model parameters and that of the residual image, but we have yet to get it right. Different from classical compression algorithms, techniques based on matching pursuit, fractal transforms and projection on to convex sets are recursive, and encode transform or projection parameters instead of either pixel or transform coefficient values. Nevertheless, they have failed so far to live up to their great expectations in terms of rate-distortion performance in practical coding systems and applications. We have long since realized that much higher compression ratios can be achieved than what is achievable by the best lossless coding techniques or the theoretical lower bound set by the information theory without noticeable distortion when viewed by human subjects. Various adaptive quantization and bit allocation techniques and algorithms have been investigated to incorporate some of the aspects of the human visual systems (HVS) [CS77, CP84, Cla85], most of which focus on spatial contrast sensitivity and masking effects. Various visually weighted distortion measures have been also explored in either performance evaluation [JN84] or rate-distortion optimization of image or video coders [Tau00].

Limited investigations have been conducted in constant quality coder design, impeded by the lack of a commonly acceptable quality metric which correlates well with subjective or perceived quality indices, such as the mean opinion score (MOS) [ITU98]. Long has the question been asked, "What's wrong with mean-squared error?" [Gir84], as well as its derivatives such as the peak signal to noise ratio (PSNR), as the quality or distortion measure. Nonetheless, obtaining credible and widely acceptable alternative perceptual based quantitative quality and/or impairment metrics have so far eluded us till most recently [LB82, VQE00]. Consequently, attempts and claims of providing users with guaranteed or constant quality visual services have been by and large unattainable or unsubstantiated. Lacking HVS-based quantitative quality or impairment metrics, more often than not we opt for a much higher bit rate for quality critical visual service applications than what is necessary, resulting in users carrying extra costs; and just as likely a coding strategy may reduce a particular type of coding distortions or artifacts at the expense of manifesting or enhancing other types of distortions. One of the most challenging questions begging for an answer is how to define "psychovisual redundancy" for lossy image and video coding, if it can ever be defined quantitatively in a similar way to the "statistical redundancy" defined for lossless coding. It would help to set the theoretical lower bound for lossy image data coding at just noticeable level compared with the original.

This book attempts to address two of the above raised issues which may form a critical part of theoretical research and practical system development in the field, i.e., HVS based perceptual quantitative quality/impairment metrics for digitally coded pictures (i.e., images and videos), and perceptual picture coding. The book consists of three parts, i.e., Part I, "Fundamentals"; Part II, "Testing and Quality Assessment of Digital Pictures" and Part III, "Perceptual Coding and Postprocessing."

Part I comprises the first three chapters, covering a number of fundamental concepts, theory, principles and techniques underpinning issues and topics addressed by this book.

Chapter 1, "Digital Picture Compression and Coding Structure," by Hwang, Wu and Rao provides an introduction to digital picture compression, covering basic issues and techniques along with popular coding structures, systems and international standards for compression of images and videos.

"Fundamentals of Human Vision and Vision Modeling" are presented by Montag and Fairchild in Chapter 2 which forms foundations of materials and discussions related to the HVS and its applications presented in Parts II and III on perceptual quality/impairment metrics, image/video coding and visual communications. The most recent achievements and findings in vision research are included, which are relevant to digital picture coding engineering practice.

Various digital image and video coding/compression algorithms and systems introduce highly structured coding artifacts or distortions, which are different from those in their counterpart analog systems. It is important to analyze and understand these coding artifacts in either subjective and objective quality assessment of digitally encoded images or video sequences. In Chapter 3, a comprehensive classification and analysis is presented by Yuen of various coding artifacts in digital pictures coded using well known techniques.

Part II of this book consists of eight chapters dealing with a range of topics regarding picture quality assessment criteria, subjective and objective methods and metrics, testing procedures, and development of international standards activities in the field.

Chapter 4, "Video Quality Testing" by Corriveau, provides an in-depth discussion on subjective assessment methods and techniques, experimental design, international standard test methods for digital video images in contrast to objective assessment methods, highlighting a number of critical issues and findings. Commonly used test video sequences are presented. The chapter also covers test criteria, test procedures and related issues for various applications in digital video coding and communications. Although subjective assessment methods have been well documented in the literature and standardized by the international standards bodies [ITU98], there has been a renewed interest in and research publications on various issues with subjective test methods and new methods, approaches or procedures which may further improve the reliability of subjective test data.

A comprehensive and up-to-date review is provided by Winkler on "Perceptual Video Quality Metrics" in Chapter 5, including both traditional measures, such as the mean square error (MSE) and the PSNR, and HVS based metrics as reported in the literature [YW00, YWWC02] as well as by international standards bodies such as VQEG [VQE00]. It discusses factors which affect human viewers' assessment on picture quality, classification of objective quality metrics, and various approaches and models used for metrics design.

In Chapter 6, Miyahara and Kawada discuss the "Philosophy of Picture Quality Scale." It provides insights into the idea and concept behind the PQS which was introduced by Miyahara, Kotani and Algozi in [MKA98], an extension of the method pioneered by Miyahara in 1988 [Miy88]. It examines applications of PQS to various digital picture services, including super HDTV, extra high quality images, and cellular video phones, in the context of international standards and activities.

Wang, Bovik and Sheikh present a detailed account on "Structural Similarity Based Image Quality Assessment" in Chapter 7. The structural similarity based quality metric is devised to complement the traditional error sensitive picture assessment methods, by targeting at perceived structural information variation, an approach which mimics high level functionality of the HVS. Quality prediction accuracy of the metric is evaluated with significant lower computational complexity than vision model based quality metrics.

"Vision Model Based Digital Video Impairment Metrics" introduced recently are described by Yu and Wu in Chapter 8 for blocking and ringing impairment assessments. In contrast with traditional vision modeling and parameterization method used in vision research, the vision model used in the impairment metrics are parameterized and optimized using subjective test data provided the VQEG where original and distorted video sequences were used instead of simple test patterns. Detailed descriptions of impairment metric implementations are provided with performance evaluation which have showed good agreements with the MOS obtained via subjective evaluations.

"Computational Models for Just-Noticeable Difference" are reviewed and closely examined by Lin in Chapter 9. It provides a systematic introduction to the field to date as well as a practical user's guide for related techniques. JND estimation techniques in both DCT subband domain and image pixel domain are discussed along with issues regarding conversions between the two domains.

In Chapter 10, Caviedes and Oberti investigate issues with "No-Reference Quality Metric for Degraded and Enhanced Video." The concept of "virtual reference" is introduced and defined. It highlights the importance of assessing picture quality enhancement as well as degradation in visual communications services and applications in absence of original pictures. A framework for the development of no-reference quality metric is described. An extensive description is provided on the no-reference overall quality metric (NROQM) which the authors have developed for digital video quality assessment.

In Chapter 11, Corriveau presents an overview of "Video Quality Experts Group" activities highlighting its goals, test plans, major findings, and future work and directions.

The next six chapters form Part III of this book, focusing on digital image and video coder designs based on the HVS, and post-filtering, restoration, error correction and concealment techniques which paly an increasing role in improvement of perceptual picture quality by reduction of perceived coding artifacts and transmission errors. A number of new perceptual coders introduced in recent years are presented in Chapters 12, 13 and 14, including rate-distortion optimization using perceptual distortion metrics and foveated perceptual coding. A noticeable feature of these new perceptual coders is that they use much more sophisticated vision models resulting in significant visual performance improvement. Discussions are included in these chapters on possible new coding architectures based on vision model as compared with existing statistically based coding algorithms and architectures predominant in current software and hardware products and systems.

Chapter 12 by Pica, Isnardi and Lubin examines critical issues associated with "HVS Based Perceptual Video Encoders." It covers an overview of perceptual based approaches, possible architectures and applications, and future directions. Architectures which support perceptual based video encoding are discussed for an MPEG-2 compliant encoder.

Tan and Wu present "Perceptual Image Coding" in Chapter 13, which provides a comprehensive review of HVS based image coding techniques to date. The review covers traditional techniques where various HVS aspects or simple vision models are used for coder designs. Until most recently, this traditional approach has dominated research on the topic with numerous publications, which forms one of, at least, four approaches to perceptual coding design. The chapter describes a perceptual distortion metric based image coder and a vision model based perceptual lossless coder along with detailed discussions on model calibration and coder performance evaluation results.

Chapter 14 by Wang and Bovik investigates novel "Foveated Image and Video Coding" techniques, which they introduced most recently. It provides an introduction to the foveation feature of the HVS, a review of various foveation techniques that have been used to construct image and video coding systems, and detailed descriptions of example foveated picture coding systems.

Chapter 15 by Chen and Wu discusses the topic of "Artifact Reduction by Post-Processing in Image Compression." Various image restoration and processing techniques have been reported in recent years to eliminate or to reduce picture coding arti-facts introduced in the encoding or transmission process to improve perceptual image or video picture quality. It becomes widely accepted that these post-filtering algorithms are an integral part of a compression package or system from a rate-distortion optimiza-tion standpoint. This chapter focuses on reduction of blocking and ringing artifacts in order to improve the visual quality of reconstructed pictures. A DCT domain deblock-ing technique is described with a fast implementation algorithm after a review of coding artifacts reduction techniques to date.

Color bleeding is a prominent distortion associated with color images encoded by block DCT based picture coding systems. Coudoux and Gazalet present in Chapter 16 a novel approach to "Reduction of Color Bleeding in DCT Block-Coded Video," which they introduced recently. This post-processing technique is devised after a thorough analysis of the cause of color bleeding. The performance evaluation results have demon-strated marked improvement in perceptual quality of reconstructed pictures.

Issues associated with "Error Resilience for Video Coding Service" are investigated by Zhang in Chapter 17. It provides an introduction to error resilient coding techniques and concealment methods. Significant improvement in terms of visual picture quality has been demonstrated by using a number of techniques presented.

Chapter 18, the final chapter of the book, highlights a number of critical issues and challenges of the field which may be beneficial to the readers for future research.

Performance measures used to evaluate objective quality/impairment metrics against subjective test data are discussed in Appendix A.

We hope that readers will enjoy reading this book as much as we have enjoyed writing it and find materials provided in it useful and relevant to their work and studies in the field.

H. R. Wu
Royal Melbourne Institute of Technology,
Australia

K. R. Rao
University of Texas at Arlington,
U.S.A.

References

[Cla85]   R. J. Clarke. *Transform Coding of Images*. London: Academic Press, 1985.

[Cla95]   R. J. Clarke. *Digital Compression of Still Images and Video*. London: Academic Press, 1995.

[CP84]   W.-H. Chen and W. K. Pratt. Scene adaptive coder. *IEEE Trans. Commun.*, COM-32:225–232, March 1984.

[CS77]   W.-H. Chen and C. H. Smith. Adaptive coding of monochrome and color images. *IEEE Trans. Commun.*, COM-25:1285–1292, November 1977.

[Cut52]   C. C. Cutler. Differential Quantization of Communication Signals, U.S. Patent No.2,605,361, July 1952.

[Gab46]   D. Gabor. Theory of communication. *Journal of IEE*, 93:429–457, 1946.

[Gir84]   B. Girod. What's wrong with mean-squared error? In A. B. Watson, Ed., *Digital Images and Human Vision*, 207–220. Cambridge, MA: MIT Press, 1984.

[Gir03]   B. Girod. Video coding for compression and beyond, keynote. In *Proceedings of IEEE International Conference on Image Processing*, Barcelona, Spain, September 2003.

[Goo51]   W. M. Goodall. Television by pulse code modulation. *Bell Systems Technical Journal*, 28:33–49, January 1951.

[Har52]   C. W. Harrison. Experiments with linear prediction in television. *Bell Systems Technical Journal*, 29:764–783, 1952.

[Hua65]   T. S. Huang. PCM picture transmission. *IEEE Spectrum*, 2:57–63, December 1965.

[Huf52]   D. A. Huffman. A method for the construction of minimum redundancy codes. *IRE Proc.*, 40:1098–1101, 1952.

[ITU98]   ITU. ITU-RBT. 500-9, methodology for the subjective assessment of the quality of television pictures. *ITU-RBT*, 1998.

[JN84]   N. S. Jayant and P. Noll. *Digital Coding of Waveforms — Principles and Applications to Speech and Video*. Upper Saddle River, NJ: Prentice Hall, 1984.

[Kel29]   R. D. Kell. Improvements Relating to Electric Picture Transmission Systems, British Patent No.341,811, 1929.

[LB82]   F. J. Lukas and Z. L. Budrikis. Picture quality prediction based on a visual model. *IEEE Transactions on Communications*, COM-30:1679–1692, July 1982.

[Miy88]   M. Miyahara. Quality assessments for visual service. *IEEE Communications Magazine*, 26(10):51–60, October 1988.

[MKA98]   M. Miyahara, K. Kotani, and V. R. Algazi. Objective picture quality scale (pqs) for image coding. *IEEE Transactions on Communications*, 46(9):1215–1226, September 1998.

[NH95]   A. N. Netravali and B. G. Haskell. *Digital Pictures — Representation, Compression and Standards*. New York: Plenum Press, 2nd ed., 1995.

[RH96]   K. R. Rao and J. J. Hwang. *Techniques and Standards for Image, Video and Audio Coding.* Upper Saddle River, NJ: Prentice Hall, 1996.

[SB65]   A. J. Seyler and Z. L. Budrikis. Detail perception after scene changes in television image presentations. *IEEE Trans. on Information Theory,* IT-11(1):31–43, January 1965.

[Sey62]   A. J. Seyler. The coding of visual signals to reduce channel-capacity requirements. *Proc. IEE, pt. C,* 109(1):676–684, 1962.

[Sey63]   A. J. Seyler. Real-time recording of television frame difference areas. *Proc. IEEE,* 51(1):478–480, 1963.

[Sha48]   C. E. Shannon. A mathematical theory of communication. *Bell System Technical Journal,* 27:379–623, 1948.

[Tau00]   D. Taubman. High performance scalable image compression with ebcot. *IEEE Trans. Image Proc.,* 9:1158–1170, July 2000.

[VQE00]   VQEG. *Final Report from the Video Quality Experts Group on the Validation of Objective Models of Video Quality Assessment.* VQEG, March 2000. Available from *ftp.its.bldrdoc.gov.*

[YW00]   Z. Yu and H. R. Wu. Human visual systems based objective digital video quality metrics. In *Proceedings of Internetional Conference on Signal Processing 2000 of 16th IFIP World Computer Congress,* 2:1088–1095, Beijing, China, August 2000.

[YWWC02] Z. Yu, H. R. Wu, S. Winkler, and T. Chen. Vision model based impairment metric to evaluate blocking artifacts in digital video. *Proc. IEEE,* 90(1):154–169, January 2002.

Contents

II Picture Quality Assessment and Metrics   123

Part I

Picture Coding and Human Visual System Fundamentals

Chapter 1

Digital Picture Compression and Coding Structure

Jae Jeong Hwang[†], Hong Ren Wu [‡] and K.R. Rao[§]

†   *Kunsan National University, Republic of Korea*
‡   *Royal Melbourne Institute of Technology, Australia*
§   *University of Texas at Arlington, U.S.A.*

1.1   Introduction to Digital Picture Coding

Digital video service has become an integral part of entertainment, education, broadcasting, communication, and business arenas [Say00, Bov05, PE02, PC00, GW02, Ric03, Gha03, Gib97]. Digital camcorders are more preferred than analog ones in the consumer market with their convenience and high quality. In fact, still image or moving video taken by digital cameras can be stored, displayed, edited, printed or transmitted via the Internet. Digital television provides strong affinities to the TV audience and is going to expel analog television receivers from the market. Digital video and image are simple alternative means of carrying the same information as their analog counterparts. An ideal analog recorder should exactly record the natural phenomena in the form of video, image or audio. An ideal digital recorder has to do the same work with a number of advantages such as interactivity, flexibility, and compressibility. Although, in the real situation, ideal conditions seldom prevail and may not be possible by means of both analog and digital techniques, digital compression is one of the techniques used to lower the cost for a video system while maintaining the same quality of service. Data compression is a process to yield a compact representation of a signal in the digital format. For delivery or transmission of information, the key issue is to minimize the bit rate that represents the number of bits per second in the real-time delivery system such as a video stream or the number of bits per picture element (pixel or pel) in the static image. Digital data contains huge amounts of information. Full motion video, e.g., in NTSC format at 30 frames per second (fps) and at 720 x 480 pixel resolution, generates data for luminance component at 10.4 Mbytes/sec, assuming 8 bits per sample quantization. If we include color components for a 4:2:2 format, data rate of 20.8 Mbytes/sec is needed, allowing only 31 seconds of video storage on a 650 Mbyte CD-ROM. The storage capacity up to 74 minutes is only possible by means of compression technology. Then, how can it be compressed? There is considerable *statistical redundancy* in the signal.

- Spatial correlation: Within a single two-dimensional image plane, there usually exists significant correlation among neighboring samples.

- Temporal correlation: For temporal data, such as moving video through temporal direction, there usually exists significant correlation among samples in adjacent frames.

- Spectral correlation: For multispectral images, such as satellite images, there usually exists significant correlation among different frequency bands.

Original video/image data containing any kind of correlation or redundancy can be compressed by appropriate techniques such as predictive or transform based coding that reduces correlation inherently. Image compression aims at reducing the number of bits needed to represent an image by removing the spatial and spectral redundancies as much as possible, while video compression is achieved by removing temporal redundancy as well. This is called redundancy reduction, the principle behind compression. Another important principle behind compression is irrelevancy reduction that will not be noticed by the signal receiver, namely the Human Visual System (HVS). Two ways of classifying compression techniques in terms of reproduction quality at the decoder are *lossless* compression and *lossy* compression. In lossless compression schemes, the reconstructed image, after compression, is numerically identical to the original image. This is also referred to as a reversible process. However lossless compression can only achieve a modest amount of compression depending on the amount of data correlation. An image reconstructed following lossy compression contains degradation relative to the original. Often this is because the compression scheme can completely discard redundant information. However, lossy schemes are capable of achieving much higher compression. Visually lossless coding is achieved if no visible loss is perceived by human viewers under normal viewing conditions. Different classes of compression techniques with respect to statistical redundancy and irrelevancy (or psychovisual redundancy) reductions are illustrated in Figure 1.1.

Another classification in terms of coding techniques is based on prediction or transformation techniques. In predictive coding, information already sent or available is used to predict future values, and the difference is coded and transmitted. Prediction can be performed in any domain, but is usually done in the image or spatial domain. It is relatively simple to implement and is readily adapted to local image characteristics. Differential Pulse Code Modulation (DPCM) is one particular example of predictive coding

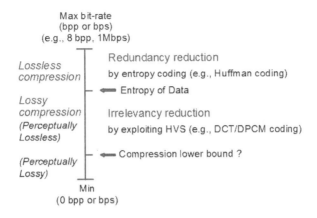

Figure 1.1: Illustration of digital picture compression fundamental concepts.

in the spatial or time domain. Transform coding, on the other hand, first transforms the image from its spatial domain representation to a different type of representation using some well-known transforms such as DCT, DWT (See details in Section 1.3) and then encodes the transformed values (coefficients). This method provides greater data compression compared to predictive methods, although at the expense of higher computational complexity.

As a result of a quantization process, inevitable errors or distortions happen in the decoded picture quality. Distortion measures can be divided into two categories: *subjective* and *objective* measures. It is said to be subjective if the quality is evaluated by humans. The use of human analysts, however, is quite impractical and may not guarantee objectivity. The assessment is not stationary, depending on their feelings. Moreover, the definition of distortion highly depends on the application, i.e. the best quality evaluation is not always made by people at all.

In the objective measures, the distortion is calculated as the difference between the original image, x_o, and the reconstructed image, x_r, by a predefined function. It is assumed that the original image is perfect. All changes are considered as occurrences of distortion, no matter how they appear to a human observer. The quantitative distortion of the reconstructed image is commonly measured by the *mean square error* (MSE), the *mean absolute error* (MAE), and the *peak-to-peak signal to noise ratio* (PSNR):

$$MSE = \frac{1}{M \times N} \sum_{m=0}^{M-1} \sum_{n=0}^{N-1} (\mathbf{x}_o[m,n] - \mathbf{x}_r[m,n])^2 \qquad (1.1)$$

$$MAE = \frac{1}{M \times N} \sum_{m=0}^{M-1} \sum_{n=0}^{N-1} |\mathbf{x}_o[m,n] - \mathbf{x}_r[m,n]| \qquad (1.2)$$

$$PSNR = 10 \log_{10} \frac{255^2}{MSE} \qquad (1.3)$$

where M and N are the height and the width of the image, respectively, and (1.3) is defined for 8 bits/pixel monochrome image representation.

These measures are widely used in the literature. Unfortunately, these measures do not always coincide with the evaluations of a human expert. The human eye, for example, does not observe small changes of intensity between individual pixels, but is sensitive to the changes in the average value and contrast in larger regions. Thus, one approach would be to calculate the local properties, such as *mean* values and *variances* of some small regions in the image, and then compare them between the original and the reconstructed images. Another deficiency of these distortion functions is that they measure only local, pixel-by-pixel differences, and do not consider global artifacts, such as blockiness, blurring, jaggedness of the edges, ringing or any other type of structural degradation.

1.2   Characteristics of Picture Data

1.2.1   Digital Image Data

Digital image is visual information represented in a discrete form, suitable for digital electronic storage and transmission. It is obtained by image sampling techniques that a discrete array $x[m, n]$ is extracted from the continuous image field at some time instant over some rectangular area $M \times N$. The digitized brightness value is called the grey level value. Each image sample is a *picture element* called a *pixel* or a *pel*. Thus, a two-dimensional (2-D) digital image is defined as:

$$x[m, n] = \begin{bmatrix} x[0, 0] & x[0, 1] & \cdots & x[0, N-1] \\ x[1, 0] & x[1, 1] & \cdots & x[1, N-1] \\ \vdots & \vdots & \ddots & \vdots \\ x[M-1, 0] & x[M-1, 1] & \cdots & x[M-1, N-1] \end{bmatrix} \qquad (1.4)$$

where its array of image samples is defined on the two-dimensional Cartesian coordinate system as illustrated in Figure 1.2. The number of bits, b, we need to store an image of size $M \times N$ with 2^q different grey levels is $b = M \times N \times q$. That is, to store a typical image of size 512×512 with 256 grey levels ($q = 8$), we need 2,097,152 bits or 262,144 bytes. We may try to reduce the factor M, N or q to save capacity of storage or bits for transmission, but it is not said to be compressed, since it results in significant loss in the quality of the picture.

$$x(m,n) = \begin{bmatrix} x(0,0) & x(0,1) & \cdots & x(0,N-1) \\ x(1,0) & x(1,1) & \cdots & x(1,N-1) \\ \vdots & \vdots & & \vdots \\ x(M-1,0) & x(M-1,1) & \cdots & x(M-1,N-1) \end{bmatrix}$$

Figure 1.2: Geometric relationship between the Cartesian coordinate system and its array of image samples.

1.2.2   Digital Video Data

A natural video stream is continuous in both spatial and temporal domains. In order to represent and process a video stream digitally it is also necessary to sample spatially and temporally as shown in Figure 1.3. An image sampled in the spatial domain is typically represented on a rectangular grid and a video stream is a series of still images sampled at regular intervals in time. In this case, the still image is usually called a *frame*. For processing video signal in a television format, a couple of fields are interlaced to construct a frame. It is called a *picture* for processing non-interlaced (frame-based) video signal. Each spatio-temporal sample, pixel, is represented as a positive digital number that describes the brightness (luminance) and color components.

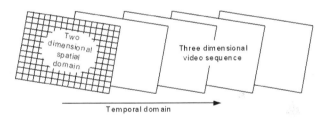

Figure 1.3: Three dimensional (spatial and temporal) domain in a video stream.

A natural video scene is captured, typically with a camera, and converted to a sampled digital representation as shown in Figure 1.4. Digital video is represented in digital color-difference format YC_1C_2 rather than in the original RGB natural color format. It may then be handled in the digital domain in a number of ways, including processing, storage and transmission. At the final output of the system, it is displayed to a viewer by reproducing it on a video monitor.

The RGB (red, green, and blue) color space is the basic choice for computer graphics and image frame buffers because color CRTs use red, green, and blue phosphors to create the desired color as the three primary additive colors. Individual components are added together to form a color and an equivalent addition of all components produces white. However, RGB is not very efficient for representing real-world images, since equal bandwidths are required to describe all the three color components. The equal

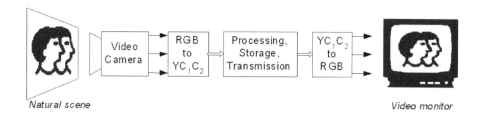

Figure 1.4: Digital representation and color format conversion of natural video stream.

bandwidths result in the same pixel depth and display resolution for each color component. Using 8 bits per component requires 24 bits information for a pixel, resulting in 3 times the capacity of the luminance component. Moreover, the sensitivity of the color component of the human eye is less than that of the luminance component. For these reasons, many image coding standards and broadcast systems use luminance and color difference signals. These are, for example, YUV and YIQ for analog television standards and YC_bC_r for their digital version.

The YC_bC_r format recommended by the ITU.R BT-601 [ITU82] as a worldwide video component standard is obtained from digital gamma-corrected RGB signals as follows:

$$Y = 0.299R' + 0.587G' + 0.114B'$$
$$C_b = -0.169R' - 0.331G' + 0.500B' \qquad (1.5)$$
$$C_r = 0.500R' - 0.419G' - 0.081B'$$

The color-difference signals are given by:

$$(B - Y) = -0.299R' - 0.587G' + 0.886B'$$
$$(R - Y) = 0.701R' - 0.587G' - 0.114B' \qquad (1.6)$$

where the values for $(B-Y)$ have a range of ±0.886 and for $(R-Y)$ a range of ±0.701, while those for Y have a range of 0 to 1.

To restore the signal excursion of the color-difference signals to unity (-0.5 to +0.5), $(B-Y)$ is multiplied by a factor 0.564 (0.5 divided by 0.886) and $(R-Y)$ is multiplied by a factor 0.713 (0.5 divided by 0.701). Thus the C_b and C_r are the re-normalized blue and red color difference signals, respectively.

Given that the luminance signal is to occupy 220 levels (16 to 235), the luminance signal has to be scaled to obtain the digital value, Y_d. Similarly, the color difference signals are to occupy 224 levels and the zero level is to be level 128. The digital representation for the three components are expressed as [NH95]:

$$Y_d = 219Y + 16$$
$$C_b = 224[0.564(B - Y)] + 128 = 126(B - Y) + 128 \qquad (1.7)$$
$$C_r = 224[0.713(R - Y)] + 128 = 160(R - Y) + 128$$

or in its vector form:

$$\begin{bmatrix} Y_d \\ C_b \\ C_r \end{bmatrix} = \begin{bmatrix} 65.481 & 128.553 & 24.966 \\ -37.797 & -74.203 & 112.000 \\ 112.000 & -93.786 & -18.214 \end{bmatrix} \begin{bmatrix} R \\ G \\ B \end{bmatrix} + \begin{bmatrix} 16 \\ 128 \\ 128 \end{bmatrix} \qquad (1.8)$$

where the corresponding level number after quantization is the nearest integer.

Video transmission bit rate is decreased by adopting lower sampling rates while pre-serving acceptable video quality. Given image resolution of 720×576 pixels represented with 8 bits each, the bit rate required is calculated as:

4:4:4 resolution: $720 \times 576 \times 8 \times 3 = 10$ Mbits/frame
10 Mbits/frame $\times 29.97$ frames/sec $= 300$ Mbits/sec

4:2:0 resolution: $(720 \times 576 \times 8) + (360 \times 288 \times 8) \times 2 = 5$ Mbits/frame
5 Mbits/frame $\times 29.97$ frames/sec $= 150$ Mbits/sec

The 4:2:0 version requires half as many bits as the 4:4:4 version but compression is still necessary for transmission and storage.

1.2.3 Statistical Analysis

The *mean* value of the discrete image array, $\mathbf{x}$ as defined in (1.4), expressed conveniently in vector-space form is given by

$$\mu_x = E\{\mathbf{x}\} = \frac{1}{M \times N} \sum_{m=0}^{M-1} \sum_{n=0}^{N-1} \mathbf{x}[m, n] = \sum_{k=0}^{2^b - 1} x_k p(x_k) \qquad (1.9)$$

where x_k denotes the k-th grey level that varies from value 0 to maximum level $2^b - 1$ defined by the quantization bits b and $p(x_k) = n_k/(M \times N)$ the probability of x_k.

The *variance* function of the image array, $\mathbf{x}$, is defined as

$$\sigma_x^2 = \frac{1}{M \times N} \sum_{m=0}^{M-1} \sum_{n=0}^{N-1} (\mathbf{x}[m, n] - \mu_x)^2 = \sum_{k=0}^{2^b - 1} (x_k - \mu_x)^2 p(x_k) \qquad (1.10)$$

that involves the squared variation of pixel levels from the image mean, μ_x. It is identical to the squared sum of all pixel values divided by the total number of pixels, called the average total element (pixel) *energy* of an image, if the mean equals zero. Variance, in general, is expanded by

$$
\begin{aligned}
\sigma_x^2 &= E\{(\mathbf{x}[m,n] - \mu_x)^2\} \\
&= E\{\mathbf{x}^2[m,n] + \mu_x^2 - 2\mathbf{x}\mu_x\} \\
&= E\{\mathbf{x}^2[m,n]\} - \mu_x^2
\end{aligned}
\tag{1.11}
$$

where $E(\cdot)$ represents the expectation operator which calculates the average over the whole image. Thus, the average total element (pixel) energy of an image, $E\{\mathbf{x}^2[m,n]\}$, is composed of the variance and the mean energy per pixel, i.e., 'AC' energy and 'DC' energy, respectively, as [Cla85]

$$
\underbrace{E\{\mathbf{x}^2[m,n]\}}_{Total\ energy} = \underbrace{\sigma^2}_{AC\ energy} + \underbrace{\mu_x^2}_{DC\ energy}
\tag{1.12}
$$

For simplicity, a random variable x is used to represent the value of an arbitrary image element, $\mathbf{x}[m,n]$, of an image sequence, (1.11) is then reduced to (1.13)

$$
\sigma_x^2 = E\{(x - \mu_x)^2\}
\tag{1.13}
$$

The correlation property of images is the most important and crucial factor in the development of image video compression techniques. From the variance representation in (1.11), we may form an equivalent operation between two different sequences, x_0 and x_1, thus

$$
\sigma_{x_0 x_1}^2 = E\{(x_0 - \mu_{x_0})(x_1 - \mu_{x_1})\}
\tag{1.14}
$$

where μ_{x_0} and μ_{x_1} denote the means of the two sequences, respectively and $\sigma_{x_0 x_1}^2$ is termed the *covariance* of the two sequences. Define the vector

$$
[\mathcal{X} - \mu_{\mathcal{X}}] = \begin{bmatrix} (x_0 - \mu_{x_0}) \\ (x_1 - \mu_{x_1}) \end{bmatrix}.
\tag{1.15}
$$

The covariance matrix of the two sequences is defined as

$$
\begin{aligned}
COV(\mathcal{X}) &= E\{(\mathcal{X} - \mu_{\mathcal{X}})(\mathcal{X} - \mu_{\mathcal{X}})^T\} \\
&= \begin{bmatrix} (x_0 - \mu_{x_0})(x_0 - \mu_{x_0}) & (x_0 - \mu_{x_0})(x_1 - \mu_{x_1}) \\ (x_1 - \mu_{x_1})(x_0 - \mu_{x_0}) & (x_1 - \mu_{x_1})(x_1 - \mu_{x_1}) \end{bmatrix}.
\end{aligned}
\tag{1.16}
$$

where T represents the transpose operation. In an image, we assume that the two sequences can be generated with a certain displacement. The approach can be extended to define the covariance between any number of sequences, i.e., any possible displacements in a sequence. Thus, we now form the vectorized covariance of K sequences as

$COV(\mathcal{X})$

$$= \begin{bmatrix} (x_0 - \mu_{x_0})(x_0 - \mu_{x_0}) & (x_0 - \mu_{x_0})(x_1 - \mu_{x_1}) & \cdots & (x_0 - \mu_{x_0})(x_\kappa - \mu_{x_\kappa}) \\ (x_1 - \mu_{x_1})(x_0 - \mu_{x_0}) & (x_1 - \mu_{x_1})(x_1 - \mu_{x_1}) & \cdots & (x_1 - \mu_{x_1})(x_\kappa - \mu_{x_\kappa}) \\ \vdots & \vdots & \ddots & \vdots \\ (x_\kappa - \mu_{x_\kappa})(x_0 - \mu_{x_0}) & (x_\kappa - \mu_{x_\kappa})(x_1 - \mu_{x_1}) & \cdots & (x_\kappa - \mu_{x_\kappa})(x_\kappa - \mu_{x_\kappa}) \end{bmatrix}$$

$$= \begin{bmatrix} \sigma^2_{x_0 x_0} & \sigma^2_{x_0 x_1} & \cdots & \sigma^2_{x_0 x_\kappa} \\ \sigma^2_{x_1 x_0} & \sigma^2_{x_1 x_1} & \cdots & \sigma^2_{x_1 x_\kappa} \\ \vdots & \vdots & \ddots & \vdots \\ \sigma^2_{x_\kappa x_0} & \sigma^2_{x_\kappa x_1} & \cdots & \sigma^2_{x_\kappa x_\kappa} \end{bmatrix}$$

(1.17)

where $\kappa = K - 1$.

Assuming that the system is symmetric and stationary, the diagonal elements represent the variances of the original sequences and off-diagonal terms represent the covariances between different sequences that can be extracted from the one original sequence. Then, the equal-step covariances are the same as shown in (1.18).

$$\begin{cases} \sigma^2_{x_0 x_0} &= \sigma^2_{x_1 x_1} &= \cdots & \cdots &= \sigma^2_{x_\kappa x_\kappa} &= \sigma^2 \\ \sigma^2_{x_0 x_1} &= \sigma^2_{x_1 x_0} &= \cdots &= \sigma^2_{x_{\kappa-1} x_\kappa} &= \sigma^2_{x_\kappa x_{\kappa-1}} &= \sigma^2_1 \\ \vdots & \vdots & \vdots & \vdots & \vdots \\ \sigma^2_{x_0 x_{\kappa-1}} &= \sigma^2_{x_{\kappa-1} x_0} &= \sigma^2_{x_1 x_\kappa} &= \sigma^2_{x_\kappa x_1} &= \sigma^2_{x_{\kappa-1}} \\ \sigma^2_{x_0 x_\kappa} &= \sigma^2_{x_\kappa x_0} &= \sigma^2_{x_\kappa} \end{cases}$$

(1.18)

Then, the K-step covariance matrix is given by

$$COV(\mathcal{X}) = \begin{bmatrix} \sigma^2 & \sigma^2_1 & \cdots & \sigma^2_\kappa \\ \sigma^2_1 & \sigma^2 & \cdots & \sigma^2_{\kappa-1} \\ \vdots & \vdots & \ddots & \vdots \\ \sigma^2_\kappa & \sigma^2_{\kappa-1} & \cdots & \sigma^2 \end{bmatrix}$$

$$= \sigma^2 \begin{bmatrix} 1 & \rho^2_1 & \cdots & \rho^2_\kappa \\ \rho^2_1 & 1 & \cdots & \rho^2_{\kappa-1} \\ \vdots & \vdots & \ddots & \vdots \\ \rho^2_\kappa & \rho^2_{\kappa-1} & \cdots & 1 \end{bmatrix}$$

$$= \sigma^2 COR(\mathcal{X})$$

(1.19)

where ρ is the k-step correlation coefficient and $COR(\mathcal{X})$ is the *correlation (normalized covariance)* matrix of the stationary first-order Markov process where each one-step elements are the same. The word correlation implies the amount of analogy, whilst covariance implies the amount of difference. However, they have a close relationship as in (1.19), representing the one-dimensional sequences. Since the two-dimensional image signal can be separately processed, the one-dimensional case can be expanded via horizontal and vertical directions. In both cases, we model the image statistics assuming that the correlation is stationary (diagonal elements are all same) and that it depends on distance from the current sample.

1.3   Compression and Coding Techniques

1.3.1   Entropy Coding

Entropy coding is based on the fact that every signal has its unique information and the average length of the code is bounded by the entropy of the information source, known as Shannon's first theorem. The entropy of the source x with m symbols, $\{x_i, i = 1, \ldots, m\}$, is defined as

$$H(x) = -\sum_{i=1}^{m} p(x_i)\log_2 p(x_i) \tag{1.20}$$

where $p(x_i)$ represents the probability of the symbol x_i. The codeword for each symbol may not be of constant length, but may be of variable length based on the amount of entropy. A shorter codeword is allocated to the symbol which has larger probability (smaller entropy), and vice versa.

To be practical in use, codes need to have some desirable characteristics. Obviously, a code has to be uniquely decodable if it is to be of use. A uniquely decodable code is said to be instantaneous if it is possible to decode each codeword in a code symbol sequence without knowing the succeeding codewords, allowing a memoryless decoding capability. Furthermore, a uniquely decodable code is said to be compact if its average length is the minimum among all other uniquely decodable codes based on the same source and codeword. A compact code is also referred to as a *minimum redundancy* code, or an *optimum* code.

Two techniques for implementing variable-length entropy coding have been used: Huffman coding and arithmetic coding. While Huffman coding is a *block-oriented* coding technique, arithmetic coding is a *stream-oriented* coding technique. Both Huffman coding and arithmetic coding are included in the international still image coding standards, JPEG (Joint Photographic Experts Group) [ISO93b] and MPEG (Moving Picture Experts Group) [ISO93a] coding. With improvements in implementation, arithmetic coding has gained increasing popularity. The adaptive arithmetic coding algorithms have been adopted by the international bilevel image coding standard JBIG (Joint Bilevel Image experts Group) [ISO92] and H.264 [ITU03]. Detailed implementation issues are referred to in [RH96, Sal00].

1.3.2 Predictive Coding

The differential coding technique has played an important role in image and video coding. In the international coding standard for still images, JPEG, differential coding is used in the lossless mode and in the DCT-based mode for coding DC coefficients. Motion-compensated (MC) coding is essentially a predictive coding technique applied to video sequences involving displacement motion vectors. MC coding has been a major development in video coding since the 1980s and has been adopted by all the international video coding standards such as H.261 [CCI90], MPEG-1 [ISO93a], MPEG-2, H.262 [ISO93c], H.263 [ITU96], MPEG-4 [ISO01], and H.264 [ITU03]. The underlying philosophy behind predictive coding is that, if the present sampled signal, say a current pixel, can be reasonably well predicted based on the previous neighborhood samples (causal predictor), then the prediction error has a smaller entropy compared to the original sampled signal [JN84, O'N76]. Hence the prediction error can be quantized with fewer quantization levels compared to the number of quantized levels of the sampled signal. The prediction can be based on the statistical distribution of the signal (the signal can be audio, speech, image, video, text, graphics, etc). The most common predictive coding is called differential pulse code modulation (DPCM) shown in Figure 1.5. The following relations are valid.

$$
\begin{aligned}
Prediction\ error: \quad & e_p[n] && = && x[n] - x_p[n] \\
Quantization\ error: \quad & q[n] && = && e_p[n] - e_{pq}[n] \\
Input\ signal: \quad & x[n] && && \\
Reconstructed\ signal: \quad & \hat{x}[n] && = && x_p[n] + e_{pq}[n] \\
Predicted\ value\ of\ x[n]: \quad & x_p[n] && && \\
Quantized\ prediction\ error: \quad & e_{pq}[n] && && \\
Reconstruction\ error: \quad & x[n] - \hat{x}[n] && = && (x_p[n] + e_p[n]) - (e_{pq}[n] + x_p[n]) \\
& && = && e_p[n] - e_{pq}[n] \\
& && = && q[n] \quad (Quantization\ error)
\end{aligned}
$$

Thus, the only factor to result in the reconstruction error is caused by the quantization error, i.e., no reconstruction error without quantization of prediction error (lossless coder). Since most international video coding standards adopt the DPCM structure to reduce interframe redundancy, in the absence of channel noise, the quantization error directly affects the picture quality resulting in a lossy coder. The feedback loop (closed-loop) avoids the accumulation of the quantization error during successive prediction cycles, while, in a feedforward predictor, the quantization error can accumulate during successive cycles. The predictor at the decoder is an exact replica of the predictor at the encoder. In practice, as only reconstructed pixels (not the original pixels) are available

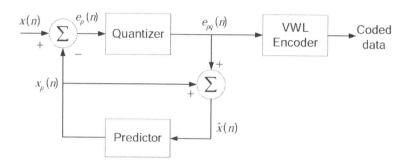

Figure 1.5: Closed-loop DPCM encoder. P (Predictor) and/or Q (Quantizer) can be adaptive. VWL: Variable Word Length.

at the decoder, the prediction is based only on the previously reconstructed pixels. Also the prediction process can be improved by adaptive prediction and/or quantization.

The order of the predictor is the number of previously reconstructed pixels used in the prediction. $x_p[n] = (\hat{x}[n-1], \hat{x}[n-2], \ldots, \hat{x}[n-k])$ is a k-th order predictor. Dimension of the predictor is dependent on the locations of the previously reconstructed pixels (causal), i.e., pixels in the same line mean 1-D predictor. Pixels in the horizontal and vertical directions imply 2-D predictor. Pixels in the horizontal, vertical and temporal domains imply 3-D predictor (Figure 1.6). Interfield or interframe prediction requires storage of field(s) or frame(s).

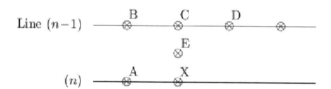

Figure 1.6: Causal pixels to be used for prediction of current pixel X on a video stream. Pixels A, B, C, and D are in spatial domain and pixel E is in temporal domain.

1.3.3   Transform Coding

1.3.3.1   Discrete cosine transform (DCT)

The DCT [RY90] is originally derived from the DFT (N-point signal is transformed to N-point frequency coefficients). While the DFT coefficients are real and imaginary, DCT coefficients are real only. An N-point sequence can be extended to a 2N-point sequence in a time-reversed order, called even extension or mirroring, as shown in Figure 1.7(c). The resulting sequence is defined as

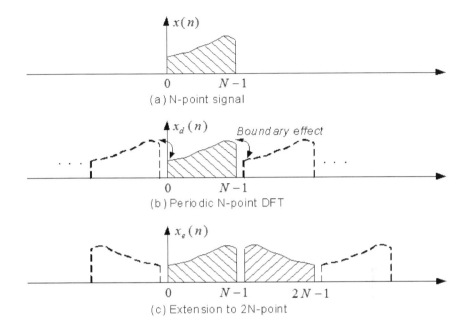

Figure 1.7: Even extension of N-point signal, a) N-point input signal, b) N-point DFT that causes end (boundary) effect, c) even extension which is symmetrical about the point (N-1).

$$
x_e[n] = \begin{cases} x[n], & n = 0, 1, \ldots, N-1 \\ x[2N-n-1], & n = N, N+1, \ldots, 2N-1 \end{cases}
\tag{1.21}
$$

The 2N-point DFT of the extended signal in (1.21) is defined as

$$
\begin{aligned}
X_{DFT}[k] &= \sum_{n=0}^{2N-1} x_e[n] e^{-j\frac{2\pi k}{2N}n} \\
&= \sum_{n=0}^{N-1} x[n] e^{-j\frac{2\pi k}{2N}n} + \sum_{n=N}^{2N-1} x[2N-1-n] e^{-j\frac{2\pi k}{2N}n}
\end{aligned}
\tag{1.22}
$$

Let $m = 2N - 1 - n$, $(n = 2N - 1 - m)$, then the 2nd term of (1.22) is

$$
\sum_{m=N-1}^{0} x[m] e^{-j\frac{2\pi k}{2N}(2N-1-m)} = \sum_{0}^{m=N-1} x[m] e^{-j\frac{2\pi k}{2N}(2N-1-m)}
\tag{1.23}
$$

Note that we assume the system is linear. The index m in (1.23) can be changed back to n. Then (1.22) can be written as

$$
\begin{aligned}
X_{DFT}[k] &= \sum_{n=0}^{N-1} x[n]e^{-j\frac{2\pi k}{2N}n} + \sum_{0}^{n=N-1} x[n]e^{-j\frac{2\pi k}{2N}(2N-1-n)} \\
&= e^{j\frac{\pi k}{2N}}\left(\sum_{n=0}^{N-1} x[n]e^{-j\frac{\pi k}{2N}(2n+1)} + \sum_{0}^{n=N-1} x[n]e^{j\frac{\pi k}{2N}(-4N+2n+1)}\right) \\
&= e^{j\frac{\pi k}{2N}}\left(\sum_{n=0}^{N-1} x[n]e^{-j\frac{\pi k}{2N}(2n+1)} + \sum_{0}^{n=N-1} x[n]e^{j\frac{\pi k}{2N}(2n+1)}\right) \\
&= e^{j\frac{\pi k}{2N}}\left(\sum_{n=0}^{N-1} x[n]\left(e^{-j\frac{\pi k}{2N}(2n+1)} + e^{j\frac{\pi k}{2N}(2n+1)}\right)\right) \\
&= 2e^{j\frac{\pi k}{2N}}\left(\sum_{n=0}^{N-1} x[n]\cos\frac{\pi k}{2N}(2n+1)\right) \\
&= 2e^{j\frac{\pi k}{2N}}X_{DCT}[k]
\end{aligned}
$$

(1.24)

where $X_{DCT}[k]$ is the type-II DCT [RY90]. Therefore N-point DCT can be derived from 2N-point even extended DFT with a weighting factor, i.e.,

$$
X_{DCT}[k] = \frac{1}{2}e^{-j\frac{\pi k}{2N}}X_{DFT}[k]
$$

(1.25)

where $k = 0, 1, \ldots, N-1$.

The effect of even extension is to turn it into an even function whose sine components are cancelled out, hence the name of the transform. Another advantage is that doubling the block length doubles the frequency resolution, so that twice as many frequency components are produced. Note that N-point signal is transformed to N-point DCT coefficients and 2N-point extension is formed by itself. The close relation to DFT reveals that a fast DCT implementation is possible via the Fast Fourier Transform algorithms (FFT).

For image processing, the two-dimensional $N \times M$-point (type-II) DCT transform is defined as

$$
X_{DCT}[u,v] = \frac{4}{N \times M}c[u]c[v]\sum_{n=0}^{N-1}\sum_{m=0}^{M-1} x[n,m]\cos\left[\frac{(2n+1)u\pi}{2N}\right]\cos\left[\frac{(2m+1)v\pi}{2M}\right]
$$

(1.26)

where $X_{DCT}[u,v]$, for $u = 0, 1, \ldots, N-1$ and $v = 0, 1, \ldots, M-1$, represents the DCT coefficient at the location (u,v), i.e., line (row, vertical axis) u and column (horizontal axis) v, and

$$
c[u], c[v] = \begin{cases} \frac{1}{\sqrt{2}} & for \ u, v = 0; \\ 1 & otherwise. \end{cases}
$$

The first cosine term is the vertical basis function generator represented by sampled cosine wave. n sets the sample number and u sets the frequency. For the same reason, the second cosine term is followed as the horizontal basis function generator. Since the DCT is separable, the two-dimensional DCT can be obtained by computing 1-D DCT in each dimension separately.

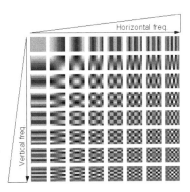

Figure 1.8: 2-D (8×8) DCT basis images. The DCT breaks up an image area into discrete frequencies in two dimensions.

The basis images for 2-D (8×8) DCT are shown in Figure 1.8. The top left image represents the mean intensity, DC value. The 2-D DCT of image data is basically a structural decomposition in terms of its corresponding basis images.

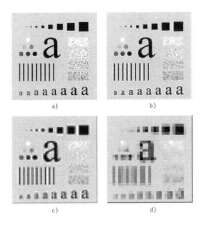

Figure 1.9: Blocking artifact by the DCT; a) a test image, b) the inverse DCT with (4×4) low-frequency coefficients, c) the inverse DCT with (2×2) low-frequency coefficients, and d) the inverse DCT with only the DC coefficient.

For the implementation view point, most of international standards favor 8×8 block size, considering its complexity and performance. It converts the image block into a form where redundancy or correlation in the image data is reordered in terms of the basis images, so that the redundancy can be easily detected and removed. The detection is possible by the virtue of orthogonal property of the basis images; non-zero coefficients are obtained if an image pattern block coincides with the basis block. Natural image data, of course, may not coincide with the rectangular shaped basis images. Although

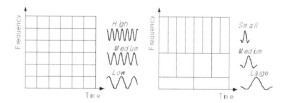

Figure 1.10: Fourier (left) and Wavelet (right) basis functions in the time-frequency plane.

the DCT shows sufficient performance of compression capability, one major disadvantage of the DCT is the block structure that dominates at very low bit rates, called *blocking artifacts* (as shown in Figure 1.9d). This feature, of course, is a characteristic of all block transforms. If the difference of quantization errors between two adjacent blocks is so large, it would be easily detectable by human eye and the block disparity occurs.

1.3.3.2 Discrete wavelet transform (DWT)

The Fourier transform represents signals as the sum of sines and cosines that have infinite time duration. It determines quantity or value at a particular frequency for signals with statistical properties which are constant over time or space. Therefore it is less suitable for the signal which is nonstationary or changing in time. The Fourier-related cosine transform has the same limitations. For example, the more limited in time a signal is (such as an impulse), the more high frequency components are needed. This is a motivation of the wavelet representation of the signal, which is localized in space.

One way to see the time-frequency resolution differences between the Fourier transform and the wavelet transform is to look at the basis function in the time-frequency plane [VH92]. Figure 1.10 shows a windowed Fourier transform, where the window is simply a square. The time resolutions are the same at all locations in the time-frequency plane and the frequency resolutions are varying in frequency.

The simplest wavelet is a Haar wavelet which has been used in Haar transform. The basis images shown in Figure 1.11 are comparable to the basis images of DCT (Figure 1.8). Frequency distribution is similar to that in DCT, increasing from top-left to bottom-right, except representing localized analysis. The Haar wavelet functions are discontinuous and compactly supported, while others are continuous as shown in Figure 1.12.

The discrete wavelet transform (DWT) is, in general, implemented by the dyadic wavelet transform as shown in Figure 1.13. The DWT is obtained by iterating a two-channel filter bank on its lowpass output. This multiresolution decomposition of a signal into its coarse detail components is useful for data compression, feature extraction, and frequency-domain filtering. The wavelet representation is well-matched to psychovi-

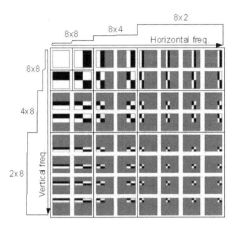

Figure 1.11: (8×8) DHT basis images. In each image, white means a positive number, black means a negative number, and grey means zero level.

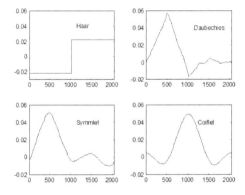

Figure 1.12: Several different wavelet functions.

sual models, and compression systems based on the wavelet transform yield perceptual quality superior to other methods at medium and high compression ratios. The basic approach to wavelet-based image processing or compression is achieved by first computing the 2D wavelet transform, and second, altering the transform coefficients, and last, computing the inverse transform. While the DCT suffers from the block effect, the DWT does not result in such an artifact, since it is based on full image transform and short time localization.

We can experiment with the blurring artifact from the DWT, when it is adopted in image data compression by discarding some low energy coefficients, i.e., high frequency details.

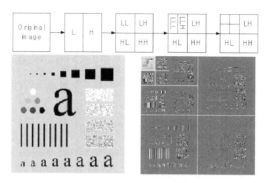

Figure 1.13: Dyadic wavelet decomposition of an image and application to a sample image. Horizontal and vertical edge components are explicitly decomposed at three levels.

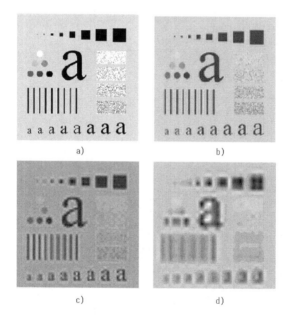

Figure 1.14: Blurring artifact by three-level wavelet transform; a) a test image, b) the inverse transform after zeroing the first-level detail coefficients, c) and d) results after zeroing second- and third-level details.

In general images, the high frequency bands have lower energy and are less important than the low frequency bands. By zeroing the first-level detail coefficients of the wavelet transformed image (as in Figure 1.14b), the resulting image shows only slightly blurred/smoothed copy. Additional blurring is present in the next result (Figure 1.14c) that shows the effect of zeroing the second level detail coefficients as well. More artifacts are found as the next level coefficients are zeroed. Figure 1.14d is obtained by only the lowest frequency band in the three-level wavelet transform.

1.4 Picture Quantization

1.4.1 Uniform/Nonuniform Quantizer

Quantization of the sampled data is performed to quantify with a finite number of levels. It is assumed that the sampling is uniform and sampling rate is above the Nyquist rate so that there is no aliasing in the frequency domain. Some criteria such as minimization of the quantizer distortion have been used for quantization of image data [Ger77]. Quantizer design includes input (decision) levels, output (representation) levels and the number of levels. The design can be enhanced by psychovisual or psychoacoustic perception.

Quantizers can be classified as memoryless or with memory. The former assumes that each sample is quantized independently with no prior knowledge of input samples whereas the latter takes into account previous samples. Another classification is uniform or nonuniform. A uniform quantizer is completely defined by the number of levels, step size and if it is a midriser or a midtreader. Our discussion also is limited to symmetric quantizers, i.e., the input and output levels in the third quadrant are negatives of the corresponding levels in the first quadrant. A nonuniform quantizer implies that the step sizes are not constant. Hence, the nonuniform quantizer has to be specified by the input and output levels and designed by using the probability density function of the input signal. In spite of the type of quantizers, a quantized output (reconstruction) value is determined in a certain interval (quantization step) where any of the input values happens. Since the reconstruction value represents the whole range of input values, quantization inherently is a lossy process and the lost information may not be recovered. Since, usually, the distribution of image data is concentrated on mean value region and image processing, including predictive coding and transform coding, produces more abundant distribution on smaller levels near zero, which means less energy or variance, the region can be quantized with fine step size, while others can be quantized with coarse step size. A nearly uniform quantizer is designed using these properties, enlarging the step size only in the mean value region, called a *deadzone*. Except for the deadzone (input range for which the output is zero), the stepsize is constant. Such a nearly uniform quantizer has been specified in H.261, MPEG-1 video, MPEG-2 video and H.263 [RH96].

1.4.2 Optimal Quantizer Design

For the development of an optimal quantizer, let x and $\hat{x}$ represent the amplitude of a signal sample and its quantized value, respectively. It is assumed that x is a random variable with probability density function $p(x)$ and lies in the range

$$a_L \leq x \leq a_U \tag{1.27}$$

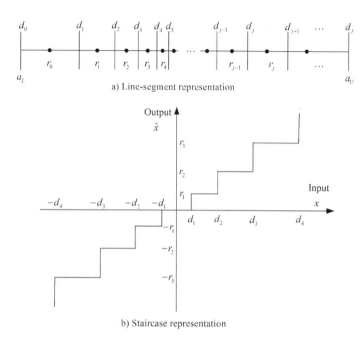

a) Line-segment representation

b) Staircase representation

Figure 1.15: Nonuniform symmetric quantization (d_j: input decision level, r_j: output reconstruction level).

where a_L and a_U represent lower and upper limits as indicated in Figure 1.15.

Quantization is performed with a set of decision levels d_j and a set of reconstruction levels r_j such that if

$$d_j \leq x \leq d_{j+1} \qquad (1.28)$$

then the sample is quantized to a reconstruction value r_j.

Given the range of input x as from a_L to a_U and the number of output levels as J, the quantizer is designed such that the MSQE (mean square quantization error), ε, is minimized, i.e.,

$$\varepsilon = E\left[(x - \hat{x})^2\right] = \int_{a_L}^{a_U} (x - \hat{x})^2 p(x) dx \qquad (1.29)$$

where E is the expectation operator. (1.29) can be rewritten, in terms of the decision and reconstruction levels, as the summation of possible errors at all output levels :

$$\varepsilon = \sum_{j=0}^{J-1} \int_{d_j}^{d_{j+1}} (x - r_j)^2 p(x) dx \qquad (1.30)$$

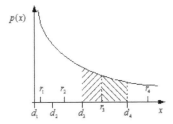

Figure 1.16: Reconstruction level by centroid between two decision levels.

Since r_j and d_j are variables, partial derivatives are set to zero to minimize the quantization error, i.e., $\frac{\partial \varepsilon}{\partial r_j} = 0$ and $\frac{\partial \varepsilon}{\partial d_j} = 0$. The optimal decision level is derived by taking only two incidents of d_j into account :

$$
\begin{aligned}
\frac{\partial \varepsilon}{\partial d_j} &= \frac{\partial}{\partial d_j}\left[\int_{d_{j-1}}^{d_j} (x - r_{j-1})^2 p(x)dx + \int_{d_j}^{d_{j+1}}(x - r_j)^2 p(x)dx\right] \\
&= (d_j - r_{j-1})^2 p(d_j) + (d_j - r_j)^2 p(d_j) \\
&= 0
\end{aligned}
\tag{1.31}
$$

which yields $(d_j - r_{j-1}) = \pm(d_j - r_j)$. Since $(d_j - r_{j-1}) > 0$ and $(d_j - r_j)$, only the condition $(d_j - r_{j-1}) = -(d_j - r_j)$ is valid. Thus, the decision level is defined between two reconstruction levels as

$$
d_j = \frac{r_j + r_{j-1}}{2}
\tag{1.32}
$$

implying that the decision level is the average of the two adjacent reconstruction levels.

The reconstruction level is defined in a similar manner, i.e.,

$$
\begin{aligned}
\frac{\partial \varepsilon}{\partial r_j} &= \frac{\partial}{\partial r_j}\left[\int_{d_j}^{d_{j+1}}(x - r_{j-1})^2 p(x)dx\right] \\
&= -2\int_{d_j}^{d_{j+1}}(x - r_j)^2 p(x)dx \\
&= 0
\end{aligned}
\tag{1.33}
$$

Hence,

$$
r_j = \frac{\int_{d_j}^{d_{j+1}} x p(x)dx}{\int_{d_j}^{d_{j+1}} p(x)dx}
\tag{1.34}
$$

implying that the output (reconstruction) level is the centroid of the adjacent input levels as illustrated in Figure 1.16.

Recursive solution of these equations for a given probability distribution provides optimum values for the decision and reconstruction levels. Max [Max60] has developed a solution for optimum design for a Gaussian density and has computed tables of optimum levels as a function of the number of quantization steps.

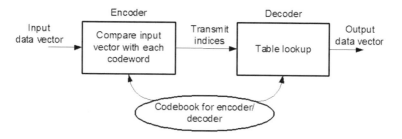

Figure 1.17: A block diagram of a simple vector quantizer.

If the decision and reconstruction levels are obtained to satisfy the optimum level conditions, it can easily be shown that the MSQE becomes

$$\varepsilon_{\min} = \sum_{j=0}^{J-1} \left[\int_{d_j}^{d_{j+1}} x^2 p(x) dx - \int_{d_j}^{d_{j+1}} r_j^2 p(x) dx \right] \qquad (1.35)$$

which would be zero if the reconstruction level is equivalent to input level. In the special case of a uniform probability density, the MSQE becomes

$$\varepsilon_{\min} = \frac{1}{12 J^2} = \frac{\Delta^2}{12} \qquad (1.36)$$

where $\Delta = 1/J$ denotes the quantization step size. The smaller number of output levels or the larger step size contributes to the larger quantization errors.

1.4.3   Vector Quantization

The concept of VQ is quite simple, i.e., identifying or representing an input vector with a member of the codebook (assume this has already been designed and stored) based on some valid criterion (best match, least distortion, etc). Basically this is a pattern matching procedure as illustrated in Figure 1.17. There are, of course, several ways of choosing the representative codevector. An index (not the codevector) is sent to the receiver where an exact replica of the codebook (look-up table) is stored. The decoder, using this index, retrieves the codevector from the look-up table and outputs it as the reconstructed vector. As such the decoder is inherently much simpler. VQ, hence, is ideal in a single encoder/multiple decoder scenario.

The result of Shannon's rate distortion theory is that for a given rate R (measured in bits/pixel), the distortion of the vector quantized image decreases with growing block dimension K [Ber71]. The limiting process (vector dimension reaching infinity) is, of course, impractical. From a practical viewpoint, a vector dimension of small size is selected. Scalar quantization is a special case of vector quantization with dimension 1. The main practical concerns in VQ are how to generate the codebook, how to scale the codebook for various rate-distortion requirements, and how much storage space is reasonable for storing the codebook. Once the codebook size is chosen, the objective is to design the codebook such that it is optimal in some fashion. One criterion, for the optimality, is that the selected codebook results in the lowest possible distortion among all the possible codebooks of the same size. Using a probabilistic model for the input vectors, an analytical technique for codebook design can be developed. This by itself is quite complex. A more reasonable and empirical approach is that formulated by Linde, Buzo and Gray [LBG80].

1.5   Rate-Distortion Theory

Assuming that X is the Gaussian distributed signal with zero mean and variance σ^2, i.e., its probability density function is given by $p_X(x) = \frac{1}{\sqrt{2\pi\sigma^2}}e^{-\frac{x^2}{2\sigma^2}}$, then the differential entropy is derived as

$$
\begin{aligned}
H(x) &= -\int_{-\infty}^{\infty} p_X(x)\ln p_X(x)\,dx \\
&= -2\int_{-\infty}^{\infty} p_X(x)\ln \frac{1}{\sqrt{2\pi\sigma^2}}e^{-\frac{x^2}{2\sigma^2}}\,dx \\
&= -\ln\frac{1}{\sqrt{2\pi\sigma^2}} + \frac{1}{2}\left(\int_{-\infty}^{\infty} p_X(x)\,dx = 1, \int_{-\infty}^{\infty} x^{2n}\,e^{-ax^2}\,dx = \frac{1\cdot 3\cdots(2n-1)}{2^{n+1}a^n}\sqrt{\frac{\pi}{a}}\right) \\
&= -\ln\frac{1}{\sqrt{2\pi\sigma^2}} + \frac{1}{2}\ln e \\
&= \frac{1}{2}\ln(2\pi e\sigma^2)
\end{aligned}
$$

(1.37)

where ln denotes the natural log.

Rate-distortion function $R(D)$ is defined as the minimum number of bits required to reproduce a memoryless source with distortion less than or equal to D, that implies the minimum mutual information between two random variables, the original and the reconstructed information, given by

$$
\begin{aligned}
R(D) &= \min I(x;\hat{x}) \leq I(x;\hat{x}) \\
&= H(x) - H(x|\hat{x})
\end{aligned}
$$

(1.38)

where $H(x|\hat{x})$ denotes the conditional entropy. Since entropy is invariant to shifting,

$$
\begin{aligned}
R(D) &= H(x) - H(x|\hat{x}) \\
&\leq H(x) - H(x - \hat{x}) \\
&= \frac{1}{2}\ln(2\pi e\sigma^2) - \frac{1}{2}\ln(2\pi eD)
\end{aligned}
$$

(1.39)

where D is defined by the mean square error (MSE) criterion, i.e., $D = E\{(x - \hat{x})^2\}$. Thus the rate-distortion lower-bound based on MSE of the Gaussian distributed signal is derived by

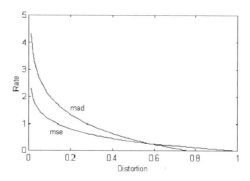

Figure 1.18: Rate-distortion curve based on distortion criteria, MSE and MAE.

$$R(D)|_{Gaussian,MSE} = \frac{1}{2}\ln\frac{\sigma^2}{D}, \qquad 0 \le D \le \sigma^2 \tag{1.40}$$

In the case of the mean absolute difference (MAD) criterion, the rate-distortion lower-bound is given by

$$\begin{aligned} R(D)|_{Gaussian,MAD} &= \tfrac{1}{2}\ln(2\pi e\sigma^2) - \ln(2eD) \\ &= \tfrac{1}{2}\ln\left(\frac{\pi\sigma^2}{2eD}\right), \quad 0 \le D \le \sqrt{\frac{\pi\sigma^2}{2e}} \end{aligned} \tag{1.41}$$

As compared in Figure 1.18, the MAD criterion with Gaussian source results in a higher rate than the MSE criterion when the distortion is less than about 0.6 [Ber71].

The encoder attempts to lower the bit rate as much as possible at the expense of greater distortion. Rate-distortion optimization has to be performed, in the practical codec, in order to maximize image quality (lower distortion), subject to transmission bit rate constraints. Practical algorithms for the control of bit rate can be judged according to how closely they approach optimum performance. Many rate control algorithms have been developed in various international standards, such as JPEG, MPEG, and ITU-T [RH96]. Sophisticated algorithms can achieve better performance, usually at the cost of increased computational complexity.

1.6   Human Visual Systems

A critical design goal for a digital video/image coding system is that the video/images produced by the system should be acceptable and pleasant to the viewer, i.e., human eye is the final observer. In order to achieve this goal it is necessary to take into account the response of the *human visual system* (HVS). The HVS is the system by which a human observer views, interprets and responds to visual stimuli. The simplified human

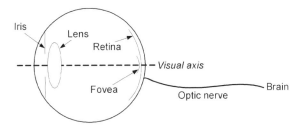

Figure 1.19: Simplified human eye cross section.

eye cross section is shown in Figure 1.19. The image is focused on the *retina* surface by the *lens* that changes shape under muscular control to perform proper focusing of near and distant objects. The *iris* controls the aperture of the lens and hence the amount of light entering the eye. The retina consists of an array of *cones* (photoreceptors sensitive to color at high light levels) and *rods* (photoreceptors sensitive to luminance at low light levels). The more sensitive cones are concentrated in a central region (the *fovea*) which means that high-resolution color vision is only achieved over a small area at the center of the field of view. Nerves connecting to the retina leave the eyeball through the *optic nerve*. The human brain processes and interprets visual information, based partly on the received information (the image detected by the retina) and partly on prior learned responses (such as known object shapes).

The human observer is the final assessor of the image's quality. Hence it is advantageous and important to incorporate the HVS in image processing applications. Therefore, the picture quality evaluation can be improved subjectively by including the HVS model with parameters like orientation and the field angle affecting the perception of the human eye. The operation of the HVS is a large and complex area of study. Some of the important features of the HVS are as follows.

1.6.1 Contrast Ratio

The response of the eye to changes in the intensity of illumination is known to be nonlinear. Consider a patch of light intensity surrounded by a background of intensity I. The just noticeable difference (JND), ΔI, of a target (object) intensity is to be determined as a function of I [Cho95], which states that the sensitivity of human eyes to discriminate between differences in intensity depends not only on the difference itself but also on the level of intensity. Over a wide range of intensities, it is found that the ratio $\Delta I / I$, called the *Weber fraction*, is nearly constant at a value of about 0.02 [Hec24, Sch86]. This result does not hold at very low or very high intensities, where the JND increases remarkably as shown in Figure 1.20. General expression $(I + \Delta I)/I$ represents the contrast ratio, which is constant in some range of intensity.

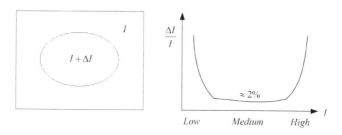

Figure 1.20: Contrast sensitivity measurements.

According to [SW96], the constant property of Weber's ratio is simulated for digital images with 8 bpp resolution and results are: The HVS is more sensitive at closer distances to the monitor for background intensity I from gray levels 0 to 50 (low gray level zone). However the viewing distances do not influence the JND values for background intensities above 50. For the influence of ambient lighting condition, the HVS is more sensitive when the room is dark, for the low gray level zone and it is independent for high gray level zone. For the influence of monitor brightness, the HVS is more sensitive to changes for higher monitor brightness and is less sensitive for lower one.

Compared to Weber's ratio which is flat (constant) in the middle range of intensity and increases in the lower and higher ranges, these experimental results show fairly flat JND values from gray levels 50 to 255 (only small increment shown from 235 to 255). This means that the dynamic range of monitor brightness does not expand so much, even though the gray level is maximum (255). From the results of contrast ratio, it is recognized that the quantization errors in the lower luminance range become insensitive and those in the middle range are sensitive.

1.6.2   Spatial Frequency

Overall measurements of the luminance threshold as a function of spatial frequency have been performed by many workers and some fruitful results are shown in [MS74], where a peak response of the *modulation transfer function* (MTF) is near 8 cyc/deg, as measured by

$$F_{MTF}(f_r) = 2.6(0.0192 + 0.144f_r)e^{-(0.114f_r)^{1.1}} \tag{1.42}$$

where f_r denotes the *spatial frequency*.

The MTF provides an easy means of integrating HVS sensitivity into image coding applications. Examples of MTF usage include the transformation of an image into perceptual domain prior to coding, estimation of image quality, and the determination of the distortion measure used in computing the bit-rate.

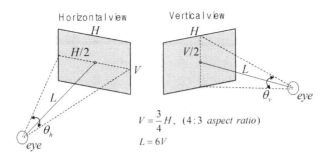

Figure 1.21: Viewing environment and spatial frequency concept.

Spatial frequency denotes a frequency normalized by viewing angle as in Figure 1.21, showing horizontal and vertical observation from constant distance, say, six times the vertical height. Then, the viewing angles are calculated as

$$
\begin{aligned}
\theta_h &= 2\arctan\left(\frac{H/2}{6V}\right) \\
\theta_v &= 2\arctan\left(\frac{V/2}{6V}\right)
\end{aligned}
\tag{1.43}
$$

Maximum spatial frequencies in horizontal direction and vertical direction are derived with the unit of cycles/degree as

$$
\begin{aligned}
f_{h,max} &= \frac{H/2}{\theta_h} \\
f_{v,max} &= \frac{V/2}{\theta_v}
\end{aligned}
\tag{1.44}
$$

Further, the spatial frequency of an object is obtained in horizontal direction $f_{h,VO}$ and in vertical direction $f_{v,VO}$ as

$$
\begin{aligned}
f_{h,VO} &= \frac{f_{h,max}}{H_{VO}/2} \\
f_{v,VO} &= \frac{f_{v,max}}{V_{VO}/2}
\end{aligned}
\tag{1.45}
$$

where H_{VO} and V_{VO} denote the horizontal and vertical sizes of an object, respectively.

Radial frequency is obtained using the spatial frequencies in both directions as

$$
f_{r,VO} = \sqrt{f_{h,VO}^2 + f_{v,VO}^2}
\tag{1.46}
$$

The corresponding sensitivity of an object can be calculated by using the empirically derived MTF such as in (1.42).

Quantization matrices are used for quantizing the transform coefficients in JPEG and MPEG as discussed in Section 1.7. The matrices have nonnegative elements that become large in the horizontal and vertical directions. The elements are divided by values of the coefficients for quantization. Larger element value or division by a large number means the coarse quantization of the high frequency coefficients, while the lower frequency coefficients are quantized with finer step sizes or division by a smaller number. Hence, the international standards have more or less utilized the HVS concept in the coder.

The fact that human eye is not equally sensitive to distorted signals by a video/image encoder can be explained in view of the spatial frequency concept of the HVS. For example, assuming that the sampling density is 64 pixels/deg and 512×512 pixels in an image, the lowest spatial frequency in an 8×8 block is about 4 cyc/deg, which lies in the most sensitive band in the MTF. Considering that the block effect of the DCT-based coder is dominant with large distortion in the lowest coefficient with the size of 8×8, the spatial frequency concept takes a role of why it is so sensitive to human eye.

1.6.3   Masking Effect

The visibility of the distortion is of primary importance in any image coding application. When an image has a high activity content, there is a loss of sensitivity to errors in these regions. This is the *masking effect* from the image activity. Researchers attempted to quantify the amount of masking by using different visibility functions to measure the sensitivity of the eye to distortions [SN77, LR78]. Due to variations in viewing distance, sampling frequency and vast differences in picture content, the subjectively measured thresholds do not apply globally in all cases. An adaptation to local picture content, using more complex activity functions to measure the masking function, was introduced by [Sch83]. These masking functions are used in the design of quantizers to achieve a lower entropy.

1.6.4   Mach Bands

Images containing sharp edges produce large prediction errors in predictive coding or large transform coefficients in transform coding, which are the main obstacles in a high compression coder. Mach (named for the Austrian physicist E. Mach) bands are frequently used for explaining the problem, defined as an illusion that emphasizes edges in a picture where the intensity is in fact changing smoothly [Gla95]. Figure 1.22 shows a set of vertical gray bars with eight edges. Within each band, the gray levels are constant. At the boundary between two edges, the right side of each band appears a bit darker than the middle of the band, and left side appears a bit lighter. The transition from one band to another is emphasized by the human visual sensitivity to the intensity.

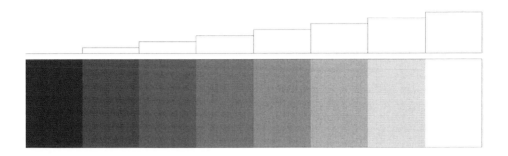

Figure 1.22: Gray scale bands in equal increments of intensity.

1.7 Digital Picture Coding Standards and Systems

1.7.1 JPEG-Still Image Coding Standard

The JPEG standard is intended to compress still continuous tone monochrome and color images. This may be achieved by one of the four main processing modes, i.e., sequential, progressive, lossless, and hierarchical mode of operations. Compression and storage are main targets of JPEG baseline system at bit rates ranging 0.25 - 0.5 bit per pixel with moderate to good image quality [LOW91].

Figure 1.23 shows the JPEG baseline system. Input source image data are level shifted to a signed two's complement representation by subtracting 2^{p-1}, where p is the precision parameter of the image intensity in bits. When $p = 8$ in baseline mode, the level shift is 128. An inverse level shift at the decoder restores the sample to the unsigned 8 bit representation. The level shifting does not affect AC coefficients or variances. It affects only the DC coefficient, shifting a neutral gray intensity to zero. Difference values between DC coefficients are also unaffected. Initial starting value at the beginning of the image and at restart interval, may be set to zero. The level shifted image is partitioned into 8×8 blocks along a raster scan fashion for the 2-D (8×8) DCT operations. The transform coefficients are quantized by using predefined quantization tables for luminance and chrominance. In spite of the transform coefficients being quantized to nonnegative values based on criteria that optimal step size should be defined by its entropy and energy and the quantization error should be less than the visual sensitivity, JPEG [ISO93b] intensively experimented for simple and easy quantization methods and suggested default tables for quantization of DC and AC coefficients. Each of the 64 resulting DCT coefficients is quantized by a different uniform quantizer.

The step sizes are based on visibility thresholds of 64-element quantization matrices for luminance and chrominance components.

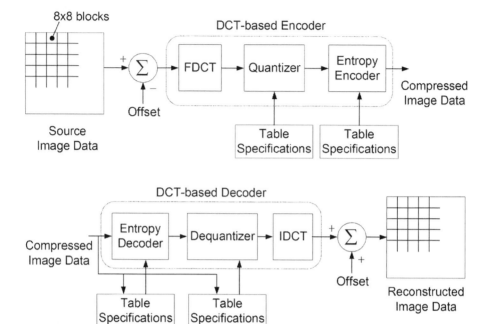

Figure 1.23: JPEG baseline encoding and decoding flow.

Quantized DCT coefficients, S_{quv}, are defined by the following equation.

$$S_{quv} = \text{Nearest Integer} \left(\frac{S_{uv}}{Q_{uv}} \right) \qquad (1.47)$$

where S_{uv} is DCT coefficient and Q_{uv} is quantization matrix. This is useful for simple and fast implementation of uniform quantizers for 8×8 coefficients, reducing complicated steps at the coder. Figure 1.24 shows the quantization matrices which are obtained empirically. The step size Q_{uv} is a function of the transform coefficient frequency. Coefficients at lower frequency are quantized with smaller step sizes and ones at higher frequency are quantized with larger step sizes, resulting in larger irreversible quantization errors.

Since the encoder and decoder should be symmetric for perfect reconstruction at the decoder, the quantization matrix elements need to be transmitted to the decoder. If the default quantization matrices are used, then there is no need to transmit them. Inverse operation is accomplished by multiplying the same elements, as shown in (1.48). For the DC coefficient, reconstruction from differential value is preceded, since differential DC value between two adjacent blocks has been quantized, while each AC coefficient has been uniformly quantized.

16	11	10	16	24	40	51	61
12	12	14	19	26	58	60	55
14	13	16	24	40	57	69	56
14	17	22	29	51	87	80	62
18	22	37	56	68	109	103	77
24	35	55	64	81	104	113	92
49	64	78	87	103	121	120	101
72	92	95	98	112	100	103	99

a)

17	18	24	47	99	99	99	99
18	21	26	66	99	99	99	99
24	26	56	99	99	99	99	99
47	66	99	99	99	99	99	99
99	99	99	99	99	99	99	99
99	99	99	99	99	99	99	99
99	99	99	99	99	99	99	99
99	99	99	99	99	99	99	99

b)

Figure 1.24: Luminance a) and chrominance b) quantization matrix example in JPEG [ISO93b].

Reconstruction of coefficients is performed by the same quantization matrix as

$$R_{uv} = S_{quv}Q_{uv} \qquad (1.48)$$

Since blocking artifacts mainly from DC coefficient are sensitive to spatial frequency response by the human visual system, DC coefficient is treated separately from the 63 AC coefficients. It is differentially coded by the following first-order prediction

$$DIFF = DC_i - DC_{i-1} \qquad (1.49)$$

where DC_i and DC_{i-1} are current (8×8)-pixel block and previous (8×8)-pixel block DC coefficients, respectively.

The quantized 63 AC coefficients are formatted as per the zigzag scan (Figure 1.25), in preparation for entropy coding. Along the zigzag scan, the DCT coefficients represent increasing spatial frequencies and in general decreasing variances. Besides, quantization by coarse step for the high frequency coefficients (Figure 1.24), responding to the HVS sensitivity results in many zero coefficients after quantization. An efficient VLC table can be developed, representing runs of zero coefficients along the zigzag scan followed by the size of the non zero coefficient.

Since the JPEG standard, the needs and requirements of advanced still image coding have grown and evolved. In March 1997, a call for contributions were launched for the development of JPEG2000 standard that became the International Standard by the end of 2000. The JPEG2000 standard provides a set of features that are important to many high-end applications. It is also based on the idea that the coefficients of a transform that decorrelates the pixels of an image can be coded more efficiently than the original data. However, the wavelet transform, instead of the DCT, packs most of the important visual information into a small number coefficients.

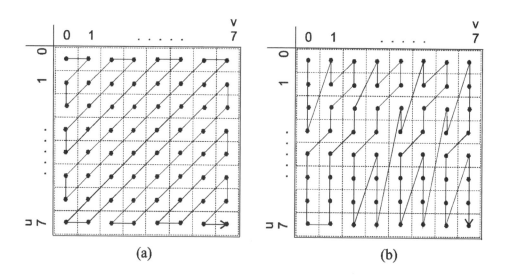

Figure 1.25: Scanning order of the DCT coefficients; a) zigzag scan adopted by most of coding standards and b) alternate scan used in MPEG-2.

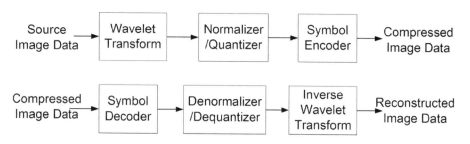

Figure 1.26: Simplified JPEG2000 encoder/decoder block diagram.

Figure 1.26 shows a simplified JPEG2000 coding system. As in the JPEG standard, level shift is first applied by subtracting 2^{m-1}, where 2^m is the number of gray levels in the image. The discrete wavelet transform is then computed. The transform coefficients are quantized and entropy coded before forming the output bitstream. The decoder is the reverse of the encoder. The compressed bitstream is first entropy decoded, dequantized and inverse wavelet transformed, thus resulting in the reconstructed image data. The forward DWT is performed by a series of 1-D subband decompositions into a lowpass band and a highpass band. Lowpass samples represent a downsampled low-resolution version of the original data set, while highpass samples represent a downsampled high-resolution version. The transform can be used for reversible (lossless, error-free) or irreversible (lossy) compression. The default reversible transform is performed by means of the (5, 3) filter, while the default irreversible transform is performed by means of the Daubechies (9, 7) filter. Note that the first number in the filter specification represents

Table 1.1: Simulation results of *Girl* image by JPEG and JPEG2000 codec.

	Compression Ratio	Error(rms)	Compression parameter
JPEG	21.85	7.63	Quality factor = 4
JPEG2000	22.53	6.26	m = 8, e = 8.6

the number of filter taps used for lowpass filtering at the analysis stage or for high-pass filtering at the synthesis stage. The filter coefficients are referred to in the reference [ISO00, CSE00]. After transformation, all coefficients are quantized. Quantization is the process by which the coefficients are reduced in precision, resulting in image data compression. It is a lossy operation that can not be reversible, unless the quantization step size is 1. Each of the transform coefficients S_{uv} in a subband is quantized to the value S_{quv} according the formula:

$$S_{quv} = S_{uv} \lfloor \frac{|S_{uv}|}{\triangle_b} \rfloor \tag{1.50}$$

where $\triangle_b$ denotes the quantization step in the subband which is defined by

$$\triangle_b = 2^{R_b - e_b}(1 + \frac{m_b}{2^{11}}) \tag{1.51}$$

where R_b is the nominal dynamic range of the subband and e_b and m_b are the number of bits allotted to the *exponent* and *mantissa* of the transform coefficients. The nominal dynamic range is the sum of the number of bits used to represent the original image and the analysis gain bits for subband b. The *exponent/mantissa* pairs (e_b, m_b) are either explicitly signaled in the bit stream syntax for every subband, which is known as *explicit quantization*, or only signaled in the bit stream for the LL band, which is known as *implicit quantization*. In the latter case, the remaining subbands are quantized using extrapolated LL subband parameters. Letting e_0 and m_0 be the number of bits allocated to the LL subband, the extrapolated parameters for subband b are

$$m_b = m_0 \tag{1.52}$$
$$e_b = e_0 nsd_b - nsd_0 \tag{1.53}$$

where nsd_b denotes the number of subband decomposition levels from the original image to subband b.

The principal difference between the DCT-based JPEG and the wavelet-based JPEG-2000 is the omission of the subblock process in the latter. Subdivision into blocks is unnecessary, since the wavelet transform is both inherently local (i.e., their basis functions are varied in duration) and computationally efficient as well. Figure 1.27 shows two

Figure 1.27: Comparison between JPEG (left column) and JPEG2000 (right column) codec. From the top, reconstruction images, enlarged images to distinguish details, and error images are shown.

column reconstruction images from the JPEG (left column) and JPEG2000 (right column) codec. From the top row, reconstructed images, enlarged partial images, and error images are shown. The removal of the subdivision process in the JPEG2000 eliminates the blocking artifact that characterizes DCT-based codec at high compression ratios. Besides decreasing reconstruction error in rms value as shown in Table 1.1, JPEG2000 increases image quality in a subjective perceptual sense.

1.7.2   MPEG-Video Coding Standards

MPEG-1 was developed for multimedia CD-ROM applications named ISO/IEC specification 11172, which is the standard for coding of moving picture and associated audio

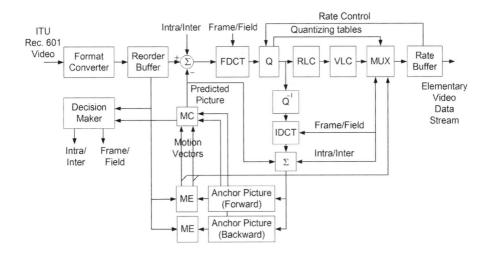

Figure 1.28: Encoder block diagram of the MPEG-1 video coder. There is not so much change in MPEG-2 coder to maintain backward compatibility with MPEG-1.

for digital storage media at up to about 1.5 Mbps. MPEG-2 is formally referred to as ISO/IEC specification 13818, which is the second phase of MPEG video coding solution for applications not originally covered by the MPEG-1 standard. Specifically, MPEG-2 was developed to provide video quality not lower than NTSC/PAL and up to HDTV quality. Its target bit rates for NTSC/PAL are about 2 to 15 Mbps. The bit rates used for HDTV signals are about 19 Mbps. In general, MPEG-2 can be seen as a superset of the MPEG-1 coding standard and is backward compatible to the MPEG-1 standard. In other words, every MPEG-2-compatible decoder is able to decode a compliant MPEG-1 bit stream [MPFL97, RH96].

MPEG-1 video coding algorithm is mainly based on the DCT coding for intraframe redundancy reduction and motion estimation/compensation for interframe redundancy reduction. In MPEG coding, the video sequence is first divided into groups of pictures (GOP). Each GOP may include three types of pictures or frames: intra (I), predictive (P), and bidirectionally (B) predictive picture. I-pictures are coded by intraframe techniques only, with no previous frame information. They are used as random access points and anchors for forward and/or backward prediction. P-pictures are coded using forward motion-compensated prediction from a previous I or P picture. The distance between two nearest I-frames is the size of GOP. The distance between two nearest anchor frames (I or P) is another parameter of GOP structure. These parameters are user-selectable parameters during the encoding. A larger GOP size increases the coding performance but also causes error propagation or drift. Usually the size is chosen from 12 to 15 and anchor distance from 1 to 3.

The typical MPEG-1 video encoder structure is shown in Figure 1.28. It uses motion compensation to remove the interframe redundancy. The concept of motion compensation is based on the estimation of motion between video frames. Motion vectors are used for replacing a corresponding 16×16-pixel macroblock by one from a previous frame at the decoder. Assuming the spatial correlation between adjacent pixels is usually very high, it is not necessary to transmit sub-block information. This would be too expensive and the coder would never be able to reach a high compression ratio. The motion vectors for any block are found within a search window that can be up to 512 pixels in each direction. Also, the matching can be done at half-pixel accuracy, where the half-pixel values are computed by averaging the full-pixel values. For interframe coding, the prediction differences or error images are coded and transmitted with motion information. A 2-D DCT is used for coding both the intraframe pixels and the predictive error pixels. The image to be coded is first partitioned into 8×8-pixel blocks, followed by 2-D 8×8-point DCT, resulting in a frequency domain representation. The goal of the transformation is to decorrelate and compress the block data so that the transform coefficients are quantized by using the quantization matrices. High frequency coefficients are quantized with coarser quantization steps that will suppress high frequencies with no subjective degradation, thus taking advantage of human visual perception characteristics. Selection of the quantization matrices and scale factors (used to control the quantization matrix) are the main elements to achieve high compression with less degradation of picture quality.

The goal of the MPEG-4 standard is to provide the core technology that allows efficient content-based coding and storage/transmission/manipulation of video, graphics, audio, and other data. It has many interesting features such as improved coding efficiency, robustness of transmission, and interactivity among end users.

It is possible to code multiple video objects (MVO), since the contents are represented in the form of primitive audio visual objects. They can be natural or synthetic scenes. The incorporation of an object- or content-based coding structure is the feature that allows MPEG-4 to provide more functionality. The receiver can receive separate bitstreams for different objects contained in the video. To achieve content-based coding, the MPEG-4 uses the concept of a video object plane (VOP). It is assumed that each frame of an input video is first segmented into a set of arbitrarily shaped regions or VOPs. Each region can cover a particular image or video object in the scene. The shape and the location of the VOP can vary from frame to frame. A sequence of VOPs is referred to as a video object (VO). The different VOs may be encoded into separate bitstreams. MPEG-4 specifies demultiplexing and composition syntax which provide the tools for the receiver to decode the separate VO bitstreams and composite them into a frame. In this way, the decoders have more flexibility to edit or rearrange the decoded video objects.

MPEG-4 provides many flexible and functional tools as follows:

- Motion estimation and compensation

- Texture coding/Shape coding
- Sprite coding
- Interlaced video coding
- Wavelet-based texture coding
- Generalized temporal, spatial, and hybrid scalabilities
- Error resilience

In late 2001, ISO/IEC MPEG and ITU-T VCEG (Video Coding Experts Group) decided on a joint venture towards enhancing video coding performance - specifically in the areas where bandwidth and/or storage capacity are limited. This joint team of both standard organizations is called JVT (Joint Video Team). The standard thus formed is called H.264/MPEG-4 Part 10 and is preferably referred to as JVT/H.26L/Advanced Video Coding (AVC). MPEG-2, H.263, MPEG-4 and H.264/MPEG-4 Part 10 are based on similar concepts and the main differences involved are the prediction signal, the block sizes used for transform coding and the entropy coding [J+02].

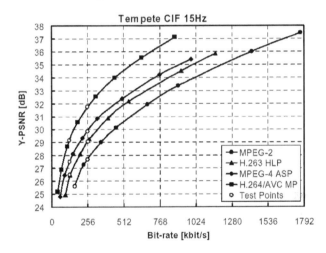

Figure 1.29: Rate distortion curves for *Tempete* sequence at 15 Hz in streaming video comparison [J+02]

H.264/MPEG-4 Part 10 has some specific additional features that distinguish it from other standards. It uses more flexible motion compensation model supporting various rectangular partitions in each macroblock. Previous standards allowed only square sized partitions in a macroblock. Multiple reference pictures also help in better prediction, though the complexity is increased. Moreover, quarter-pel accuracy provides high spatial accuracy. Due to these enhanced properties, the H.264 provides better performance than existing standards as shown in Figure 1.29. Table 1.2 provides a comparison of MPEG-1, MPEG-2, MPEG-4 and H.264/MPEG-4 Part 10.

Since the H.264/MPEG-4 Part 10 includes enhanced features like multiple reference frames, CABAC and various profiles, it is suitable for both video streaming and video

Table 1.2: Comparison of standards MPEG-1, MPEG-2, MPEG-4 and H.264/MPEG-4 Part 10.

Feature/ Standard	MPEG-1	MPEG-2	MPEG-4	H.264/MPEG-4 Part 10
Macroblock size	16x16	16x16 (frame mode) 16x 8 (field mode)	16x16	16x16
Block size	8x8	8x8	16x16, 8x8, 16x8 8x8, 16x8	8x8, 16x8, 8x16, 16x16, 4x8, 4x4, 4x4
Transform	DCT	DCT	DCT/Wavelet transform	4x4 Integer transform
Transform size	8x8	8x8	8x8	4x4
Quantization step size	Increases with constant increment	Increases with constant increment	Vector Quantization	Step sizes increase at rate of 12.5%
Entropy coding	VLC	VLC(different VLC tables for Intra and Inter modes)	VLC	VLC, CAVLC and CABAC
Motion estimation & compensation	Yes	Yes	Yes	Yes, more flexible up to 16 MVs per MB
Pel accuracy	Integer 1/2-pel	Integer 1/2-pel	Integer 1/2-pel, 1/4-pel	Integer 1/2-pel, 1/4-pel
Profiles	No	5 profiles. Several levels within a profile	8 profiles. Several levels within a profile	3 profiles. Several levels within a profile
Reference frame	Yes, One frame	Yes, One frame	Yes, One frame	Yes, Multiple frames(as many as 5 frames allowed)
Picture types	I, P, B, D	I, P, B	I, P, B	I, P, B, SI, SP
Playback & random access	Yes	Yes	Yes	Yes
Error robustness	Synchronization concealment	Data partitioning, redundancy, FEC for important packet transmission	Synchronization, data partitioning, header extension, reversible VLCs	Deals with packet loss and bit errors in error-prone wireless network
Transmission rate	Up to 1.5Mbps	2-15 Mbps	64Kbps-2Mbps	64Kbps-150Mbps
Encoder complexity	Low	Medium	Medium	High
Compatible with previous standards	Yes	Yes	Yes	No

conferencing applications. Error resilience is achieved by using parameter sets, which can be either transmitted in-band or out-of-band. It is more flexible compared to the previous standards and this enables improved coding efficiency as shown in Table 1.3. However, it should be noted that this is at the expense of increased complexity and is not backward compatible to the previous standards. One way to reduce the level of complexity of the decoder is to design specifically for a profile and a level.

1.8 Summary

A simple introduction to digital video/image compression and coding is presented. This is followed by insights to various compression and coding techniques. Theoretical

Table 1.3: Average bit rate savings for video streaming [J$^+$02].

	Average bit rate savings relative to		
Coder	MPEG-4 ASP	H.263	MPEG-2
H.264/MPEG-4 Part 10	39%	49%	64%
MPEG-4 ASP	-	17%	43%
H.263 HLP	-	-	31%

background is closely incorporated in international standards such as JPEG, MPEG, and H.26x series. The emphasis is on general concepts and quality of reconstructed video/image rather than detailed and in-depth description of the different compression algorithms. References at the end of the chapter dealing with the ITU/ISO/IEC standards, basic books on data compression, digital image processing, etc., provide the readers with a full familiarization with this ever-expanding field.

References

[Ber71]   T. Berger. *Rate-Distortion Theory*. Upper Saddle River, NJ: Prentice Hall, 1971.

[Bov05]   A. Bovik (Ed.). *Handbook of Image and Video Processing*. Orlando, FL: Academic Press, 2nd ed., 2005.

[CCI90]   CCITT Study Group XV. Recommendations of the H–Series, CCITT Study Group XV - Report R37. Technical report, CCITT, August 1990.

[Cho95]   C. H. Chou. Adaptive transform coding of images based on removing just noticeable distortion. In *Proceedings of SPIE Visual Com. Im. Proc.*, 2501:607–618, Taipei, Taiwan, May 1995.

[Cla85]   R. J. Clarke. *Transform Coding of Images*. New York: Academic Press, 1985.

[CSE00]   C. Christopoulos, A. Skodras, and T. Ebrahimi. The JPEG2000 still image coding system: An overview. *IEEE Trans. Consumer Elect.*, 46:1103–1127, November 2000.

[Ger77]   A. Gersho. Quantization. *IEEE Communications Society Magazine*, 15:16–29, September 1977.

[Gha03]   M. Ghanbari. *Standard Codecs: Image Compression to Advanced Video Coding*. London, UK: IEE, 2003.

[Gib97]   J. D. Gibson (Ed.). *Multimedia Compression: Applications and Standards*. San Francisco, CA: Morgan Kaufmann, 1997.

[Gla95]   A. S. Glassner. *Principles of Digital Image Synthesis*. San Francisco, CA: Morgan Kaufman, 1995.

[GW02]   R. C. Gonzalez and R. E. Woods. *Digital Image Processing*. Upper Saddle River, NJ: Prentice–Hall, 2nd ed., 2002.

[Hec24]   S. Hecht. The visual discrimination of intensity and the weber-fechner law. *Journal General Physiology*, 7:241, September 1924.

[ISO92]   ISO/IEC JTC1/SC2/WG9. CD11544, Progressive Bi-Level Image Compression, Revision 4.1. Technical report, ISO/IEC, September 1992.

[ISO93a]   ISO/IEC. ISO/IEC 11172 Information Technology: Coding of Moving Pictures and Associated Audio for Digital Storage Media at up to about 1.5 Mbit/s. Technical report, ISO/IEC, 1993.

[ISO93b]   ISO/IEC JTC1 10918-1. ITU-T Rec. T.81, Information Technology — Digital Compression and Coding of Continuous-Tone Still Images: Requirements and Guidelines. Technical report, ISO/IEC, 1993.

[ISO93c]   ISO/IEC JTC1/SC29/WG11. CD 13818, Generic Coding of Moving Pictures and Associated Audio. Technical report, ISO/IEC, November 1993.

[ISO00]   ISO/IEC FCD 15444-1. JPEG 2000 Image Coding System. Technical report, ISO/IEC, March 2000.

[ISO01]   ISO/IEC. ISO/IEC 14496, Information Technology — Coding of Audio-Visual Objects. Technical report, ISO/IEC, 2001.

[ITU82]   ITU-R. ITU-R Recommendation BT.601, Encoding Parameters of Digital Television for Studios. Technical report, ITU-R, 1982.

[ITU96]   ITU-T. ITU-T Recommendation H.263, Video Coding for Low Bit Rate Communication. Technical report, ITU-T, 1996.

[ITU03]   ITU-T. ITU-T Recommendation H.264, Advanced Video Coding for Generic Audiovisual Services. Technical report, ITU-T, 2003.

[J$^+$02]   A. Joch *et al.* Performance comparison of video coding standards using lagrangian coder control. In *Proceedings of International Conference on Image Processing*, 2:501–504, September 2002. Special Session: The emerging JVT/H.26L video coding standard.

[JN84]   N. S. Jayant and P. Noll. *Digital Coding of Waveforms, Principles and Applications to Speech and Video.* Upper Saddle River, NJ: Prentice Hall, 1984.

[LBG80]   Y. Linde, A. Buzo, and R. M. Gray. An algorithm for vector quantization design. *IEEE Trans. on Commun.*, 28:84–95, January 1980.

[LOW91]   A. Leger, T. Omachi, and G. K. Wallace. JPEG still picture compression algorithm. *Optical Engineering*, 30:949–954, July 1991.

[LR78]   J. O. Limb and C. B. Rubinstein. On the design of quantizer for DPCM coders : A functional relationship between visibility, probability and masking. *IEEE Trans. on Commun.*, 26:573–578, May 1978.

[Max60]   J. Max. Quantizing for minimum distortion. *IRE Trans. Information Theory*, IT-6:16–21, March 1960.

[MPFL97]   J. L. Mitchell, W. B. Pennebaker, C. E. Fogg, and D. J. LeGall. *MPEG Video Compression Standard.* New York: Chapman Hall, 1997.

[MS74]   J. L. Mannos and D. J. Sakrison. The effects of a fidelity criterion on the encoding of images. *IEEE Trans. Information Theory*, IT-20:525–536, July 1974.

[NH95]   A. N. Netravali and B. G. Haskell. *Digital Pictures-Representation, Compression, and Standards.* New York and London: Plenum Press, 2nd ed., 1995.

[O'N76] J. B. O'Neal Jr. Differential pulse code modulation with entropy coding. *IEEE Trans. Information Theory*, IT-21:169–174, March 1976.

[PC00] A. Puri and T. Chen (Eds.). *Multimedia Systems, Standards and Networks*. New York: Marcel Dekker, 2000.

[PE02] F. Pereira and T. Ebrahimi (Eds.). *The MPEG-4 Book*. Upper Saddle River, NJ: IMSC Press, 2002.

[RH96] K. R. Rao and J. J. Hwang. *Techniques and Standards for Image, Video, and Audio Coding*. Upper Saddle River, NJ: Prentice Hall, 1996.

[Ric03] I. E. G. Richardson. *H.264 and MPEG-4 Video Compression*. Hoboken, NJ: Wiley, 2003.

[RY90] K. R. Rao and P. Yip. *Discrete Cosine Transform: Algorithms, Advantages, Applications*. San Diego, CA: Academic Press, 1990.

[Sal00] D. Saloman. *Data Compression : The Complete Reference*. Berlin: Springer–Verlag, 2000.

[Say00] K. Sayood. *Introduction to Data Compression*. San Francisco, CA: Morgan Kaufmann, 2000.

[Sch83] R. Schafer. Design of adaptive and nonadaptive quantizers using subjective criteria. *Signal Processing*, 5:333–345, July 1983.

[Sch86] W. F. Schreiber. *Fundamentals of Electronic Imaging Systems*. Berlin: Springer–Verlag, 1986.

[SN77] D. K. Sharma and A. N. Netravali. Design of quantizers for DPCM coding of picture signals. *IEEE Trans. on Commun.*, 25:1267–1274, November 1977.

[SW96] D. Shen and S. Wang. Measurements of JND property of HVS and its applications to image segmentation, coding and requantization. In *Proceedings of SPIE Dig. Comp. Tech. Sys. for Video Com.*, 2952:113–121, October 1996.

[VH92] M. Vetterli and C. Herley. Wavelets and filter banks: Theory and design. *IEEE Trans. on Sig. Proc.*, 40:2207–2232, November 1992.

Chapter 2

Fundamentals of Human Vision and Vision Modeling

Ethan D. Montag and Mark D. Fairchild
Rochester Institute of Technology, U.S.A.

2.1 Introduction

Just as the visual system evolved to best adapt to the environment, video displays and video encoding technology is evolving to meet the requirements demanded by the visual system. If the goal of video display is to veridically reproduce the world as seen by the mind's eye, we need to understand the way the visual system itself represents the world. This understanding involves finding the answer to questions such as: What characteristics of a scene are represented in the visual system? What are the limits of visual perception? What is the sensitivity of the visual system to spatial and temporal change? How does the visual system encode color information?

In this chapter, the limits and characteristics of human vision will be introduced in order to elucidate a variety of the requirements needed for video display. Through an understanding of the fundamentals of vision, not only can the design principles for video encoding and display be determined but methods for their evaluation can be established. That is, not only can we determine what is essential to display and how to display it, but we can also decide on the appropriate methods to use to measure the quality of video display in terms relevant to human vision. In addition, an understanding of the visual system can lead to insight into how rendering video imagery can be enhanced for aesthetic and scientific purposes.

2.2 A Brief Overview of the Visual System

Our knowledge of how the visual system operates derives from many diverse fields ranging from anatomy and physiology to psychophysics and molecular genetics, to name a few. The physical interaction of light with matter, the electro-chemical nature of nerve conduction, and the psychology of perception all play a role. In his *Treatise on Phys-iological Optics* [Hel11], Helmholtz divides the study of vision into three parts: 1) the theory of the path of light in the eye, 2) the theory of the sensations of the nervous mech-anisms of vision, and 3) the theory of the interpretation of the visual sensations. Here we will briefly trace the visual pathway to summarize its parts and functions.

Light first encounters the eye at the cornea, the main refractive surface of the eye. The light then enters the eye through the pupil, the hole in the center of the circular pigmented iris, which gives our eyes their characteristic color. The iris is bathed in aqueous humor, the watery fluid between the cornea and lens of the eye. The pupil diameter, which ranges from about 3 to 7 mm, changes based on the prevailing light level and other influences of the autonomic nervous system. The constriction and dilation of the pupil changes its area by a factor of 5, which as we will see later contributes little to the ability of the eye to adapt to the wide range of illumination it encounters. The light then passes through the lens of the eye, a transparent layered body that changes shape with accommodation to focus the image on the back of the eye. The main body of the eyeball contains the gelatinous vitreous humor maintaining the eye's shape. Lining the back of the eye is the retina, where the light sensitive photoreceptors transduce the electromagnetic energy of light into the electro-chemical signals used by the nervous system. Behind the retina is the pigment epithelium that aids in trapping the light that is not absorbed by the photoreceptors and provides metabolic activities for the retina and photoreceptors.

An inverted image of the visual field is projected onto the retina. The retina is consid-ered an extension of the central nervous system and in fact develops as an outcropping of the neural tube during embryonic development. It consists of five main neural cell types organized into three cellular layers and two synaptic layers. The photoreceptors, spe-cialized neurons that contain photopigments that absorb and initiate the neural response to light, are located on the outer part of the retina meaning that light must pass through all the other retinal layers before being detected. The signals from the photoreceptors are processed via the multitude of retinal connections and eventually exit the eye by way of the optic nerve, the axons of the ganglion cells, which make up the inner cellular layer of the retina. These axons are gathered together and exit the eye at the optic disc forming the optic nerve that projects to the lateral geniculate nucleus, a part of the thalamus in the midbrain. From here, there are synaptic connections to neurons that project to the primary visual cortex located in the occipital lobe of the cerebral cortex. The human cortex contains many areas that respond to visual stimuli (covering 950 cm^2 or 27% of the cortical surface [VE03]) where processing of the various modes of vision such as form, location, motion, color, etc. occur. It is interesting to note that these various areas maintain a spatial mapping of the visual world even as the responses of neurons become more complex.

There are two classes of photoreceptors, rods and cones. The rods are used for vision at very low light levels (scotopic) and do not normally contribute to color vision.

The cones, which operate at higher light levels (photopic), mediate color vision and the seeing of fine spatial detail. There are three types of cones known as the short wavelength sensitive cones (S-cones), the middle wavelength sensitive cones (M-cones), and the long wavelength sensitive cones (L-cones), where the wavelength refers to the visible region of the spectrum between approximately 400 - 700 nm. Each cone type is color-blind; that is, the wavelength information of the light absorbed is lost. This is known as the Principle of Univariance so that the differential sensitivity to wavelength by the photoreceptors is due to the probability of photon absorption. Once a photopigment molecule absorbs a photon, the effect on vision is the same regardless of the wavelength of the photon. Color vision derives from the differential spectral sensitivities of the three cone types and the comparisons made between the signals generated in each cone type. Because there are only three cone types, it is possible to specify the color stimulus in terms of three numbers indicating the absorption of light by each of the three cone photoreceptors. This is the basis of trichromacy, the ability to match any color with a mixture of three suitably chosen primaries.

As an indication of the processing in the retina, we can note that there are approximately 127 million receptors (120 million rods and 7 million cones) in the retina yet only 1 million ganglion cells in the optic nerve. This overall 127:1 convergence of information in the retina belies the complex visual processing in the retina. For scotopic vision there is convergence, or spatial summation, of 100 receptors to 1 ganglion cell to increase sensitivity at low light levels. However, in the rod-free fovea, cone receptors may have input to more than one ganglion cell. The locus of light adaptation is substantially retinal and the center-surround antagonistic receptive field properties of the ganglion cells, responsible for the chromatic-opponent processing of color and lateral inhibition, are due to lateral connections in the retina.

Physiological analysis of the receptive field properties of neurons in the cortex has revealed a great deal of specialization and organization in their response in different regions of the cortex. Neurons in the primary visual cortex, for example, show responses that are tuned to stimuli of specific sizes and orientations (see [KNM84]). The area MT, located in the medial temporal region of the cortex in the macaque monkey, is an area in which the cells are remarkably sensitive to motion stimuli [MM93]. Damage to homologous areas in humans (through stroke or accident, for example) has been thought to cause defects in motion perception [Zek91]. Deficits in other modes of vision such as color perception, localization, visual recognition and identification, have been associated with particular areas or processing streams within the cortex.

2.3 Color Vision

As pointed out by Newton, color is a property of the mind and not of objects in the world. Color results from the interaction of a light source, an object, and the visual system. Although the product of the spectral radiance of a source and the reflectance of an object (or the spectral distribution of an emissive source such as an LCD or a CRT display) specify the spectral power distribution of the distal stimulus to color vision, the color signal can be considered the product of this signal with the spectral sensitivity of the three cone receptor types. The color signal can therefore be summarized as three numbers that express the absorption of the three cone types at each "pixel" in the scene. Unfortunately, a standard for specifying cone signals has not yet been agreed upon, but the basic principles of color additivity have led to a description of the color signal that can be considered as linearly related to the cone signals.

2.3.1   Colorimetry

Trichromacy leads to the principle that any color can be matched with a mixture of three suitable chosen primaries. We can characterize an additive match by using an equation such as:

$$C1 = r_1 R + g_1 G + b_1 B, \tag{2.1}$$

where the Red, Green, and Blue primaries are mixed together to match an arbitrary test light C by adjusting their intensities with the scalars r, g, and b. The amounts of each primary needed to make the match are known as the tristimulus values. The behavior of color matches follows the linear algebraic rules of additivity and proportionality (Grassmann's laws) [WS82]. Therefore, given another match:

$$C2 = r_2 R + g_2 G + b_2 B, \tag{2.2}$$

the mixture of lights C1 and C2 can be matched by adding the constituent match components:

$$C3 = C1 + C2 = (r_1 + r_2)R + (g_1 + g_2)G + (b_1 + b_2)B. \tag{2.3}$$

Scaling the intensities of the lights retains the match (at photopic levels):

$$kC3 = kC1 + kC2. \tag{2.4}$$

The color match equations can therefore be scaled so that the coefficients of the primaries add to one representing normalized mixtures of the primaries creating unit amounts of the matched test light. These normalized weights are known as chromaticities and can be calculated by dividing the tristimulus value of each primary by the sum of all three tristimulus values. In this way matches can be represented in a trilinear coordinate system (a Maxwell diagram) as shown in Figure 2.1 [KB96]. Here a unit amount of C1 is matched as follows:

$$C1 = 0.3R + 0.45G + 0.25B. \tag{2.5}$$

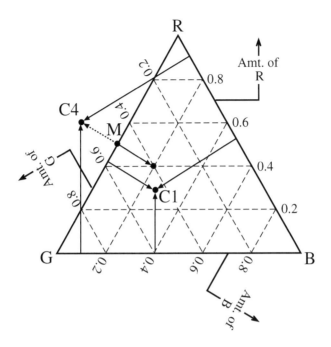

Figure 2.1: Maxwell diagram representing the concept of color-matching with three primaries, R, G, and B.

As shown in Equation (2.1) and Figure 2.1, the range of colors that can be produced by adding the primaries fill the area of the triangle bounded by unit amounts of the primaries. However if the primaries are allowed to be added to the test light, colors outside of this gamut can be produced, for example:

$$C4 + bB = rR + gG = M, \tag{2.6}$$

where M is the color that appears in the matched fields. This can be rewritten as:

$$C4 = rR + gG - bB, \tag{2.7}$$

where a negative amount of a primary means the primary is added to the test field. In Figure 2.1, an example is shown where:

$$C4 = 0.6R + 0.6G - 0.2B. \tag{2.8}$$

Colorimetry is based on such color matching experiments in which matches to equal energy monochromatic test lights spanning the visible spectrum are made using real R,

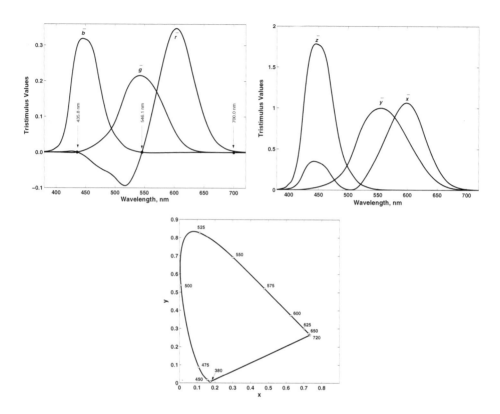

Figure 2.2: The CIE 1931 2° Standard Observer System of colorimetry. A) (Top left) The color-matching functions based on real primaries R, G, and B. B) (Top right) The transformed color-matching functions for the imaginary primaries, X, Y, and Z. C) (Bottom) The (x, y) chromaticity diagram.

G, and B primaries. In this way, color can be specified based on the matching behavior of a standard, or average observer. One such case is the CIE 1931 (**R,G,B**) Primary System shown in Figure 2.2A in which monochromatic primaries of 700.0 (**R**), 536.1 (**G**), and 435.8 (**B**) were used to match equal-energy monochromatic lights along the visible spectrum in a 2° bipartite field. The color matching functions, $\bar{r}(\lambda)$, $\bar{g}(\lambda)$, and $\bar{b}(\lambda)$, shown in Figure 2.2A, represent the average tristimulus values of the matches.

This system of matches was transformed to a set of matches made with imaginary primaries, **X**, **Y**, and **Z**, so that the color-matching functions, $\bar{x}(\lambda)$, $\bar{y}(\lambda)$, and $\bar{z}(\lambda)$, shown in Figure 2.2B, are all positive and the $\bar{y}(\lambda)$ color-matching function is the luminous efficiency function, $V(\lambda)$, to facilitate the calculation of luminance. As a result of additivity and proportionality, one can calculate the tristimulus values, X, Y, Z, for any color by integrating the product of the spectral power of the illuminant, the reflectance of the surface, and each of the color-matching functions:

$$X = \int_\lambda S_\lambda R_\lambda \bar{x}(\lambda)d\lambda, \quad Y = \int_\lambda S_\lambda R_\lambda \bar{y}(\lambda)d\lambda, \quad Z = \int_\lambda S_\lambda R_\lambda \bar{z}(\lambda)d\lambda \quad (2.9)$$

where S_λ is the spectral power distribution of the illuminant and R_λ is the spectral reflectance of the object. In practice, tabulated color-matching functions are used in summations to calculate the tristimulus values:

$$X = k\sum_\lambda S_\lambda R_\lambda \bar{x}(\lambda)\Delta\lambda, \quad Y = k\sum_\lambda S_\lambda R_\lambda \bar{y}(\lambda)\Delta\lambda, \quad Z = k\sum_\lambda S_\lambda R_\lambda \bar{z}(\lambda)\Delta\lambda$$

$$(2.10)$$

where k is a scaling factor used to define relative or absolute colorimetric values. Chromaticities can then be determined using: $x = X/(X+Y+Z)$, $y = Y/(X+Y+Z)$, and $z = Z/(X+Y+Z) = 1 - x - y$. The CIE 1931 (x,y) chromaticity diagram showing the spectral locus is shown in Figure 2.2C. Typically a color is specified by giving its (x,y) chromaticities and its luminance, Y.

Ideally, the color-matching functions (the CIE has adopted two standard observers, the 1931 2° observer, above, and the 1964 10° observer based on color matches made using a larger field) can be considered a linear transform of the spectral sensitivities of the L-, M-, and S- cones. It is therefore possible to specify color in terms of L-, M- and S-cone excitation.

Currently, work is underway to reconcile the data from color matching, luminous efficiency, and spectral sensitivity into a system that can be adopted universally. Until recently, specification of the cone spectral sensitivities has been a difficult problem because the overlap of their sensitivities makes it difficult to isolate these functions. A system of colorimetry based on L, M, and S primaries would allow a more direct and intuitive link between color specification and the early physiology of color vision [Boy96].

2.3.2 Color Appearance, Color Order Systems and Color Difference

A chromaticity diagram is useful for specifying color and determining the results of additive color mixing. However, it gives no insight into the appearance of colors. The appearance of a particular color depends on the viewer's state of adaptation, both globally and locally, the size, configuration, and location of the stimulus in the visual field, the color and location of other objects in the scene, the color of the background and surround, and even the colors of objects presented after the one in question. Therefore, two colored patches with the same chromaticity and luminance may have wildly different appearances.

Although color is typically thought of as being three-dimensional due to the trichromatic nature of color matching, five perceptual attributes are needed for a complete specification of color appearance [Fai98]. These are: *brightness*, the attribute according to which an area appears to be more or less intense; *lightness*, the brightness of an area relative to a similarly illuminated area that appears to be white; *colorfulness (chromaticness)*, the attribute according to which an area appears to be more or less chromatic; *chroma*, the colorfulness of an area relative to a similarly illuminated area that appears to be white; and *hue*, the attribute of a color denoted by its name such as blue, green, yellow, orange, etc. One other common attribute, saturation, defined as the colorfulness of an area relative to its brightness, is redundant when these five attributes are specified. Brightness and colorfulness are considered sensations that indicate the absolute levels of the corresponding sensations while lightness and chroma indicate the relative levels of these sensations. In general, increasing the illumination increases the brightness and colorfulness of a stimulus while the lightness and chroma remain approximately constant. Therefore a video or photographic reproduction of a scene can maintain the relative attributes even though the absolute levels of illumination are not realizable.

In terms of hue, it is observed that the colors red, green, yellow, and blue, are unique hues that do not have the appearance of being mixtures of other hues. In addition, they form opponent pairs so that perception of red and green cannot coexist in the same stimulus nor can yellow and blue. In addition, these terms are sufficient for describing the hue components of any unrelated color stimulus. This idea of opponent channels, first ascribed to Hering [Her64], indicates a higher level of visual processing concerned with the organization of color appearance. Color opponency is observed both physiologically and psychophysically in the chromatic channels with L- and M-cone (L-M) opponency and opponency between S-cones and the sum of the L- and M-cones (S-(L+M)). However, although these chromatic mechanisms are sure to be the initial substrate of red-green and yellow-blue opponency, they do not account for the phenomenology of Hering's opponent hues.

Even under standard viewing conditions of adaptation and stimulus presentation, the appearances of colors represented in the CIE diagram do not represent colors in a uniform way that allows specification of colors in terms of appearance. Various color spaces have been developed as tools for communicating color, calculating perceptual color difference, and specifying the appearance of colors.

The *Munsell Book of Color* [Nic76] is an example of a color-order system used for education, color communication, and color specification. The system consists of a denotation of colors and their physical instantiation arranged according to the attributes of value (lightness), chroma and hue. Figure 2.3A shows the arrangement of the hues in a plane of constant value in the Munsell Book of Color (R = red, YR = yellow-red, P = purple, etc.). Fig 2.3B shows a hue leaf with variation of value and chroma. The space is designed to produce a visually uniform sampling of perceptual color space when the samples are viewed under a standard set of conditions. The Munsell notation of a color

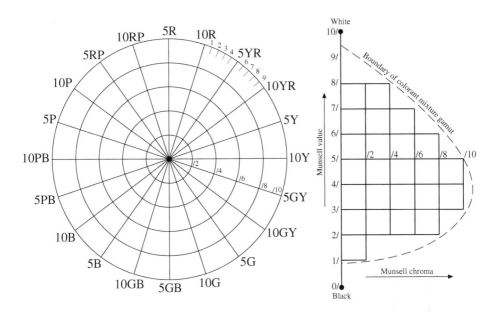

Figure 2.3: The organization of the *Munsell Book of Color*. A) (Left) Arrangement of the hues in a plane of constant value. B) (Right) A hue leaf with variation in value and chroma.

is HV/C, where H represents hue, V represents value, and C represents chroma. The relationship between relative luminance (scaling Y of the white point to a value of 100) and the Munsell value, V, is the 5th-order polynomial:

$$Y = 1.2219V - 0.23111V^2 + 0.23951V^3 - 0.021009V^4 + 0.0008404V^5 \qquad (2.11)$$

The CIE 1976 L^*, a^*, b^*, (CIELAB), color space was adopted as an empirical uniform color space for calculating the perceived color difference between two colors [CIE86]. The Cartesian coordinates, L^*, a^*, and b^*, of a color correspond to a lightness dimension and two chromatic dimensions that are roughly red-green and blue-yellow. The coordinates of color are calculated from its tristimulus values and the tristimulus values of a reference white so that the white is plotted at $L^* = 100$, $a^* = 0$, and $b^* = 0$. The space is organized so that the polar transformations of the a^* and b^* coordinates give the chroma, C^*_{ab}, and the hue angle, h_{ab}, of a color. The equations for calculating CIELAB coordinates are:

$$
\begin{aligned}
L^* &= 116(Y/Y_n)^{1/3} - 16 \\
a^* &= 500[(X/X_n)^{1/3} - (Y/Y_n)^{1/3}] \\
b^* &= 200[(Y/Y_n)^{1/3} - (Z/Z_n)^{1/3}] \\
C^*_{ab} &= [a^{*2} + b^{*2}]^{1/2} \\
h_{ab} &= \arctan(\tfrac{b^*}{a^*})
\end{aligned}
\qquad (2.12)
$$

where (X,Y,Z) are the color's CIE tristimulus values and (X_n, Y_n, Z_n) are the tristimulus values of a reference white. For values of Y/Y_n, X/X_n, and Z/Z_n less than 0.01 the formulae are modified (see [WS82] for details). Lightness, L^*, is an exponential function of luminance. There is also an "opponent-like" transform for calculating a^* and b^*. Figure 2.4 shows a typical CRT gamut plotted in CIELAB space.

Color differences, ΔE_{ab}^*, can be calculated using the CIELAB coordinates:

$$\Delta E_{ab}^* = \left[(\Delta L^*)^2 + (\Delta a^*)^2 + (\Delta b^*)^2 \right]^{1/2} \qquad (2.13)$$

Or in terms of the polar coordinates:

$$\Delta E_{ab}^* = \left[(\Delta L^*)^2 + (\Delta C_{ab}^*)^2 + (\Delta H_{ab}^*)^2 \right]^{1/2} \qquad (2.14)$$

where ΔH_{ab}^* is defined as:

$$\Delta H_{ab}^* = \left[(\Delta E_{ab}^*)^2 - (\Delta L^*)^2 - (\Delta C_{ab}^*)^2 \right]^{1/2} \qquad (2.15)$$

More recent research [LCR01] has led to modification of the CIELAB color difference equation to correct for observed nonuniformities in CIELAB space. A generic form of these advanced color difference equations [Ber00] is given as:

$$\Delta E = \frac{1}{k_E} \left[\left(\frac{\Delta L^*}{k_L S_L} \right)^2 + \left(\frac{\Delta C_{ab}^*}{k_C S_C} \right)^2 + \left(\frac{\Delta H_{ab}^*}{k_H S_H} \right)^2 \right]^{1/2} \qquad (2.16)$$

where k_E, k_L, k_C, and k_H are parametric factors that are adjusted according to differences in viewing conditions and sample characteristics and S_L, S_C, and S_H are weighting functions for lightness, chroma and hue that depend on the position of the samples in CIELAB color space. It should be noted that the different color difference formulae in the literature may have different magnitudes on average so that comparisons of performance based on ΔE values should be based on the relative rather than absolute values of the calculated color differences.

Color appearance models [Fai98] attempt to assign values to the color attributes of a sample by taking into account the viewing conditions under which the sample is observed so that colors with corresponding appearance (but different tristimulus values) can be predicted. These models generally consist of a chromatic-adaptation transform that adjusts for the viewing conditions (e.g., illumination, white-point, background, and surround) and calculations of at least the relative color attributes. More complex models include predictors of brightness and colorfulness and may predict color appearance phenomena such as changes in colorfulness and contrast with luminance [Fai98]. Color spaces can then be constructed based on the coordinates of the attributes derived in the model. The CIECAM02 color appearance model [MFH$^+$02] is an example of a color ap-

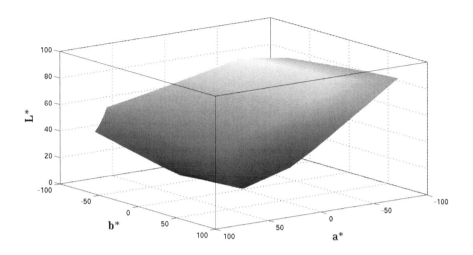

Figure 2.4: A CRT gamut plotted in CIELAB space.

pearance model that predicts the relative and absolute color appearance attributes based on specifying the surround conditions (average, dim, or dark), the luminance of the adapting field, the tristimulus values of the reference white point, and the tristimulus values of the sample.

2.4 Luminance and the Perception of Light Intensity

Luminance is a term that has taken on different meanings in different contexts and therefore the concept of luminance, its definition, and its application can lead to confusion. Luminance is a photometric measure that has loosely been described as the "apparent intensity" of a stimulus but is actually defined as the effectiveness of lights of different wavelengths in specific photometric matching tasks [SS99]. The term is also used to label the achromatic channel of visual processing.

2.4.1 Luminance

The CIE definition of luminance [WS82] is a quantity that is a constant times the integral of the product of radiance and $V(\lambda)$, the photopic spectral luminous efficiency function.

$V(\lambda)$ is defined as the ratio of radiant flux at wavelength λ_m to that of wavelength λ, when the two fluxes produce the same luminous sensations under specified condition with a value of 1 at $\lambda_m = 555$ nm. Luminance efficiency is therefore tied to the tasks that are used to measure it. Luminance is expressed in units of candelas per square meter (cd/m^2). To control for the amount of light falling on the retina, the troland (td) unit is defined as cd/m^2 multiplied by the area of the pupil.

The $V(\lambda)$ function in use by the CIE was adopted by the CIE in 1924 and is based on a weighted assembly of the results of a variety of visual brightness matching and minimum flicker experiments. The goal of the $V(\lambda)$ in the CIE system of photometry and colorimetry is to predict brightness matches so that given two spectral power distributions, $P_{1\lambda}$ and $P_{2\lambda}$, the expression:

$$\int_\lambda P_{1\lambda} V(\lambda) d\lambda = \int_\lambda P_{2\lambda} V(\lambda) d\lambda \qquad (2.17)$$

predicts a brightness match and therefore additivity applies to brightness.

In direct heterochromatic brightness matching, it has been shown that this additivity law, known as Abney's law, fails substantially [WS82]. For example when a reference "white," **W**, is matched in brightness to a "blue" stimulus, **C$_1$**, and to a "yellow" stimulus, **C$_2$**, the additive mixture of **C$_1$** and **C$_2$** is found to be less bright than 2**W**, a stimulus that is twice the radiance of the reference white [KB96]. Another example of the failure of $V(\lambda)$ to predict brightness matches is known as the *Helmholtz-Kohlrausch effect* where chromatic stimuli of the same luminance as a reference white stimulus appear brighter than the reference. In Figure 2.5, we see the results from Sanders and Wyszecki [SW64] and presented in [WS82] that show how the ratio, B/L, of the luminance of a chromatic test color, B, to the luminance of a white reference, L, varies as a function of chroma and hue. Direct hetereochromatic brightness matching experiments also demonstrate large inter-observer variability.

Other psychophysical methods for determining "matches" have derived luminance efficiency functions that are more consistent between observers and obey additivity. Heterochromatic flicker photometry (HFP) is one such technique in which a standard light, λ_S, at a fixed intensity is flickered in alteration with a test light, λ_T, at a frequency of about 10 to 15 Hz [KB96]. The observer's task is to adjust the intensity of λ_T to minimize the appearance of flicker in the alternating lights. The normalized reciprocal of the radiances of the test lights measured across the visible spectrum needed to minimize flicker is the resulting luminous efficiency function. Comparable results are obtained in the minimally distinct border method (MDB) in which the radiance of λ_T is adjusted to minimize the distinctness of the border between λ_T and λ_S when they are presented as precisely juxtaposed stimuli in a bipartite field [KB96]. Modifications of the CIE $V(\lambda)$ function are aimed at achieving brightness additivity and proportionality by developing functions that are in better agreement to HFP and MDB data.

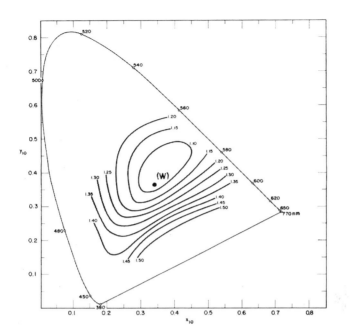

Figure 2.5: The *Helmholtz-Kohlrausch effect* in which the apparent brightness of isoluminant chromatic colors increases with chroma. The loci of constant ratios of the luminance of chromatic colors to a white reference (B/L) based on heterochromatic brightness matches of Sanders and Wyszecki [WS82, SW64] are shown in the CIE 1964 (x_{10}, y_{10}) chromaticity diagram. From [WS82].

The shape of the luminous efficiency function is often considered as the spectral sensitivity of photopic vision and as such it is a function of the sensitivity of the separate cone types. However, unlike the spectral sensitivities of the cones, the luminosity function changes shape depending on the state of adaptation, especially chromatic adaptation [SS99]. Therefore the luminosity function only defines luminance for the conditions under which it was measured. As measured by HFP and MDB, the luminosity function can be modeled as a weighted sum of L- and M-cone spectral sensitivities with L-cones dominating. The S-cones contribution to luminosity is considered negligible under normal conditions. (The luminosity function of dichromats, those color-deficient observers who lack either the L- or M-cone photopigment, is the spectral sensitivity of the remaining longer wavelength sensitive receptor pigment. Protanopes, who lack the L-cone pigment, generally see "red" objects as being dark. Deuteranopes, who lack the M-cone pigment, see brightness similarly to color normal observers.)

2.4.2 Perceived Intensity

As the intensity of a background field increases, the intensity required to detect a threshold increment superimposed on the background increases. Weber's law states that the ratio of the increment to the background or adaptation level is a constant so that $\Delta I/I = k$, where I is the background intensity and ΔI is the increment threshold (also known as the just noticeable difference or JND) above this background level. Weber's law is a general rule of thumb that applies across different sensory systems.

To the extent that Weber's Law holds, it states that threshold contrast is constant at all levels of background adaptation so that a plot of ΔI versus I will have a constant slope. In vision research, plots of $\log(\Delta I)$ versus $\log(I)$, known as threshold versus intensity or t.v.i. curves, exhibit a slope of one when Weber's law applies. The curve that results when the ratio $\Delta I/I$, the Weber fraction, is plotted against background intensity, has a slope of zero when Weber's law applies.

The Weber fraction is a measure of contrast and it should be noted that contrast is a more important aspect of vision than the detection of absolute intensity. As the illumination on this page increases from dim illumination to full sunlight, the contrast of the ink to the paper remains constant even as the radiance of the reflected light off the ink in sunlight surpasses the radiance reflected off the white paper under dim illumination.

If the increment threshold is considered a unit of sensation, one can build a scale of increasing sensation by integrating JNDs so that $S = K\log(I)$, where S is the sensation magnitude and K is a constant that depends on the particular stimulus conditions. The logarithmic relationship between stimulus intensity and the associated sensation is known as Fechner's Law. The increase in perceptual magnitude is characterized by a compressive relationship with intensity as shown in Figure 2.6. Based on applicability of Weber's law and the assumptions of Fechner's Law, a scale of perceptual intensity can be constructed based on the measurement of increment thresholds using classical psychophysical techniques to determine the Weber fraction.

Weber's law, in fact, does not hold for the full range of intensities that the visual system operates. However, at luminance levels above 100 cd/m^2, typical of indoor light-ning, Weber's Law holds fairly well. The value of the Weber fraction depends upon the exact stimulus configuration with a value as low as 1% under optimal conditions [HF86]. In order to create a scale of sensation using Fechner's Law, one would need to choose a Weber fraction that is representative of the conditions under which the scale would be used.

Logarithmic functions have been used for determining lightness differences such as in the BFD color difference equation [LR87]. It should be noted, however, that ΔE values are typically larger than one JND. Since they do not represent true increment thresholds, Weber's law may not apply.

The validity of Fechner's law has been questioned based on direct brightness scal-ing experiments using magnitude estimation techniques. Typically these experiments yield scales of brightness that are power functions of the form $S = kI^a$, where S is the sensation magnitude, I is the stimulus intensity, a is exponent that depends on the stimulus conditions, and k is a scaling constant. This relationship is known as Stevens' Power Law. For judgments of lightness, the exponent, a, typically has a value less than one, demonstrating a compressive relationship between stimulus intensity and perceived magnitude.

The value of the exponent is dependent on the stimulus conditions of the experiment used to measure it. Therefore, the choice of exponent must be based on data from experiments that are representative of the conditions in which the law will be applied. As shown above, the *Munsell Book of Color*, Equation (2.11), and the L* function in CIELAB, Equation (2.12), are both exponential functions.

Figure 2.6 shows a comparison of various lightness vs luminance factor functions that are summarized in [WS82]. Functions based on Fechner's Law (logarithmic functions) and power functions have been used for lightness scales, as shown in the figure. Although the mathematical basis of a lightness versus luminance factor function may be theoretically important, it is the compressive nonlinearity observed in the data that must be modeled.

As seen in Figure 2.6, different functional relationships between luminance and lightness result depending on the exact stimulus conditions and psychophysical techniques used to collect the data. For example, curve 2 was observed with a white background, curves 1, 3, and 4 were obtained using neutral backgrounds with a luminance factor of approximately 20% (middle-gray), curve 6 was obtained on a background with a luminance factor of approximately 50%, and curve 5 was based on application of Fechner's law applied to observations with backgrounds with luminance factors close to the gray chips being compared.

The effect of the background influences the appearance of lightness in two ways. Due to simultaneous contrast, a dark background will cause a patch to appear lighter while a light background will cause a patch to look darker. Sensitivity to contrast increases when the test patch is close to lightness of the background. Known as the *crispening effect*, this phenomenon leads to nonuniformities in the shapes of the luminance vs lightness function with different backgrounds and increased sensitivity to lightness differences when patches of similar lightness to the background are compared [WS82, Fai98]. In addition, the arrangement and spatial context of a pattern can influence the apparent lightness. White's illusion [And03] is an example where a pattern of identical gray bars appears lighter (darker) when embedded in black (white) stripes although simultaneous contrast would predict the opposite effect due to the surrounding region. Figure 2.7 shows examples of simultaneous contrast, crispening, and White's illusion.

2.5   Spatial Vision and Contrast Sensitivity

Computational analysis has been the dominant paradigm for the study of visual processing of form information in the cortex replacing a more qualitative study based on identi-

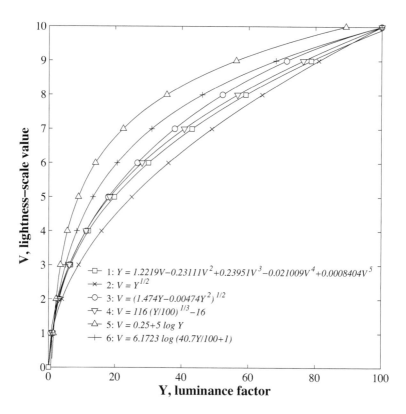

Figure 2.6: A variety of luminance versus lightness functions: 1) Munsell renotation system, observations on a middle-gray background; 2) Original Munsell system, observations on a white background; 3) Modified version of (1) for a middle-gray background; 4) CIE L* function, middle-gray background; 5) Gray scale of the *Color Harmony Manual*, background similar to the grays being compared; 6) Gray scale of the *DIN color chart*, gray background. Equations from [WS82].

fying the functional specialization of cortical modularity and feature detection in cortical neurons. The response of the visual system to harmonic stimuli has been used to analyze the processing of the visual system to see whether the response to more complex stimuli can be modeled from the response of simpler stimuli in terms of linear systems analysis. Because the visual system is not a linear system, these methods reveal where more comprehensive study of the visual system is needed.

2.5.1   Acuity and Sampling

The first two factors that need to be accounted for in the processing of spatial information by the visual system are the optics of the eye and the sampling of the visual scene

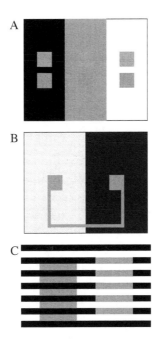

Figure 2.7: Various lightness illusions. A) Simultaneous contrast: The appearance of the identical gray squares changes due to the background. B) Crispening: The difference between the two gray squares is more perceptible when the background is close to the lightness of the squares. C) White's illusion: The appearance of the identical gray bars changes when they are embedded in the white and black stripes.

by the photoreceptor array. Once the scene is sampled by the photoreceptor array, neural processing determines the visual response to spatial variation.

Attempts to measure the modulation transfer function (MTF) of the eye have relied on both physical and psychophysical techniques. Campbell and Gubisch [CG66], for example, measured the linespread function of the eye by measuring the light from a bright line stimulus reflected out of the eye and correcting for the double passage of light through the eye for different pupil sizes. MTFs derived from double-pass measurements correspond well with psychophysical measurements derived from laser interferometry in which the MTF is estimated from the ratio of contrast sensitivity to conventional gratings and interference fringes that are not blurred by the optics of the eye [WBMN94]. The MTF of the eye demonstrates that for 2 mm pupils, the modulation of the signal falls off very rapidly and is approximately 1% at about 60 cpd. For larger pupils, 4–6 mm, the modulation falls to 1% at approximately 120 cpd [PW03].

In the fovea the cone photoreceptors are packed tightly in a triangular arrangement with a mean center-to-center spacing of 32 arc min [Wil88]. This corresponds to a sampling rate of approximately 120 samples per degree or a Nyquist frequency of around 60 cpd. Because the optics of the eye sufficiently degrade the image above 60 cpd we are spared the effects of spatial aliasing in normal foveal vision. The S-cone packing in the retina is much more sparse than the M- and L-cone packing so that its Nyquist limit is approximately 10 cpd [PW03].

Visual spatial acuity is therefore considered to be approximately 60 cpd although under special conditions, for example, peripheral vision, large pupil sizes, and laser interferometry, higher spatial frequencies can be either directly resolved or seen via the effects of aliasing. This would appear to set a useful limit for the design of displays, for example. However, in addition to the ability of the visual system to detect spatial variation, there also exists the ability to detect spatial alignment known as Vernier acuity or hyperacuity. In these tasks, observers are able to judge whether two stimuli (dots or line segments) separated by a small gap are misaligned with offsets as small as 2–5 arc sec, which is less than one-fifth the width of a cone [WM77]. It is hypothesized that it is the blurring of the optics of eye that contributes to this effect by distributing the light over a number of receptors. By comparing the distributions of the responses to the two stimuli, the visual system can localize the position of the stimuli at a resolution finer than the receptor spacing [Wan95, DD90].

2.5.2   Contrast Sensitivity

It can be argued that the visual system evolved to discriminate and identify objects in the world. The requirements for this task are different from those needed if the purpose of the visual system was meant to measure and to record the variation of light intensity in the visual field. In this regard, characterization of the visual system's response to variations in contrast as a function of spatial frequency is studied using harmonic stimuli. The contrast sensitivity function (CSF) plots the sensitivity of the visual system to sinusoids of varying spatial frequency.

Figure 2.8 shows the contrast sensitivity of the visual system based on Barten's empirical model [Bar99, Bar04]. The contrast sensitivity is a plot of the reciprocal of the threshold (ordinate) contrast needed to detect sinusoidal gratings of different spatial frequency (abscissa) in cycles per degree of visual angle. Contrast is usually given using Michelson contrast: $(L_{max} - L_{min})/(L_{max} + L_{min})$, where L_{max} and L_{min} are the peak and trough luminance of the grating, respectively. Typically plotted on log-log coordinates, the CSF reveals the limits of detecting variation in intensity as a function of size.

Each curve in Figure 2.8 shows the change in contrast sensitivity for sinusoidal gratings modulated around mean luminance levels ranging from 0.01 cd/m^2 to 1000 cd/m^2. As the mean luminance increases to photopic levels, the CSF takes on its characteristic band-pass shape. As luminance increases, the high frequency cut-off, indicating spatial acuity, increases. At scotopic light levels, reduced acuity is due to the increased spatial pooling in the rod pathway which increases absolute sensitivity to light.

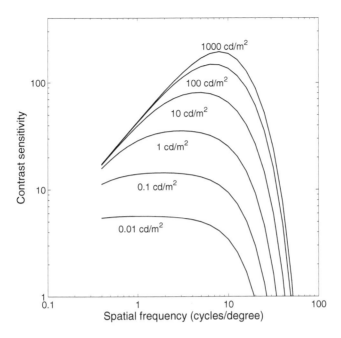

Figure 2.8: Contrast sensitivity function for different mean luminance levels. The curves were generated using the empirical model of the achromatic CSF in [Bar04].

As the mean luminance level increases, the contrast sensitivity to lower spatial frequencies increases and then remains constant. Where these curves converge and overlap, we see that the threshold contrast is constant despite the change in mean luminance. This is where Weber's Law holds. At higher spatial frequencies, Weber's Law breaks down.

The exact shape of the CSF depends on many parametric factors so that there is no one canonical CSF curve [Gra89]. Mean luminance, spatial position on the retina, spatial extent (size), orientation, temporal frequency, individual differences, and pathology are all factors that influence the CSF. (See [Gra89] for a detailed bibliography of studies related to parametric differences.)

Sensitivity is greatest at the fovea and tends to fall off linearly (in terms of log sensitivity) with distance from the fovea for each spatial frequency. This fall off is faster for higher spatial frequencies so that the CSF becomes more low-pass in the peripheral retina. As the spatial extent (the number of periods) in the stimulus increases, there is an increase in sensitivity up to a point at which sensitivity remains constant. The change in sensitivity with spatial extent changes as the distance from the fovea increases so that the most effective stimuli in the periphery are larger than in the fovea. The *oblique effect* is the term applied to the reduction in contrast sensitivity to obliquely oriented gratings compared to horizontal and vertical ones. This reduction in sensitivity (a factor of 2 or 3) occurs at high spatial frequencies.

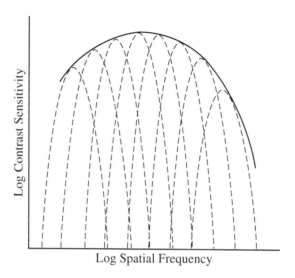

Figure 2.9: The contrast sensitivity function represented as the envelope of multiple, more narrowly tuned spatial frequency channels.

2.5.3   Multiple Spatial Frequency Channels

Psychophysical, physiological, and anatomical evidence suggest that the CSF represents the envelope of the sensitivity of many more narrowly tuned channels as shown in Figure 2.9. The idea is that the visual system analyzes the visual scene in terms of multiple channels each sensitive to a narrow range of spatial frequencies. In addition to channels sensitive to narrow bands of spatial frequency, the scene is also decomposed into channels sensitive to narrow bands of orientation. This concept of multiresolution representations forms the basis of many models of spatial vision and pattern sensitivity (see [Wan95, DD90]).

The receptive fields of neurons in the visual cortex are size specific and therefore show tuning functions that correspond with multiresolution theory and may be the physiological underpinnings of the psychophysical phenomena described below. Physiological evidence points to a continuous distribution of peak frequencies in cortical cells [DD90], although multiresolution models have been built using a discrete number of spatial channels and orientations (e.g., Wilson and Regan [WR84] suggested a model with six spatial channels at eight orientations). Studies have shown that six channels seem to model psychophysical data well without ruling out a larger number [WLM+90]. Both physiological and psychophysical evidence indicate that the bandwidths of the underlying channels are broader for lower frequency channels and become narrower at higher frequencies as plotted on a logarithmic scale of spatial frequency. On a linear scale, however, the low-frequency channels have a much narrower bandwidth than the high-frequency channels.

2.5.3.1 Pattern adaptation

Figure 2.10A is a demonstration of the phenomenon of pattern adaptation first described by Pantle and Sekuler [PS68] and Blakemore and Campbell [BC69]. The two patterns in the center of Figure 2.10A, the test patterns, are of the same spatial frequency. The adaptation patterns on the left are higher (top) and lower (bottom) spatial frequencies. By staring at the fixation bar between the two adaptation patterns for a minute or so, the channels that are tuned to those spatial frequencies adapt, or become fatigued, so that their response is suppressed for a short period of time subsequent to adaptation. After adaptation, the appearance of the two test patterns is no longer equal. The top pattern now appears to be of a lower spatial frequency and the bottom appears higher.

The response of the visual system depends on the distribution of responses in the spatial channels. Before adaptation, the peak of the response is located in the channels tuned most closely to the test stimulus. Adaptation causes the channels most sensitive to the adaptation stimuli to respond less vigorously. Upon subsequent presentation of the test pattern, the distribution of response is now skewed from its normal distribution so that higher-frequency adaptation pattern will lead to relatively more response in the lower-frequency channels and vice versa. Measurement of the CSF after contrast adaptation reveals a loss of sensitivity in the region surrounding the adaptation frequency as shown in Figure 2.10B. This effect of adaptation has been shown to be independent of phase [JT75] demonstrating that phase information is not preserved in pattern adaptation.

Adaptation to the two right hand patterns in Figure 2.10A leads to a change in the apparent orientation of the test patterns in the center. As with spatial frequency, the response of the visual system to orientation depends on the pattern of response in multiple channels tuned to different orientations. Adaptation to a particular orientation will fatigue those channels that are more closely tuned to that particular orientation so that the pattern of response to subsequent stimuli will be skewed [BC69].

2.5.3.2 Pattern detection

Campbell and Robson [CR68] measured the contrast detection thresholds for sine wave gratings and a variety of periodic complex waveforms (such as square waves and saw-tooth gratings) and found that the detection of complex patterns were determined by the contrast of the fundamental component rather than the contrast of the overall pattern. In addition, they observed that the ability to distinguish a square wave pattern from a sinusoid occurred when the third harmonic of the square wave reached its own threshold contrast.

Graham and Nachmias [GN71] measured the detection thresholds for gratings composed of two frequency components, f and $3f$, as a function of the relative phase of

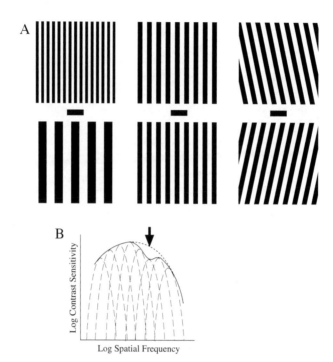

Figure 2.10: Pattern and orientation adaptation. A) After adapting to the two patterns on the left by staring at the fixation bar in the center for a minute or so, the two identical patterns in the center will appear to have different widths. Similarly, adaptation to the tilted patterns on the right will cause an apparent orientation change in the central test patterns. B) Adaptation to a specific spatial frequency (arrow) causes a transient loss in contrast sensitivity in the region of the adapting frequency and a dip in the CSF.

the two gratings. When these two components are added in phase so that their peaks coincide, the overall contrast is higher than when they are combined so that their peaks subtract. However, the thresholds for detecting the gratings were the same regardless of phase. These results support a spatial frequency analysis of the visual stimulus as opposed to detection based on the luminance profile.

2.5.3.3 Masking and facilitation

The experiments on pattern adaptation and detection dealt with the detection and inter-action of gratings at their contrast threshold levels. The effect of suprathreshold patterns on grating detection is more complicated. In these cases a superimposed grating can either hinder the detection (masking) of a test grating or it can lead to a lower detection threshold (facilitation), depending on the properties of the two gratings. The influence of the mask on the detectability of the test depends on the spatial frequency, orientation,

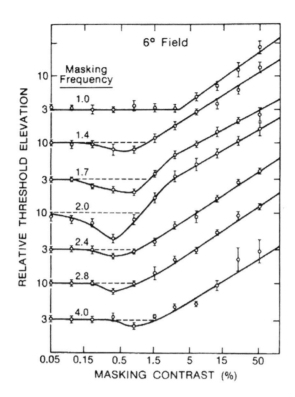

Figure 2.11: Contrast masking experiment showing masking and facilitation by the masking stimulus. The threshold contrasts (ordinate) for detecting a 2 cpd target grating are plotted as a function of the contrast of a superimposed masking grating. Each curve represents the results for a mask of a different spatial frequency. Mask frequencies ranged from 1 to 4 cpd. The curves are displaced vertically for clarity. The dashed lines indicate the unmasked contrast threshold. From [LF80].

and phase of the mask relative to the test. This interaction increases as the mask and target become more similar. For similar tests and masks, facilitation is seen at low mask contrasts and as the contrast increases the test is masked.

In the typical experimental paradigm a grating (or a Gabor pattern) of a fixed contrast called the "pedestal" or "masking grating" is presented along with a "test" or "signal" grating (or Gabor). The threshold contrast at which the test can be detected (or discriminated in a forced-choice) is measured as a function of the contrast of the mask. The resulting plot of contrast threshold versus the masking contrast is sometimes referred to as the threshold versus contrast, or TvC, function.

Figure 2.11 shows the results from Legge and Foley [LF80]. In these experiments both the mask and test were at the same orientation. As the spatial frequency of the mask approaches the same frequency of the test we see facilitation at low contrasts. At higher mask contrasts, the detection of the target is masked leading to an elevation in the contrast threshold. The initial decrease and subsequent increase seen in these curves is known as the "dipper effect." In Figure 2.11, we see that this masking is slightly diminished when the mask frequency is much different from that of the test. Other studies (e.g., [SD83, WMP83]) have shown that the effectiveness of the masking is reduced as the mask and test frequencies diverge showing a characteristic band-limited tuning response. These masking effects support the multiresolution theory.

There is no effect of the phase relationship between the mask and test at higher masking contrasts where masking is observed. However, the relative phase of the mask and test are critical at low mask contrasts where facilitation is seen [BW94, FC99]. Foley and Chen [FC99] describe a model of contrast discrimination that explains both masking and facilitation and the dependence of phase on facilitation in which spatial phase is an explicit parameter of the model. Yang and Makous' [YM95] model of contrast discrimination models the phase relationship seen in the dipper effect without specific phase parameters. Bowen and Wilson [BW94] attribute the dipper effect to local adaptation preceding spatial filtering.

Both adaptation and masking experiments have been used to determine the bandwidths of the underlying spatial frequency channels. The estimates of bandwidth, although variable, correspond to the bandwidths of cells in the primary visual cortex of animals studied using physiological techniques [DD90, WLM$^+$90].

Experiments studying the effect of orientation of the mask on target detection have shown that as the orientation of the mask approaches that of the target, the threshold contrast for detecting the target increases. Phillips and Wilson [PW84], for example, measured the threshold elevation caused by masking for various spatial frequencies and masking contrasts. Their estimates of the orientation bandwidths show a gradual decrease in bandwidth of about $\pm 30°$ at 0.5 cpd to $\pm 15°$ at 11.3 cpd which agrees well with physiological results from macaque striate cortex. No differences were found for tests oriented at $0°$ and $45°$ indicating that the oblique effect is not due to differences in bandwidth at these two orientations but rather is more likely due to a predominance of cells with orientation specificity to the horizontal and vertical [PW84]. The same relationships found between the spatial and orientation tuning of visual mechanisms elucidated using the psychophysical masking paradigm agree with those revealed in recordings from cells in the macaque primary cortex [WLM$^+$90].

2.5.3.4 Nonindependence in spatial frequency and orientation

The multiresolution representation theory postulates independence among the various spatial frequency and orientation channels that are analyzing the visual scene. Models of spatial contrast discrimination are typically based on the analysis of the stimulus through independent spatial frequency channels followed by a decision stage in which the outputs from these channels are compared (e.g., [FC99, WR84]).

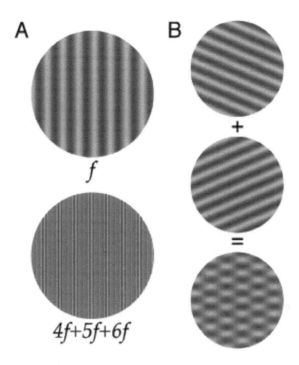

Figure 2.12: (A) The top grating has a spatial frequency of f. A perceived periodicity of f is seen in the complex grating on the bottom that is the sum of the frequencies of $4f$, $5f$, and $6f$. Redrawn from [DD90]. (B) The sum of two gratings at $\pm 67.5°$ has no horizontal or vertical components yet these orientations are perceived.

However, there is considerable evidence of nonlinear interactions between separate spatial frequency and orientation channels that challenge a strong form of the multiresolution theory (see [Wan95]). Both physiologically and psychophysically, there is evidence of interactions between channels that are far apart in their tuning characteristics. Figure 2.12 shows two perceptual illustrations of nonlinear interaction for spatial frequency (Figure 2.12A) and orientation (Figure 2.12B).

Figure 2.12A shows a grating of frequency f, top, and a grating that is the sum of gratings of $4f$, $5f$, and $6f$. The appearance of the complex grating has a periodicity of frequency f although there is no component at this frequency. Even though the early cortical mechanisms tuned to f are not responding to the complex grating, the higher frequency mechanisms tuned to the region of $5f$ somehow produce signals that fill in the "missing fundamental" during later visual processing (see [DD90]).

Figure 2.12B shows the grating produced by the sum of two gratings at $\pm 67.5°$ from the vertical. Although there are no grating components in the horizontal or vertical direction, there is the appearance of such stripes. Derrington and Henning [DH89] performed a masking experiment in which individual gratings at orientations of $\pm 67.5°$ did not mask the detection of a vertical grating of the same spatial frequency but the mixture of the two gratings produced substantial masking. De Valois [DD90] presents further examples of interactions between spatial frequency channels from adaptation experiments, physiological recordings, and anatomical studies in animals. These studies point to mutual inhibition among spatial frequency and orientation selective channels.

2.5.3.5   Chromatic contrast sensitivity

Sensitivity to spatial variation for color has also been studied using harmonic stimuli. Mullen [Mul85] measured contrast sensitivity to red-green and blue-yellow gratings using counterphase monochromatic stimuli for the chromatic stimuli. The optics of the eye produce chromatic aberrations and magnification differences between the chromatic gratings. These artifacts were controlled by introducing lenses to independently focus each grating and scaling the size of the gratings. In addition, the relative luminances of the gratings were matched at the different spatial frequencies using flicker photometry. The wavelengths for the chromatic gratings were chosen to isolate the two chromatic-opponent mechanisms.

Figure 2.13 shows the resulting luminance and isolumant red-green and blue-yellow CSFs redrawn from one observer in Mullen. The chromatic CSFs are characterized by a low-pass shape and have high frequency cut-offs at much lower spatial frequencies than the luminance CSF. The acuity of the blue-yellow mechanism is limited by the sparse distribution of S-cones in the retinal mosaic, however the lower acuity in the red-green mechanism is not limited by retinal sampling and is therefore imposed by subsequent neural processing [Mul85, SWB93b].

Sekiguchi et al. [SWB93b, SWB93a] used laser interferometry to eliminate chromatic aberration with drifting chromatic gratings moving in opposite directions to simultaneously measure the chromatic and achromatic CSFs. They found a close match between the two chromatic mechanisms and slightly higher sensitivity than Mullen's results.

Poirson and Wandell [PW95] used an asymmetric color-matching task to explore the relationship between pattern vision (spatial frequency) and color appearance. In this task, square-wave gratings of various spatial frequencies composed of pairs of complementary colors (colors whose additive mixture is neutral in appearance) were presented to observers. The observers' task was to match the color of a two-degree square patch to the colors of the individual bars of the square-wave grating. They found that the color appearance of the bars depended on the spatial frequency of the pattern. Using a pattern-color separable model, Poirson and Wandell derived the spectral sensitivities and spatial contrast sensitivities of the three mechanisms that mediated the color appearance judgments. These mechanisms showed a striking resemblance to the luminance and two chromatic-opponent channels identified psychophysically and physiologically.

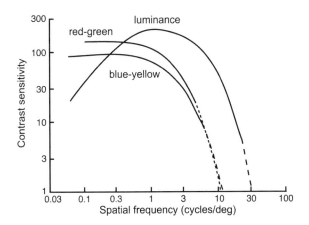

Figure 2.13: Contrast sensitivity functions of the achromatic luminance channel and the red-green, and blue yellow opponent chromatic channels from [Mul85]. Solid lines indicate data from one subject. Dashed lines are extrapolated from the measured data to show the high-frequency acuity limit for the three channels in this experiment. Redrawn from Mullen [Mul85].

One outcome of the difference in the spatial response between the chromatic and luminance channels is that the sharpness of an image is judged based on the sharpness of the luminance information in the image since the visual system is incapable of resolving high-frequency chromatic information. This has been taken advantage of in the compression and transmission of color images, since the high spatial frequency chromatic information in an image can be removed without a loss in perceived image quality (e.g., [MF85]). The perceptual salience of many geometric visual illusions is severely diminished or eliminated when they are reproduced using isoluminant patterns due to this loss of high spatial frequency information. The minimally distinct border method for determining luminance matches is also related to the reduced spatial accuity of the chromatic mechanisms.

2.5.3.6 Suprathreshold contrast sensitivity

The discussion so far has focused on threshold measurements of contrast. Our sensitivity to contrast at threshold is very dependent on spatial frequency and has been studied in-depth to understand the limits of visual perception. The relationship between the perception of contrast and spatial frequency at levels above threshold will be briefly discussed here. It should be noted that there is increasing evidence that the effects seen at threshold are qualitatively different from those at suprathreshold levels so that models of detection and discrimination may not be applicable (e.g., [FBM03]).

Figure 2.14 presents data for one subject redrawn from Georgeson and Sullivan [GS75] showing the results from a suprathreshold contrast matching experiment. In this experiment, observers made apparent contrast matches between a standard 5 cpd

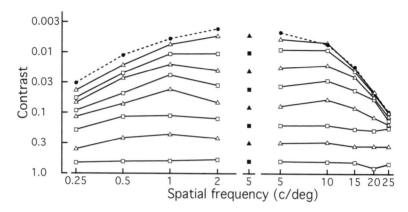

Figure 2.14: Results from Georgeson and Sullivan [GS75] showing equal-contrast contours for suprathreshold gratings of various spatial frequencies matched to 5 cpd gratings.

grating and test gratings that varied from 0.25 to 25 cpd. The uppermost contour reflects the CSF at threshold. As the contrast of the gratings increased above threshold levels, the results showed that the apparent contrast matched when the physical contrasts were equal. This flattening of the equal-contrast contours in Figure 2.14 is more rapid at higher spatial frequencies. The flattening of the contours was termed "contrast constancy." Subjects also made contrast matches using single lines and band-pass filtered images. Again it was shown that the apparent contrast matched the physical contrast when the stimuli were above threshold. Further experiments demonstrated that these results were largely independent of mean luminance level and position on the retina.

Georgeson and Sullivan suggest that an active process is correcting the neural and optical blurring seen at threshold for high spatial frequencies. They hypothesize that the various spatial frequency channels adjust their gain independently in order to achieve contrast constancy above threshold. This process, they suggest, is analogous to deblurring techniques used to compensate for the modulation transfer function in photography. Vimal [Vim00] found a similar flattening effect and chromatic contrast constancy for isoluminant chromatic gratings, however, he proposed that different mechanisms underlie this process.

Switkes and Crognale [SC99] compared the apparent contrast of gratings that varied in color and luminance in order to gauge the relative magnitude of contrast perception and the change of contrast perception with increasing contrast. In these experiments, contrast was varied in 1 cpd gratings modulated in five different directions in color space: luminance (lum), isoluminant red-green (LM), isoluminant blue-yellow (S), L-cone excitation only (L), and M-cone excitation only (M). A threshold procedure was used to determine equivalent perceptual contrast matches between pairs of these gratings at various contrast levels. The results are shown in Figure 2.15. Contrast was measured in terms of cone contrast, which is the deviation in cone excitation from the mean level.

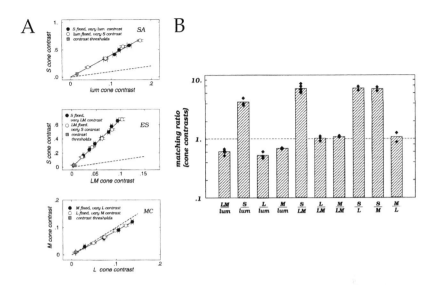

Figure 2.15: (A) Contrast matches between stimuli that vary in different dimensions of color space: luminance (lum), isoluminant red-green (LM), isoluminant blue-yellow (S), L-cone excitation only (L), and M-cone excitation only (M). The data show proportionality in contrast matches (solid lines). The dashed lines, slope equals 1, indicate the prediction for contrast based on equal cone contrast. (B) Average contrast-matching ratios across observers for the different pairwise comparisons of contrast. Diamonds indicate individual observer's ratios. A ratio of 1 (dashed line) indicates prediction based on cone-contrast. From [SC99].

Although the authors considered that the cross-modal matching (color against luminance) task might be difficult due to the problem of comparing "apples vs. oranges," they found very good inter- and intra- observer consistency. The results of the experiment showed that the contrast matches between pairs of stimuli that varied in the different dimensions were proportional so that a single scaling factor could be used to describe the contrast matches between the stimuli as they varied in physical contrast. In Figure 2.15A, the matches between luminance and blue-yellow (top), red-green and blue-yellow (middle), and L-cone and M-cone (bottom) gratings are shown for increasing levels of contrast. Matches are determined symmetrically so that error bars are determined in both dimensions. Looking at the top panel of Figure 2.15A, we see that as the physical contrast of the luminance grating and the blue-yellow grating increases, the perceived contrast match is always at the same ratio of physical contrast (a straight line can be fit to the data). Much more S-cone contrast is needed to match the L- and M- cone contrast in the luminance grating.

Figure 2.15B summarizes the results of the experiment by showing the ratio of cone-contrasts needed to produce contrast matches for the different pairs of color and luminance contrasts. For example, for the data in the top panel of Figure 2.15A, the slope of the line fit to the data corresponds to the matching ratio labeled S/lum. These results indicate the physical scaling needed to match luminance and chromatic gratings at least at the spatial frequency tested. These data shed light on the performance of the visual system along different axes in color space by providing a metric for scaling the salience of these mechanisms [SC99].

2.5.3.7 Image compression and image difference

Despite the fact that studies of the ability of the human visual system to detect and discriminate patterns do not give a complete understanding of how the visual system works to produce our final representation of the world, this information on perceptual limits can have practical use in imaging applications. For image compression, we can use this information to help guide the development of perceptually lossless encoding schemes. Similarly, we can use this information to eliminate information from imagery to which the visual system is insensitive in order to compute image differences based on what is visible in the scene.

In his book, *Foundations of Vision*, Wandell [Wan95] introduces the concepts of the discrete cosine transformation in JPEG image compression, image compression using Laplacian pyramids, and wavelet compression. He points out the similarities between these algorithms and the multiresolution representation theory of pattern vision. Perceptually lossless image compression is possible because of the ability to quantize high spatial frequency information in the transformed image. Glenn [Gle93] enumerates and quantifies how the visual effects based on thresholds, which have been described in part here, such as the luminance and chromatic CSFs, pattern masking, the oblique effect, and temporal sensitivity (see below) can contribute to perceptually lossless compression. Zeng et al. [ZDL02] summarize the "visual optimization tools" based on spatial frequency sensitivity, color sensitivity, and visual masking that are used in JPEG 2000 in order to optimize perceptually lossless compression.

It was pointed out above that the reduced resolution of the chromatic channels had been taken advantage of in image compression and transmission. This difference in the processing of chromatic and achromatic vision has been used for describing image differences in Zhang and Wandell's extension of CIELAB for digital image reproduction called S-CIELAB for Spatial-CIELAB (Zhang and Wandell [ZW96, ZW88]). This method is intended to compute the visible color difference for determining errors in color image reproduction.

In this procedure, an original color image and a reproduction are each converted into three bands corresponding to the luminance and two opponent chromatic channels. From here each band is convolved with a spatial filter that removes the information beyond the limits of contrast visibility. The images are then converted into CIELAB coordinates where pixel-by-pixel color differences can be computed.

Johnson and Fairchild [JF02, JF03] explored the use of different CSF functions, including the standard S-CIELAB filters, on their ability to predict image differences.

They found that the use of filters that were normalized to maintain the DC component of the image performed better [JF02]. This also has an effect of boosting the gain for certain spatial frequencies since the function produces a filter with coefficients greater than one when the DC component is fixed at a value of one. In addition, they found that anisotropic filters (with parameters that predict the oblique effect), such as the one used in Daly's Visible Difference Predictor [Dal93], improve the predictions of the image difference model. Johnson and Fairchild [JF02] also applied spatial frequency adaptation, both a generic statistical model based on the $1/f$ approximation of spatial frequency content in natural scenes and an image dependent model, in their image difference predictor and found improvement. This adaptation tends to shift the peak of the CSF to higher frequencies.

2.6 Temporal Vision and Motion

The sensitivity of the visual system to change over time and the perception of object motion are often linked together because the stimuli for motion produce temporal variations in light intensity falling on the retina. It follows that the perception of motion, at the early stages of visual processing, is driven by the same mechanisms that detect changes in intensity over time. In this section, the response of the visual system to temporally periodic stimuli will be briefly presented, followed by a short discussion of real and apparent motion.

2.6.1 Temporal CSF

Before the application of linear systems theory to the study of temporal vision, the literature was dominated by studies that examined the temporal sensitivity of the eye in terms of the critical flicker frequency (CFF) of periodic waveforms modulated in amplitude [Wat86]. CFF refers to the temporal frequency (Hz) at which the perception of flicker in the stimulus changes to a steady or fused appearance. Two concepts that emerged from this work are the Ferry-Porter law and the Talbot-Plateau law. The Ferry-Porter law states that CFF rises linearly with the logarithm of the time-average background intensity. The Talbot-Plateau law states that the perceived intensity of a fused periodic stimulus is the same as a steady stimulus of the same time-average intensity.

De Lange [de 54, de 58] introduced the ability to modulate temporal contrast while keeping the time-average background constant, which allowed the measurement of CFF and temporal contrast sensitivity at various mean intensities and temporal frequencies. His application of linear systems theory produced a coherent framework for interpreting the literature on temporal sensitivity [Wat86].

In a typical experiment for the measurement of temporal contrast sensitivity, a grating is modulated temporally in counterphase so that the contrast reverses sinusoidally over time (see Figure 2.16). The contrast threshold for detecting a pattern is measured as a function of both the temporal frequency of the modulation and the spatial frequency of the pattern.

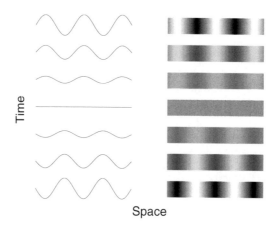

Space

Figure 2.16: Temporally counterphase modulated gratings change contrast over time sinusoidally producing a pattern that flickers. Each plot on the left shows the amplitude at a point in time. The luminance profile is simulated on the right.

Figure 2.17A shows the results from Kelly [Kel61] of the temporal contrast sensitivity of a large 68° field presented for frequencies from 1.6 to 75 Hz and mean luminances ranging from 0.06 to 9300 trolands. For these low spatial frequency stimuli, the shape of the temporal CSF is bandpass with a peak sensitivity at approximately 20 Hz at the highest luminance levels. As the mean luminance level decreases, curves become more low pass as the peak sensitivity decreases and shifts to lower frequencies. The overlapping sections of the curves for the moderate to high luminance levels indicate the region where Weber's law holds. The CFF shifts to higher frequencies with increased luminance.

In Figure 2.17B the data are replotted so that the reciprocal of threshold amplitude, rather than contrast, is plotted on the ordinate. When plotted in this manner, the high frequency portions of the curves at all luminance levels converge. This convergence indicates that at high flicker rates, the visual system cannot adapt to the signal and the response is linear.

Robson [Rob66] first explored the relationship between spatial and temporal contrast sensitivity using temporally modulated counterphase gratings (see Figure 2.16) at a variety of spatial and temporal frequencies. The results of this experiment are shown in Figure 2.18. Here the mean luminance was 20 cd/m². The data are presented two ways. Figure 2.18A shows the spatial CSFs for gratings flickered from 1 to 22 Hz. We see that as the temporal frequency increases the spatial CSF changes from band-pass to low-pass. This indicates that the process responsible for the low spatial frequency atten-

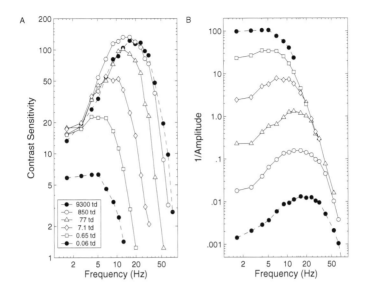

Figure 2.17: (A) Temporal CSF as a function of mean luminance for a large flickering field. (B) The data in A are replotted in terms of amplitude rather than contrast. Replotted from [Kel61].

uation is time dependent. Figure 2.18B shows the temporal CSFs for gratings varying in spatial frequency from 0.5 to 22 cpd. Here we see that the temporal CSF is bandpass at low spatial frequencies. We can infer here that the mechanism responsible for suppressing temporal contrast sensitivity is size dependent. It is remarkable to note the similarities between the spatial and temporal mechanism seen in Figure 2.18.

Families of curves such as those in Figure 2.18 can be accumulated to produce a spatiotemporal contrast-sensitivity contour for a given mean luminance level. Here contrast sensitivity is plotted as a function of both temporal and spatial frequency. Figure 2.19 is a characterization of this surface [BK80]. Similar results are obtained with drifting gratings moving at various velocities.

The spatiotemporal contrast response to chromatic stimuli has also been measured (see [Kel83, Kel89]). The response of the visual system to isoluminant red-green gratings exhibits much less attenuation at low temporal and spatial frequencies so that the shape of the spatiotemporal contrast surface is much more low-pass, especially above 0.5 Hz. The CFFs of the chromatic mechanisms occur at a lower temporal frequency than that of the achromatic channel.

2.6.2 Apparent Motion

There are a number of ways in which the appearance of motion occurs without any real movement in the stimulus. For example, adaptation to motion in a particular direction will lead to a motion aftereffect in the opposite direction. The moon can appear to move

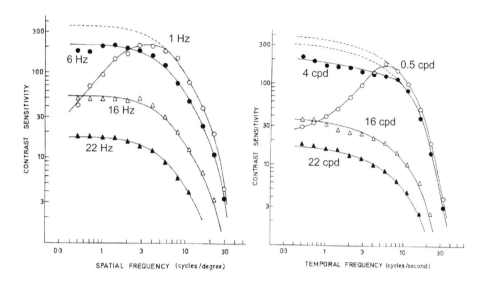

Figure 2.18: (A) Spatial contrast sensitivity measured for gratings modulated at various temporal frequencies. (B) Temporal contrast sensitivity for gratings of various spatial frequencies. The mean luminance was held constant at 20 cd/m². From [Rob66].

behind the clouds due to induced motion because of a misjudged frame of reference. Apparent motion, or stroboscopic motion, is the most common case of perceived motion without a real moving stimulus.

Consider an object at position 1 at time 1. At some point in time, time 2, the object disappears. After a certain interstimulus interval (ISI), the object reappears at time 3 in a new position, position 2. Depending on the distance and ISI, the object appears to have moved from position 1 to position 2. Interestingly, if the shape of the object is changed (within limits), the visual system "fills in" or "morphs" the object's shape during the ISI. Changes in the object's configuration are perceived as movement of the object's parts as well as a global position change. No such filling in is observed when the object's color is changed; an abrupt color change is perceived during the ISI.

Typically, the example of marquee lights is used when introducing the concept of apparent motion. However, in virtually all display systems, the perception of motion is produced with sequential static frames. Ironically, the study of motion in the laboratory is done using static stimuli presented in succession without any real motion!

The early Gestalt psychologist, Wertheimer, studied the relationship of perceived motion and ISI and classified the perception of motion based on the ISI duration [Pal99]. For very short ISI's (< 30 msec) no motion is observed. As the ISI increases (30–60 msec), there is a perception that the object moved from position 1 to position 2, however there is no impression of the intermediate position of the objects. This "disembodied

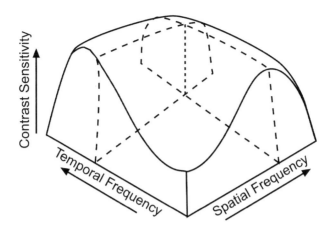

Figure 2.19: A representation of a spatiotemporal contrast sensitivity surface. Redrawn from [BK80].

motion" has been referred to as *phi motion*. At ISI's of approximately 100 msec, continuous motion is perceived. This has been referred to as *beta motion*. At longer ISI's no motion is observed. We see a sequential alternation of the object. Additional observations [Pal99] known as Korte's laws have been enumerated for beta motion: higher contrast is needed for larger positional changes; higher contrast is needed for shorter ISI's, and larger separations require longer ISI's. One can see that there is a relationship between temporal frequency, spatial frequency, and contrast revealed by Korte's laws.

Watson [Wat86] summarizes the reexamination of apparent motion in terms of the spatiotemporal sensitivity of the visual system. If we consider a 1-dimensional stationary grating of spatial frequency u and contrast c, we can plot its location in Figure 2.19, as a point on the spatial frequency (u) axis with zero temporal frequency ($w = 0$) so that we can define its contrast as $c(u, 0)$. For a moving grating, it can be derived that the contrast is $c(u, w + ru)$ where r is the grating's speed. The location of the grating plotted in the spatiotemporal plane is shifted in the temporal frequency dimension by $-ru$, the product of the speed and spatial frequency. The components of the frequency spectrum of a complex grating are sheared around the origin of the (u, w) plane [WA85]. For stroboscopic stimuli, the spatiotemporal contrast spectrum is the same as that for the smoothly moving stumulus with the addition of parallel replicas at intervals of the strobe frequency [WAF86].

For stroboscopic vision to appear smooth and real, Watson et al. [WAF86] predicted that the contrast of parallel replicas of the stimuli must be beyond the sensitivity of the visual system. That is, the temporal and spatial frequency spectra of these replications should be of sufficiently low contrast or beyond the CFF and acuity limits. Psychophysical tests of this prediction were confirmed. These results have implication for the design of motion capture and display systems to reduce the visibility of the aliasing components.

2.7 Visual Modeling

Visual modeling in one form or another has become an integral part of the study of the visual system. Rarely are basic psychophysical and physiological findings not put in the context of a model of visual processing in order to provide an understanding of the data in context of what is known about the visual system. This is necessary in order to assemble a full picture of the complexity of visual processing given that discoveries are made in small steps looking at specific aspects of visual function. In addition to the goal of understanding how vision works, what other purposes can visual modeling serve?

One possible goal, machine vision, is to simulate human vision in order to create systems that can replace humans for certain tasks. Another reason to study human vision is to discover the processing algorithms used by the visual system and apply these algorithms in other contexts. Wandell [Wan95] points out that until we fully understand why the visual system works the way it does, it is more likely that vision scientists will learn more from engineers than engineers will learn from vision scientists.

One other goal of modeling visual processing is in order to improve our ability to reproduce the visual world. Imagine a camera that can accurately capture, pixel for pixel, the complete spectral and radiometric information in a scene. Now imagine a device that can play it back with complete fidelity (perhaps some virtual reality environment). If we had such devices, there would be no need to understand vision from an application point of view. However, because we are limited in our ability to capture and display imagery, we need to discover the relevant information that must be captured and how to display it. Our reproductions are generally presented at a different size, resolution, adaptation level, dynamic range and surround environment, to name a few. We must therefore understand how vision processes the original scene information in order to best reproduce the scene.

2.7.1 Image and Video Quality Research

There has been significant research applying visual models to video quality and video quality metrics, often aimed at the creation and optimization of encoding/compression/decoding algorithms such as MPEG-2 and MPEG-4. The static-image visible differences predictor of Daly [Dal93] is one such model used to predict the visibility of artifacts introduced into still images by JPEG image compression, for example. The Daly model was designed to predict the probability of detecting an artifact (i.e., is the artifact above the visual threshold).

Recently, Fairchild has been developing an image appearance model, iCAM (for image color appearance model) that has had a different objective with respect to image quality [FJ02, FJ03]. Instead of focusing on threshold differences in quality, this work has focused on the prediction of image quality scales (e.g., scales of sharpness, contrast and graininess) for images with changes well above threshold. The question has not been "can a difference be perceived?" but rather "how large is the perceived difference?" Such suprathreshold image differences are a new domain of image quality research based on image appearance that separate the iCAM model from previous image quality models.

A similar situation exists in the area of video quality metrics. Metrics have been published to examine the probability of detection of artifacts in video (i.e., threshold metrics). Two well-known video image quality models are the Sarnoff JND model and the NASA DVQ model.

The Sarnoff JND model, a proprietary model of human vision [ATI01], is based on the multi-scale model of spatial vision published by Lubin [Lub93, Lub95] with some extensions for color processing and temporal variation. The Lubin model is similar in nature to the Daly model mentioned above in that it is designed to predict the probability of detection of artifacts in images. These are threshold changes in images often referred to as just-noticeable differences, or JNDs. The Sarnoff JND model has no mechanisms of chromatic and luminance adaptation, as are included in the iCAM model [FJ02]. The input to the Sarnoff model must first be normalized (which can be considered a very rudimentary form of adaptation). The temporal aspects of the Sarnoff model are aimed at predicting the detectability of temporal artifacts. The model only uses two frames (four fields) in its temporal processing. Thus, it is capable of predicting the perceptibility of relatively high frequency temporal variation in the video. It cannot predict the visibility of low frequency variations that would require an appearance-oriented, rather than JND-oriented model. The Sarnoff model also cannot be used for rendering video, but rather is limited to the prediction of difference-visibility maps on a frame-by-frame basis. While it is well accepted in the vision science literature that JND predictions are not linearly related to suprathreshold appearance differences, it is certainly possible to use a JND model to try to predict suprathreshold image differences and the Sarnoff JND model has been applied with some success to such data.

A similar model, the DVQ (Digital Video Quality) metric has been published by Watson and colleagues [Wat98, WHM01]. The DVQ metric is similar in concept to the Sarnoff JND model, but significantly different in implementation. Its spatial decomposition is based on the coefficients of a discrete cosine transform (DCT) making it amenable to hardware implementation and likely making it particularly good at detecting artifacts introduced by DCT-based video compression algorithms. It also has a more robust temporal filter that should be capable of predicting a wider array of temporal artifacts. Like the Sarnoff model, the DVQ metric is aimed at predicting the probability of detection of threshold image differences.

2.8 Conclusions

In this chapter we presented a basic outline of some of the psychophysical aspects of vision focusing on color vision, spatial vision, and temporal vision. At first glance, it may seem that our knowledge of the visual system based on the rather simple stimuli used in the laboratory may be quite rudimentary when we compare these stimuli to the complexity seen in the visual world. However, it is not the case that simple stimuli are used in visual psychophysics because they are easy. Rather, they are chosen in order to build up our knowledge of visual perception based on first principles. Before we can understand how we see objects, we must understand how the visual system responds to patterns. And before we can fully understand this we must know how the visual system responds to intensity. Although the stimuli used in psychophysical experiments may appear artificial, they are designed to measure specific aspects of visual function.

Notwithstanding, the gains in understanding of visual processing have had successful application to practical imaging. In this regard, the use of vision models for image compression and image quality has been briefly discussed. In order for there to be continued opportunity for the application of vision research and color science to technical challenges in video and imagery there must be continued interaction among the basic research community and the engineering and technical community in industry. By providing a brief summary of human vision, it is hoped that this chapter will facilitate this interaction.

References

[And03]   B. L. Anderson. Perceptual organization and white's illusion. *Perception*, 32(3):269–284, 2003.

[ATI01]   ATIS. Objective perceptual video quality measurement using a JND-based full reference technique. Technical Report T1.TR.PP.75-2001, Alliance for Telecommunications Industry Solutions Technical Report, 2001.

[Bar99]   P. G. J. Barten. *Contrast Sensitivity of the Human Eye and Its Effects on Image Quality*. Bellingham, WA: SPIE Press, 1999.

[Bar04]   P. G. J. Barten. Formula for the contrast sensitivity of the human eye. In Y. Miyake and D. R. Rasmussen, Eds., *Image Quality and System Performancei, Proceedings of SPIE-IS&T Electronic Imaging*, 5294:231–238, 2004.

[BC69]   C. Blakemore and F. W. Campbell. On the existence of neurons in the human visual system selectively sensitive to the orientation and size of retinal images. *J. Physiol.*, 203:237–260, 1969.

[Ber00]   R. S. Berns. *Billmeyer and Salzman's Principles of Color Technology*. New York: John Wiley & Sons, Inc., 2000.

[BK80]   C. A. Burbeck and D. H. Kelly. Spatiotemporal characteristics of visual mechanisms: Excitatory-inhibitory model. *J. Opt. Soc. Am.*, 70(9):1121–1126, 1980.

[Boy96]   R. M. Boynton. History and current status of a physiologically based system of photometry and colorimetry. *J. Opt. Soc. Am. A*, 13(8):1609–1621, 1996.

[BW94]   R. W. Bowen and H. R. Wilson. A two-process analysis of pattern masking. *Vision Research*, 34:645–658, 1994.

[CG66]   F. W. Campbell and R. W. Gubisch. Optical quality of the human eye. *J. Physiol.*, 186:558–578, 1966.

[CIE86]   CIE No. 15.2. *Colorimetry.* Commission Internationale de l'Éclairage, Vienna, Austria, 2nd ed., 1986.

[CR68]   F. W. Campbell and J. G. Robson. Application of fourier analysis to the visibility of gratings. *J. Physiol.*, 197:551–566, 1968.

[Dal93]   S. Daly. The visible differences predictor: An algorithm for the assessment of image fidelity. In A. W. Watson, Ed., *Digital Images and Human Vision*, 179–206. Cambridge, MA: MIT Press, 1993.

[DD90]   R. L. De Valois and K. K. De Valois. *Spatial Vision.* New York: Oxford University Press, 1990.

[de 54]   H. de Lange. Relationship between critical flicker frequency and a set of low frequency characteristics of the eye. *J. Opt. Soc. Am.*, 44:380–389, 1954.

[de 58]   H. de Lange. Relationship between critical flicker frequency and a set of low frequency characteristics of the eye. *J. Opt. Soc. Am.*, 48:777–784, 1958.

[DH89]   A. M. Derrington and G. B. Henning. Some observations on the masking effects of two-dimensional stimuli. *Vision Research*, 28:241–246, 1989.

[Fai98]   M. D. Fairchild. *Color Appearance Models.* Reading, MA: Addison–Wesley, 1998.

[FBM03]   J. Fiser, P. J. Bex, and W. Makous. Contrast conservation in human vision. *Vision Research*, 43(25):2637– 2648, 2003.

[FC99]   J. M. Foley and C.-C. Chen. Pattern detection in the presence of maskers that differ in spatial phase and temporal offset: Threshold measurements and a model. *Vision Research*, 39:3855–3872, 1999.

[FJ02]   M. D. Fairchild and G. M. Johnson. Meet iCAM: A next-generation color appearance model. In *Proceedings of IS&T/SID 10th Color Imaging Conference*, 33–38, Scottsdale, AZ, 2002.

[FJ03]   M. D. Fairchild and G. M. Johnson. Image appearance modeling. In B. R. Rogowitz and T. N. Pappas, Eds., *Human Vision and Electronic Imaging VIII, Proceedings of SPIE-IS&T Electronic Imaging*, 5007:149–160, 2003.

[Gle93]   W. E. Glenn. Digital image compression based on visual perception. In A. W. Watson, Ed., *Digital Images and Human Vision*, 63–72. Cambridge, MA: MIT Press, 1993.

[GN71]   N. Graham and J. Nachmias. Detection of grating patterns containing two spatial frequencies: A comparison of single channel and multichannel models. *Vision Research*, 11:251–259, 1971.

[Gra89]   N. V. S. Graham. *Visual Pattern Analyzers.* New York: Oxford University Press, 1989.

[GS75]   M. A. Georgeson and G. D. Sullivan. Contrast constancy: Deblurring in human vision by spatial frequency channels. *J. Physiol.*, 252(3):627–656, 1975.

[Hel11]   H. v. Helmholtz. *Treatise on Physiological Optics.* Rochester, New York: Optical Society of America, 1866/1911. Translated from 3rd German ed., J. P. Southall, Ed.

[Her64]   E. Hering. *Outlines of a Theory of the Light Sense.* Cambridge, MA: Harvard University Press, 1920/1964.

[HF86]   D. C. Hood and M. A. Finkelstein. Sensitivity to light. In K. R. Boff, L. Kaufman, and J. P. Thoma, Eds., *Handbook of Perception and Human Performance, Vol. 1: Sensory Processes and Perception*, (5–1) – (5–66), Vienna, 1963, 1986. New York: John Wiley & Sons.

[JF02]   G. M. Johnson and M. D. Fairchild. On contrast sensitivity in an image difference model. In *Proceedings of the IS&T PICS Conference*, 18–23, Portland, OR, 2002.

[JF03]   G. M. Johnson and M. D. Faichild. Measuring images: Differences, quality, and appearance. In B. R. Rogowitz and T. N. Pappas, Eds., *Human Vision and Electronic Imaging VIII, Proceedings of SPIE*, 5007:51–60, 2003.

[JT75]   R. M. Jones and U. Tulunay-Keesey. Local retinal adaptation and spatial frequency channels. *Vision Research*, 15:1239–1244, 1975.

[KB96]   P. K. Kaiser and R. M. Boynton. *Human Color Vision.* Washington, DC: Optical Society of America, 2nd ed., 1996.

[Kel61]   D. H. Kelly. Visual responses to time-dependent stimuli, I. Amplitude sensitivity measurements. *J. Opt. Soc. Am.*, 51(4):422–429, 1961.

[Kel83]   D. H. Kelly. Spatiotemporal variation of chromatic and achromatic contrast thresholds. *J. Opt. Soc. Am.*, 73(11):1784–1793, 1983.

[Kel89]   D. H. Kelly. Opponent-color receptive-field profiles determined from large-area psychophysical measurements. *J. Opt. Soc. Am. A*, 6(11):1784–1793, 1989.

[KNM84]   S. W. Kuffler, J. G. Nicholls, and A. R. Martin. *From Neuron to Brain.* Sunderland, MA: Sinauer Associates Inc., 2nd ed., 1984.

[LCR01]   M. R. Luo, G. Cui, and B. Rigg. The development of the CIE 2000 colour-difference formula: CIEDE2000. *Color Research and Application*, 26(5):340–350, 2001.

[LF80]   G. E. Legge and J. M. Foley. Contrast masking in human vision. *J. Opt. Soc. Am.*, 70(12):1458–1471, 1980.

[LR87]   M. R. Luo and B. Rigg. BFD (l:c) colour-difference formula, Part 1 — Development of the formula. *J. Soc. Dyers. Colour*, 103:86–94, 1987.

[Lub93]   J. Lubin. The use of psychophysical data and models in the analysis of display system performance. In A. W. Watson, Ed., *Digital Images and Human Vision*, 163–178. Cambridge, MA: The MIT Press, 1993.

[Lub95]   J. Lubin. A visual discrimination model for imaging system design and evaluation. In E. Peli, Ed., *Vision Models for Target Detection and Recognition*, 245–283. Singapore: World Scientific, 1995.

[MF85]   N. M. Moroney and M. D. Fairchild. Color space selection for JPEG image compression. *J. Electronic Imaging*, 4(4):373–381, 1985.

[MFH⁺02]   N. Moroney, M. D. Fairchild, R. W. G. Hunt, C. Li, M. R. Luo, and T. Newman. The CIECAM02 color appearance model. In *Proceedings of IS&T/SID Tenth Color Imaging Conference*, 23–27, Scottsdale, AZ, 2002.

[MM93]   W. H. Merigan and J. H. R. Maunsell. How parallel are the primate visual pathways. *Annual Review of Neuroscience*, 16:369–402, 1993.

[Mul85]   K. T. Mullen. The contrast sensitivity of human colour vision to red-green and blue-yellow chromatic gratings. *J. Physiol.*, 359:381–400, 1985.

[Nic76]   D. Nickerson. History of the munsell color system. *Color Research and Application*, 1:121–130, 1976.

[Pal99]   S. E. Palmer. *Vision Science: Photons to Phenomenology.* Cambridge, MA: The MIT Press, 1999.

[PS68]   A. Pantle and R. W. Sekular. Size-detection mechanisms in human vision. *Science*, 162:1146–1148, 1968.

[PW84]   G. C. Phillips and H. R. Wilson. Orientation bandwidths of spatial mechanisms measured by masking. *J. Opt. Soc. Am. A*, 1(2):226–232, 1984.

[PW95]   A. B. Poirson and B. A. Wandell. The appearance of colored patterns: Pattern-color separability. *J. Opt. Soc. Am. A*, 12:2458–2471, 1995.

[PW03]   O. Packer and D. R. Williams. Light, the retinal image, and photoreceptors. In S. K. Shevell, Ed., *The Science of Color*, 41–102. New York: Optical Society of America, 2003.

[Rob66]   J. G. Robson. Spatial and temporal contrast-sensitivity functions of the visual system. *J. Opt. Soc. Am.*, 56(8):1141–1142, 1966.

[SC99]   E. Switkes and M. A. Crognale. Comparison of color and luminance contrast: Apples versus oranges? *Vision Research*, 39(10):1823–1832, 1999.

[SD83]   E. Switkes and K. K. DeValois. Luminance and chromaticity interactions in spatial vision. In J. D. Mollon and L. T. Sharpe, Eds., *Colour Vision*, 465–470. London: Academic Press, 1983.

[SS99]   A. Stockman and L. T. Sharpe. Cone spectral sensitivities and color matching. In K. R. Gegenfurtner and L. T. Sharpe, Eds., *Color Vision*, 53–88. Cambridge, UK: Cambridge University Press, 1999.

[SW64]   C. L. Sanders and G. Wyszecki. Correlate for brightness in terms of CIE color matching data. In *CIE Proceedings 15th Session*, Paper P–63.6. CIE Central Bureau, Paris, 1964.

[SWB93a]   N. Sekiguchi, D. R. Williams, and D. H. Brainard. Aberration-free measurements of the visibility of isoluminant gratings. *J. Opt. Soc. Am. A*, 10(10):2105–2117, 1993.

[SWB93b]   N. Sekiguchi, D. R. Williams, and D. H. Brainard. Efficiency in detection of isoluminant and isochromatic interference fringes. *J. Opt. Soc. Am. A*, 10(10):2118–2133, 1993.

[VE03]   D. C. Van Essen. Organization of the visual areas in macaque and human cerebral cortex. In L. M. Chalupa and J. S. Werner, Eds., *The Visual Sciences*, 507–521. Cambridge, MA: The MIT Press, June 2003.

[Vim00]   R. L. P. Vimal. Spatial color contrast matching: Broad-bandpass functions and the flattening effect. *Vision Research*, 40(23):3231–3244, 2000.

[WA85]   A. B. Watson and A. J. Ahumada, Jr. Model of human visual-motion sensing. *J. Opt. Soc. Am. A*, 2(2):322–342, 1985.

[WAF86]   A. B. Watson, A. J. Ahumada, Jr., and J. E. Farrell. Window of visibility: A psychophysical theory of fidelity in time-sampled visual motion displays. *J. Opt. Soc. Am. A*, 3(3):300–307, 1986.

[Wan95]   B. A. Wandell. *Foundations of Vision*. Sunderland, MA: Sinauer Associates, Inc., 1995.

[Wat86]   A. B. Watson. Temporal sensitivity. In K. R. Boff, L. Kaufman, , and J. P. Thoma, Eds., *Handbook of Perception and Human Performance, Vol. 1: Sensory Processes and Perception*, (6–1) – (6–43). New York: John Wiley & Sons, 1986.

[Wat98]   A. B. Watson. Toward a perceptual video quality metric. In *Human Vision and Electronic Imaging III*, 3299:139–147. Bellingham WA: SPIE Press, 1998.

[WBMN94] D. R. Williams, D. H. Brainard, M. J. McMahon, and R. Navarro. Double-pass and interferometric measures of the optical quality of the eye. *J. Opt. Soc. Am. A*, 11(12):3123–3135, 1994.

[WHM01]   A. B. Watson, J. Hu, and J. F. McGowan. DVQ: A digital video quality metric based on human vision. *J. Electronic Imaging*, 10:20–29, 2001.

[Wil88]   D. R. Williams. Topography of the foveal cone mosaic in the living human eye. *Vision Research*, 28(3):433–454, 1988.

[WLM$^+$90] H. R. Wilson, D. Levi, L. Maffei, J. Rovamo, and R. DeValois. The perception of form: Retina to striate cortex. In L. Spillman and J. Werner, Eds., *Visual Perception: The Neurophysiological Foundations*, 231–272. New York: Academic Press, Inc., 1990.

[WM77]   G. Westheimer and J. McKee. Spatial configurations for visual hyperacuity. *Vision Research*, 17:941–947, 1977.

[WMP83]   H. R. Wilson, D. K. McFarlane, and G. C. Phillips. Spatial frequency tuning of orientation selective units estimated by oblique masking. *Vision Research*, 23:873–882, 1983.

[WR84]   H. R. Wilson and D. Regan. Spatial frequency adaptation and grating discrimination: Predictions of a line element model. *J. Opt. Soc. Am. A*, 1:1091–1096, 1984.

[WS82]   G. Wyszecki and W. S. Stiles. *Color Science Concepts and Methods, Quantitative Data and Formulae*. New York: John Wiley and Sons, 2nd ed., 1982.

[YM95]   J. Yang and W. Makous. Modeling pedestal experiments with amplitude instead of contrast. *Vision Research*, 35(14):1979–1989, 1995.

[ZDL02]   W. Zeng, S. Daly, and S. Lei. An overview of the visual optimization tools in JPEG 2000. *Signal Processing: Image Communication*, 17:85–104, 2002.

[Zek91]   S. Zeki. Cerebral akinetopsia (visual motion blindness). *Brain*, 114(2):811–824, 1991.

[ZW88]   X. Zhang and B. A. Wandell. Color image fidelity metrics evaluated using image distortion maps. *Signal Processing*, 70(3):201–214, 1988.

[ZW96]   X. Zhang and B. A. Wandell. A spatial extension to CIELAB for digital color image reproduction. In *Society for Information Display Symposium Technical Digest*, 27:731–734, 1996.

Chapter 3

Coding Artifacts and Visual Distortions

Michael Yuen
ESS Technology, Inc., P.R. China

3.1   Introduction

Evaluation and classification of coding artifacts produced by the use of the hybrid MC/DPCM/DCT video coding algorithm [ITU90, ITU95, Sec91, Sec92], as described in Chapter 1, is essential in order to evaluate the performance of various video coding software and hardware products proliferating the telecommunications, multimedia and consumer electronics markets. A comprehensive classification will also assist in the design of more effective adaptive quantization algorithms, coding mechanisms and post-processing techniques under the current constant bit-rate schemes, and, therefore, improve video codec performance. The classification of coding artifacts is of equal importance in the evaluation and minimization of these artifacts to achieve constant "quality" video compression.

In this chapter, a comprehensive characterization is presented of the numerous coding artifacts which are introduced into reconstructed video sequences through the use of the hybrid MC/DPCM/DCT algorithm. The isolation of the individual artifacts, which exhibit consistent identifying attributes, will be conducted with the aim of obtaining descriptions of the artifacts' visual manifestations, causes, and relationships. Additionally, the spatial and temporal characteristics of a video sequence that are susceptible to each artifact, and in which the artifacts are visually prominent, will be noted. To account for the adaptive quantization mechanisms and human visual system (HVS) masking properties utilized in practical coding applications, an MPEG-1 compliant codec, utilizing source content-based adaptive bit-allocation, will be employed as the origin of the coding distortions. Finally, because of the growing ubiquity of DVD/VCD players, a discussion will be presented of the distortions caused by various post-processing techniques which are common in such consumer electronic audio/video devices.

It is important to note that the task of quantifying the cause and effect of the various coding artifacts is hampered by the transformation of the spatial data to the transform domain, where the main source of the distortion—quantization—originates. Consequently, the total error introduced into an individual pixel does not have a direct relationship with the quantization operation; rather, the error introduced into each pixel is a sum of functions of the quantization error of each of the DCT coefficients.

Quantifying the artifacts in terms of their visual impact is an even more difficult task. Due to the complexity of the HVS (see Chapter 2), which has not yet been satisfactorily modeled [Cla95, 6], the perceived distortion is not directly proportional to the absolute quantization error, but is also subject to the local and global spatial, and temporal characteristics of the video sequence.

As a consequence of these factors, at a local level it is not possible to provide a definitive value indicating the level of quantization that will induce any one artifact to a certain visual degree, or when an artifact is perceived to even exist (the so-called *just noticeable distortion* or *supra-threshold* distortion) [Jay92]. Consequently, at a global level it is not possible to indicate a specific bit-rate at which any one artifact manifests. This is exacerbated by the different varieties of bit-allocation techniques that have been proposed which may, or may not, exploit the masking effects of the HVS.

Therefore, the discussion will be limited to descriptions of the artifacts' visual manifestations, causes, and relationships. We also describe the spatial and temporal characteristics of video sequences which are susceptible to each artifact and where the artifacts are visually prominent. The characteristics of some of the artifacts may be directly attributable to aspects of the MPEG-1 coding process (such as the use of bidirectional prediction); however, the overall discussion relates to the hybrid MC/DPCM/DCT algorithm in general. Where possible, pictures affected by the artifacts have been extracted from MPEG-1 coded sequences and presented to illustrate the artifacts' visual effect; however, as noted by Clarke [Cla95, 14–15], a true visual appreciation of the artifacts can only be provided by viewing the affected sequences on a high quality video monitor.

Due to the detailed discussion of the causes of the various artifacts, it is assumed that the reader has some background knowledge of the coding structure, as discussed in Chapter 1, in the H.261/MPEG-1/MPEG-2 standards: such as picture types, GOP and macroblock structures, and intra- and inter-frame coding techniques.

3.2   Blocking Effect

The *blocking effect* is, perhaps, the most widely studied distortion found in block-based picture coding. Visually, the effect is seen as a discontinuity between adjacent blocks in a picture. A number of examples of the effect are shown in Figure 3.1, where it is most evident around the face and the van in the foreground. The straight edges induced by the block boundary discontinuities, and the regular spacing of the blocks, result in the blocking effect being highly conspicuous.

Figure 3.1: Examples of the blocking effect. It is most evident in the areas which are of low-to-medium luminance and are smoothly textured.

The blocking effect is defined here as the discontinuities found at the boundaries of adjacent blocks in a reconstructed picture [RL83, Tzo88], since it is a common practice to measure and to reduce the blocking effect by only taking into account the pixels at the block boundaries[1] [KK95, AH94, YGK95]. However, the term can be generalized to mean an overall difference between adjacent blocks [Plo89, 46], but we have categorized these differences as individual artifacts.

One of the main reasons for the coding of pixels as block units is to exploit the high local inter-pixel correlation in a picture. Unfortunately, coding a block as an independent unit does not take into account the possibility that the correlation of the pixels may extend beyond the borders of a block into adjacent blocks, thereby leading to the border discontinuities. Even if an attempt is made by a basic MPEG-1 encoder to reduce the blocking effect, it is hampered by the fact that regulation of the quantization step-sizes, for intraframe coded blocks, is the only method available for adjusting the spatial content of the reconstructed blocks (a more appropriate MC prediction may be available to predictive coded macroblocks). Unfortunately, due to the use of transform coding, simply ensuring a similarity between the quantization step-sizes of adjacent blocks does not, necessarily, mean that the blocking effect will be eliminated.

---

[1]Ramamurthi and Gersho in [RG86] use the term "grid noise" to describe these block-edge discontinuities.

3.2.1 Intraframe Coded Macroblocks

The severity of the blocking effect is subject to the coarseness of the quantization of the DCT coefficients of either one or both of the adjacent blocks. The threshold level of the quantization, in either of the blocks, above which the blocking effect is noticeable to the HVS, relies on the content of the blocks as well as the masking effects of the HVS. Generally, the effect is hidden in either the more spatially active areas, or the bright or very dark areas [Gir89].

Since the blocking effect is more visible in the smoothly textured sections of a picture, the lower order DCT coefficients play the most significant role in determining the visibility of the blocking effect; this is especially true for the DC coefficient [NH89, 418]. However, the blocking effect may occur in spatially active areas as a result of very coarse quantization. With the medium to higher order AC coefficients quantized to zero, an originally spatially active block will have a smoothly textured reconstruction which, in terms of the blocking effect, is subject to the same visibility concerns as originally smooth blocks.

3.2.2 Predictive Coded Macroblocks

The occurrence of the blocking effect in predictive coded macroblocks can be categorized into two different forms: one relating to the external boundary of the macroblock, and the other found between the four constituent luminance 8×8-element blocks of the macroblock.

Since the MC prediction, generally, provides a good prediction of the lower-frequency information of a macroblock [Le 91], the internal blocking effect, typically, does not occur for macroblocks with a smoothly textured content.[1] In this situation, the prediction error is minimal and, therefore, after quantization the prediction error is reduced to zero. If the prediction error for all the constituent blocks is quantized to zero, the internal blocking effect will not occur. The internal blocking effect mainly occurs in mildly textured areas where the prediction error is sufficiently large such that it is not quantized to zero. The combination of a relatively smooth prediction and coarsely quantized prediction error results in visible discontinuities between the internal blocks. The areas within a picture which produce very large prediction errors usually contain high spatial activity, which masks the internal blocking effect.

Discontinuities induced between the internal blocks will result in discontinuities between adjacent macroblocks. However, the external blocking effect is most visibly significant around the borders of moving objects, and is a consequence of poor MC pre-

---

[1]The copying of the blocking effect to the current picture from the MC reference pictures will be discussed in Section 3.10

diction, which is typical around moving areas in the recorded scene. This phenomenon will be discussed in more detail in Section 3.11. Suffice to say that the task of the MC prediction is made more difficult in these areas since it must produce a single motion vector for the whole macroblock where, in some situations, the constituent pixels of a macroblock may have originated from separate objects within the scene with divergent motion. The result is a predicted macroblock that is ill-fitting in one or more of the enclosed moving objects, with disparities occurring along the boundary of the macroblock. The contribution of the quantized prediction error to the reconstruction is, typically, a high-frequency noise-like effect, resulting from the presence of the edge(s) between the boundaries of the moving objects in the prediction, and the high energy of the prediction error needed to compensate for the poor prediction. This tends to mask the blocking effect, although it produces a temporal effect which will be discussed in Section 3.12.

The external block effect may also occur where the boundary edge of a macroblock coincides with the boundary of a moving object; the straight well-defined edge of the macroblock produces an unnatural visual sharpness to the object's boundary.

3.3 Basis Image Effect

The visual prominence of the blocking effect is primarily a consequence of the regularity of the size and spacing of the block-edge discontinuities. Each of the DCT basis images have a distinctive regular horizontally- or vertically-oriented pattern which make them visually conspicuous [RY90]. This leads to the *basis image effect* which manifests as blocks bearing a distinct likeness to one of the sixty-three AC DCT basis images. The regular pattern of the basis images, and their fixed size, makes them visually prominent; examples are given in Figure 3.2.

The effect is caused by coarse quantization of the AC DCT coefficients in areas of high spatial activity within a picture, resulting in the nullification of low-magnitude DCT coefficients. In a situation where a single AC basis image is prominent in the representation of a block, the result after coarse quantization is the reduction of all, except the most prominent, basis images to insignificance. This results in an emphasis of the pattern contributed by the prominent basis image, since the combined energy of the other AC basis images are insufficient to mute its contribution to the aggregated reconstruction.

The visual characteristics of the DCT basis images may also produce other coding artifacts when adjacent blocks are taken into consideration. For example, blocks suffering from the basis image effect invariably do not fit well, in terms of appearance, with the surrounding blocks; therefore, this results in the mosaic pattern effect, which will be described in Section 3.8. Additionally, for basis images which have a zero, or very low, frequency content in either the horizontal or vertical direction, the blocking effect may result along the sections of the block boundary which contain limited spatial activity.

Figure 3.2: Examples of the basis image effect, extracted from the background of the *Table-Tennis* sequence (see the example given for the ringing effect in Section 3.7).

3.3.1 Visual Significance of Each Basis Image

The MPEG-1 intra quantization weighting matrix was derived from experiments on the psychovisual thresholding of quantization distortion for each of the DCT basis images [Sec91]. This thresholding relates to the ability of the HVS to discern changes of contrast in each of the basis images. The use of different weights for each of the basis images, and other research into the calculation of visually optimum quantization matrices [WSA94, PAW93, KSC92, Ahu92], demonstrate that the HVS does not perceive each of the basis images with equal significance. Therefore, the visual impact of the basis image effect is subject to the proportion of AC energy concentrated into any one coefficient, as well as the visual significance of the respective DCT basis image.

3.3.2 Predictive Coded Macroblocks

As with intraframe coded macroblocks, the basis image effect in predictive coded macroblocks occurs in high spatial activity areas. This is especially evident when the prediction offered by the MC reference contains little or no spatial detail. The attempt to compensate for the loss of high frequency information with coarsely quantized prediction error may result in the emergence of a single AC basis image.

Similar to the mosaic pattern effect (see Section 3.8), if the same high spatial activity area is visible in a number of pictures of a sequence, the basis image effect will decrease over time in this area. This is a consequence of the accumulation and refinement of higher order AC component information for each succeeding P-type picture (MC reference picture) in the sequence.

3.3.3 Aggregation of Major Basis Images

Although blocks containing the basis image effect resemble a single DCT basis image, it is possible that a significant proportion of the AC energy is concentrated into more than one AC coefficient. This is most evident when the primary basis images are of a similar directional orientation and frequency, resulting in an enhancement of parts of the contributed visual patterns. This will retain the frequency of the contributing basis images, as well as a perceived regularity in the aggregated pattern.

3.4 Blurring

Blurring manifests as a loss of spatial detail in moderate to high spatial activity regions of pictures, such as in roughly textured areas or around scene object edges.

For intraframe coded macroblocks, blurring is directly related to the suppression of the higher order AC DCT coefficients through coarse quantization, leaving only the lower order coefficients to represent the contents of a block. In many respects, blurring, as a consequence of transform coding, can be considered as a specific case of the basis image effect where the prominent AC basis images after quantization are of a lower frequency, resulting in a reconstructed block with low spatial activity. Also similar to the basis image effect, the blurring of blocks in areas of high spatial activity may coincide with both the blocking effect and the mosaic pattern effect. The result of blurring of the chrominance information will be discussed in Section 3.5.

For predictive-coded macroblocks, blurring is mainly a consequence of the use of a predicted macroblock with a lack of spatial detail. However, blurring can also be induced in bidirectionally predicted macroblocks, where the interpolation of the backward and forward predictions results in an averaging of the contents of the final bidirectional prediction. In both these cases, the blurred details are supplemented by the prediction error which supplies some higher frequency information to the reconstruction, thereby reducing the blurring effect.

3.5 Color Bleeding

The blurring of the luminance information, as discussed in Section 3.4, results in the smoothing of spatial detail. The corresponding effect for the chrominance information results in a smearing of the color between areas of strongly contrasting chrominance.

As with blurring, *color bleeding* is caused by the quantization to zero of the higher order AC coefficients, resulting in the representation of the chrominance components with only the lower frequency basis images. Since the chrominance information is typically subsampled, the bleeding is not limited to an 8×8-pixel area, as for the luminance information, but extends to the boundary of the macroblock.

For chrominance edges of very high contrast, or where the quantization of the higher frequency AC coefficients does not result in their truncation, the color artifact corresponding to the ringing effect occurs. This will be discussed in Section 3.7.

Figures 3.3 and 3.4 show the C_b and C_r chrominance components, respectively, of an I-type picture from the *Table-Tennis* sequence, coded at 0.6Mbps. The blurring of the chrominance component, which leads to color bleeding, is most evident along the top edge of the arm in Figure 3.3, as well as around the table-tennis paddle. The source of the chrominance ringing is seen along the edge of the table-tennis table in both Figures 3.3 and 3.4. The corresponding color picture is shown in Figure 3.5.

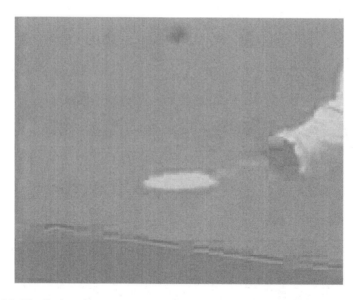

Figure 3.3: The C_b chrominance component of a coded picture from the *Table-Tennis* sequence.

It is interesting to note that strong chrominance edges are accompanied by strong luminance edges; however, the existence of a strong luminance edge does not necessarily coincide with a strong chrominance edge [NH89, 230]. Therefore, color bleeding is not necessarily found at blurred edges in a reconstructed color picture.

Figure 3.4: The C_r chrominance component of a coded picture from the *Table-Tennis* sequence.

Figure 3.5: Reproduction of a picture suffering from color bleeding and "color ringing" from the *Table-Tennis* sequence, corresponding with the C_b and C_r chrominance components shown in Figs. 3.3 and 3.4. Note the high frequency changes in color around the table's edge, corresponding to the ringing, and the gradual bleeding around the arm.

3.6   Staircase Effect

The DCT basis images are not attuned to the representation of diagonal edges and features [RY90]. Consequently, more of the higher frequency basis images are required to
satisfactorily represent diagonal edges or significant diagonally-oriented features. After
coarse quantization, the truncation of the contributions made by the higher order basis
images results in the failure to mute the contributions made by the lower order basis images to the reconstructed block. For a diagonal edge angled towards the horizontal, this
will result in a reconstruction with a horizontal orientation, and vice versa for blocks
angled towards the vertical. The ringing effect, which is also found in blocks containing
an edge, will be discussed in Section 3.7.

The *staircase effect* is related to both the blocking and mosaic pattern effects in terms
of the manifestation of discontinuities between adjacent blocks. When a diagonal edge
is represented within a string of consecutive blocks, the consequence of coarse quantization is the reconstruction of the diagonal edge as a number of horizontal or vertical
steps. These individual steps do not merge smoothly at a block's boundary with the con-
tinuance of the edge in adjacent blocks. A number of horizontally oriented examples
of the staircase effect are shown in Figure 3.6; note that the step-wise discontinuities
occur at block boundaries. Figure 3.6 also exhibits significant ringing effect, especially
around the edges near the top of the image.

Figure 3.6: Examples of the staircase effect.

3.7 Ringing

The poor energy compacting properties of the DCT for blocks containing diagonal edges, as previously described, also extends to vertical and horizontal edges. However, whereas the energy is distributed generally throughout all the AC coefficients for diagonal edges, for horizontal edges the energy tends to be limited to the coefficients which represent little or no vertical energy; and vice-versa for vertical edges.

The representation of a block can be considered as a carefully balanced weighted combination of each of the DCT basis images, such that the feature contributed by any one basis image is either enhanced or muted by the contribution of the other basis images [RY90]. Therefore, quantization of an individual coefficient results in the generation of an error in the contribution made by the corresponding basis image. Since the higher frequency basis images play a significant role in the representation of an edge, the quantized reconstruction of the block will include high frequency irregularities.[1]

The *ringing effect* is most evident along high contrast edges in areas of generally smooth texture in the reconstruction, and appears as a shimmering or rippling outwards from the edge up to the encompassing block's boundary. The higher the contrast of the edge, the greater the level of the peaks and troughs of the rippling. Examples of this are shown in Figure 3.7, where it is most evident along the edge of the table-tennis table and the bottom of the player's arm. The generally smooth texture in the surrounding blocks results in a greater visibility of the ringing effect, where otherwise a masking effect would be introduced.

Figure 3.8 shows a section of an I-type picture from the *Claire* sequence coded at 1.0 Mbps. The PSNR of the picture is approximately 45 dB. Note that at this relatively high bit-rate (for the spatial/temporal characteristics of sequence), a slight ringing effect is still produced along the high contrast edge of the character's arm.

The discussion and background of the ringing effect that has been presented so far also applies to the chrominance components; consequently, a similar effect to ringing occurs for the chrominance information at strong chrominance edges (see Figures 3.3 and 3.4). The ringing of the chrominance information appears as wave-like transitions of color away from the chrominance edge up to the boundary of the encompassing macroblock. The colors produced as a result of ringing often do not correspond to the colors of the surrounding area. Due to the subsampling of the chrominance information, the *chrominance ringing* spans a whole macroblock.

---

[1]If most of the higher-order AC coefficients are nullified as a result of very coarse quantization then, depending on the energy and distribution of the AC coefficients, the basis image effect or blurring may result instead.

Figure 3.7: Example of the ringing effect, where it is most evident around the bright table-edge and the boundary of the arm.

Figure 3.8: Example of the ringing effect in a sequence coded at a relatively high bit-rate. Most prominent along the edge formed by the upper-arm in the scene.

3.8 Mosaic Patterns

The general nature of the *mosaic pattern effect* is the apparent mismatch between all, or part, of the contents of adjacent blocks; this has a similar effect to using visually ill-fitting square tiles in a mosaic. This may mean a block with a certain contour or texture dissimilar to the surrounding blocks, or a block used in the representation of an object which does not blend satisfactorily with the other constituent blocks.

The mosaic pattern, typically, coincides with the blocking effect; however, the existence of the blocking effect between two blocks does not necessarily imply the manifestation of the mosaic pattern between the same two blocks. For example, a smoothly textured area is highly susceptible to the blocking effect, but the smooth characteristics of the adjacent blocks would not induce a mosaic pattern. Measures to reduce the blocking effect in these low activity areas by smoothing the block boundaries have been shown to be effective [RL83, Tzo88], with no visually apparent mosaic pattern. Figure 3.9 shows the result of such a filter proposed by Tzou [Tzo88] applied to the reconstructed picture depicted in Figure 3.1. Compare the smoothly textured areas of the window and hood (bonnet) of the vehicle in Figures 3.1 and 3.9. Examples of the mosaic pattern effect can be seen on the face of the human character in Figure 3.9. The mosaic pattern may also be introduced by the various artifacts discussed in previous sections. This is most evident with blocks suffering from the basis image effect, as can be seen in Figure 3.2.

Figure 3.9: Example of the mosaic pattern effect, where it is most evident around the character's face and adjacent to the horizontal edges near the van's window.

3.8.1   Intraframe Coded Macroblocks

For intraframe coded blocks, the mosaic pattern typically occurs in areas of the reconstructed picture where high spatial activity existed in the original picture. The typical quantization weightings and probability distributions associated with each of the AC coefficients [Sec91, Sec92] means that very coarse quantization will result in the truncation of a significant proportion of the higher frequency AC coefficients to zero. Consequently, upon reconstruction, the blocks will contain textures constructed only from the lower frequency AC basis images, which may be of a dissimilar texture or contour than their neighbours. It is important to note that even if the blocks contained a similar texture in the original picture, it cannot be guaranteed that the coarsely quantized reconstruction will be the same for all the blocks. There cannot even be a guarantee of a similarity of the general texture between the reconstructed blocks. An example of this can be seen by comparing Figures 3.10 and 3.11, which show the background of the first picture of the *Table-Tennis* sequence, before and after transform coding,[1] respectively.

Figure 3.10: A section of the background from the original (uncoded) *Table-Tennis* sequence.

From an examination of the lower-frequency AC basis images, it can be seen that their general orientations can be placed into one of three categories: horizontal, vertical, and indeterminate.[2] One of the major factors affecting the visibility of the mosaic pattern effect is the general directional orientation of the adjacent blocks. For example, if the

---

[1] For this example, the same quantization weighting matrix and scaler were used for all the 8×8-pixel blocks to code the picture in Figure 3.11.

[2] An attempt to define a set of diagonally oriented DCT basis images can be found in [Plo89, 101].

Figure 3.11: The same section of the background from the *Table-Tennis* sequence, shown in Fig. 3.10, after transform coding. An identical quantization matrix and scaler were used for all the 8×8-pixel blocks.

contents of two adjacent blocks are of dissimilar directional orientation, then the mosaic effect will be more pronounced; this is especially true for a combination of horizontally and vertically oriented blocks.

Even if adjacent blocks do have a similar orientation, the visibility of the mosaic pattern will also be affected by the prominence of the major frequency component of each of the adjacent blocks. The worst case would be if a block is reconstructed using only a single AC basis image (see Section 3.3). A simple example of this can be seen by placing the lowest order horizontal AC basis image adjacent to any of the higher order horizontal basis images. Therefore, a similar major frequency component between the adjacent blocks would reduce the perceived mosaic pattern.

A number of aspects of the mosaic pattern effect for intraframe coded blocks have similarities to the basis image effect, which was described in Section 3.3. However, the coarseness of the quantization necessary for the mosaic pattern to occur is less than that for the basis image effect, and the resulting patterns of the adjacent blocks need only be sufficiently dissimilar, i.e., they need not resemble DCT basis images.

3.8.2 Predictive-Coded Macroblocks

The typical consequence of the low bit-rate coding of I-type pictures is the suppression of the higher order AC components in favour of the lower order components. For the predictive coded macroblocks of succeeding pictures in a sequence, the MC prediction macroblocks originating from I-type pictures typically contain a good prediction of the lower order components [Le 91]. Therefore, for areas of high spatial activity, the main task of the prediction error coding is to reconstruct the higher frequency information.

As with intraframe coded macroblocks, the mosaic pattern effect in predictive coded macroblocks typically occurs in high spatial activity areas of a picture. The effect results from the attempt to compensate for the loss of high frequency information in an MC predicted macroblock with coarsely quantized prediction error. This may produce differing higher frequency AC components in the reconstruction of adjacent blocks within a macroblock, or in neighbouring macroblocks. In smooth areas, however, the prediction is of a reasonable quality[1] and the likeness between adjacent macroblocks preclude the mosaic pattern effect.

It is interesting to note that if the same high spatial activity area is visible in a number of consecutive pictures of a sequence, the mosaic pattern gradually decreases over time. This is as a result of the accumulation and refinement of the higher frequency AC component information for each successive MC reference picture. However, the visibility of the mosaic pattern in B-type pictures is subject to the type of prediction used for each macroblock, as well as which reference pictures were used for the MC prediction. Figures 3.12–3.14 show the same section from the reconstruction of the *Table-Tennis* sequence for an I-type picture and the following two P-type pictures,[2] respectively. A comparison between these three reconstructions demonstrates the gradual refinement of high activity spatial information for each successive MC reference picture. However, it is important to note that the degree to which the high frequency information may be refined is subject to the coarseness of quantization of the related prediction error. Note that the I-type picture was filtered [YWR95] for the blocking effect prior to it being used as an MC reference; this was done to separate out the effect of the propagation of the blocking effect to the following predictive coded pictures (false edges, see Section 3.10).

3.9 False Contouring

The simple artifact of *false contouring* often results from the direct quantization of pixel values. It occurs in smoothly textured areas of pictures containing a gradual transition in the value of the pixels over a given area. The many-to-one mapping operation of quantization effectively restricts the allowable pixel values to a subset of the original range, resulting in a series of step-like gradations in the reconstruction over the same given area. This artifact is directly attributed to either an inadequate number of quantization levels for the representation of the area, or their inappropriate distribution [GW92, 34].

---

[1] A poor prediction for a smooth area would typically result in the encoder intraframe coding the macroblock [Sec91].

[2] Due to the structure of the GOP used in the MPEG-1 coding process for this example, the P-type pictures are the 5th and 9th pictures in the sequence.

Figure 3.12: Section of the reconstruction of an I-type picture from the *Table-Tennis* sequence, coded at 0.6Mbps, with the application of a filter for the blocking effect.

Figure 3.13: Section of the reconstruction of the first P-type picture from the *Table-Tennis* sequence, coded at 0.6Mbps, using the I-type picture in Fig. 3.12 as the MC reference.

Figure 3.14: Section of the reconstruction of the second P-type picture from the *Table-Tennis* sequence, coded at 0.6Mbps, using the P-type picture in Fig. 3.13 as the MC reference.

False contouring can also occur in block-based transform coding with similar circumstances and causes to that of direct quantization. The artifact is a consequence of the inadequate quantization of the DC coefficient and the lower-frequency AC coefficients in smoothly textured areas. The effect appears in the reconstruction as step-like gradations in areas of originally smooth transition, similar to the effect described above. However, it is important to note that the gradations tend to affect a whole block. Examples of this are shown in the background of Figure 3.15, which contains a section of an I-type picture from the *Claire* sequence.[1] The original picture had contained a gradual reduction of the luminance radially away from the speaker's head.

3.10   False Edges

The exploitation of interframe redundancies relies on the transfer of previously coded information from MC reference pictures to the current predictive coded picture. The transferred information, unfortunately, also includes the coding artifacts formed in the reconstruction of the MC reference. *False edges* are a consequence of the transfer of the block-edge discontinuities formed by the blocking effect from the reference picture.

The generation of false edges is explained in Figure 3.16, where the predicted macroblock copied from the MC reference picture contains the block boundary disconti-

---

[1] Note that, for purposes of demonstration, the image has been filtered to exaggerate the distortion.

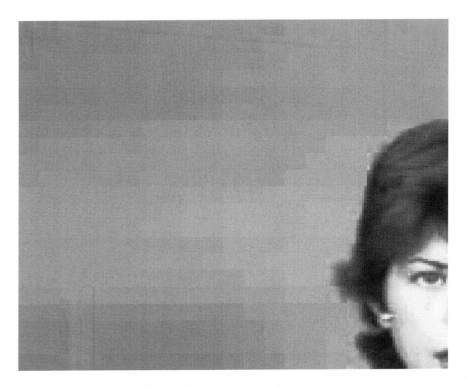

Figure 3.15: Example of false contouring.

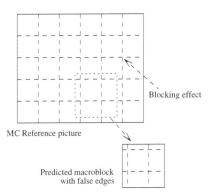

Figure 3.16: Example of the propagation of the blocking effect to false edges.

nuities formed by the blocking effect. These false edges may, or may not, coincide with the boundaries of macroblocks/blocks in the current predictive coded picture, but for skipped macroblocks the boundaries will always coincide if the MC reference is an I-type picture.

As with the blocking effect, false edges are mainly visible in smooth areas of predictive coded pictures. The prediction error in such areas would typically be minimal, or quantized to zero, and therefore the false edges would not be masked. In comparison, significant false edges may also be contained in the prediction for areas of higher spatial activity, however, a masking effect will be produced by the high level of the quantized prediction error. As a consequence of these masking effects, the manifestation of false edges typically do not propagate to subsequent predictive coded pictures, especially in areas of high spatial activity.

3.11   MC Mismatch

Motion compensation between pictures of a sequence is commonly conducted using a full-search block-matching technique, where the contents of a macroblock to be coded is compared to all the possible macroblock-sized regions within a limited search window of the MC reference pictures. The common metric used for the comparison is a disparity measure, such as the mean squared error or mean absolute error. The final product is a motion vector indicating the spatial displacement of the current macroblock from its prediction.

This simple, yet computationally intensive, operation models the motion within a sequence as being composed of translations of rigid objects. This has, generally, been found to provide a close approximation of the true motion within a sequence [DM95], and the use of a disparity measure for the search criterion has the benefit of determining the prediction, within the search window, which produces the minimum prediction error energy. However, a problem arises as a result of assuming that the true motion of all the constituent pixels of a macroblock are identical. This is most evident around the boundaries of moving scene objects, where a macroblock may encompass pixels forming part of a moving object as well as pixels representing sections of the scene not connected with the same object. In this situation, the motion of the pixels within a macroblock would be better defined as a collection of multiple sub-macroblock motion vectors [DM95].

For a better understanding of the *MC mismatch effect* we examine the case where only two objects are involved, i.e. a macroblock straddling the boundary between two objects in a scene whose motions are divergent with relation to each other. Examples of such macroblocks are illustrated in 'Picture K' of Figure 3.17. In this situation, it is unlikely for there to exist the same congruence between the two objects within the MC reference pictures as in the current picture; therefore, the chosen MC predicted mac-

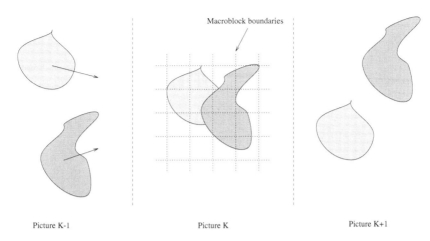

Picture K-1   Picture K   Picture K+1

Figure 3.17: Illustration of macroblocks encompassing the regions of two scene objects with divergent motion.

roblock would contain an unsatisfactory representation of one, or possibly even both, of the objects it is intended to represent. This reveals one of the disadvantages of the disparity measure of the block-matching operation: it does not seek to retain the integrity of any one of the objects' contents, nor the boundary dividing the two objects. The limited search window of the block-matching operation also exacerbates the misrepresentation of the MC prediction for very high motion of the constituent objects.

Since the macroblock is situated on an object boundary, it will contain a significant amount of high-frequency detail, such as a scene edge; therefore, it would not be advantageous to intraframe code the macroblock, even though the poor prediction will result in a large prediction error energy. This corresponds with the intra/predictive coding decision mechanism used in MPEG-1, based on the variances of the macroblock and the prediction error.

If a macroblock is reconstructed solely from a poor quality prediction, then, visually, the effect of MC mismatch would be seen as an ill-fitting square section of pixels with relation to either one or more of the scene objects that the macroblock encompasses. Consequently, a relationship would exist with the previously discussed artifacts of the mosaic pattern effect and the blocking effect. However, due to the changing spatial characteristics induced by the moving objects, a large prediction error energy will result. After quantization of the DCT coefficients, the contribution of the prediction error to the reconstruction is typically a high-frequency noise-like effect. This helps to mask the mismatch of the prediction, although it also produces a temporal effect which will be discussed in Section 3.12. The consequence of quantizing the higher order AC coefficients in blocks containing a high contrast edge was discussed in Section 3.7 with relation to ringing. Examples of the high-frequency noise induced by MC mismatch can be seen in Figure 3.18 around the boundary between the arm and the background.

This example was taken from a P-type picture, and it is important to note that all the macroblocks representing the arm and its boundary were forward predicted.

Figure 3.18: Example of the high-frequency noise induced by MC mismatch around the boundaries of moving objects.

3.12 Mosquito Effect

The *mosquito effect* is a temporal artifact seen mainly in smoothly textured regions as fluctuations of luminance/chrominance levels around high contrast edges, or moving objects, in a video sequence [KHK87]. This effect is related to the high frequency distortions introduced by both the ringing effect, and the prediction error produced by the MC mismatch artifact; both of which have been described previously. Generally, the degree and visibility of the fluctuations is less for the ringing-related effect than that resulting from the MC mismatch prediction error. The mosquito effect, regardless of the origin, is a consequence of the varied coding of the same area of a scene in consecutive pictures of a sequence. This may mean a difference in the type of prediction (forward, backward, bidirectional, or skipped), quantization level, MC prediction (motion vector), or a combination of these factors.

3.12.1 Ringing-Related Mosquito Effect

In predictive coded pictures, a macroblock containing a high contrast edge would most likely be predictive coded due to the high cost of intraframe coding such a spatial feature. If available, the macroblock's MC prediction would also contain a high contrast edge corresponding to the contents of the current macroblock. Assuming that the I- or P-type MC reference picture was poorly quantized, this prediction would also be suffering from the ringing effect due to the presence of the high contrast edge.

The mosquito effect occurs in such a situation, generally, as a result of varying attempts, from picture to picture, to correct the ringing effect in the MC prediction, or the use of a different prediction. This may be a consequence of the use of a different prediction type, a differently positioned MC prediction, or simply a change in the level of quantization of the prediction error. Any of these factors will cause a difference in the reconstruction of the same area of a picture. This assumes that the quantization is sufficiently coarse to prevent the satisfactory correction of the ringing effect, in which case under- or over-correction of the high frequency ringing noise may be introduced by the coarsely quantized prediction error.

As discussed in Section 3.7, the higher frequency DCT basis images play a significant role in the representation of a high contrast edge, and the ringing effect results from the quantization error of the higher order AC coefficients. To correct the ringing effect in the predicted macroblock, the prediction error will require the use of a significant level of the higher frequency basis images; therefore, differences in the quantized prediction error, from picture to picture, will result in high frequency fluctuations in areas around high contrast edges during the display of the decoded video sequence.

Figure 3.19 shows the difference between the same section of five consecutive coded pictures from the *Claire* sequence, i.e. the difference between pictures 0 and 1, 1 and 2, etc.[1] The speaker's left arm remains relatively stationary over the duration of the five pictures. High frequency noise-like blocks can be seen around the high contrast edge formed by the speaker's arm and the background. Note that this noise is the difference between consecutive pictures, therefore, it is equivalent to the temporal change in the area between consecutive pictures.

As an indication of the different MC prediction types used by each of these predictive coded pictures, each macroblock within the same Co-sited region is delineated in Figure 3.20 for pictures 1 through 4 of the coded *Claire* sequence, with the MC prediction method shown for each macroblock. These displayed regions correspond with those presented in Figure 3.19.

3.12.2 Mismatch-Related Mosquito Effect

As discussed in Section 3.11, the divergent motion around the boundaries of moving scene objects is a hindrance to the selection of a satisfactory MC prediction. The presence of the boundary edge, and the significant level of the prediction error energy, results in a high frequency noise-like effect in the area around the boundary.

---

[1]Note that, for purposes of demonstration, the images have been filtered to exaggerate the differences.

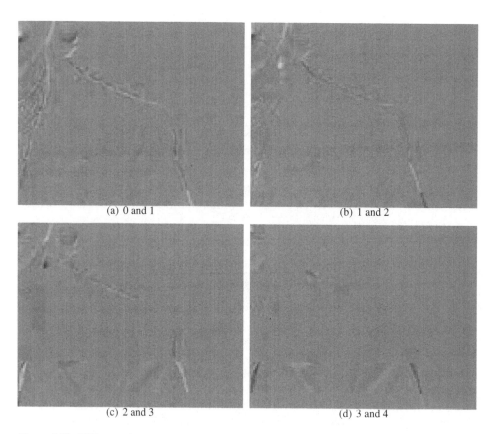

Figure 3.19: Difference between the same section of five consecutive coded pictures from the *Claire* sequence. The jaw of the speaker can be seen at the top-left of the difference images. To enable their display, the differences were offset by +128 and clipped to be within 0 and 255.

The causes of the mosquito effect in these areas are similar to those originating from the ringing effect, as discussed above. However, as a consequence of the changing spatial characteristics of either the moving object or its background, the differences would be more pronounced between the predicted macroblocks used in consecutive pictures for the same section of the moving object. Also, if the object is moving with relation to the camera, the section of the object that is encompassed by any one macroblock will change from picture to picture. The different locations of the discontinuities caused by the changing position of the macroblock boundaries, with reference to the moving object, will result in an increased mosquito effect.

3.13   Stationary Area Fluctuations

Similar fluctuations like that produced around high contrast edges and moving objects have also been seen in stationary areas containing significant spatial activity. This is

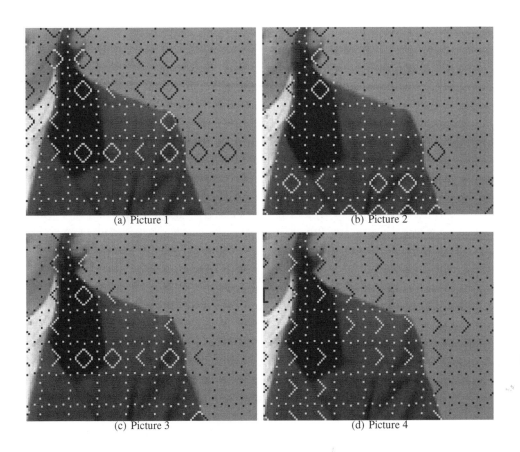

(a) Picture 1   (b) Picture 2

(c) Picture 3   (d) Picture 4

Figure 3.20: The MC prediction types used to code each of the macroblocks in the same spatial region of pictures 1 through 4 of the *Claire* sequence. This region corresponds to that used in the difference images in Fig. 3.19. Chevrons pointing right and left represent forward and backward prediction, respectively. Bidirectional (interpolated) prediction is represented by a diamond-like symbol, and the letter "I" indicates that the macroblock was intraframe coded. The absence of a symbol indicates that the macroblock was skipped.

despite the masking effect provided by the high spatial detail. These fluctuations tend not to be perceived in similarly textured areas that are also undergoing significant motion, where it would be difficult to discern any minor difference between a section of a scene in one picture and the same section in its new position in the next picture of the sequence.

As before, the fluctuation's causes relate to the varied coding of the same area of a picture for consecutive pictures of a sequence. The high spatial activity in the area makes it unlikely that a macroblock will be intraframe coded, especially where there is no movement; therefore, the fluctuations result from the differing types of prediction, quantization levels, MC prediction, or a combination of these factors.

Since a relatively significant level of quantization error is introduced into the higher order AC coefficients in I-type pictures, which propagates to the succeeding predictive coded pictures of the sequence, the main task of the prediction error coding is to compensate for the loss of the higher frequency information. Consequently, the difference in the prediction error from picture to picture causes high spatial frequency fluctuations.

Figure 3.21 show two consecutive pictures of the coded *Sons and Daughters* sequence where there is an absence of motion in the background, and slight motion by the two characters at the center of the scene. As an indication of the flickering in the background, Figure 3.22 shows the difference between the pictures in Figure 3.21. Note that the difference in the background consists of a large number of blocks containing either high frequency information, or no information at all (skipped).

(a) Picture 40   (b) Picture 41

Figure 3.21: *Sons and Daughters sequence.* The absolute difference of these pictures is shown in Fig. 3.22.

3.14 Chrominance Mismatch

As was mentioned in Section 3.11, motion compensation between pictures of a sequence is commonly conducted using a full-search block-matching technique, where the contents of the macroblock to be coded is compared to all the possible macroblock-sized groups of pixels within a limited search window of the MC reference pictures. One important factor of this search is that it is conducted, more often than not in practical implementations, only with the luminance information, and the same motion vector is used by all the components (luminance and chrominance) of the macroblock.

Figure 3.22: Difference between pictures 40 and 41 of the *Sons and Daughters* sequence. For the purpose of display, the difference is offset by +128 and clipped to be within 0 and 255.

Although a mismatch of the chrominance information occurs with the general MC mismatch artifact described in Section 3.11, the chrominance mismatch described in this section does not necessarily occur at object boundaries, and the quality of the associated luminance prediction is generally satisfactory. *Chrominance mismatch* appears as a misplacement of a macroblock with respect to its own general color and the color of the surrounding area within the picture.

The use of only the luminance information for the block-matching operation results in the selection of the macroblock-sized region within the search window which has the highest luminance correlation to the macroblock currently being coded. This correlation may not extend to the chrominance information, and may even be totally disparate to the chrominance information of the current macroblock. Figure 3.23 shows a photograph of a predictive coded picture from the *Flowers* sequence. Examples of the manifestation of chrominance mismatch can be seen within the tree-trunk, where sections of macroblocks containing a red-orange color (presumably, from the nearby roof tiles) are situated within the grey-brown trunk. It must be noted that the color disparity can be reduced by the coded prediction error, but its effectiveness is a function of the coarseness of the quantization.

3.15 Video Scaling and Field Rate Conversion

Advanced or high definition, 16:9 aspect ratio, televisions are slowly gaining in popularity, and are now being sold along side traditional 4:3 aspect ratio televisions. Addi-

Figure 3.23: Macroblocks located centrally in the trunk of the tree in this picture provide examples of chrominance mismatch within the *Flowers* sequence.

tionally, with the huge popularity of DVDs, source material typically isolated within one geographic region (and TV standard, such as NTSC) is being distributed worldwide and are expected to be displayed on TVs which are not of the same TV standard or aspect ratio.

To allow for the correct rendition of various source formats on any given display, there is a need to perform correct scaling and frame-rate conversion. Following is a discussion of the various artifacts which are caused by both these tasks.

3.15.1   Video Scaling

The simplest method of scaling down is sample decimation, but such an approach introduces aliasing and severely distorts details in the direction of scaling especially with increasing scaling ratios. Similar artifacts are caused by sample repetition when scaling up. The crude nature of such approaches are sometimes necessitated by limitations in processing capacity.

More sophisticated approaches utilize convolution implemented using standard digital filter design techniques based on the scaling requirements. However, despite convolution providing an improved outcome compared to decimation/repetition, it is inevitable that some degree of blurring will result for large up-scaling factors. For down-scaling, the main issue is aliasing, with the level of aliasing generally a function of filter length and the filter's ability to suppress the high frequency components in the source that cannot be accurately represented in the down-scaled (lower sampling rate) output.

The other consideration that needs to be taken into account is the aspect ratio. The consequence of not preserving the original aspect ratio is the visible squeezing or stretching of the image; for example, circles would appear as ovals. Generally, this can be prevented by scaling vertically and horizontally by the same factor. However, special consideration must be taken when source material intended for widescreen, 16:9 aspect, TVs is displayed on standard, 4:3 aspect, TVs. Typically, this 16:9 to 4:3 conversion is performed by vertically scaling the source down by 4-to-3, i.e. every 4 lines are scaled down to 3 lines. This produces the well-known "letterbox" effect with horizontal black bands at the top and bottom of the TV display.

Lastly, due care must be taken with interlaced sequences. Scaling filters must account for the possible inter-field disparity of interlaced source material. Filters which produce adequate results for progressive sequences may cause severe artifacts for interlaced sequences. Further discussion of the distortions caused by processing interlaced sequences can be found in Section 3.16 which covers deinterlacing.

3.15.2 Field Rate Conversion

In addition to correctly scaling the source material to fit the output display dimensions and preserve the aspect ratio, the output field rate must be converted appropriately. The simplest approach to this problem is the duplication or removal of fields. For example, one in every six fields is dropped when converting from a 60Hz NTSC source to a 50Hz PAL display. However, though simple to implement, the dropped/repeated fields can be seen as regular abrupt jerks in sequences containing persistent, steady motion, such as panning. This jerky-motion artifact can be eliminated by using temporal interpolation to implement the rate conversion. However, this, in turn, produces ghosting artifacts.

More sophisticated rate conversion schemes employ motion compensation to avoid these ghosting artifacts. With this approach, moving objects can be tracked, and an intelligent decision can be made as to the source of the weighted combination that will form the interpolation, including the source field and the location within the field. However, for very fast motion, newly exposed areas within the scene may be outside the motion estimation search area, thereby resulting in the inability to produce an adequate prediction. Depending on the sophistication of the rate conversion scheme, this can cause inconsistent spatial details to be generated in the interpolated sequence. Similar distortions may be caused at scene changes and can persist for a number of fields after the change.

Though not necessarily directly related to rate conversion, *tearing* can result if care is not taken when updating a frame buffer at the same time it is being displayed. This occurs only if there is a difference between the field-line update and raster display scan rates. Tearing typically appears as the transitory display of two frames simultaneously: the upper part of one, and the bottom part of the other. This is caused by the buffer update overtaking the position in the buffer being displayed, or vice-versa. If the buffer update is bursty, then multiple horizontal bands can appear. This can be eliminated by using multiple buffers and ensuring that the buffer currently being displayed is not updated simultaneously. Alternatively, "beam racing" can be enforced to prevent the update address in the frame buffer from overtaking the display address.

3.16   Deinterlacing

The ever-growing prevalence of progressive-scan displays, in the form of HDTV, PCs, and streaming-video, has spurred an interest in finding a robust and efficient means of converting traditionally interlaced sources to progressive output. The fundamental objective of such *deinterlacers* is to synthesize a complete frame (both odd and even fields) in every field period. This must be accomplished by utilizing the information provided in the odd or even fields of the source. Additionally, there are strict time constraints since the deinterlacing is typically performed in the display device.

As detailed by De Haan and Bellers [HB98] in their comprehensive discussion and review of deinterlacing methods: due to a lack of an anti-aliasing filter prior to inter-lacing the source material, deinterlacing is not a straightforward, linear sampling rate up-conversion problem. The application of a basic linear up-conversion filter is insuf-ficient to synthesize a distortion-free, progressive-scan (frame) output that confidently represents the pre-interlaced source.

The main challenge facing researchers is the need to devise a deinterlacer which is applicable to all types of spatio-temporal source characteristics. Early deinterlacing attempts were found to be generally successful for some forms of source characteristics, but caused highly visible distortions for other characteristics.

Hybrid combinations of these relatively basic methods have also been proposed that utilize the favourable characteristics of each method. The choice of which method is to be applied at any given instant is governed by the local spatio-temporal characteris-tics. This assumes that it is possible to quantify the source characteristics to enable the mapping to a specific deinterlacing method.

This leads to the main complication of such hybrid methods: there needs to exist a reliable method of quantifying the source characteristics to enable the mapping to a specific deinterlacing method. This in itself is not a trivial task; indeed, for some hybrid schemes the quantifying of the source characteristics is more computationally intensive than the individual deinterlacing methods; the main complication being the quantification of the temporal characteristics which necessitates some form of motion estimation. The overall success of such hybrid schemes can be generally associated with the accuracy and robustness of the source characteristic quantifier.

It is beyond the scope of this section to describe each and every deinterlacing method and the visual distortions they produce. Therefore, the objective of this section is to illustrate the distortions produced by some of the more prevalent, basic deinterlacing methods. Since these fundamental methods are typically incorporated into hybrid schemes, this overview will cover most distortions that will be encountered as a result of poor deinterlacing.

3.16.1 Line Repetition and Averaging

Line repetition is a purely spatial deinterlacing technique which simply repeats line n of a given source field to line $2n + 1$ of the progressive output frame for $n = 0 \ldots N - 1$, where N is the number of lines in the source field. Line repetition exploits the correlation between vertical intrafield samples, but causes severe aliasing artifacts in the presence of high vertical frequencies. Blurring is also caused, though to a lesser visual extent. The main advantage of such a technique is the low implementation cost since all that is required is the storage, and repetition, of the last source line.

To address the issue of aliasing, the synthesized lines can be interpolated rather than simply repeated. The application of a first order interpolator, or line averager, is commonly referred to as *bob*. Although such a deinterlacing approach does suppress aliasing it is at the cost of increased blurring of the vertical detail and the need for an extra line buffer.

3.16.2 Field Repetition

Field repetition, commonly referred to as *weave*, exploits temporal correlation within the source, and is the temporal analog of line repetition. Given a sequence of fields:

$$\mathbf{B}_0 \; \mathbf{T}_1 \; \mathbf{B}_2 \; \mathbf{T}_3 \; \mathbf{B}_4 \; \mathbf{T}_5 \; \cdots ,$$

where the T and B represent top- and bottom-fields, respectively, the output progressive frame sequence as a result of field repetition is as follows:

$$[\mathbf{T}_1 : \mathbf{B}_0]_0 \; [\mathbf{T}_1 : \mathbf{B}_2]_1 \; [\mathbf{T}_3 : \mathbf{B}_2]_2 \; [\mathbf{T}_3 : \mathbf{B}_4]_3 \; \cdots ,$$

where each $[\mathbf{T}_x : \mathbf{B}_y]_z$ represents an interleaving of fields $\mathbf{T}_x$ and $\mathbf{B}_y$ to produce a progressive frame for field period z. Note that the parity, either top or bottom, of the source fields are preserved when forming the progressive frame, i.e. a top field remains top and bottom remains bottom.

Such an approach requires the buffering of an entire field. It has a spatial all-pass characteristic and, therefore, no spatial degradation is caused for sequences depicting stationary scenes. However, in the presence of motion, field repetition causes the highly annoying artifact of *serration* around the edges of moving objects. This is due to the sampling delay between the two interleaved fields and the distance traveled in this intervening time.

3.16.3   Motion Adaptivity

Many hybrid deinterlacing schemes attempt to differentiate between the stationary and moving areas of a sequence, thereby allowing, for example, the application of field repetition in the stationary areas and intraframe processing (such as line averaging) elsewhere.

The success of these hybrid schemes is directly related to the accuracy and applicability of the motion detection mechanism that is employed. The range of motion detectors that have been proposed is large and varied, from simple inter-field, pixel-by-pixel luminance difference evaluators, to computationally intensive, pixel- or block-based, motion compensation schemes.

3.16.3.1   Luminance difference

A simple means of determining local motion is to calculate the luminance difference of a pixel with that of two fields earlier.[1] The luminance difference provides a rough indication of the presence of motion: if the difference exceeds a given threshold then motion is assumed, or if it is below a certain threshold then stationarity is assumed.

Such motion detectors need to account for noise that will influence the difference value, and, hence, the motion/stationarity decision. Additionally, abrupt switching between inter- and intra-field deinterlacing schemes is visible as high-frequency temporal noise. So, a degree of cross-fading is required when transitioning.

3.16.3.2   Median filters

One of the more popular motion adaptive deinterlacers is the 3-point, spatio-temporal median filter [HLS93]. This approach has the added advantage of incorporating the motion detection mechanism as a part of the median filter, thereby greatly reducing the computational burden.

Given the target, synthesized pixel p, the 3 points of this median filter consist of the two original pixels above and below p, a and b, respectively, as well as the pixel in the previous field, c, which is positioned equivalently to p.

---

[1]The two field delay is required due to the vertical phase difference between top and bottom fields. Additionally, motion for the top field needs to be determined separately from the bottom field.

For a stationary pixel, it is assumed that c will have a value which is somewhere in between the values of a and b, so c will be the output of the median filter, thereby achieving field repetition. For pixels in motion, the value of c will typically be outside the range bounded by the values of a and b, and, therefore, either a or b will be used to represent the target pixel.

The main artifacts produced by such a deinterlacer were found to be distortions of vertical details and the introduction of aliasing. Correspondingly, flickering of vertical details and lines were also evident.

3.16.3.3 Motion compensation

The ability of line repetition to produce a progressive frame without spatial degradation for stationary sequences has spurred interest in producing a similar outcome for moving sequences. Line repetition achieves its favorable result through the fact that the pixel data is not spatially filtered in any way: the data is blindly copied from a previous field. However, this is only possible for stationary sequences.

To achieve a similar result for moving sequences by copying pixels between fields, it is necessary to accurately determine the pixel positions from one field to another, i.e., determine the motion vector field for the sequence. This is where motion compensation (MC) comes into play.

As with the use of MC for video coding, robustness to MC errors is a critical issue. However, with MC-based deinterlacing schemes there is no supplemental error signal to hide MC errors. Such errors are seen as highly discordant patches within a scene, and are typically found in areas of complex motion.

Additionally, secondary spatial deinterlacing methods are needed for areas which cannot be adequately predicted from previous fields, such as newly exposed scene content. Some of these spatial deinterlacing methods have been described previously, and produce the same distortions when applied in this circumstance.

3.17 Summary

Due to the non-linearity of the quantization process, and the energy distributing effects of the inverse DCT, it is not possible to predict the visual distortionary outcomes of the quantization error. Therefore, the aim of this chapter was to identify the distortions which have a consistent identifying characteristic and, hence, may be labeled as a consequence of the quantization process rather than a natural characteristic of the source.

The quantization error introduced into the DCT coefficients can be distinguished into two basic processes. The first relates to fluctuations of a DCT coefficient around its original value, and the second results in the complete nullification of a DCT coefficient:

the truncation effect. These two basic error processes, which affect the DCT coefficients, are the root causes of most of the distortions described in this chapter.

The spatial distortions may be divided into two general categories, one relating to the actual distortions introduced into a block by the quantization process, and the other regarding the association and continuity between blocks. The latter distortion category is more a function of the independent coding of each of the blocks than the actual quantization error. The distortions in this group include the blocking effect, the mosaic pattern effect, the staircase effect, and false contouring.

The distortions in the first general category can be further divided into two groups related to the two basic quantization error processes. The combination of fluctuations to the contribution of certain basis images by the quantization error, and the DCT's poor efficiency in representing high contrast edges and diagonal features, results in the distortions of ringing and the staircase effect. Although the quantization truncation effect also plays a role in the manifestation of these two distortions, truncation is mainly associated with blurring, color bleeding, and the basis image effect.

The temporal artifacts, in the main, result from the inconsistency in the coding of the same area of a scene in consecutive pictures of a sequence. This includes variations in the type of predictions, quantization levels, MC predictions, or a combination of these factors. MC mismatch is associated with the inability to provide a satisfactory MC prediction for boundaries of moving scene objects. Whereas, the mosquito effect is a general manifestation of the inability by the prediction error to satisfactorily correct for the loss of high frequency information in MC predictions.

Finally, with the growing diversity of display media and the world-wide distribution of various source formats, post-processing of the source material has become a ubiquitous component in digital video products. As such, there is a growing need to understand and overcome the distortions associated with such post-processing techniques.

References

[AH94]   A. J. Ahumada and R. Horng. De-blocking DCT compressed images. In *Proc. SPIE Human Vision, Visual Processing, and Digital Display*, 2179:109–116, May 1994.

[Ahu92]   A. J. Ahumada. Luminance-model-based DCT quantization for color image compression. In *Proc. SPIE Human Vision, Visual Processing, and Digital Display*, 1666:365–374, 1992.

[Cla95]   R. J. Clarke. *Digital Compression of Still Images and Video*. London: Academic Press, 1995.

[DM95]   F. Dufaux and F. Moscheni. Motion estimation techniques for digital TV: A review and a new contribution. *Proc. IEEE*, 83(6):858–876, June 1995.

[Gir89]   B. Girod. The information theoretical significance of spatial and temporal masking in video signals. In *Proc. SPIE Human Vision, Visual Processing, and Digital Display*, 1077:178–187, October 1989.

[GW92]   R. C. Gonzales and R. E. Woods. *Digital Image Processing*. Reading, MA: Addison–Wesley, 1992.

[HB98]   G. D. Haan and E. B. Bellers. Deinterlacing—an overview. *Proc. IEEE*, 86(9), September 1998.

[HLS93]   H. Hwang, M. H. Lee, and D. I. Song. Interlaced to progressive scan conversion with double smoothing. *IEEE Trans. Consumer Electronics*, 39:241–246, August 1993.

[ITU90]   ITU. *ITU-T Recommendation H.261: Video Codec for Audiovisual Services at $p \times 64$ kbits/s*, 1990.

[ITU95]   ITU. *Draft ITU-T Recommendation H.263: Video Coding for Low Bitrate Communication*, 1995.

[Jay92]   N. Jayant. Signal compression: Technology targets and research directions. *IEEE J. Sel. Areas Commun.*, 10(5):796–818, June 1992.

[KHK87]   M. Kaneko, Y. Hatori, and A. Koike. Improvements of transform coding algorithm for motion-compensated interframe prediction errors—DCT/SQ coding. *IEEE J. Sel. Areas Commun.*, SAC-5(7):1068–1078, August 1987.

[KK95]   S. A. Karunasekera and N. G. Kingsbury. A distortion measure for blocking artifacts in images based on human visual sensitivity. *IEEE Trans. Image Processing*, 4(6):713–724, June 1995.

[KSC92]   S. A. Klein, A. D. Silverstein, and T. Carney. Relevance of human vision to JPEG-DCT compression. In *Proc. SPIE Human Vision, Visual Processing, and Digital Display*, 1666:200–215, 1992.

[Le 91]   D. Le Gall. MPEG: A video compression standard for multimedia applications. *Communications of the ACM*, 34(4):47–57, April 1991.

[NH89]   A. N. Netravali and B. G. Haskell. *Digital Pictures: Representation and Compression*. New York: Plenum Press, 1989.

[PAW93]   H. A. Peterson, A. J. Ahumada, and A. B. Watson. Improved detection model for DCT coefficient quantization. In *Proc. SPIE Human Vision, Visual Processing, and Digital Display*, 1913:191–201, September 1993.

[Plo89]   R. Plompen. Motion video coding for visual telephony, 1989.

[RG86]   B. Ramamurthi and A. Gersho. Nonlinear space-variant postprocessing of block coded images. *IEEE Trans. Acoust., Speech, Signal Processing*, ASSP-34(5):1258–1268, October 1986.

[RL83]   H. C. Reeve and J. S. Lim. Reduction of blocking effects in image coding. In *Proc. IEEE ICASSP*, 1212–1215, Boston, MA, April 1983.

[RY90]   K. R. Rao and P. Yip. *Discrete Cosine Transform: Algorithms, Advantages, Applications*. San Diego, CA: Academic Press, 1990.

[Sec91]   Secretariat ISO/IEC JTC1/SC29. *ISO CD11172-2: Coding of moving pictures and associated audio for digital storage media at up to about 1-5 Mbit/s*. ISO, November 1991. MPEG-1.

[Sec92]   Secretariat ISO/IEC JTC1/SC29/WG11. *Test Model 1*. ISO, May 1992. Description of MPEG-2 test model.

[Tzo88]   K.-H. Tzou. Post-filtering of transform-coded images. In *Proc. SPIE Applications of Digital Image Processing XI*, 974:121–126, 1988.

[WSA94]   A. B. Watson, J. A. Solomon, and A. J. Ahumada. DCT basis-function visibility: effects of viewing distance and contrast masking. In *Proc. SPIE Human Vision, Visual Processing, and Digital Display*, 2178:99–108, May 1994.

[YGK95]   Y. Yang, N. P. Galatsanos, and A. K. Katsaggelos. Projection-based spatially-adaptive reconstruction of block-transform compressed images. *IEEE Trans. Image Processing*, 4(7):896–908, July 1995.

[YWR95]   M. Yuen, H. R. Wu, and K. R. Rao. Performance evaluation of POCS loop filtering in generic MC/DPCM/DCT video coding. In *Proc. SPIE Visual Commun. and Image Processing*, 2501:65–75, Taipei, Taiwan, May 1995.

Part II

Picture Quality Assessment and Metrics

Chapter 4

Video Quality Testing

Philip Corriveau
Intel Media and Acoustics Perception Lab, U.S.A.

This chapter offers a perspective on video quality, its evolution and how it is being measured now. There are many aspects to video quality and how it is perceived from a provider and a consumer points of view. What is presented here is a balance of different perspectives on the topic of quality evaluation.

4.1   Introduction

What is video? For the purpose of this chapter and Chapter 11, video is defined as material presented to an end-user in a predefined video file. For the average person this would mean terrestrial broadcast. Programs that are broadcast and received over some type of antenna and then displayed to a consumer's television or it could represent the presentation of visual material over a cable system or satellite service. Until a few years ago, these were the predominant ways of delivering video to the consumer. With the deployment of DVD, personal computer and communications technologies, video files are disseminated in many different formats from CD to AVI (Audio Video Interleave) and other file types.

To put things into perspective, video technology has changed drastically over the generations. Society has switched from black and white to color television and now a person can create recordable media with high resolutions in his/her own living room. Video material, today, can be transferred from place to place on multiple types of media. The providers and consumers of video material have kept pace with the times by increasing their expectations of quality. The problem that the industry faces now is how we measure video quality, guarantee delivery of high quality video and prove that the quality is delivered at the level promised.

Migration from analog signals to digital signals also added another wrinkle into the problem. Classic methods that had been developed to measure video quality rapidly fell to the wayside and new ones were devised to take their place. Reliable methods were developed to test certain video types in certain situations and volumes of research are still needed to identify and standardize methods in other situations [ZC96].

There are two primary ways to measure video quality. The first is Subjective Quality Assessment. This method is a psychologically based method using structured experimental designs and "human" participants to evaluate the quality of the video presented when compared with a fixed reference. The second is Objective Quality Assessment. This measures physical aspects of a video signal and considers the physical aspects and psychological issues. Both types of testing are far from an exact science but they have proven to be very useful tools.

4.2   Subjective Assessment Methodologies

Subjective assessment uses human subjects (real end users) to evaluate, compare or assess the quality of images under test. Subjective assessment is the most reliable way to determine actual image quality, and cannot be replaced with objective testing. In the realm of subjective assessment, there are many different methodologies, categories of subjects and rules for designing and defining tests. The next few sections will describe these methods and the proper procedures for conducting reliable and repeatable tests.

Before considering a design for a particular assessment, the question to be answered must be well defined. For example, you might want to determine the proper operational bit-rate range for transmitting a video signal at a desired quality over a distribution path whose bandwidth is limited to 4.5 Mbs. A properly defined question allows the implementation of the right procedures, manipulation of the proper variables and choice of the appropriate testing scale.

4.3   Selection of Test Materials

Every variable in a test design has the potential to affect image quality at the point where it is viewed. It is well known that decreasing the bit rate of an encoded video stream from 8Mbs to 1.5 Mbs will introduce major visible artifacts in the picture that the end user will see. Or is this "so-called" well-known fact really true? What engineers, algorithm developers and others sometimes forget is that besides compression, transmission and image manipulation, the largest factor that must be considered is content!

Image content can have a large effect on the results of a test. For example, if one reduces the video bit rate from 8 to 1.5 Mbs the image may be perfectly acceptable, if all that was tested were "head and shoulder shots" with little or no motion. For example,

Figure 4.1: Standard test sequence — *Susie*

the test sequence *Susie* as shown in Figure 4.1 can be encoded at low rates with little apparent loss in quality.

There have been tests where the sequence *Susie* showed no visible quality impairment from such a bit rate reduction while other standard sequences, such as those shown in Figure 4.2, were significantly impaired and unacceptable to the viewer. *Waterfall* and *Mobile & Calendar* are exceptionally good test sequences because they highlight the loss in visual quality caused by reducing the video bit rate. *Tree* and *Tempete* stress other quality aspects of compression and transmission techniques.

It is very important that multiple types of video material be used in any image quality test involving motion video systems or equipment. Multiple sources should also be considered since film, studio and graphic material would also have different behaviors. In any given test, at least 8 to 16 different test material types should be used. Good test sequences can be obtained from various standards bodies such as SMPTE, IEEE and others. There are also open sourced sequences available from the Video Quality Experts Group (VQEG) at *www.vqeg.org*.

Figure 4.2: Examples of other Standard Test Sequences: (a) *Waterfall*, (b) *Tree*, (c) *Tempete*, and (d) *Mobile & Calendar*.

4.4   Selection of Participants — Subjects

Test subjects can be either experts or non-experts. All subjects should be screened for acuity, color blindness and other visual defects. To produce reliable results, a large number of participants should be used. It is generally accepted that between 16 and 24 subjects will provide a statistically valid result. The more participants the more statistically sound the results will be. Also ensuring that the subjects meet some minimum requirement by performing proper screening ensures that the data is valid and carries more weight in the research and consumer community.

4.4.1   Experts

Experts in the video field have extensive experience in evaluating, producing, distributing or designing systems that deal with video. These people have a specialized way of looking at images. The advantage of using experts is it can be a fast test and they know what they are looking for. The disadvantage is that some experts cannot remove themselves from their own way of thinking. Results from these studies are beneficial for video algorithm development but not good for generalization to the average consumer. Experts also are very good at providing technical performance assessments of video and they perform better in free form viewing.

4.4.2 Non-Experts

Non-experts represent the general public, the average user and someone who will consume the product. Non-experts can sometimes see artifacts that an expert might miss since they have no pre-determined way of looking at a video sequence. Non-experts will generally provide excellent data when the tests are structured properly. When selecting participants, one should strive for a representative sample with respect to age, gender and other factors that might influence the tests.

4.4.3 Screening

Regardless of the subjects that you chose to employ in your test, you must ensure that you screen them for two main factors: color blindness and visual acuity. There are standardized tests that can be used for this. Subjects should be 20/20 normal or corrected to normal vision and the Snellen Eye Chart$^©$ can be used (Figure 4.3a). For color blindness you can use the Ishihara$^©$ test, which is a set of plates you show subjects and ask if they see the numbers or pictures inside (Figure 4.3b). Other tests may also be conducted including a subject's spatial acuity, although this is rarely necessary.

4.5   Experimental Design

There are several factors an experimenter should try to control when conducting a subjective test. For example, carefully considering the number of sequences and the different variables desired is essential when you realize that the average subject can watch no more than 90 minutes of material. These sessions are usually divided into 30-minute sections with warm-up and re-set trails at the beginning of each section. Industry and research bodies have settled on using test stimuli that range in 5 to 12 seconds in duration. The time duration will also affect the trial structure.

Tests can be either double or single presentation. This choice will affect the length of the test and might determine the number of conditions covered. Repeating test conditions is possible, but in the past it has not shown to add much to the stability or repeatability of a test result. Most experimenters prefer to cover more test conditions and

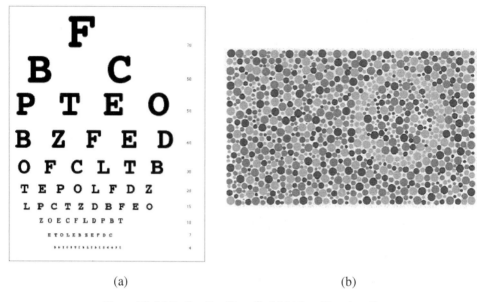

(a)   (b)

Figure 4.3: (a) Snellen Eye Chart ©; (b) Ishihara Test plate ©

use a hidden reference; reference conditions are employed to judge the reliability and repeatability of any given test subject.

The data analysis associated with collecting and reporting subjective data can be found in journal articles or regular statistical textbooks and International Telecommunications Union Radiocommunications Broadcast Technology 500 (ITU-R Recommendation BT-500) [ITU98].

4.5.1 Test Chamber

The test environments in different facilities around the world vary from extremely complex to very simple. One of the most advanced video quality test facilities is the Communications Research Centre (CRC)/Industry Canada in Ottawa, Canada. Figure 4.4 shows a typical room layout at CRC for a dual monitor evaluation. This facility meets and exceeds the standards defined in the ITU-R Recommendation BT 500.7. Other facilities such as those at NTIA/ITS (National Telecommunications and Information Administration/Institute of Telecommunications Sciences) and Intel Corporation use soundproof chambers that are designed to meet industry standards, but accommodate much smaller displays than that at the CRC.

Three major factors that must be considered are lighting, ambient noise and the quality and calibration of the display. The most important information is contained in Recommendation BT-500, which one should read and understand before conducting

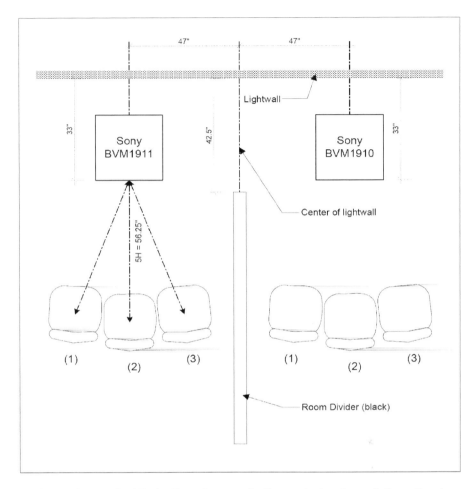

Figure 4.4: Example of Testing Room Layout at the Communications Research Centre Canada.

subjective testing. These will affect the experimental results if these factors are not controlled carefully.

4.5.2 Common Experimental Mistakes

One of the most common mistakes is the use of a limited set of sequences; another is not covering the range of conditions you are trying to test. Each subjective test method has the ability to answer certain questions. If one is conducting a test to examine small hand held devices with a limited processing power and storage, the encoding rates for broadcast television would not apply, thus one must ensure that they are not setting the reference out of the range of possible variables to be tested. Also if the purpose of the device is to be able to display regular television content then the experimenter better sample from all possible sources for video material, from film, sports, graphics and news channels.

Another mistake is designing an experiment that takes a participant too long to complete. Since these tests are designed with small numbers of test material, the subject will see the same sequences over and over again. Fatigue and boredom are larger factors leading to poor data from a subject. Previously, it was suggested that a maximum of 90 minutes be employed. A balance must be reached to ensure that subjects do not get bored or tired. Some experts believe that 1 hour is the maximum that you should have a subject rate video.

The last mistake that is commonly made in designing a subjective test is to not include the warm-up and re-set trials in a session. Once a solid pseudo-randomized order has been developed for the trial presentations five warm-up trials must be inserted at the beginning of the test. These trials span the breadth of the test conditions from best to worst to ground the subjects to the scale and allow them some time to adjust to the pace of the evaluations. Then before resuming after each break, roughly every 30 minutes, three re-set trials need to be inserted for the same purpose. The data from all of these trials are discarded and not used in the final data analysis.

4.6   International Test Methods

Test methods defined in the following sections are internationally accepted and used by private industry, public sector and standards bodies for conducting subjective tests [ITU98]. Some of the methods are relatively new, such as the continuous evaluation methods but others have been in use for many years. One must take care to instruct the subjects very carefully for each method to ensure that the participant will perform the task requested. Please note that statistical analysis methods will not be covered but typical representation of results will be.

4.6.1   Double Stimulus Impairment Scale Method

Impairment implies that there will be a visible degradation in the image that the subject will be asked to evaluate. It is common to use a five point ordinal scale. Ordinal scale implies that it is not continuous; the participant is forced to vote one of five options and nothing in-between. The five choices depicted in Figure 4.5 are Imperceptible (no difference); Perceptible, but not annoying (they see something there but it does not bother them); Slightly Annoying (something is there and it bothers them); Annoying (the artifacts are visible and bothersome enough to stop consumption of the material and Very Annoying (the subject would not watch the material under any condition). This scale is internationally recognized and accepted for use in this type of test.

Figure 4.5: Five point Impairment Rating Scale.

The best use of the impairment rating scale is to determine a failure characteristic. For example, if a test is designed to investigate a variable like bit-rate, the experimenter will determine the highest rate achievable and set this as the reference. Every processed sequence is then compared against this reference and can be judged as possessing picture quality which is equal to or less than that of the reference. This allows one to chart the decline in quality as one reduces the rate off of the reference point. If this test is designed correctly, the subject will choose the top and bottom boxes occasionally, if they do not there is a design flaw in the experimental procedure.

Subjects should always be provided with instructions like those as shown in Figure 4.6 on the next page.

The trial structure as shown in Figure 4.7 is always presented such that the "Reference" is presented first followed by the "Test." The reference sequence shows the video material at the highest quality that the experimenter wants to use for the benchmark. The reason for saying the highest quality and not the best quality is that there might be a certain ceiling already imposed on the process above which one cannot go; therefore, so we start the failure characterization from that point. The test presentations are presented after the reference. The subject is instructed to compare the test to the reference and judge the visibility and severity of impairments introduced into the video material based on the five point scale described above. In this case a double presentation is used, giving the subject two chances to examine the reference and test sequences prior to providing a response.

INSTRUCTIONS FOR IMPAIRMENT TESTS

In this test, we ask you to judge the visibility and severity of impairments in the television picture resulting from imperfections or interference in broadcasting.

Each trial will be announced verbally by number. The first presentation of a trial will be announced as "Reference", and the second presentation will be announced as "Test".

The test consists of a series of judgement trials, each consisting of two presentations of the same piece of video material. The first presentation shows the video material from a Reference signal (a broadcast not subjected to interference). The second presentation shows the same material from a Test signal (a broadcast subjected to a measured amount of interference).

We will now show you four demonstration trials. Across the four demonstrations, the test material will exhibit different levels of visible impairment to illustrate the kinds of impairments you will see during the test session.

DEMONSTRATION TRIALS

You are asked to examine the Reference presentation and note any deficiencies present, and then to assess the visibility and severity of any additional impairment in the Test material.

In judging the visibility and severity of transmission impairments, we ask you to use judgement scales like the sample shown below.

IMPERCEPTIBLE ☐

PERCEPTIBLE, BUT NOT ANNOYING ☐

SLIGHTLY ANNOYING ☐

ANNOYING ☐

VERY ANNOYING ☐

SAMPLE JUDGEMENT SCALE

The judgement scale offers five response options. These options are: "imperceptible", "perceptible, but not annoying", "slightly annoying", "annoying", and "very annoying". You are asked to place a single "X" in the box that best corresponds to your judgement (as shown in the example).

IMPERCEPTIBLE ☐

PERCEPTIBLE, BUT NOT ANNOYING ☒

SLIGHTLY ANNOYING ☐

ANNOYING ☐

VERY ANNOYING ☐

SAMPLE JUDGEMENT SCALE

We ask you to use the judgement scales to express your judgement of the visibility and severity of the incremental impairment in the Test material. Please note, in some trials, the Reference material itself may have visible impairments. These are not of interest in this test.

Please refrain from recording your judgement until the final grey period of the trial.

Figure 4.6: Instructions for Impairment Test.

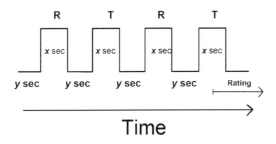

Figure 4.7: Trial Structure Double Stimulus Impairment Test.

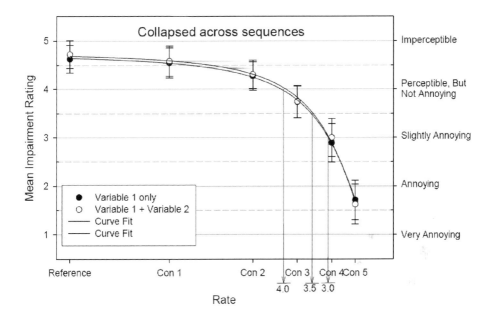

Figure 4.8: Overall Impairment Results.

Subjects are always instructed to wait for the "Vote/Rating" period that occurs at the end of each trial prior to registering a response. This is important to ensure that the subject watches the entire two sets of reference/test presentations. Under the assumption that the test was designed correctly and your subjects performed reliably, the mock results shown in Figure 4.8 show a typical outcome. Figure 4.8 allows the experimenter to look closely at the overall results to see if the conditions imposed on the reference had the expected effect. The plot is arranged such that the mean subject rating is plotted on the "y" axis and the conditions tested are on the "x" axis, with the lines representing data collapsed over all sequences tested. As shown in the figure, the results start near 5.0

for the reference and fall off smoothly to reach condition 5, reaching approximately 1.5 on the impairment scale. The failure shape is one of an experiment that was designed perfectly. In this particular case, there were two variables tested inside the design. The results show that when the second variable was added there was statistically no effect on the results. In this case it proved that adding a variable that might improve the perceived quality when coupled with the first variable, actually achieved nothing.

When determining the desired condition or parameter based on the condition variable one usually uses drop-down lines from three major points on the Rating Scale. The points of 4.0, 3.5 and 3.0 are of most interest. It can be considered that if one were to design to the value corresponding to 4.0 it would be over engineering the solution whereas designing to the value corresponding to 3.0 could be risky depending on the content. Very often, the experimenter will opt to go with the value that corresponds to 3.5.

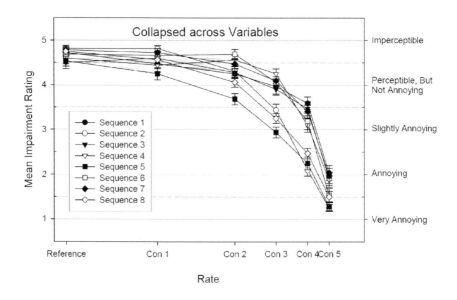

Figure 4.9: Individual Sequence Results.

In some cases, the experimenter must be more aware of the content used in the test since the results have the potential to vary greatly. In this example, Figure 4.9 shows the individual sequence results that contributed to the collapsed graph in Figure 4.8. It can be observed that sequence 5 has a much sharper failure characteristic compared to several of the other sequences used. If one was to determine operating points based on sequence 6, the system would fail most of the time for all other types of material. On the other hand, if operating points were set based on sequence 5, the system would never fail. Since either way is an extreme, the next step is to determine roughly what percent of the time these classes of sequences will be presented and work from that number.

In most cases, operational points are determined based on the collapsed data in Figure 4.8 with close attention paid to the failure cases identified in Figure 4.9.

4.6.2   Double Stimulus Quality Scale Method

Quality implies that the subject will be asked to assess the raw perceived quality of any given presentation. A difference between the impairment method described above and this one is that in this case, "double" not only means two presentations to the subject but it also requires two responses for each trial. The double stimulus quality method uses a "double" scale depicted in Figure 4.10, which when used must be exactly 10cm in length and needs the internationally accepted set of adjectives to the left.It can be noted that the scale is divided into five equal intervals with the following terms from top to bottom; Excellent (100–80), Good (79–60), Fair (59–40), Poor (39–20) and Bad (19-0). The scale presented to participants is identical to that in Figure 4.10, except that the numbers to the right are not present; these are used in the analysis only.

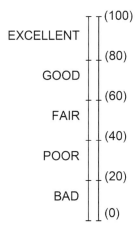

Figure 4.10: Double Stimulus Quality Scale.

The 10 cm scale provides the experimenter with a "continuous" rating system that helps reduce quantization errors in participant responses.

The best use of the quality scale is to tease out small differences in video quality such that the system developed performs better, equal or worse than the reference being compared. When comparing the impairment method and the quality method, one major difference is that the results can yield cases where the system under test can be rated higher than the reference against which it is being compared. In the impairment method, they can only be equal or less where, in this case, the reference can be rated lower than the test. There is a two-stage analysis process to this method to ensure reliable data and to determine what the true differences are. If the results yield huge differences between the reference and test sequences, the test was not designed properly. On this scale results are usually a max of 40 units different and often stay under 25 units of 100.

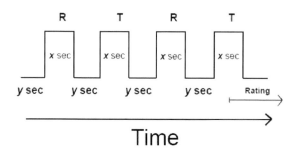

Figure 4.11: Trial Structure Double Stimulus Quality Scale.

The trial structure shown in Figure 4.11 is different from the previous method such that the "Reference" and "Test" presentations are blind to the subject. That means the participant has no idea from trial to trial if the reference or the test is presented first or second. But instead, the viewer is requested to rate both sequences. The advantage of them not knowing the presentation order is that the participant will not always mark A or B at 100.

The participant sees presentation A then presentation B twice and then is required to vote on each during the rating period at the end of each trial. Like in all methods, subjects are instructed to watch both sets of presentations fully before rating the sequences to ensure thorough examination of the video clips.

Subjects should always be given explicit instructions prior to testing similar to those shown in Figure 4.12.

One of the facts the experimenter will have to deal with is that of human nature, subjects will tend to avoid the ends of the scales. It has been noted in past experience that Expert Viewers are not shy to rate at both extremes of the scales, however, non-experts tend to convince themselves that there might always be something better or worse to come and stay away from the 100 and the 0 on the scale. One trick to help reduce this binding effect is to educate your subjects on what your reference quality is and what the worst test case looks like.

INSTRUCTIONS FOR QUALITY TESTS

In this test, we ask you to evaluate the <u>overall</u> quality of the video material you see. By overall quality, we mean the quality of the appearance of the video, <u>not</u> the desirability of the material itself.

Possible problems in quality include:

- poor, or inconsistent, reproduction of detail;
- poor reproduction of colours, brightness, or depth;
- poor reproduction of motion;
- imperfections, such as false patterns, or "snow".

The test consists of a series of judgement trials, each consisting of four presentations of the same piece of video material.

Each trial will be announced verbally by number. The first presentation of a trial will be announced as "A", and the second as "B". This pair of presentations will then be repeated, thereby completing a single trial. Please note that one of the presentations, A *or* B, will be a reference picture that shows the video material in a high quality format, with the other being a test picture that shows the same video material in the format being tested.

We will now show you four demonstration trials.

DEMONSTRATION TRIALS

In judging the overall quality of the presentations, we ask you to use judgement scales like the samples shown below.

SAMPLE QUALITY SCALE

As you can see, there are two scales for each trial, one for the "A" presentation and one for the "B" presentation, since both the "A" and "B" presentations are to be judged.

The judgement scales are continuous vertical lines that are divided into five segments. As a guide, the adjectives "excellent", "good", "fair", "poor", and "bad" have been aligned with the five segments of the scales. You are asked to place a <u>single horizontal</u> line at the point on the scale that best corresponds to your judgement of the overall quality of the presentation (as shown in the example).

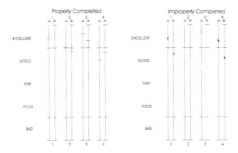

You may make your mark at any point on the scale which most precisely represents your judgement.

In making your judgements, we ask you to use the first pair of presentations in the trial to form an impression of the quality of each presentation, but to refrain from recording your judgements. You may then use the second pair of presentations to confirm your first impressions and to record your judgements in your Response Booklet.

Figure 4.12: Instructions for Quality Tests.

Typical quality results are shown in Figure 4.13 and the mean opinion scores are used for two purposes. First to check the relative stability of the reference responses and second to get a quick glance at the differences between the reference and test sequences, inversions or not. If we use the results below, this test was looking at a system that performed extremely close to the reference. In this case the reference was always rated higher but at times almost equal to the test. It must be stressed that a full statistical analysis usually accompanies these types of figures.

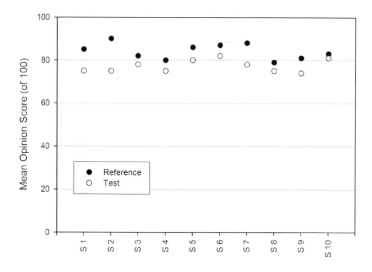

Figure 4.13: Quality (Reference/Test) Mean Opinion Scores.

Having reviewed the relative placement of the reference and test sequences the next step is to extract the difference score. Since the actual placement on the quality scale is arbitrary for reference, meaning that it will not always be at the 100 mark, and in reality the participant is performing a comparison between the reference and test sequences, the difference is the most valuable piece of data. Figure 4.14 shows an example of a difference data set, from an experiment different from that shown in Figure 4.13.

In this case there were two test variables compared to the same reference but using the quality scale and run in the same test. The subjects were unaware of the presentation order and unaware of the variable being tested. If the figure is examined it can be seen that all of the differences calculated "Test-Reference" is below zero, with one sequence fairly close.

The convention of test minus reference allows for rapid identification of cases where the test was rated higher than the reference. One can see the effect each processing method had on each of the sequences used. Both of the variables in this case were almost equal, save the first case. It is clear from the results that the visual performance

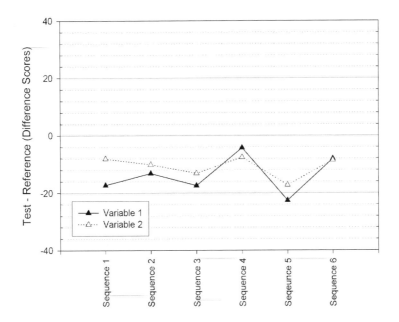

Figure 4.14: Quality Difference Scores.

changes depending on the sequence. This strengthens the fact that one must use as many sequences as possible to further explore the different types of video material that will ultimately be dealt with. The end result of this test method is the experimenter will be able to state a relative level of quality with a specific performance gap between the reference and test cases.

4.6.3 Comparison Scale Method

This method is best suited for comparing two systems that the experimenter feels are on equal footing with respect to capability. In this case, the participant is instructed to place a mark on the scale that best represents his/her preference and magnitude. The comparison scale is a continuous scale, as shown in Figure 4.15 that has three adjective markers. "A is much better", "A=B" and "B is much better."

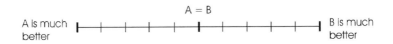

Figure 4.15: Comparison Rating Scale.

The numerical range of this scale is 100 points in total and is 10 cm in length just like the quality scale. This allows for a fifty-point spread between the two systems being tested, where zero is when they are perceived equal.

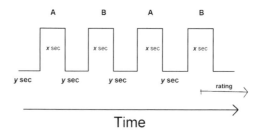

Figure 4.16: Trial Structure for Comparison Test.

The trial structure for this test, shown in Figure 4.16, is similar to the quality test such that the reference system and the test system are blind to the subject. They do not know if the A or B presentation in the trial is the reference or test system. This allows the experimenter to swap them and see if this changes the placement on the scale. Like the other methods, this can be a double or single presentation. A double presentation style allows the subject to view the video material twice before rating. Subjects are always instructed to wait for the vote screen before registering their rating.

4.6.4   Single Stimulus Methods

Figure 4.17 depicts the overall results that would be typical for this type of test. The "y" axis presents the difference scale from 50 to -50, which corresponds to the adjectives

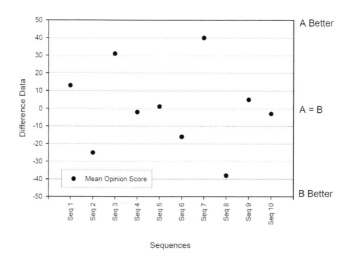

Figure 4.17: Overall Comparison Results.

used in the test. The mean opinion score is shown for each of the sequences tested. This allows the experimenter to see the placement of sequences and preference towards one processing way or another. It is interesting to note the results in Figure 4.17. Once again, these results show the importance of using a wide variety of video test material. The sequences range in rating from 40 for one sequence to almost -40 for another. This shows that certain systems have a huge effect on different types of video material. Once it is determined where the system under test ranks with respect to the reference, problems can be identified and solved.

Any of the internationally accepted methods described above can be conducted using a single stimulus methodology.

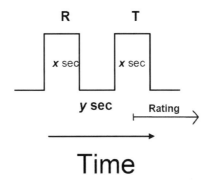

Figure 4.18: Single Stimulus Trial Structure.

The only thing that changes in a single stimulus presentation is the trial structure. Figure 4.18 shows the single presentation of the reference and test conditions. For methods where the reference and test are blind, the labels A/B would be used.

The advantage is it allows an experimenter to cover a lot of test conditions in a shorter period of time, maximizing the use of the subjects' time and collecting the most data possible. The disadvantage is that the participant only has one shot to examine the two conditions and must remain more vigilant for rating the sequences. One can repeat a sub-set of the same test conditions scattered throughout the session to double check a subject's response. This allows for a thread of results that are used to determine if the subjects are responding relatively the same way to the same set of stimuli presentations.

4.6.5 Continuous Quality Evaluations

All of the previously discussed methods have separate presentations of the reference and test material such that the participant knows there are two things to be rated or compared. In the continuous quality evaluations there is no such reference present. Depending on how the experiment is designed, the subject has his/her own internal reference or, the experimenter can provide a reference example to the subject prior to the start of the sessions.

This method is called the Single Stimulus Continuous Quality Evaluation method (SSCQE). The SSCQE method uses computer equipment to record a continuous rating of the video material seen by the viewers throughout the entire duration of the test.

One advantage to the continuous evaluation is the ability to measure variations in picture quality within sequences as well as between sequences. This is not possible with the overall sequence ratings generated with DSCQS. Automated voting devices ("sliders"), fitted with a moveable indicator and a single version of the scale used in the DSCQS method, are used to collect data from the viewers. The sequences are rated independently of one another and are only shown once. At the beginning of the sequence, a computer activates the sliders and the viewers enter their opinion scores by moving the indicator along the scale during the test. The control computer samples ratings from the sliders at a predetermined rate (e.g., every 0.5 seconds) throughout the test; this is what creates the viewers rating trace, used in the analysis.

Figure 4.19 has three panels, which display three parts of the typical SSCQE set-up.

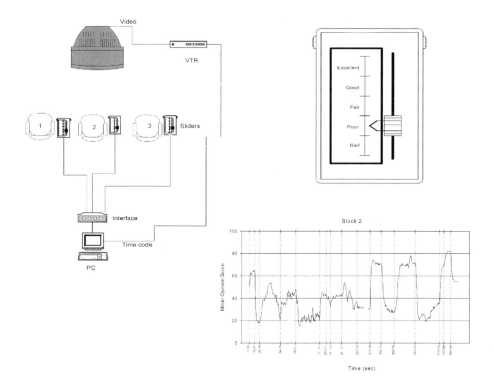

Figure 4.19: Single Stimulus Continuous Evaluation Layout, Hardware and Results.

The left side of the figure shows the room layout with three subjects taking the test at the same time. Each viewer watches the same video presentation on the monitor and uses sliders, like that on the top right side of the diagram, to continuously register his/her perception of the video quality.

The scale on the front of the slider is the same scale used in the double stimulus quality method. The slider length must be exactly 10cm in length and meets the internationally accepted set of adjectives to the left. It should be noted that the scale is divided into five equal intervals with the following terms from top to bottom; Excellent (100-80), Good (79-60), Fair (59-40), Poor (39-20) and Bad (19-0).

The lower right side of Figure 4.19 shows a typical trace of a viewer's response. To determine the effectiveness of the SSCQE method several experiments have been conducted to compare results with those obtained from DSCQS.

4.6.6   Discussion of SSCQE and DSCQS

Due to its continuous nature, the SSCQE methodology possesses unique distinguishing characteristics. With the changing nature of the video industry, these characteristics make SSCQE a more applicable and efficient choice for subjective assessment methodology, since it has been shown to be valid against the current benchmark of DSCQS. The longer sequences utilized with this method are more representative of actual broadcast programming, which is the ultimate focus of video quality assessment. To implement DSCQS with longer sequences would be difficult and inefficient; the length of such a test, due to the reference/test repetition, would undoubtedly cause viewer fatigue. Secondly, the continuous nature of the data with its corresponding time code allows accurate pinpointing of the impact of individual errors on the subjective quality ratings. This is impossible with the single vote given by the DSCQS method. SSCQE has very valuable implications for the growing field of real-time in-service testing. In short, the SSCQE method provides much more detailed and realistic information about viewers' opinions of image quality and is therefore a more accurate subjective assessment methodology for many of today's video applications. Its values correlate very well with those of the traditional DSCQS method.

Another important asset of the SSCQE methodology is its use of the sliders (the sliders can, however, also be implemented in DSCQS methodology, in a non-continuous mode). Difficulties can be encountered when using the pen and paper data collection used by the traditional DSCQS method. Though instructions were given to place only a single horizontal tick to mark their score, some viewers completed their response booklets improperly, causing their data to be discarded. The use of the slider-voting device, such as the one used in the SSCQE experiment, effectively overcomes this problem by eliminating the need for pen and paper data collection.

Once a DSCQS test is complete, the paper response booklets are collected and the data is entered by hand into the computer using a digital tablet. It is only after these data files are created by hand that computer data analysis can begin. Entering all the data by hand leaves considerable room for human error, another difficulty that can be overcome by using the sliders. They automatically collect the viewer's opinion data according to the commands given in its programming and store the data in computer files, which are automatically created by the software. These data files can be used immediately for processing and analysis.

The use of the sliders significantly reduces preparation time, since no booklets need to be formatted, and eliminates the need to transfer all the data to files using the digital tablet. This cuts down on unnecessary data input time, and reduces the possibility of human error corrupting the results during data input. Therefore the sliders effectively facilitate data analysis, making it less vulnerable to human error, as well as more time-efficient.

When the subjective results of the two runs were compared with one another, they were found to be statistically indistinguishable. There is one excellent study that was published that investigated the SSCQE and the DSCQS methods and is referenced for further reading [PW03].

In a more advanced stage of analyzing the results of SSCQE data, concepts for collapsing subjective scores to achieve an overall program score have been proposed/suggested. It has been observed that subjects behave in certain general ways when evaluating image quality. The first concept is called Primacy and Recency effects. This is the effect of time on the presentation and perception of image quality. Primacy is the situation where a subject sees some impairments early on in the stimulus set, nonetheless has forgotten what he/she saw once a few seconds to a few minutes of video have passed. This causes subjects to behave on the Recency principles, which means what they have seen in the last 5 to 10 seconds, will weigh heavily on their overall perception of picture quality. In the DSCQS test session, these two effects are eliminated since the presentation generally runs 10 to 15 seconds in length.

Another issue that needs to be addressed in SSCQE is the habit of subjects to react sharply to degradations in image quality and to recover slower when good images are presented. This type of reaction has to be modeled mathematically to be most effective. There are several factors that will influence a viewer's judgment such as the severity and duration of the impairment and the placement in time. A good temporal modeling formula has yet to be developed to address all of these issues with this testing methodology.

It might be a few more years before cognitive modeling is complete but research laboratories are working on it.

4.6.7 Pitfalls of Different Methods

A common problem that can arise when conducting subjective assessments is contextual effect. One type of contextual effect is when there are fluctuations in the subjective rating of sequences based on the impairment present in the preceding video sequences. For example, a sequence with moderate impairment that follows a set of sequences with weak impairment may be judged lower in quality than if it followed sequences with strong impairment. A common method used to try and counter-balance this type of contextual effect is the randomization of the test trial presentation order [Cea99, CW99].

Another type of contextual effect is a range effect. This is when the entire set of responses can be compressed due to a limited range of impairments displayed. For example, if the impairment range is from weak to moderate, subjects might use only a small portion of the scale to express differences in image quality that in reality should be greater. This type of contextual effect has been referred to as a ceiling or floor effect [Ris86]. Some experimenters have addressed this problem by including anchor conditions corresponding to impairment levels at the extremes of the range examined. These anchor conditions are used to allow subjects to establish a subjective quality range and the results of these trials are not included in the data set [Sar67]. Currently, it is common practice to provide such anchors in subjective tests as is recommended in ITU-R Recommendation 500 [ITU98]. However, this procedure may not fully eliminate the occurrence of contextual effects.

The severity of contextual effects on subjective results will vary depending on the type of rating scale implemented in the experiment. Several studies have investigated the stability and reliability of different testing methods [Nar94a, NS97, Nar97, Nar94b, MD82, All80, WA80]. In a joint effort to investigate contextual effects for the ITU-R WP11E, four international labs: CRC (Communications Research Centre, Canada), CCETT (France Telecom), IRT (Institut für Rundfunktecknik, Germany) and SPTT (Swiss Telecom) designed and completed a study on contextual effects. To the best of our knowledge, the DSCQS method, DSIS II method and the Comparison method have never been directly compared using the same video material. These methods are described in greater detail in the International Telecommunication Union (ITU-R) Recommendation 500 [ITU98].

Classically, each of the methods mentioned have been used for different applications. For example, the DSCQS method has been used to evaluate the system performance (i.e., compressed video versus original video) of the digital HDTV Grand Alliance System, which was used as the basis for the new North American standard for digital TV broadcasting. The DSIS II method was used to evaluate the failure characteristic of the same system under degraded channel conditions, such as co-channel interference. The Comparison method is used mostly for direct head-to-head comparisons of different technologies. These studies used four anchor conditions embedded into a series of test trials for both high and low impairment ranges. The results of this work are covered below.

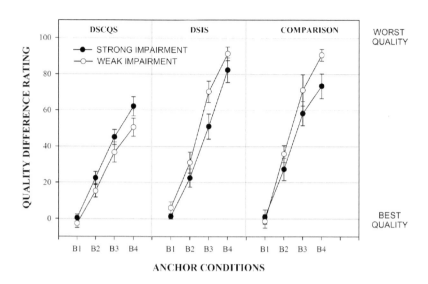

Figure 4.20: Communications Research Centre's Results for Anchor Conditions (B1, B2, B3 and B4) for the DSCQS, DSIS II and Comparison Methods.

Figure 4.20 presents the Communications Research Centre's results for the four anchor conditions (B1 to B4) for all three methods (DSCQS, DSIS II and Comparison). For this figure, the mean values for all methods were transformed to a common scale where 0 denotes the best quality and 100 the worst. From this figure, it is clear that all three methods tracked the results in the same direction, however, two effects are noticeable. One effect is that each method has varying usage levels of the entire quality range and the other is that each method demonstrates different levels of contextual effects.

The varying use of quality range is demonstrated by the results of DSCQS only reaching a maximum difference of around 70, while both DSIS II and Comparison methods have differences in the order of 90-95.

When using the DSIS II method, the scale offers the viewer five choices. If the quality of the reference and test picture is identical the viewer has only one choice, imperceptible (5). Likewise, if the difference in quality is extreme the viewer will choose very annoying (1). This occurs because the available choices are limited. With the DSCQS method, viewers tend to avoid the extremities of the scale, 100–85 and 25–0, thus there is a range effect leading to different scores that do not span the range. It is not uncommon, and is almost expected, that ratings for the DSIS II method will extend over the range.

A similar effect appears with the Comparison scale. Since only half of the range is available, viewers are more likely to use the extremities of the scale. Thus, the same effect is visible in these results also.

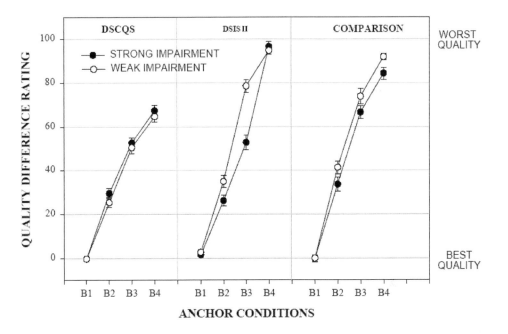

Figure 4.21: Collapsed Results over CRC, CCETT, IRT and SPTT, Taken from [ITU97], for Anchor Conditions (B1, B2, B3 and B4) for the DSCQS, DSIS II and Comparison Methods.

Contextual effects are represented by the differences between the strong and weak impairment curves for the four anchor conditions. It is evident from Figure 4.20 that there are slight contextual effects for the DSCQS method, a moderate effect for the DSIS II method and a small effect for the Comparison method. Logic would state that since these conditions (B1 to B4) are identical in both test sets (weak and strong), they should be rated at the same quality levels at either time. However, due to the level of global impairment presented, the quality fluctuates greatly for B2 to B4 and very little for B1.

If the results from testing facilities that replicated the work are taken into account, then the contextual effects become more evident [ITU97]. Figure 4.21 presents the results collapsed over all testing facilities [ITU97]. Anchor condition B1 for all methods had no effect of content. For conditions B2 to B4, it has been found that contextual effects are not always present for the DSCQS method but clearly present for the DSIS II and the Comparison methods. This means that the DSCQS method produces the most consistent results when the range of impairments is varied.

This study shows just one example of how methods can vary and their results impacted. Despite the results of this study, it is a well-known fact that each of these methods perform extremely well when the experimenter pays close attention to the design of the experiment. If that is not the case, then the results can be suspect and susceptible to problems like contextual effects.

It can not be stressed enough that in order to obtain good results, one must follow all the steps in designing, selecting subjects, conducting and controlling for variables to ensure valid and repeatable results.

4.7   Objective Assessment Methods

4.7.1   History

Objective video quality measurements traditionally have been performed by measuring the physical characteristics of a video signal, such as signal amplitude, timing, signal to noise ratio and Peak Signal to Noise Ratio (PSNR) [IL93]. There are many standardized test signals that can be used to determine what is happening to the video signal. Companies have long marketed equipment designed to do such testing.

Objective assessment refers to methods that assess video quality by taking measurements from the electrical video signal. They should therefore be repeatable and, one would expect, easier to make than performing a subjective test. They can be standardized and implemented in video test equipment. Objective methods can be generally classified into analog methods and digital methods.

Analog methods for the objective assessment of video quality are well known and have long been used for the measurement of the analog video signal [IL93]. These methods include the use of video test signals such as color bars, multiburst, and various ramp patterns to measure signal to noise ratio. For analog systems, this type of test method is sufficient since signal quality and video image quality are highly correlated. In the new world of digital video, these test signals, while still necessary, are not enough to accurately measure video quality. New methods for measuring the quality of the video image (i.e., picture quality) are now needed for digitally compressed video.

The use of digital compression techniques on the signal can result in video that is clearly impaired and yet still maintains a high signal to noise ratio and measures well against static test patterns. Confounding the assessment of digitally compressed video is the fact that the quality of the compressed video is highly dependent on the nature of the source video being compressed. For example, a scene of a newscaster showing only a head and shoulders shot will be easier to compress (and will likely be perceived as being of higher quality) than a fast action football scene. To accurately measure digitally compressed video, aspects of the human visual system must be taken into account.

Ideally, an objective measurement method will utilize natural test scenes (scenes much like those for which the system will be used) and be able to measure video quality while the video system is in service (i.e., being used for its intended purpose). There will likely be trade-offs between the amount of information available at the test instrument location and the accuracy of the measurement. These issues will ultimately be sorted out according to the needs of the video industry.

4.7.2   Requirement for Standards

The industry has been pushing for standards to be created such that products that solve problems of in service quality monitoring which are internationally accepted can be offered on the market. There is a requirement for standards in three main areas of services: full reference, reduced reference – no reference and multimedia. The Video Quality Experts Group discussed in Chapter 11 has only designed and reported on testing covering the Full Reference category [VQE03].

4.8   Summary

In all the years that I have been conducting subjective assessment, it never ceases to amaze me the things which one must truly try and control or, at least, take into account. If there is anything that should be taken away from this chapter, it should be some of the following.

Decide very carefully the question that you are trying to answer with the test and experimental design in which you are preparing to invest so much time and energy. Once you have sorted that out, decide on the standardized scale and method that will make sure you have an answer to that question, i.e., whether it is a quality, impairment or comparison/discrimination question, whether your material requires a double or single presentation and if you are required to have a hidden reference. Once you determine this, the next stage is selecting your test material.

I have taught several courses on Subjective Assessment and the number one largest factor that escapes most experimenters is the visual stimuli selected. People forget that encoding difficulty changes with the input scene and thus the output will vary greatly depending on the content. I cannot stress enough the need to cover a wide range of video material to ensure that you have covered the spectrum of the delivered content. If you only test with still images or sequences such as *Susie* (reference above in Figure 4.1) you will under design your system and your subjective testing will cause you to have false results and fail in deployment. So please heed the need for content from still, to fast action and others with high special detail to those with saturated colors. Content is extremely important, and once that is set, select your viewers.

Once you have sorted out the majority of the above, you then determine your subjects. In the past few years, my thought on this has changed from using non experts to targeted groups to experts. But in general, use non-experts for published data and experts where the work does not have wide distribution and impact. People have different reasons for using different subjects. You need to determine your correct set for your purpose.

Lastly, I would like to encourage everyone to continue to study and develop methods of subjective assessment. I cannot emphasize this enough. End users are the ones who set quality levels and expectations, and subjective testing allows us to determine what that is. Remember, subjective assessment is a truly scientific, repeatable and reliable way to determine visual quality. Let us keep developing it.

References

[All80]   J. W. Allnatt. Subjective assessment method for television digital codecs. *Electronics Letters*, 16(12):450–451, June 1980.

[Cea99]   P. Corriveau and et al. All subjective scales are not created equal: The effects of context on different scales. *Signal Processing*, 77:1–9, 1999.

[CW99]   P. Corriveau and A. Webster. VQEG evaluation of objective methods of video quality assessment. *SMPTE Journal*, 108:645–648, 1999.

[IL93]   A. F. Inglis and A. C. Luther. *Video Engineering*. New York: McGraw–Hill, 1993.

[ITU97]   ITU-R. *Investigation of Contextual Effects*. ITU-R, Canada, France, Germany, Switzerland, March 1997. ITU-R Document 11E/34-E, 24.

[ITU98]   ITU. ITU-R BT. 500-9, methodology for the subjective assessment of the quality of television pictures. *ITU-R BT*, 1998.

[MD82]   I. F. MacDiarmid and P. J. Darby. Double-stimulus assessment of television picture quality. *EBU Technical Review*, E78-A(192):70–78, April 1982.

[Nar94a]   N. Narita. Consideration of subjective evaluation method for quality of image coding. *Electronics and Communications in Japan, Part 3*, 77(7):84–103, 1994.

[Nar94b]   N. Narita. Proposal of a modified ebu method for reliable evaluation of picture quality. *NHK Laboratories Note*, E78-A(431):1–10, December 1994.

[Nar97]   N. Narita. Effect of impairment ranges on reliability of the modified ebu method. *IEICE Trans. Fundamentals*, E78-A(11):1553–1555, November 1997.

[NS97]   N. Narita and Y. Sugiura. On an absolute evaluation method of the quality of television sequences. *IEEE Transactions on Broadcasting*, 43(1):26–35, 1997.

[PW03]   M. H. Pinson and S. Wolf. Comparing subjective video quality testing methodologies. In *SPIE Video Communications and Image Processing Conference*, 146–157, Lugano, Switzerland, July 2003.

[Ris86]   D. R. Riskey. Use and abuses of category scales in sensory measurement. *Journal of Sensory Studies I*, 217–236, 1986.

[Sar67]   V. Sarris. Adaptation-level theory: Two critical experiments on Helson's weighted average model. *American Journal of Psychology*, 80:331–334, 1967.

[VQE03] VQEG. *Final Report from the Video Quality Experts Group on the Validation of Objective Models of Video Quality Assessment, Phase II.* VQEG, August 2003. (Accessible at anonymous ftp site: *ftp://ftp.its.bldrdoc.gov/dist/ituvidq/*).

[WA80] T. White and J. Allnatt. Double-stimulus quality rating method for television digital codecs. *Electronics Letters*, 16(18):714–716, 1980.

[ZC96] W. Zou and P. Corriveau. Methods for evaluation of digital television picture quality. In *SMPTE 138th Technical Conference and World Media Expo*, 146–157, October 1996.

Chapter 5

Perceptual Video Quality Metrics — A Review

Stefan Winkler
Genista Corporation, Montreux, Switzerland

5.1 Introduction

Evaluating and optimizing the quality of digital imaging systems with respect to the capture, display, storage and transmission of visual information is one of the biggest challenges in the field of image and video processing. This is rooted in the complexity of human visual perception and its properties, which determine the visibility of distortions and thus perceived quality.

Subjective experiments, which to date are the only widely recognized method of quantifying the actual perceived quality, are complex and time-consuming, both in their preparation and execution. They have to be carefully designed and controlled in order to achieve meaningful and reproducible results. Chapter 4 gives an overview of the most frequently used methods for subjective tests. Pixel-based difference measures such as the mean-squared error (MSE) or the peak signal-to-noise ratio (PSNR) on the other hand may be simple and very popular, but they cannot be expected to be reliable predictors of perceived quality, because they are ignorant of how vision works. These problems necessitate better objective methods for video quality assessment.

In this chapter, we review the history and state of the art of perceptual quality metrics for video. As any such review, it is bound to be incomplete — work on picture quality metrics goes back almost 50 years, and a vast number of metrics have been proposed over time. The large majority of them are quality metrics for still images — see for example [Ahu93, PS00] for an overview. Work on video quality metrics has flourished only in recent years.

The chapter is organized as follows. Section 5.2 discusses the factors influencing our perception of visual quality. Section 5.3 proposes a classification of video quality metrics according to their design philosophy. Based on this classification, three different types of metrics are reviewed, namely pixel-based metrics (Section 5.4), metrics based on the psychophysical approach (Section 5.5), and metrics based on the "engineering" approach (Section 5.6). Section 5.7 discusses comparative studies of metric prediction performance. Section 5.8 concludes the chapter with a summary of the state of the art and an outlook on future developments in the field.

5.2   Quality Factors

In order to be able to design reliable visual quality metrics, we have to understand what "quality" means to the viewer [AN93, Kle93, SEL00]. A viewer's enjoyment when watching a video depends on many factors:

- Individual interests and expectations: Everyone has their favorite programs, which implies that a football fan who attentively follows a game may have very different quality requirements than someone who is only marginally interested in the sport. We have also come to expect different qualities in different situations, e.g., the quality of watching a feature film in the cinema versus a short clip on a mobile phone. At the same time, advances in technology such as the highly successful DVD have also raised the quality bar — a VHS recording that nobody would have objected to a few years ago is now considered inferior quality by someone who has a DVD player at home.

- Display type and properties: There is a wide variety of displays available today — traditional CRT screens, LCD's, plasma displays, front and back projection using various technologies — with different characteristics in terms of brightness, contrast, color rendition, response time etc. For any type of display, the resolution and size (together with the viewing distance) also influence perceived quality [WR89, Lun93].

- Viewing conditions: the viewing distance determines the effective size and resolution of the displayed image on the retina. The ambient light also affects our perception. Even though we are able to adapt to a wide range of light levels and to discount the color of the illumination, high ambient light levels decrease our sensitivity to small contrast variations. Furthermore, exterior light can lead to veiling glare due to reflections on the screen that again reduce the visible luminance and contrast range [SW04].

- The fidelity of the reproduction. On one hand, we want the "original" video to arrive at the end-user with a minimum of distortions introduced along the way.

On the other hand, video is not necessarily about capturing and reproducing a scene as naturally as possible — considering animations, special effects or artistic "enhancements." For example, sharp images with high contrast are usually more appealing to the average viewer [Rou89]. Likewise, subjects prefer slightly more colorful and saturated images despite realizing that they look somewhat unnatural [dRBF95, YBdR99].

- Finally, the accompanying *soundtrack* also has a great influence on perceived quality. It is important that the sound be synchronized with the video; this is most noticeable for speech and lip synchronization, for which time lags of more than approximately 100 ms are considered very annoying [Ste96]. Additionally, the quality of the accompanying audio has a direct influence on the perceived video quality: subjective quality ratings are generally higher when the test scenes are accompanied by a good quality sound program [Rih96].

Of all the factors in the above list, individual interests in certain types of content are perhaps the most elusive for objective quality metrics; however, it is usually factored out in subjective experiments by averaging the ratings of all observers. Different quality expectations are less of a problem, because subjective experiments are typically constrained to a certain application or setting, and so are metrics.

Display and viewing conditions are sometimes parameters of a quality metric, but often the metrics simply assume "standard" or optimal conditions, under which viewers would have the highest discrimination and be most critical.

Indeed, most of today's metrics are oblivious to the majority of the factors listed above and focus on measuring the visual fidelity of the video, in particular any distortions introduced by lossy compression methods and errors or losses during transmission.

5.3 Metric Classification

Aside from pixel-based fidelity metrics such as the MSE and the PSNR, which are discussed in Section 5.4, two approaches have been taken in metric design:

- The "psychophysical approach" (see Section 5.5), where metric design is primarily based on modeling the human visual system (HVS). Such metrics try to incorporate aspects of human vision deemed relevant to picture quality, such as color perception, contrast sensitivity and pattern masking, using models and data from psychophysical experiments. Due to their generality, these metrics can be used in a wide range of video applications. Models based on neurobiology have been designed as well, but are less useful in real-world applications because of their overwhelming complexity.

- The "engineering approach" (see Section 5.6), which is based primarily on the extraction and analysis of certain features or artifacts in the video. These can be either structural elements such as contours, or specific artifacts that are introduced by a particular compression technology or transmission link. The metrics look for the strength of these features in the video to estimate overall quality. This is not to say that such metrics disregard human vision, as they often consider psychophysical effects as well, but image analysis rather than fundamental vision modeling is the conceptual basis for their design.

Both approaches typically require some degree of parameter fitting to be able to match subjective quality ratings.

Quality metrics can be further classified into the following categories based on the amount of information required about the reference video:

- Full-reference (FR) metrics – sometimes referred to as fidelity metrics – perform a frame-by-frame comparison between a reference video and the video under test; they require the entire reference video to be available, usually in uncompressed form, which is quite an important restriction on the practical usability of such metrics.

- No-reference (NR) metrics look only at the video under test and have no need of reference information. This makes it possible to measure video quality anywhere in an existing compression and transmission system, for example at the receiver side of TV broadcasts or a video streaming session. The difficulty here lies in telling apart distortions from actual content, a distinction humans are usually able to make from experience.

- Reduced-reference (RR) metrics lie between the above two extremes. They extract a number of features from the reference video (e.g. the amount of motion or spatial detail), and the comparison with the video under test is then based only on those features. This makes it possible to avoid some of the pitfalls of pure no-reference metrics while keeping the amount of reference information manageable.

Pixel-based metrics and metrics based on the psychophysical approach typically belong to the full-reference class. Reduced- and no-reference metrics almost exclusively follow the engineering approach.

One important aspect related to the use of reference information is registration and alignment. Full-reference metrics generally impose a precise spatial and temporal alignment of the two videos so that every pixel in every frame can be assigned its counterpart in the reference clip. Especially the temporal alignment is quite a strong restriction and can be very difficult to achieve in practice (even for short clips), because encoders may drop frames or vary the frame rate of the encoded video, and additional delays may be introduced during transmission. Aside from the issue of spatio-temporal alignment, full-reference metrics usually do not respond well to global changes in luminance, chrominance or contrast and require a corresponding calibration. For reduced-reference metrics, the restrictions are less severe, as only the extracted features need to be aligned. Since these features are likely subsampled compared to the original, alignment need not be as precise as with full-reference metrics. No-reference metrics are completely free from alignment issues.

5.4 Pixel-Based Metrics

The mean squared error (MSE) and the peak signal-to-noise ratio (PSNR) are the most popular difference metrics in image and video processing. The MSE is the mean of the squared differences between the gray-level values of pixels in two pictures or sequences $\mathbf{x}_o$ and $\mathbf{x}_r$

$$\text{MSE} = \frac{1}{TMN} \sum_t \sum_m \sum_n [\mathbf{x}_o(m,n,t) - \mathbf{x}_r(m,n,t)]^2, \tag{5.1}$$

for pictures of size $M \times N$ and T frames in the sequence. The average difference per pixel is thus given by the root mean squared error $\text{RMSE} = \sqrt{\text{MSE}}$.

The PSNR in decibels is defined as:

$$\text{PSNR} = 10 \log_{10} \frac{I^2}{\text{MSE}}, \tag{5.2}$$

where I is the maximum value that a pixel can take (e.g., 255 for 8-bit images). Note that the MSE and the PSNR are well-defined only for luminance information; once color comes into play, there is no general agreement on the computation of these measures.

Technically, the MSE measures image difference, whereas the PSNR measures image fidelity, i.e. how closely an image resembles a reference image, usually the uncorrupted original. The popularity of these two metrics is due to the fact that minimizing the MSE is very well understood from a mathematical point of view. Besides, computing MSE and PSNR is very easy and fast. Because they are based on a pixel-by-pixel comparison of images, however, they only have a limited, approximate relationship with the distortion or quality perceived by human observers. In certain situations, the subjective image quality can be improved by adding noise and thereby reducing the PSNR. Dithering of color images with reduced color depth, which adds noise to the image to remove the perceived banding caused by the color quantization, is a common example of this. Furthermore, the visibility of distortions depends to a great extent on the image content, a property known as masking (see Section 5.5.1). Distortions are often much more disturbing in relatively smooth areas of an image than in textured regions with a lot of activity, an effect not taken into account by pixel-based metrics. Therefore the perceived quality of images with the same PSNR can actually be very different (see Figure 5.1).

Figure 5.1: The same amount of noise has been added to these two images, such that their PSNR is identical. High-frequency noise was inserted into the bottom region of the left image, whereas band-pass filtered noise was inserted into the top region of the right image. The noise is hardly visible in the left image due to our low sensitivity to high-frequency stimuli and the strong masking by highly textured content in the bottom region. The smooth sky represents a much weaker masker, and the structured (low-frequency) noise is clearly visible. The PSNR is oblivious to both of these effects.

A number of additional pixel-based metrics have been proposed and tested [EF95]. It was found that although some of these metrics can predict subjective ratings quite successfully for a given compression technique or type of distortion, they are not reliable for evaluations across techniques. MSE was found to be a good metric for additive noise, but is outperformed by more complex HVS-related techniques for coding artifacts [ASS02]. Another study concluded that even perceptual weighting of MSE does not give consistently reliable predictions of visual quality for different pictures and scenes [Mar86]. These results indicate that pixel-based error measures are not accurate for quality evaluations across different scenes or distortion types.

5.5 The Psychophysical Approach

5.5.1 HVS Modeling Fundamentals

This section briefly introduces processes and properties of the human visual system to provide a common ground for the ensuing discussion of HVS-based metrics. More details on the fundamentals of human vision and modeling can be found in Chapter 2.

Models of the human visual system (HVS) account for a number of psychophysical effects [Win99a] which are typically implemented in a sequential process as shown in Figure 5.2.

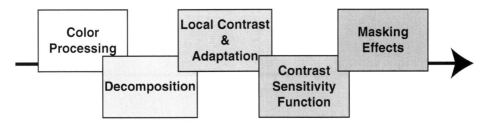

Figure 5.2: Block-diagram of a typical HVS-model.

Color Processing: The first stage in the processing chain of HVS-models concerns the transformation into an adequate perceptual color space, usually based on opponent colors. After this step the image is represented by one achromatic and two chromatic channels carrying color difference information.

This stage can also take care of the so-called *luminance masking* or *lightness nonlinearity* [Sch93], the non-linear perception of luminance by the HVS. Such a nonlinearity is inherent to more sophisticated color spaces such as the CIE $L^*a^*b^*$, but needs to be added to simple linear color spaces.

Multi-Channel Decomposition: It is widely accepted that the HVS bases its perception on multiple channels which are tuned to different ranges of spatial frequencies and orientations. Measurements of the receptive fields of simple cells in the primary visual cortex revealed that these channels exhibit approximately a dyadic structure [Dau80]. This behavior is well matched by a multi-resolution filter bank or a wavelet decomposition. An example for the former is the cortex transform [Wat87], a flexible multi-resolution pyramid, whose filters can be adjusted within a broad range. Wavelet transforms on the other hand offer the advantage that they can be implemented in a computationally efficient manner by a lifting scheme [DS98].

It is believed that there are also a number of channels processing different object velocities or temporal frequencies. These include one temporal low-pass and one, possibly two, temporal band-pass mechanisms in the human visual system [HS92, FH98], which are generally referred to as sustained and transient channels, respectively.

Contrast and Adaptation: The response of the HVS depends much less on the absolute luminance than on the relation of its local variations to the surrounding background, a property known as *Weber-Fechner law* [Sch93]. Contrast is a measure of this relative variation, which is commonly used in vision models. While it is quite simple to define a contrast measure for elementary patterns, it is very difficult to model human contrast perception in complex images, because it varies with the local image content [Pel90, Pel97, WV99]. Furthermore, the adaptation to a specific luminance level or color can influence the perceived contrast.

Contrast Sensitivity: One of the most important issues in HVS-modeling concerns the decreasing sensitivity to higher spatial frequencies. This phenomenon is parameterized by the contrast sensitivity function (CSF). The correct modeling of the CSF is especially difficult for color images. Typically, separability between color and pattern sensitivity is assumed, so that a separate CSF for each channel of the color space needs to be determined and implemented. Achromatic CSF's are summarized in [Bar99], and color CSF measurements are described in [vdHB69, GH73, Mul85].

The human contrast sensitivity also depends on the temporal frequency of the stimuli. Similar to the spatial CSF, the temporal CSF has a low-pass or slightly band-pass shape. The interaction between spatial and temporal frequencies can be described by spatio-temporal contrast sensitivity functions, which are commonly used in vision models for video [Dal98]. For easier implementation, they may be approximated by combinations of components separable in space and time [Kel83, YM94].

Masking: Masking occurs when a stimulus that is visible by itself cannot be detected due to the presence of another. Sometimes the opposite effect, facilitation, occurs: a stimulus that is not visible by itself can be detected due to the presence of another. Within the framework of image processing it is helpful to think of the distortion or coding noise being masked (or facilitated) by the original image or sequence acting as background. Masking explains why similar distortions are disturbing in certain regions of an image while they are hardly noticeable elsewhere (cf. Figure 5.1).

Several different types of spatial masking can be distinguished [KCBST97, NRK02, WBT97], but this distinction is not clear-cut. The terms *contrast masking, edge masking,* and *texture masking* are often used to describe masking due to strong local contrast, edges, and local activity, respectively. *Temporal masking* is a brief elevation of visibility thresholds due to temporal discontinuities in intensity, e.g., at scene cuts [SB65]. It can occur not only after a discontinuity, but also before [ABE98].

Pooling: It is believed that the information represented in various channels within the primary visual cortex is integrated in the subsequent brain areas. This process can be simulated by gathering the data from these channels according to rules of probability or vector summation, also known as pooling. However, little is known about the nature of the actual integration taking place in the brain, and there is no firm experimental evidence that these rules are a good description of the pooling mechanism in the human visual system [Qui74, FH98, MW00].

Often this summation is carried out over all dimensions in order to obtain a single distortion rating for an image or video, but in principle any subset of dimensions can be used depending on what kind of result is desired. For example, pooling over pixel locations may be omitted to produce a distortion map for every frame.

5.5.2 Single-Channel Models

The first models of human vision adopted a single-channel approach. Single-channel models regard the human visual system as a single spatial filter, whose characteristics are defined by the contrast sensitivity function. The output of such a system is the filtered version of the input stimulus, and detectability depends on a threshold criterion.

The first image quality metric for luminance images was developed by Mannos and Sakrison [MS74]. They realized that simple pixel-based distortion measures were not able to accurately predict the quality differences perceived by observers. On the basis of psychophysical experiments on the visibility of gratings, they inferred some properties of the human visual system and came up with a closed-form expression for the contrast sensitivity as a function of spatial frequency, which is still widely used in HVS-models. The input images are filtered with this CSF after a lightness nonlinearity. The squared difference between the filter output for the two images is the distortion measure. It was shown to correlate quite well with subjective ranking data. Despite its simplicity, this metric was one of the first works in engineering to recognize the importance of applying vision science to image processing.

The first color image quality metric was proposed by Faugeras [Fau79]. His model computes the cone absorption rates and applies a logarithmic nonlinearity to obtain the cone responses. An achromatic and two chromatic color difference components are calculated from linear combinations of the cone responses to account for the opponent-color processes in the human visual system. These opponent-color signals go through individual filtering stages with the corresponding CSF's. The squared differences between the resulting filtered components for the reference image and for the distorted image are the basis for an image distortion measure.

The first video quality metric was developed by Lukas and Budrikis [LB82]. It is based on a spatio-temporal model of the CSF using an excitatory and an inhibitory path. The two paths are combined in a nonlinear way, enabling the model to adapt to changes in the level of background luminance. Masking is also incorporated into the model by means of a weighting function derived from the spatial and temporal activity in the reference sequence. In the final stage of the metric, an L_p-norm of the masked error signal is computed over blocks in the frame whose size is chosen such that each block covers the size of the foveal field of vision. The resulting distortion measure was shown to outperform MSE as a predictor of perceived quality.

More recently, Tong et al. [THvdBL99] proposed an interesting single-channel video quality metric called ST-CIELAB (spatio-temporal CIELAB). ST-CIELAB is an extension of the spatial metric S-CIELAB for still image quality [ZW96]. Both are backward compatible to the CIELAB standard, i.e. they reduce to CIE $L^*a^*b^*$ for uniform color fields. The ST-CIELAB metric is based on a spatial, temporal, and chromatic model of human contrast sensitivity in an opponent-colors space. The resulting data are transformed to CIE $L^*a^*b^*$ space, whose difference formula (ΔE) is used for pooling.

Single-channel models are still in use today because of their relative simplicity and computational efficiency, and a variety of extensions and improvements have been proposed. However, they are intrinsically limited in prediction accuracy. They are unable to cope with more complex patterns and cannot account for empirical data from masking and pattern adaptation experiments. These data can be explained quite successfully by a multi-resolution theory of vision, which assumes a whole set of different channels instead of just one. The corresponding multi-channel models and metrics are discussed in the next section.

5.5.3   Multi-Channel Models

Multi-channel models assume that each band of spatial frequencies is dealt with by an independent channel. The CSF is just the envelope of the sensitivities of these channels. Detection occurs independently in any channel when the signal in that band reaches a threshold criterion.

A well-known image distortion metric, the Visual Differences Predictor (VDP), was proposed by Daly [Dal93]. The underlying vision model includes an amplitude nonlinearity to account for the adaptation of the visual system to different light levels, an orientation-dependent CSF, and a hierarchy of detection mechanisms. These mechanisms involve a decomposition similar to the cortex transform [Wat87] and a simple intra-channel masking function. The responses in the different channels are converted to detection probabilities by means of a psychometric function and finally combined according to rules of probability summation. The resulting output of the VDP is a visibility map indicating the areas where two images differ in a perceptual sense.

Lubin [Lub95] designed an elaborate model for measuring still image fidelity, which is also known as the Sarnoff Visual Discrimination Model (VDM). First the input images are convolved with an approximation of the point spread function of the eye's optics. Then the sampling by the cone mosaic in the retina is simulated. The decomposition stage implements a Laplacian pyramid for spatial frequency separation, local contrast computation, as well as directional filtering, from which a phase-independent contrast energy measure is calculated. This contrast energy measure is subjected to a masking stage, which comprises a normalization process and a sigmoid nonlinearity. Finally, a distance measure or JND map is computed as the L_p-norm of the masked responses. The VDM is one of the few models that take into account the eccentricity of the images in the observer's visual field. With some modifications, it was later extended to the Sarnoff JND (just noticeable differences) metric for color video [LF97].

Van den Branden Lambrecht proposed a number of video quality metrics based on multi-channel vision models [vdBL96]. The Moving Picture Quality Metric (MPQM) is based on a local contrast definition and Gabor-related filters for the spatial decomposition, two temporal mechanisms, as well as a spatio-temporal CSF and a simple intra-channel model of contrast masking [vdBLV96]. Due to the MPQM's purely frequency-domain implementation of the spatio-temporal filtering process and the resulting huge memory requirements, it is not practical for measuring the quality of sequences with a duration of more than a few seconds, however. The Normalization Video Fidelity Metric (NVFM) [LvdBL96] avoids this shortcoming by using a steerable pyramid transform for spatial filtering and discrete time-domain filter approximations of the temporal mechanisms. It is a spatio-temporal extension of Teo and Heeger's image distortion metric [TH94].

The author presented a perceptual distortion metric (PDM) for color video [Win99b]. It is based on the NVFM and a model for color images [Win98]. After conversion of the input to an opponent colors space, each of the resulting three components is subjected to a spatio-temporal decomposition by the steerable pyramid, yielding a number of perceptual channels. They are weighted according to spatio-temporal contrast sensitivity data and subsequently undergo a contrast gain control stage for pattern masking [WS97], which is realized by an excitatory nonlinearity that is normalized by a pool of inhibitory responses. Finally, the sensor differences are combined by means of an L_p-norm into visibility maps or a distortion measure. The performance of the metric is discussed in [Win00].

Masry and Hemami [MH04] designed a metric for continuous video quality evaluation (CVQE) of low bitrate video. It is one of the few metrics (especially in the psychophysical category) used on video with high distortion levels. The metric works with luminance information only. It uses temporal filters and a wavelet transform for the perceptual decomposition, followed by CSF-weighting of the different bands, a gain control model, and pooling by means of an L_3-norm. Recursive temporal summation takes care of the low-pass nature of subjective quality ratings. They also outline a method to compare the time-series data generated by the metric with the corresponding continuous MOS data from SSCQE experiments.

5.6 The Engineering Approach

Metrics based on multi-channel vision models such as the ones presented above in Subsection 5.5.3 are the most general and potentially the most accurate ones [Win99a]. However, quality metrics need not necessarily rely on sophisticated general models of the human visual system; they can exploit a priori knowledge about the compression algorithm and the pertinent types of artifacts (see also Chapter 3) using ad-hoc techniques or specialized vision models. While such metrics are not as versatile, they can perform very well in a given application area. Their main advantage lies in the fact that they generally permit a computationally more efficient implementation.

5.6.1 Full-Reference Metrics

One example of such specialized metrics is DCTune[1] [Wat95, Wat97]. It was developed as a method for optimizing JPEG image compression, but can also be used as a quality metric. DCTune computes the JPEG quantization matrices that achieve the maximum compression for a specified perceptual distortion given a particular image and a particular set of viewing conditions. It considers visual masking by luminance and contrast techniques. DCTune can also compute the perceptual difference between two images. This latter aspect was extended to video [Wat98]. In addition to the spatial sensitivity and masking effects considered in DCTune, this so-called Digital Video Quality (DVQ) metric relies on measurements of the visibility thresholds for temporally varying DCT quantization noise. It also models temporal forward masking effects by means of a masking sequence, which is produced by passing the reference through a temporal low-pass filter. A report of the DVQ metric's performance is given in [WHM01].

Hamada et al. [HMM99] proposed a picture quality assessment system based on a perceptual weighting of the coding noise. In their three-layered design, typical noise types from the compression are classified and weighted according to their characteristics. The local texture is analyzed to compute the local degree of masking. Finally, a gaze prediction stage is used to emphasize noise visibility in and around objects of interest. The PSNR computed on the weighted noise is used as distortion measure. This metric has been implemented in a system that permits real-time video quality assessment.

Tan et al. [TGP98] presented an objective measurement tool for MPEG video quality. It first computes the perceptual impairment in each frame based on contrast sensitivity and masking with the help of spatial filtering and Sobel-operators, respectively. Then the PSNR of the masked error signal is calculated and normalized. The interesting part of this metric is its second stage, a "cognitive emulator." It simulates the delay and temporal smoothing effect of observer responses, the nonlinear saturation of perceived quality, and the asymmetric behavior with respect to quality changes from bad to good and vice versa. The cognitive emulator was shown to improve the predictions of subjects' SSCQE ratings.

Hekstra et al. [H+02] proposed a perceptual video quality measure (PVQM). It uses a linear combination of three features, namely the loss of edge sharpness, the color error normalized by the saturation, and the temporal variability of the reference video. Spatio-temporal pooling of these features is carried out using an L_7-norm. The PVQM was one of the best predictors in the VQEG FR-TV Phase I test (see Section 5.7.2).

Wang et al. [WLB04] presented a video quality assessment method based on a structural similarity (SSIM) index (see also Chapter 7). It computes the mean, variance and

---

[1] A demonstration version of DCTune can be downloaded from
http://vision.arc.nasa.gov/dctune/.

covariance of small patches inside a frame and combines the measurements into a distortion map. It is applied to color video by computing the SSIM index for each color channel. Motion estimation is used for a weighting of the SSIM index of each frame. The metric was evaluated with the data from the VQEG FR-TV Phase I test (see Section 5.7.2) and showed a slight outperformance of the other metrics in this test.

5.6.2 Reduced-Reference Metrics

Wolf and Pinson [WP99] designed a video quality metric (VQM) that uses reduced reference information in the form of low-level features extracted from spatio-temporal blocks of the sequences.[1] These features were selected empirically from a number of candidates so as to yield the best correlation with subjective data. First, horizontal and vertical edge enhancement filters are applied to facilitate gradient computation in the feature extraction stage. The resulting sequences are divided into spatio-temporal blocks. A number of features measuring the amount and orientation of activity in each of these blocks are then computed from the spatial luminance gradient. To measure the distortion, the features from the reference and the distorted sequence are compared using a process similar to masking. This metric was one of the best performers in the latest VQEG FR-TV Phase II evaluation (see Section 5.7.2).

Horita et al. [HMG+03] proposed a reduced-reference metric based on 26 low-level spatial features computed from the luminance image and the Sobel edge filters. Squared frame differences are used as the temporal feature. Additionally, the global motion due to camera movements (pan, tilt and zoom) is estimated in the received video. Similar to the FR metric by Tan et al. described above, an asymmetric tracking function is applied to the instantaneous measurements to simulate the delay and memory effects seen in human observer behavior.

5.6.3 No-Reference Metrics

In contrast to full- and reduced-reference metrics, no-reference (NR) metrics always have to make assumptions about the video content and/or the distortions of interest. With this comes the risk of confusing actual content with distortions (e.g., a chessboard may be interpreted as block artifacts in the extreme case).

It should be noted that NR metrics typically use estimates of the missing reference video or the corresponding features, which means that NR metrics can also be used in an RR scenario by replacing these estimates with the actual features computed from the reference video.

---

[1] An evaluation version of the VQM can be downloaded from
http://www.its.bldrdoc.gov/n3/video/vqmsoftware.htm.

NR metrics for block-DCT encoded video

The majority of NR metrics are based on estimating blockiness, which is the most prominent artifact of block-DCT based compression methods such as H.26x, MPEG and their derivatives. Different algorithms have been developed to measure blockiness in the absence of the original [WSM01]. They all assume a fixed block structure, e.g., an 8×8 grid, on which they look for signs of blockiness. Several commercial quality assessment systems are based on this approach.

Wu and Yuen introduced the first NR blocking impairment metric [WY97], which measures the horizontal and vertical differences between the columns and rows at block boundaries. Weights for taking into account masking effects are derived from the means and standard deviations of the blocks adjacent to each boundary. The resulting measure is normalized by an average of the same measures computed at non-boundary columns and rows.

Baroncini and Pierotti [BP99] proposed an approach based on multiple filters to extract significant vertical and horizontal edge segments due to blockiness, but not limited to a regular block structure. Taking into account the temporal variations of these edge features, an overall quality measure is computed.

Wang et al. [WBE00] model the blocky image as a non-blocky image interfered with a pure blocky signal. They apply 1-D FFTs to horizontal and vertical difference signals or rows and columns in the image to estimate the average horizontal and vertical power spectra. Spectral peaks due to 8×8 block structures are identified by their locations in the frequency domain. The power spectra of the underlying non-blocky images are approximated by median-filtering these curves. The overall blockiness measure is then computed as the difference between these power spectra at the locations of the peaks.

Vlachos [Vla00] used an algorithm based on the cross-correlation of subsampled images. The sampling structure is chosen such that every sub-image contains one specific pixel from each block. Four sub-images are constructed from the four corner pixels of each block. Four more sub-images are constructed from four neighboring pixels in the top left corner of each block. Finally, the cross-correlations among the former four sub-images are normalized by the cross-correlations among the latter four sub-images to yield a measure of blockiness.

NR metrics in the DCT domain

Instead of analyzing the blockiness artifacts in the spatial (decoded) domain, it can be beneficial to carry out this analysis using the DCT coefficients directly from the encoded bitstream. This is especially useful in situations where computing power and data throughput are limited, or where a decoder is not available. The fact that the DCT coefficients already represent a frequency decomposition of the blocks can also help in modeling artifact visibility. However, the bitstream has to be parsed to find the relevant information, and partial decoding may be required.

Coudoux et al. [CGDC01] combine the detection of vertical block edges with a number of sensitivity and masking models which are applied in the DCT domain. The models consider luminance masking, intra- and inter-coefficient spatial activity masking, as well as temporal activity masking. The authors present substantial data from psychophysical experiments on these masking effects, and also evaluate their quality metric using a few test clips.

Gastaldo et al. [GZR02] present an NR measurement approach using a neural network. Features are extracted directly from the bitstream without any a priori knowledge or assumptions about human perception. They include statistics on motion vectors and macroblock type, quantizer scale and DCT coefficient energy. These features are then used as input to a circular back-propagation network. They report a correlation of 85% with SSCQE data on 12 MPEG-2 sequences.

Other NR metrics

Not all compression algorithms are based on block-DCTs, and not all video distortions are compression artifacts. For these other types of distortions, additional NR metrics are needed.

Marziliano et al. [MDWE02] proposed a blurriness metric that is based on the assumption that object boundaries are usually represented by sharp edges. The algorithm estimates the spread of significant edges in the image and derives an overall measure of blurriness. Since blurriness is inherent not only to block-DCT based compression, but most other compression schemes as well, it is more versatile than a blockiness metric. Many block-DCT based encoders and decoders today use deblocking filters, which reduce the blocking artifacts, but increase the blur. Furthermore, a blurriness metric can not only be used for measuring compression artifacts, but also for the detection of scenes that are out of focus, for example. Blurriness and ringing metrics were used successfully for the evaluation of JPEG2000 compression artifacts [MDWE04].

Winkler et al. [WC03, WD03] combined blockiness, blurriness and jerkiness artifact metrics for real-time NR quality assessment. Windows Media, Real Media, MPEG-4 and Motion JPEG2000 codecs were used to encode video and the video transmission was simulated over packet networks and wireless networks. Subjective experiments were carried out using the Single Stimulus Continuous Quality Evaluation (SSCQE) method to obtain instantaneous quality ratings, which were subsequently used to validate the metric's prediction performance. Correlations of up to 90% were achieved with the subjective ratings, compared with 40% for the PSNR.

Caviedes and Oberti [CO03] proposed a no-reference video quality metric based on several artifacts and image features, including blocking, ringing, clipping, noise, contrast, sharpness, and their interactions. They tested it on video degraded by MPEG compression as well as video enhanced through noise reduction and sharpening.

5.7   Metric Comparisons

5.7.1   Overview

While video quality metric designs and implementations abound, only few comparative studies exist that have investigated the prediction accuracy of metrics in relation to others. The prediction performance of a quality metric is usually evaluated with the help of subjective ratings for a certain test set. A number of different criteria can be considered in such an evaluation, e.g. prediction accuracy (the average error), monotonicity (the ordering of images according to their quality), and consistency (the number of outliers) — see Appendix A for more details. These criteria can be quantified with mathematical tools such as regression analysis; correlations are probably the most commonly used performance indicators.

Ahumada [Ahu93] reviewed more than 30 visual discrimination models for still images from the application areas of image quality assessment, image compression and halftoning. However, only a comparison of the implementations of their computational models is given; the performance of the metrics is not evaluated.

Comparisons of several image quality metrics with respect to their prediction performance were carried out in [EAB98, FBC95, Jac95, LMK98, MEC98]. These studies consider various pixel-based metrics as well as a number of single-channel and multi-channel models from the literature. Summarizing their findings and drawing overall conclusions is made difficult by the fact that test images, testing procedures, and applications differ greatly between studies. It can be noted that certain pixel-based metrics in the evaluations correlate quite well with subjective ratings for some test sets, especially for a given type of distortion or scene. They can be outperformed by vision-based metrics, where more complexity usually means more generality and accuracy. However, the observed gains are often so little that the computational overhead does not seem justified.

Objective measures of MPEG video quality were validated by Cermak et al. [CWT$^+$98]. However, this comparison does not consider entire quality metrics, but only a number of low-level features such as edge energy or motion energy and combinations thereof.

5.7.2   Video Quality Experts Group

The most ambitious and comprehensive performance evaluations of video quality metrics to date have been undertaken by the Video Quality Experts Group (VQEG) — refer to Chapter 11 for details. The group was formed in 1997 with the objective to collect reliable subjective ratings for a well-defined set of test sequences and to evaluate the performance of different video quality assessment systems with respect to these sequences.

In the first phase (1997-2000), the emphasis was on full-reference testing of production- and distribution-class video for television ("FR-TV"). Therefore, the test conditions comprised mainly MPEG-2 encoded sequences, including conversions between analog digital video or transmission errors. In total, 20 scenes were subjected to 16 test conditions each. Subjective ratings for these were collected using the DSCQS method (see Chapter 4) from ITU-R Rec. BT.500 [ITU02]. Both reference and processed sequences are available publicly on the VQEG web site, [1] together with all the data from the subjective experiments.

Ten different video quality metrics were submitted for evaluation on the test sequences, which were disclosed to the proponents only after the submission of their metrics. The results of the data analysis showed that the prediction performance of most models as well as the PSNR were statistically equivalent for four different statistical criteria, leading to the conclusion that no single model outperforms the others in all situations on this test material [VQE00, R$^+$00].

As a follow-up to this first phase, VQEG decided to carry out a second round of tests for full-reference metrics ("FR-TV Phase II"); the final report was finished recently [VQE03]. In order to obtain more discriminating results, this second phase was designed with a stronger focus on secondary distribution of digitally encoded television quality video and a wider range of distortions. New source sequences and test conditions were defined, and a total of 128 test sequences was produced. Subjective ratings for these sequences were again collected using the DSCQS method. Unfortunately, the test sequences of the second phase are not publicly available.

Six video quality metrics were submitted for evaluation. In contrast to the first phase, registration and calibration with the reference video had to be performed by each metric individually. Seven statistical criteria were defined to analyze the prediction performance of the metrics. These criteria all produced the same ranking of metrics, therefore only correlations are quoted here. The best metrics in the test achieved correlations as high as 94% with the MOS, thus significantly outperforming the PSNR with correlations of around 70%.

VQEG is currently preparing an evaluation of reduced- and no-reference metrics for television as well as an evaluation of metrics in a "multimedia" scenario targeted at lower bitrates and smaller frame sizes.

5.7.3 Limits of Prediction Performance

Perceived visual quality is an inherently subjective measure and can only be described statistically, i.e., by averaging over the opinions of a sufficiently large number of observers. Therefore, the question is also how well subjects agree on the quality of a given

---

[1] http://www.vqeg.org/

image or video. In the first round of VQEG tests (see above), the correlations obtained between the average ratings of viewer groups from different labs are in the range of 90–95%. While the exact figures certainly vary depending on the application and the range of the test set, this gives an indication of the limits of prediction performance for quality metrics. In the same study, the best-performing metrics only achieved correlations in the range of 80–85%. This shows that there is still room for improvement before quality metrics can replace subjective tests.

In the FR-TV Phase II tests, a statistically optimal model was defined based on the subjective data to provide a quantitative upper limit on prediction performance. Despite the generally good performance of metrics in this test, none of the submitted metrics achieved a prediction performance statistically equivalent to the optimal model.

5.8   Conclusions and Perspectives

Reliable perceptual quality metrics have applications in many fields of image processing. Pixel-based difference metrics, such as the MSE, are still widely used today, but significant improvements in prediction performance and/or versatility can be achieved by perceptual quality metrics. As this review of the state-of-the-art shows, there has been a lot of work on full-reference metrics and television broadcasting applications. Much remains to be done in the area of no-reference quality assessment, especially for video compressed at low bitrates and transmission error artifacts. Here the development of reliable metrics is still in its infancy, and many issues remain to be investigated and solved.

More comparative analysis is necessary in order to evaluate the prediction performance of metrics in realistic conditions and to determine the most promising modeling approaches. The collaborative efforts of Modelfest [C+00, C+02] or VQEG [VQE00, VQE03] represent important steps in the right direction. Even if the former concerns low-level vision and the latter entire video quality assessment systems, both share the idea of applying different models to the same set of subjective data.

Furthermore, more psychophysical experiments (especially on masking) need to be done with natural images. The use of simple test patterns like Gabor patches or noise patterns may be appropriate for elementary experiments, but they are insufficient for the modeling of more complex phenomena that occur in natural images.

Similarly, most psychophysical experiments focus on the threshold of visibility, whereas quality metrics and compression are often applied to clearly visible distortions. This obvious discrepancy has to be overcome by intensified efforts with supra-threshold experiments.

Another important aspect in video quality evaluation is the fact that people only focus on certain regions of interest in the video, e.g., persons, faces or some moving objects. Outside of the region of interest, our sensitivity is significantly reduced. Most quality assessment systems ignore this and weight distortions equally over the entire frame. Some recent metrics attempt to model the focus of attention and consider it for computing the overall video quality [CS04, LPB02, OR01, LLO+03].

An important shortcoming of existing metrics is that they measure image fidelity instead of perceived quality (cf. Section 5.2). The accuracy of the reproduction of the original on the display, even considering the characteristics of the human visual system, is not the only quality benchmark. For example, colorful, well-lit, sharp pictures with high contrasts are considered attractive, whereas low-quality, dark and blurry pictures with low contrasts are often rejected [SEL00]. Quantitative metrics of this "image appeal" were indeed shown to improve quality prediction performance [Win01].

Finally, vision may be the most essential of our senses, but it is certainly not the only one. We rarely watch video without sound, for example. Focusing on visual quality alone cannot solve the problem of evaluating a multimedia experience. Therefore, comprehensive audiovisual quality metrics are required that analyze both video and audio as well as their interactions [Han04, WF05].

As this list shows, the remaining tasks in HVS-research are challenging and need to be solved in close collaboration of experts in human vision, color science and video processing.

References

[ABE98]   A. J. Ahumada, Jr., B. L. Beard, and R. Eriksson. Spatio-temporal discrimination model predicts temporal masking function. In *Proc. SPIE*, 3299:120–127, San Jose, CA, Jan. 26–29, 1998.

[Ahu93]   A. J. Ahumada, Jr. Computational image quality metrics: A review. In *SID Symposium Digest*, 24:305–308, 1993.

[AN93]   A. J. Ahumada, Jr. and C. H. Null. Image quality: A multidimensional problem. In A. B. Watson, Ed., *Digital Images and Human Vision*, 141–148. Cambridge, MA: The MIT Press, 1993.

[ASS02]   İ. Avcıbaş, B. Sankur, and K. Sayood. Statistical evaluation of image quality measures. *J. Electronic Imaging*, 11(2):206–223, April 2002.

[Bar99]   P. G. J. Barten. *Contrast Sensitivity of the Human Eye and Its Effects on Image Quality*. Bellingham, WA: SPIE, 1999.

[BP99]   V. Baroncini and A. Pierotti. Single-ended objective quality assessment of DTV. In *Proc. SPIE*, 3845:244–253, Boston, MA, Sep. 19–22, 1999.

[C+00]   T. Carney *et al.* Modelfest: Year one results and plans for future years. In *Proc. SPIE*, 3959:140–151, San Jose, CA, Jan. 23–28, 2000.

[C+02]   T. Carney *et al.* Extending the Modelfest image/threshold database into the spatio-temporal domain. In *Proc. SPIE*, 4662:138–148, San Jose, CA, Jan. 20–25, 2002.

[CGDC01]　F. X. Coudoux, M. G. Gazalet, C. Derviaux, and P. Corlay. Picture quality measurement based on block visibility in discrete cosine transform coded video sequences. *J. Electronic Imaging*, 10(2):498–510, April 2001.

[CO03]　J. Caviedes and F. Oberti. No-reference quality metric for degraded and enhanced video. In *Proc. SPIE*, 5150:621–632, Lugano, Switzerland, July 8–11, 2003.

[CS04]　A. Cavallaro and S.Winkler. Segmentation-driven perceptual quality metrics. In *Proc. ICIP*, 3543–3546, Singapore, Oct. 24–27, 2004.

[CWT⁺98]　G. W. Cermak, S. Wolf, E. P. Tweedy, M. H. Pinson, and A. A. Webster. Validating objective measures of MPEG video quality. *SMPTE J.*, 107(4):226–235, April 1998.

[Dal93]　S. Daly. The visible differences predictor: An algorithm for the assessment of image fidelity. In A. B. Watson, Ed., *Digital Images and Human Vision*, 179–206. Cambridge, MA: The MIT Press, 1993.

[Dal98]　S. Daly. Engineering observations from spatiovelocity and spatiotemporal visual models. In *Proc. SPIE*, 3299:180–191, San Jose, CA, Jan. 26–29, 1998.

[Dau80]　J. G. Daugman. Two-dimensional spectral analysis of cortical receptive field profiles. *Vision Res.*, 20(10):847–856, 1980.

[dRBF95]　H. de Ridder, F. J. J. Blommaert, and E. A. Fedorovskaya. Naturalness and image quality: Chroma and hue variation in color images of natural scenes. In *Proc. SPIE*, 2411:51–61, San Jose, CA, Feb. 6–8, 1995.

[DS98]　I. Daubechies and W. Sweldens. Factoring wavelet transforms into lifting steps. *J. Fourier Anal. Appl.*, 4(3):245–267, 1998.

[EAB98]　R. Eriksson, B. Andrén, and K. Brunnström. Modelling the perception of digital images: A performance study. In *Proc. SPIE*, 3299:88–97, San Jose, CA, Jan. 26–29, 1998.

[EF95]　A. M. Eskicioglu and P. S. Fisher. Image quality measures and their performance. *IEEE Trans. Comm.*, 43(12):2959–2965, Dec. 1995.

[Fau79]　O. D. Faugeras. Digital color image processing within the framework of a human visual model. *IEEE Trans. Acoust. Speech Signal Process.*, 27(4):380–393, Aug. 1979.

[FBC95]　D. R. Fuhrmann, J. A. Baro, and J. R. Cox, Jr. Experimental evaluation of psychophysical distortion metrics for JPEG-coded images. *J. Electronic Imaging*, 4(4):397–406, Oct. 1995.

[FH98]　R. E. Fredericksen and R. F. Hess. Estimating multiple temporal mechanisms in human vision. *Vision Res.*, 38(7):1023–1040, 1998.

[GH73]　E. M. Granger and J. C. Heurtley. Visual chromaticity-modulation transfer function. *J. Opt. Soc. Am.*, 63(9):1173–1174, Sep. 1973.

[GZR02]　P. Gastaldo, R. Zunino, and S. Rovetta. Objective assessment of MPEG-2 video quality. *J. Electronic Imaging*, 11(3):365–374, July 2002.

[H⁺02]　A. P. Hekstra *et al.* PVQM – a perceptual video quality measure. *Signal Processing: Image Communication*, 17(10):781–798, Nov. 2002.

[Han04]　D. S. Hands. A basic multimedia quality model. *IEEE Trans. Multimedia*, 6(6):806–816, Dec. 2004.

[HMG⁺03] Y. Horita, T. Miyata, I. P. Gunawan, T. Murai, and M. Ghanbari. Evaluation model considering static-temporal quality degradation and human memory for SSCQE video quality. In *Proc. SPIE*, 5150:1601–1611, Lugano, Switzerland, July 8–11, 2003.

[HMM99] T. Hamada, S. Miyaji, and S. Matsumoto. Picture quality assessment system by three-layered bottom-up noise weighting considering human visual perception. *SMPTE J.*, 108(1):20–26, Jan. 1999.

[HS92] R. F. Hess and R. J. Snowden. Temporal properties of human visual filters: Number, shapes and spatial covariation. *Vision Res.*, 32(1):47–59, 1992.

[ITU02] ITU-R Recommendation BT.500-11. Methodology for the subjective assessment of the quality of television pictures. International Telecommunication Union, Geneva, Switzerland, 2002.

[Jac95] R. E. Jacobson. An evaluation of image quality metrics. *J. Photographic Sci.*, 43(1):7–16, 1995.

[KCBST97] S. A. Klein, T. Carney, L. Barghout-Stein, and C. W. Tyler. Seven models of masking. In *Proc. SPIE*, 3016:13–24, San Jose, CA, Feb. 8–14, 1997.

[Kel83] D. H. Kelly. Spatiotemporal variation of chromatic and achromatic contrast thresholds. *J. Opt. Soc. Am.*, 73(6):742–750, June 1983.

[Kle93] S. A. Klein. Image quality and image compression: A psychophysicist's viewpoint. In A. B. Watson, Ed., *Digital Images and Human Vision*, 73–88. Cambridge, MA: The MIT Press, 1993.

[LB82] F. X. J. Lukas and Z. L. Budrikis. Picture quality prediction based on a visual model. *IEEE Trans. Comm.*, 30(7):1679–1692, July 1982.

[LF97] J. Lubin and D. Fibush. Sarnoff JND vision model. T1A1.5 Working Group Document #97-612, ANSI T1 Standards Committee, 1997.

[LLO⁺03] Z. Lu, W. Lin, E. Ong, X. Yang, and S. Yao. PQSM-based RR and NR video quality metrics. In *Proc. SPIE*, 5150:633–640, Lugano, Switzerland, July 8–11, 2003.

[LMK98] B. Li, G. W. Meyer, and R. V. Klassen. A comparison of two image quality models. In *Proc. SPIE*, 3299:98–109, San Jose, CA, Jan. 26–29, 1998.

[LPB02] S. Lee, M. S. Pattichis, and A. C. Bovik. Foveated video quality assessment. *IEEE Trans. Multimedia*, 4(1):129–132, March 2002.

[Lub95] J. Lubin. A visual discrimination model for imaging system design and evaluation. In E. Peli, Ed., *Vision Models for Target Detection and Recognition*, 245–283. Singapore: World Scientific Publishing, 1995.

[Lun93] A. M. Lund. The influence of video image size and resolution on viewing-distance preferences. *SMPTE J.*, 102(5):407–415, May 1993.

[LvdBL96] P. Lindh and C. J. van den Branden Lambrecht. Efficient spatio-temporal decomposition for perceptual processing of video sequences. In *Proc. ICIP*, 3:331–334, Lausanne, Switzerland, Sep. 16–19, 1996.

[Mar86] H. Marmolin. Subjective MSE measures. *IEEE Trans. Systems, Man, and Cybernetics*, 16(3):486–489, May 1986.

[MDWE02]   P. Marziliano, F. Dufaux, S. Winkler, and T. Ebrahimi. A no-reference perceptual
            blur metric. In *Proc. ICIP*, 3:57–60, Rochester, NY, Sep. 22–25, 2002.

[MDWE04]   P. Marziliano, F. Dufaux, S. Winkler, and T. Ebrahimi. Perceptual blur and ringing
            metrics: Application to JPEG2000. *Signal Processing: Image Communication*,
            19(1), Jan. 2004.

[MEC98]   A. Mayache, T. Eude, and H. Cherifi. A comparison of image quality models and
            metrics based on human visual sensitivity. In *Proc. ICIP*, 3:409–413, Chicago,
            IL, Oct. 4–7, 1998.

[MH04]   M. A. Masry and S. S. Hemami. A metric for continuous quality evaluation of
            compressed video with severe distortions. *Signal Processing: Image Communi-
            cation*, 19(1):133–146, Feb. 2004.

[MS74]   J. L. Mannos and D. J. Sakrison. The effects of a visual fidelity criterion on the
            encoding of images. *IEEE Trans. Inform. Theory*, 20(4):525–536, July 1974.

[Mul85]   K. T. Mullen. The contrast sensitivity of human colour vision to red-green and
            blue-yellow chromatic gratings. *J. Physiol.*, 359:381–400, 1985.

[MW00]   T. S. Meese and C. B. Williams. Probability summation for multiple patches of
            luminance modulation. *Vision Res.*, 40(16):2101–2113, July 2000.

[NRK02]   M. J. Nadenau, J. Reichel, and M. Kunt. Performance comparison of masking
            models based on a new psychovisual test method with natural scenery stimuli.
            Signal Processing: Image Communication, 17(10):807–823, Nov. 2002.

[OR01]   W. Osberger and A. M. Rohaly. Automatic detection of regions of interest in
            complex video sequences. In *Proc. SPIE*, 4299:361–372, San Jose, CA, Jan. 21–
            26, 2001.

[Pel90]   E. Peli. Contrast in complex images. *J. Opt. Soc. Am. A*, 7(10):2032–2040, Oct.
            1990.

[Pel97]   E. Peli. In search of a contrast metric: Matching the perceived contrast of Gabor
            patches at different phases and bandwidths. *Vision Res.*, 37(23):3217–3224, 1997.

[PS00]   T. N. Pappas and R. J. Safranek. Perceptual criteria for image quality evaluation.
            In A. Bovik, Ed., *Handbook of Image & Video Processing*, 669–684. New York:
            Academic Press, 2000.

[Qui74]   R. F. Quick, Jr. A vector-magnitude model of contrast detection. *Kybernetik*,
            16:65–67, 1974.

[R+00]   A. M. Rohaly *et al.* Video Quality Experts Group: Current results and future
            directions. In *Proc. SPIE*, 4067:742–753, Perth, Australia, June 21–23, 2000.

[Rih96]   S. Rihs. The influence of audio on perceived picture quality and subjective audio-
            video delay tolerance. In *MOSAIC Handbook*, 183–187, Jan. 1996.

[Rou89]   J. A. J. Roufs. Brightness contrast and sharpness, interactive factors in perceptual
            image quality. In *Proc. SPIE*, 1077:209–216, Los Angeles, CA, Jan. 18–20, 1989.

[SB65]   A. J. Seyler and Z. L. Budrikis. Detail perception after scene changes in television
            image presentations. *IEEE Trans. Inform. Theory*, 11(1):31–43, 1965.

[Sch93]   W. Schreiber. *Fundamentals of Electronic Imaging Systems*. New York: Springer,
            1993.

[SEL00]   A. E. Savakis, S. P. Etz, and A. C. Loui. Evaluation of image appeal in consumer photography. In *Proc. SPIE*, 3959:111–120, San Jose, CA, Jan. 23–28, 2000.

[Ste96]   R. Steinmetz. Human perception of jitter and media synchronization. *IEEE J. Selected Areas in Comm.*, 14(1):61–72, Jan. 1996.

[SW04]   S. Süsstrunk and S. Winkler. Color image quality on the Internet. In *Proc. SPIE*, 5304:118–131, San Jose, CA, Jan. 19–22, 2004. Invited paper.

[TGP98]   K. T. Tan, M. Ghanbari, and D. E. Pearson. An objective measurement tool for MPEG video quality. *Signal Processing*, 70(3):279–294, Nov. 1998.

[TH94]   P. C. Teo and D. J. Heeger. Perceptual image distortion. In *Proc. SPIE*, 2179:127–141, San Jose, CA, Feb. 8–10, 1994.

[THvdBL99] X. Tong, D. Heeger, and C. J. van den Branden Lambrecht. Video quality evaluation using ST-CIELAB. In *Proc. SPIE*, 3644:185–196, San Jose, CA, Jan. 23–29, 1999.

[vdBL96]   C. J. van den Branden Lambrecht. *Perceptual Models and Architectures for Video Coding Applications*. Ph.D. thesis, École Polytechnique Fédérale de Lausanne, Switzerland, 1996.

[vdBLV96] C. J. van den Branden Lambrecht and O. Verscheure. Perceptual quality measure using a spatio-temporal model of the human visual system. In *Proc. SPIE*, 2668:450–461, San Jose, CA, Jan. 28–Feb. 2, 1996.

[vdHB69]   G. van der Horst and M. A. Bouman. Spatiotemporal chromaticity discrimination. *Journal of the Optical Society of America*, 59:1482–1488, 1969.

[Vla00]   T. Vlachos. Detection of blocking artifacts in compressed video. *Electronics Letters*, 36(13):1106–1108, June 2000.

[VQE00]   VQEG. Final report from the Video Quality Experts Group on the validation of objective models of video quality assessment, April 2000. Available at http://www.vqeg.org/.

[VQE03]   VQEG. Final report from the Video Quality Experts Group on the validation of objective models of video quality assessment — Phase II, Aug. 2003. Available at http://www.vqeg.org/.

[Wat87]   A. B. Watson. The cortex transform: Rapid computation of simulated neural images. *Computer Vision, Graphics, and Image Processing*, 39(3):311–327, 1987.

[Wat95]   A. B. Watson. Image data compression having minimum perceptual error. U.S. Patent 5,426,512, 1995.

[Wat97]   A. B. Watson. Image data compression having minimum perceptual error. U.S. Patent 5,629,780, 1997.

[Wat98]   A. B. Watson. Toward a perceptual video quality metric. In *Proc. SPIE*, 3299:139–147, San Jose, CA, Jan. 26–29, 1998.

[WBE00]   Z. Wang, A. C. Bovik, and B. L. Evans. Blind measurement of blocking artifacts in images. In *Proc. ICIP*, 3:981–984, Vancouver, Canada, Sep. 10–13, 2000.

[WBT97]   A. B. Watson, R. Borthwick, and M. Taylor. Image quality and entropy masking. In *Proc. SPIE*, 3016:2–12, San Jose, CA, Feb. 8–14, 1997.

[WC03]   S. Winkler and R. Campos. Video quality evaluation for Internet streaming applications. In *Proc. SPIE*, 5007:104–115, Santa Clara, CA, Jan. 21–24, 2003.

[WD03]   S. Winkler and F. Dufaux. Video quality evaluation for mobile applications. In *Proc. SPIE*, 5150:593–603, Lugano, Switzerland, July 8–11, 2003.

[WF05]   S. Winkler and C. Faller. Audiovisual quality evaluation of low-bitrate video. In *Proc. SPIE*, 5666:139–148, San Jose, CA, Jan. 16–20, 2005.

[WHM01]   A. B. Watson, J. Hu, and J. F. McGowan III. DVQ: A digital video quality metric based on human vision. *J. Electronic Imaging*, 10(1):20–29, Jan. 2001.

[Win98]   S. Winkler. A perceptual distortion metric for digital color images. In *Proc. ICIP*, 3:399–403, Chicago, IL, Oct. 4–7, 1998.

[Win99a]   S. Winkler. Issues in vision modeling for perceptual video quality assessment. *Signal Processing*, 78(2):231–252, Oct. 1999.

[Win99b]   S. Winkler. A perceptual distortion metric for digital color video. In *Proc. SPIE*, 3644:175–184, San Jose, CA, Jan. 23–29, 1999.

[Win00]   S. Winkler. Quality metric design: A closer look. In *Proc. SPIE*, 3959:37–44, San Jose, CA, Jan. 23–28, 2000.

[Win01]   S. Winkler. Visual fidelity and perceived quality: Towards comprehensive metrics. In *Proc. SPIE*, 4299:114–125, San Jose, CA, Jan. 21–26, 2001.

[WLB04]   Z. Wang, L. Lu, and A. C. Bovik. Video quality assessment based on structural distortion measurement. *Signal Processing: Image Communication*, 19(2):121–132, Feb. 2004.

[WP99]   S. Wolf and M. H. Pinson. Spatial-temporal distortion metrics for in-service quality monitoring of any digital video system. In *Proc. SPIE*, 3845:266–277, Boston, MA, Sep. 19–22, 1999.

[WR89]   J. H. D. M. Westerink and J. A. J. Roufs. Subjective image quality as a function of viewing distance, resolution, and picture size. *SMPTE J.*, 98(2):113–119, Feb. 1989.

[WS97]   A. B. Watson and J. A. Solomon. Model of visual contrast gain control and pattern masking. *J. Opt. Soc. Am. A*, 14(9):2379–2391, Sep. 1997.

[WSM01]   S. Winkler, A. Sharma, and D. McNally. Perceptual video quality and blockiness metrics for multimedia streaming applications. In *Proc. International Symposium on Wireless Personal Multimedia Communications*, 547–552, Aalborg, Denmark, Sep. 9–12, 2001. Invited paper.

[WV99]   S. Winkler and P. Vandergheynst. Computing isotropic local contrast from oriented pyramid decompositions. In *Proc. ICIP*, 4:420–424, Kobe, Japan, Oct. 25–28, 1999.

[WY97]   H. R. Wu and M. Yuen. A generalized block-edge impairment metric for video coding. *IEEE Signal Processing Letters*, 4(11):317–320, Nov. 1997.

[YBdR99]   S. N. Yendrikhovskij, F. J. J. Blommaert, and H. de Ridder. Towards perceptually optimal colour reproduction of natural scenes. In L. W. MacDonald and M. R. Luo, Eds., *Colour Imaging: Vision and Technology*, 363–382. New York: John Wiley & Sons, 1999.

[YM94]   J. Yang and W. Makous. Spatiotemporal separability in contrast sensitivity. *Vision Res.*, 34(19):2569–2576, 1994.

[ZW96]   X. Zhang and B. A. Wandell. A spatial extension of CIELAB to predict the discriminability of colored patterns. In *SID Symposium Digest*, 27:731–735, 1996.

Chapter 6

Philosophy of Picture Quality Scale

Makoto Miyahara[†] and Ryoichi Kawada[‡]

†   *Japan Advanced Institute of Science and Technology, Japan*
‡   *KDDI R&D Laboratories Inc., Japan*

This chapter discusses the philosophy behind the design of a digital picture quality measure known as PQS (Picture Quality Scale). First, we describe this objective measure of image quality degradation caused by coding errors in Section 1. Although the picture format which we consider here is mainly that defined by the ITU Rec. BT.601-5 [ITU00a], we surmise that the PQS is applicable to a wide range of digital picture formats. The application of the PQS and overviews of various types of electronic images, problems and recent trends are discussed in Sections 2, 3 and 4.

6.1   Objective Picture Quality Scale for Image Coding

6.1.1   PQS and Evaluation of Displayed Image

The PQS, i.e., the Picture Quality Scale, is defined as an objective evaluation measure for quality of a coded image as a function of the difference between a coded signal and its original, and represents a numerical degree of absolute quality (by a 5-grade scale) or a numerical degree of deterioration or improvement (by a 7-grade scale or a 5-grade scale with a reference). In this section, an abbreviated description of the PQS is presented based on the work by Miyahara, Kotani and Algazi [MKA98].

6.1.2   Introduction

The evaluation of picture quality is indispensable in image coding. Subjective assessment tests are widely used to evaluate the picture quality of coded images [LA65, ITU00b, Pro95]. However, careful subjective assessments of quality are experimentally difficult and lengthy, and the results obtained may vary depending on the test conditions. Furthermore, subjective assessments provide no constructive methods for coding performance improvement, and are difficult to use as a part of the design process.

Objective measures of picture quality will not only alleviate the difficulties described above, but also help to expand the field of image coding. This expansion results from both the systematic determination of objective measures for the comparison of coded images and the possibility of successive adjustments to improve or optimize the picture quality for a desired quality of service [JJS93]. The objective simulation of performance with respect to both bit-rate and image quality will also lead to a more systematic design of image coders.

It is important that an objective scale mirror the perceived image quality. For instance, simple distortion scales (i.e., measures), such as the signal to noise ratio (PSNR), or even the weighted mean square error (WMSE), are good distortion indicators for random errors, but not for structured or correlated errors. However, such structured errors are prevalent in images compressed by image coders, and degrade local features and perceived quality much more than random errors [Fuj73]. Hence, the PSNR and the WMSE alone are not suitable objective scales to evaluate compressed images. There have been many studies of the construction of objective scales that represent properties of the human visual system [Bed79, LB82, Lim79, Sak77]. Other models and applications of perception to coding have been reported [O$^+$87, C. 65, K$^+$85]. An additional discussion of more recent work on perceptual models and their application to coding performance evaluation is given later in the chapter.

In this section, we describe a new methodology for the determination of objective quality metrics, and apply it to obtain a PQS for the evaluation of coded achromatic still images [MKA98]. The properties of human visual perception suggest the transformation of images and coding errors into perceptually relevant signals. First, we transform the image signal into one which is proportional to the visual perception of luminance using Weber-Fechner's Law and the contrast sensitivity for achromatic images. Second, we apply spatial frequency weighting to the errors. Third, we describe perceived image disturbances and the corresponding objective quality factors that quantify each image degradation. Fourth, we describe the experimental method for obtaining the PQS based on these distortion factors, and assess how accurate the approximation is between the obtained PQS and the Mean Opinion Score (MOS). In the discussion section, limitations, extensions and refinements of the PQS methodology are examined. Finally, applications of PQS are briefly discussed [AFMN93, LAE96, LAE95].

6.1.3   Construction of a Picture Quality Scale

The PQS methodology is illustrated in Figure 6.1. Given the original image $x_o[m, n]$ and a distorted or reconstruction of the compressed image $x_r[m, n]$, we compute local distortion maps $\{f_i[m, n]\}$ from which the distortion factors $\{F_i\}$ are computed. We then use regression methods to combine these factors into a single number representative of

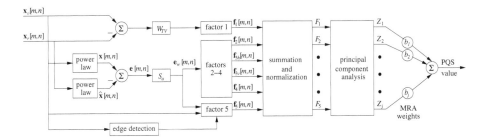

Figure 6.1: The construction of PQS.

the quality of a given image. The methodology that is presented here applies to a CRT or other types of electronic displays, as well as to print. In each case, we assume that the nonlinear characteristic of the image display system has been compensated, so that the image signal $\mathbf{x}_o[m, n]$ is equal to the luminance of the display at each pixel.

In what follows, we discuss in detail the various components of this framework, starting with the computation of contrast adjusted error images, and concluding with explicit formulae for the computation of the PQS factors and then the PQS itself.

6.1.3.1 Luminance coding error

We now mathematically account for several components of a simple model of visual perception. To provide a more uniform perceptual scale, we transform the images using a power law

$$\mathbf{x}[m, n] = k \cdot \mathbf{x}_o[m, n]^{1/2.2} \tag{6.1}$$

that approximates Weber-Fechner's Law for brightness sensitivity, where k is a scaling constant that allows for the adjustment of the range of the variable $\mathbf{x}$. Note that the exponent $1/2.2$ is only one of several approximate values commonly used [Jai89].

The contrast adjusted error image is then computed as

$$\mathbf{e}[m, n] = \mathbf{x}[m, n] - \hat{\mathbf{x}}[m, n] \tag{6.2}$$

where $\hat{\mathbf{x}}[m, n]$ is the contrast adjusted version of $\mathbf{x}_r[m, n]$. Most of the distortion factors are defined as functions of $\mathbf{e}[m, n]$.

6.1.3.2 Spatial frequency weighting of errors

The contrast sensitivity function of vision suggests a spatial frequency distortion weighting. Based in part on a measured contrast sensitivity function, the spatial frequency response is modeled approximately by

$$S(\omega) = 1.5e^{-\sigma^2\omega^2/2} - e^{-2\sigma^2\omega^2} \tag{6.3}$$

where

$$\sigma = 2, \ \omega = \frac{2\pi f}{60}, \ f = \sqrt{u^2 + v^2}, \tag{6.4}$$

u and v are the horizontal and vertical spatial frequencies, respectively, in cycles per degree. As compared to the contrast sensitivity function (CSF), modeled for instance in [Sak77], the response of (6.3) gives more weight to low frequencies, which is critical to the reproduction of edges [YH68]. We have found that high frequencies also need to be attenuated as compared with the measured CSF of human vision, to account for the transfer function of the CRT.

At higher spatial frequencies, the frequency response is anisotropic [CKL66, PW84] so that a better model [HM87] is given by

$$S_a(\omega) = S(\omega)O(\omega, \theta) \tag{6.5}$$

with

$$O(\omega, \theta) = \frac{1 + e^{\beta(\omega - \omega_o)}\cos^4 2\theta}{1 + e^{\beta(\omega - \omega_o)}} \tag{6.6}$$

where ω is defined in (6.4), $\theta = \tan^{-1}(v/u)$ is the angle with respect to the horizontal axis, $\omega_o = 2\pi f_o$,

$$\beta = 8, \ \text{and} \ f_o = 11.13 \ \text{cycles/degree}. \tag{6.7}$$

$O(\omega, \theta)$ blends in a $\cos^4 2\theta$ anisotropy, fairly quickly, for frequencies $f > f_o$. The frequency weighted error $e_w[m, n]$, then, is just the contrast adjusted error, filtered with $S_a(\omega)$.

In our presentation, perceived disturbances are first described verbally, for example, as random or structured errors. Next, these disturbances are quantified locally, resulting in distortion maps or factor images $\{f_i[m, n]\}$. A single numerical value, or distortion factor F_i, is then computed from each factor image. We now describe the perceived disturbances and define numerical measures of the corresponding distortion factors $\{F_i\}$. We make use of five such factors and then analyze their relative importance.

6.1.3.3 Random errors and disturbances

All coding techniques will produce random errors. The perceived random disturbances will be in the form of incremental noise in slowly varying regions of the image and, generally, will not be perceived in active areas of the image. We use integral square measures to compute distortion factors F_1 and F_2.

Because the CCIR has adopted a standard to quantify the effect of noise, we follow that standard in defining F_1 [CCI82].

Distortion Factor F_1: The CCIR television noise weighting standard does not take into account Weber's law. Thus, we define a direct reconstruction error image as

$$\mathbf{e}_i[m,n] = \mathbf{x}_o[m,n] - \mathbf{x}_r[m,n] \tag{6.8}$$

where $\mathbf{x}_o[m,n]$ and $\mathbf{x}_r[m,n]$ represent the original and the reconstructed images, respectively. This error is used to compute the first factor image,

$$\mathbf{f}_1[m,n] = (\mathbf{e}_i[m,n] * \mathbf{w}_{TV}[m,n])^2 \tag{6.9}$$

where $\mathbf{w}_{TV}[m,n]$ is the isotropic spatial domain weighting corresponding to the frequency weighting defined by the CCIR 567-1,

$$W_{TV}(f) = \frac{1}{1 + (f/f_c)^2}, \tag{6.10}$$

where $f = \sqrt{u^2 + v^2}$ with a 3 dB cutoff frequency $f_c = 5.56$ cycles/degree at a viewing distance of 4 times picture height ($4H$), and $*$ represents the convolution.

The distortion factor F_1 is then computed as the summation of the square of the local frequency weighted errors,

$$F_1 = \frac{\sum_{m,n} \mathbf{f}_1[m,n]}{\sum_{m,n} \mathbf{x}_o^2[m,n]} \tag{6.11}$$

where the sums are computed over all pixels in the $M \times N$ images. Note that F_1, as defined, is normalized by the weighted energy of the image. It is therefore a normalized noise to signal ratio.

Distortion Factor F_2:   Distortion factor F_2 includes a more complete single channel model of visual perception. A correction for Weber's Law (6.1) and the frequency weighting factor of (6.5) are now used. In addition, F_2 ignores values of $e_w[m, n]$ which are below a perceptual threshold T. Thus,

$$f_2[m, n] = I_T[m, n](e[m, n] * s_a[m, n])^2 \qquad (6.12)$$

and

$$F_2 = \frac{\sum_{m,n} f_2[m, n]}{\sum_{m,n} x_r^2[m, n])} \qquad (6.13)$$

where $s_a[m, n]$ is the spatial domain counterpart of the the the frequency response $S_a(\omega)$ as defined in (6.5), $I_T(m, n)$ is an indicator function for a perceptual threshold of visibility, and $T = 1$.

6.1.3.4   Structured and localized errors and disturbances

Because the perception of structured patterns is more acute, and structured and correlated errors are prevalent in coded images, we now define three additional factors to evaluate the contribution of correlated errors.

Distortion Factor F_3 (End of Block Disturbances):   We are specially sensitive to linear features in images and, therefore, to such features in errors as well. Such structured disturbances are quite apparent in most coders. These disturbances occur in particular at the end of blocks for transform coders and are due to error discontinuities.

We define distortion factor F_3 as a function of two factor images, which represent the horizontal and the vertical block error discontinuities, respectively. Thus,

$$f_{3h}[m, n] = I_h[m, n]\Delta_h^2[m, n] \qquad (6.14)$$

where

$$\Delta_h[m, n] = e_w[m, n] - e_w[m, n + 1] \qquad (6.15)$$

and $I_h[m, n]$ is an indicator function whose elements are set to 1 to select only those differences which span horizontal block boundaries, otherwise set to 0. Now,

$$F_{3h} = \frac{1}{N_h} \sum_{m,n} f_{3h}[m, n] \qquad (6.16)$$

where $N_h = \sum_{m,n} \mathbf{I}_h[m,n]$ is the number of pixels selected by the corresponding indicator function. The distortion factor F_3 is defined as

$$F_3 = \sqrt{F_{3h}^2 + F_{3v}^2} \qquad (6.17)$$

where F_{3v} is defined to account for block error discontinuities at vertical block boundaries in a similar way to F_{3h}. Note that more elaborate models for end of block errors have also been proposed [KK95].

Distortion Factor F_4 (Correlated Errors): Even if they do not occur at the end of a block, image features and textures with strong spatial correlation are much more perceptible than random noise. In order to evaluate structured errors, we make use of their local spatial correlation. The distortion factor F_4 is thus defined as a summation over the entire image of local error correlations.

We compute locally the factor image

$$\mathbf{f}_4[m,n] = \sum_{(k,l)\in W;\; l=k\neq 0} |r(m,n,k,l)|^{0.25} \qquad (6.18)$$

where W is the set of lags to include in the computation, and the local correlation

$$r(m,n,k,l) = \frac{1}{\mathcal{W}-1}\left[\sum_{i,j}\mathbf{e}_w[i,j]\mathbf{e}_w[i+k,j+l] - \frac{1}{\mathcal{W}}\sum_{i,j}\mathbf{e}_w[i,j]\sum_{i,j}\mathbf{e}_w[i+k,j+l]\right]$$

$$(6.19)$$

and $\mathcal{W}$ is the number of pixels in the window, the sums are computed over the set of pixels where (i,j) and $(i+k,j+l)$ both lie in a 5×5 window centered at (m,n). We include all unique lags with $|k|,|l| \le 2$, except for $(0,0)$, which is the error variance. Due to symmetry, only 12 lags are included in the sum. Note that the 0.25 exponent is used to de-emphasize the relative magnitude of the errors, as compared to their correlation or structure. By summation of the local distortions we obtain the distortion factor

$$F_4 = \frac{1}{MN}\sum_{m,n}\mathbf{f}_4[m,n] \qquad (6.20)$$

Distortion Factor F_5 (Errors in the vicinity of high contrast image transitions): Two psychophysical effects affect the perception of errors in the vicinity of high contrast transitions: visual masking, which refers to the reduced visibility of disturbances in active areas, and enhanced visibility of misalignments, even when they are quite small. Here, we account only for the masking effect. The second effect, denoted edge busyness is prevalent for very low bit-rate DPCM coders but is not a major effect for other commonly used coding techniques, such as those considered in this chapter.

Although visual masking reduces the visibility of impairments in the vicinity of transitions, coding techniques will introduce large errors and major visual disturbances in the same areas. Thus, even though these disturbances are masked, they will still be very important. Distortion factor F_5 measures all disturbances in the vicinity of high contrast transitions. In contrast to the other factors, F_5 is based on an analysis of the original image, as well as on the contrast enhanced error image $\mathbf{e}_w[m, n]$.

A horizontal masking factor [Lim79]

$$\mathbf{S}_h[m, n] = e^{\{-0.04\mathbf{V}_h[m,n]\}} \tag{6.21}$$

is defined in terms of a horizontal local activity function

$$\mathbf{V}_h[m, n] = \frac{|\mathbf{x}_o[m, n-1] - \mathbf{x}_o[m, n+1]|}{2} \tag{6.22}$$

Defining the vertical masking factor $\mathbf{S}_v[m, n]$ similarly, we compute the masked error at each pixel as

$$\mathbf{f}_5[m, n] = \mathbf{I}_M[m, n] \cdot |\mathbf{e}_w[m, n]| \cdot (\mathbf{S}_h[m, n] + \mathbf{S}_v[m, n]) \tag{6.23}$$

where $\mathbf{I}_M[m, n]$ is an indicator function which selects pixels *close* to high intensity transitions. Note that the masking factors can be substantially less than 1 in highly active regions of the image.

The final factor is now computed as

$$F_5 = \frac{1}{N_K} \sum_{m,n} \mathbf{f}_5[m, n] \tag{6.24}$$

where N_K is the number of pixels whose 3×3 Kirsch edge response $k(m, n) \geq K$, for a threshold $K = 400$. The indicator function $\mathbf{I}_M[m, n]$ of (6.23) selects the set of all pixels within $l = 4$ pixels of those pixels detected by the Kirsch operator. F_5 thus measures, with a suitable weight to account for visual masking [Lim79], these large localized errors.

6.1.3.5   Principal component analysis

The distortion factors were defined so as to quantify specific types of impairments. Clearly, some of the local image impairments will contribute to several or all factors, and the factors $\{F_1, \ldots, F_5\}$ will be correlated.

A principal component analysis is carried out to quantify this correlation between distortion factors. We compute the covariance matrix

$$\mathbf{C_F} = COV(\mathbf{F}) = E\{(\mathbf{F} - \bar{\mu}_{\mathbf{F}})(\mathbf{F} - \bar{\mu}_{\mathbf{F}})^T\} \qquad (6.25)$$

where $\mathbf{F} = [F_1, F_2, F_3, F_4, F_5]^T$ is the vector of distortion factors and the vector $\bar{\mu}_{\mathbf{F}}$ is its mean.

The matrix of eigenvectors will diagonalize the covariance matrix $\mathbf{C_F}$. The eigenvalues $\{\lambda_j, j = 1, 2, \ldots, 5\}$ also indicate the relative contributions of the transformed vectors, or principal components, to the total energy of the vector $\mathbf{F}$. The "eigenfactors" are now uncorrelated and, as we shall show, are more effective and robust in the objective assessment of image quality.

6.1.3.6   Computation of PQS

We compute the PQS quality metric as a linear combination of principal components Z_j as

$$PQS = b_0 + \sum_{j=1}^{J} b_j Z_j \qquad (6.26)$$

where the b_j are the partial regression coefficients which are computed using Multiple Regression Analysis (MRA) [Ken75] of the PQS given by (6.26) with the mean opinion scores obtained experimentally in quality assessment tests.

In order to illustrate visually the five distortion factor images, they have been evaluated for the image *Lena* and a low quality (quality factor $= 15$) JPEG encoded version of it using a software codec [Ind]. To illustrate the relative spatial contributions of $\mathbf{f}_i[m, n]$ to F_i, and to differentiate the factors from one another, we show in Figure 6.2 the original image and each of the $\mathbf{f}_i[m, n]$ suitably magnified. We observe that, as compared to $\mathbf{f}_1[m, n]$, $\mathbf{f}_2[m, n]$ discards a number of small errors which occur in the flat portions of the image. $\mathbf{f}_3[m, n]$, restricted to block boundaries, is quite high in active portions of the image and also near high contrast intensity transitions, where it is most visible. The structured error $\mathbf{f}_4[m, n]$, compared to $\mathbf{f}_5[m, n]$, shows that structured errors are very common, and that they do not consistently coincide with image regions where visual masking occurs.

6.1.4   Visual Assessment Tests

We now turn to the experimental determination of the subjective mean opinion score for each of the encoded images.

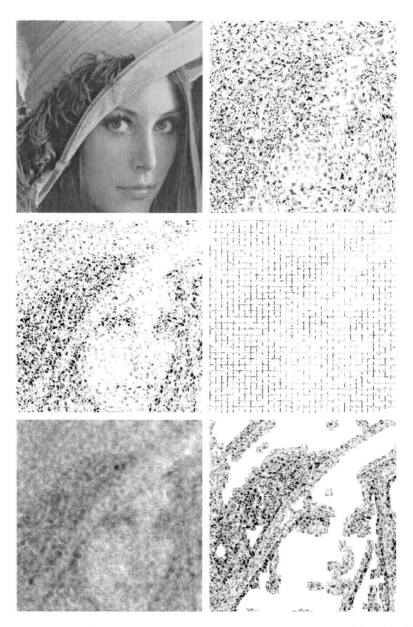

Figure 6.2: PQS factor images. Left to right, then top to bottom: original, $\mathbf{f}_1[m, n]$, $\mathbf{f}_2[m, n]$, $\sqrt{\mathbf{f}_{3h}^2[m, n] + \mathbf{f}_{3v}^2[m, n]}$, $\mathbf{f}_4[m, n]$, and $\mathbf{f}_5[m, n]$.

Table 6.1: MOS grading scale

Scale	Impairment
5	Imperceptible
4	Perceptible, but not annoying
3	Slightly annoying
2	Annoying
1	Very annoying

6.1.4.1 Methods

The visual or subjective evaluation of image quality has drawn the attention of a number of researchers for many years, principally in relation to the evaluation of new transmission or coding schemes, and in the development of advanced television standards. As it applies to television, an excellent presentation of the complex issues involved has been given by Allnatt [All83]. The standardization committees of the ISO and, in particular, the ITU-R (formerly the CCIR), have published recommendations on methods for the assessment of picture quality in television. In our work, we follow closely the ITU-R 500 recommendations with respect to subjective scales and experimental conditions [ITU00b]. Note that we are making use of the numerical scores associated with the impairment descriptors, or categories, of Table 6.1, for the computation of average MOS scores and regression analysis. This requires that the subjective numerical scale also provide equal perceptual intervals. This property holds for the ITU-R MOS impairment scale of Table 6.1 [JM86]. The general issue of subjective numerical category scaling as applied to image coding is reviewed in [vDMW95].

Table 6.2: Conditions of the subjective assessment tests

Ratio of viewing distance to picture height	4
Room illumination	None
Peak luminance on the screen	42.5 (cd/m^2)
Lowest luminance on the screen	0.23 (cd/m^2)
Time of observation	unlimited
Number of observers	9 (expert observers)

In Table 6.1, we show the 5-point (MOS) impairment scale, and in Table 6.2, the conditions for subjective assessment tests recommended in [ITU00b]. The specific conditions used were as follows:

Figure 6.3: PQS test images. From left to right, then top to bottom: Church, Hairband, Weather, Barbara, and Cameraman.

1. The pictures used were all 256×256 pixels and were viewed at 4 times the picture height ($4H$).

2. The selected observers, principally graduate students at JAIST, the Japan Advanced Institute of Science and Technology, received limited training. The selection of subjects was based on consistency in the evaluation of picture quality. The training consisted of the description and illustration of the types of distortions that the subjects would observe.

3. The subjects were instructed to grade the image quality in 1/2 step increments.

A number of coded images were evaluated informally at various times during the study, with more than 800 coded image evaluations performed, in the preliminary and final assessment tests. The results reported are based on the 675 quality evaluations of 75 encoded images by the pool of nine subjects.

6.1.4.2   Test pictures

The image impairments represented by the distortion factors F_1 and F_2 are mainly global distortions, while F_3, F_4, and F_5 characterize local distortions. Local distortions are apparent only in portions of the images and depend on the density of transitions and flat regions in the images. Five test images were used that represent a range of characteristics. These images, shown in Figure 6.3, include the ITE (Institute of Television Engineers of Japan) test images "Church," "Hairband" and "Weather" [ITE87], and the widely used "Barbara" and "Cameraman" images.

6.1.4.3 Coders

There are a large variety of coders, such as the DPCM, Orthogonal Transform Coders (OTC), the VQ, subband coders, etc. We have concentrated on the block DCT based JPEG, as well as widely used wavelet and subband coding techniques. The subband coder used in our experiments is a 28 band decomposition using Johnston's 16 tap filters [Joh80] and the wavelet coder, a 10 band decomposition with 8 tap Daubechies' filters [Dau88].

6.1.4.4 Determination of MOS

The observers are asked to assign a score $A(i, k)$ to each encoded image, where $A(i, k)$ is the score given by the ith observer to image k. Each score in the range of 1 to 5, according to the impairment scale of Table 6.1, is assigned in 1/2 step increments. For each encoded image, the scores are averaged to obtain the MOS value for a specific image,

$$MOS(k) = \frac{1}{n} \sum_{i=1}^{n} A(i, k) \tag{6.27}$$

where n denotes the number of observers. Note that the possibility of assigning a half step has no implication on the accuracy of the subjective data. We found that observers like to have the option of half step scoring when they were uncertain about a full step. Repetition of the experiment by the same observer may result in a score that may differ by more than a half step from the previous one.

6.1.5 Results of Experiments

The experiments resulted in a set of five images coded with one of three types of coders and for the entire range of quality. A total of seventy five encoded images were assessed by nine observers as described, and the average MOS score was computed for each encoded image. The set of error images were then analyzed.

6.1.5.1 Results of principal component analysis

The set of error images was first used to compute the distortion factors. From this set of distortion factors, we compute the covariance matrix $\mathbf{C_F}$ of Table 6.3. A principal component analysis of $\mathbf{C_F}$ is then carried out. In Table 6.4, we show the eigenvalues and the eigenvectors of matrix $\mathbf{C_F}$.

Table 6.3: Covariance matrix

	F_1	F_2	F_3	F_4	F_5
F_1	1.0000	0.9967	0.9753	0.8391	0.5823
F_2	0.9967	1.0000	0.9743	0.8231	0.5682
F_3	0.9753	0.9743	1.0000	0.8749	0.5696
F_4	0.8391	0.8231	0.8749	1.0000	0.6714
F_5	0.5823	0.5682	0.5696	0.6714	1.0000

Table 6.4: Eigenvalues and corresponding eigenvectors

λ_1	λ_2	λ_3	λ_4	λ_5
4.19165	0.59144	0.19021	0.02392	0.00278

ℓ_1	ℓ_2	ℓ_3	ℓ_4	ℓ_5
0.47500	-0.24449	0.27083	-0.38709	0.70100
0.47198	-0.26829	0.32384	-0.31044	-0.70993
0.47526	-0.24723	0.01921	0.84271	0.04966
0.45058	0.10598	-0.86219	-0.20065	-0.04604
0.35030	0.89213	0.27936	0.05793	-0.00216

Note the very high correlation between F_1 and F_2. This is expected, since these factors both evaluate random errors, with some changes in their weighting. The high correlation of F_3 with F_1 and F_2 is more surprising. Although the spatial contributions to these factors are distinct, the high correlation indicates that, when aggregated into a single number, they do track each other as the coding parameters change.

The resulting eigenvalues have a wide spread of values, and the largest 3 eigenvalues amount to 99.5% of the total energy. The transform from F_i to Z_i, for $i = 1, 2, \ldots, 5$, is given by (6.28):

$$
\begin{bmatrix} Z_1 \\ Z_2 \\ Z_3 \\ Z_4 \\ Z_5 \end{bmatrix} = \begin{bmatrix} 0.47500 & 0.47198 & 0.47526 & 0.45058 & 0.35030 \\ -0.24449 & -0.26829 & -0.24723 & 0.10598 & 0.89213 \\ 0.27083 & 0.32384 & 0.01921 & -0.86219 & 0.27936 \\ -0.38709 & -0.31044 & 0.84271 & -0.20065 & 0.05793 \\ 0.70100 & -0.70993 & 0.04966 & -0.04604 & -0.00216 \end{bmatrix} \begin{bmatrix} F_1 \\ F_2 \\ F_3 \\ F_4 \\ F_5 \end{bmatrix}
$$
(6.28)

The space spanned by the 5 distortion factors is essentially three-dimensional. The eigenvectors ℓ_1, ℓ_2, ℓ_3 provide a useful first transformation of the F_i into an effective principal component representation (Z_1, Z_2, Z_3). To obtain a numerical distortion value, we carry out a multiple regression analysis between the principal component vector and the measured MOS values.

6.1.5.2 Multiple regression analysis

The partial regression coefficients b_0 and $\{b_j\}$ for the three principal components, Z_1 to Z_3, have been evaluated, so that PQS for any coded image is given by

$$PQS = 5.632 - 0.068Z_1 - 1.536Z_2 - 0.0704Z_3 \tag{6.29}$$

Note that the PQS, that is derived using principal components, can be also be expressed in terms of the distortion factors, F_i.

$$\begin{aligned} PQS &= 5.797 + 0.035F_1 + 0.044F_2 + 0.01F_3 \\ &\quad -0.132F_4 - 0.135F_5 \end{aligned} \tag{6.30}$$

It is important to use the principal components in the multiple regression analysis and the determination of the PQS. Since the covariance matrix is nearly singular, the results obtained by multiple regression with the F_i or with the entire set of Z_i are unstable. Although a slightly better fit of the specific test data is then obtained, the regression coefficients are not robust, and the results may not be usable for images outside the test set [CP77].

6.1.5.3 Evaluation of PQS

A fairly good agreement between the PQS and the MOS is achieved, as shown in the scatter diagram of Figure 6.4. Note that the goodness of fit is better in the middle of the quality range than at its extremes. This observation is elaborated in Subsection 6.1.6. In order to describe the degree of approximation of the PQS to the MOS quantitatively, the correlation coefficient ρ [Ken75] between PQS and MOS is evaluated. The correlation coefficient $\rho = 0.928$, which is a great improvement when compared to the correlation of $\rho = 0.57$ of the conventional WMSE scale which is calculated using F_1 alone. We also analyzed the errors in the PQS values about the regression line. We find that the absolute error is within 0.5 with a 70% probability. These results are quite reliable and consistent and have been verified by us and other researchers in several additional studies. It has been used in the comparative evaluation of wavelet coders for alternative choices of wavelet basis and of quantizers [LAE96]. The PQS methodology was also used at the high end of the quality range in comparing wavelet and JPEG coders [AV98]. We have found that PQS is very relevant and useful for such parametric studies. In particular, it captures differences in image quality between JPEG and wavelet coders

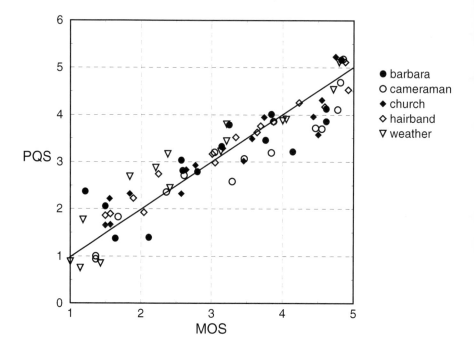

Figure 6.4: PQS versus MOS scatter diagram.

that are missed by the PSNR metric [AV98]. Some of the limitations of the PQS are discussed in Subsection 6.1.6.

6.1.5.4   Generality and robustness of PQS

We have already discussed the importance of using principal component analysis in formulation of the PQS. Other issues are related to generality and robustness of the PQS in its use for image data outside the test set, which are critical to any application. In order to assess this, the complete PQS evaluation was conducted on each set of four images, and the resulting formulas were then used for the fifth image. We find that the results with respect to regression coefficients and correlation with the MOS are very close to the values reported. We also evaluated the effect of the encoding technique on the results. We find that the PQS evaluation is only slightly dependent on the specific encoding technique.

6.1.6 Key Distortion Factors

We have defined the distortion factors F_i as measures of perceived disturbances which are common and basic to coding techniques. We now consider the interpretation of the combinations of F_is into principal components, and whether each F_i, as defined, is a key distortion factor so as to further remove redundancies or to rank the importance of the factors in the set $\{F_i\}$.

6.1.6.1 Characteristics of the principal components

The eigenvectors $\{\ell_k\}$ which were used to compute $\{Z_j\}$ are given in Table 6.4. We also found that the characterization of the overall distortion is concentrated into three principal components. Let us consider the characteristics of ℓ_1, ℓ_2 and ℓ_3.

- Z_1 may be reasonably named "the amount of error," since the entries in the first eigenvector, ℓ_{i1} for $i = 1, 2, \ldots, 5$, in Table 6.4 are almost equal to each other.
- Z_2 may be named "the location of error," because ℓ_{52} is very large (ℓ_{52}=0.89 as shown in Table 6.4), which measures the contribution of errors in the vicinity of image edges or transitions.
- Z_3 may be named "the structure of error," because of the weight given to the factor F_4 in the third eigenvector.

Therefore, it can be said that the PQS is a linear combination of three distortion factors: "the amount of error," "the location of error" and "the structure of error." The last two factors again emphasize the importance of non-random errors in the quality evaluation of coded images.

6.1.6.2 Contribution of the distortion factors

We now evaluate the importance of the distortion factors, taken singly and in combination. For all sets of combinations of F_i, the correlation coefficients ρ and ρ^* are shown in Table 6.5, where ρ^* is the modified ρ [Ken75] adjusted for the degrees of freedom. Specifically,

$$\rho^* = \sqrt{\frac{\rho^2(n-1) - p}{n - p - 1}} \tag{6.31}$$

where n is the total number of coded images in the test set, and p is the number of factors F_i retained. Note that, as done in the previous section, a principal component analysis was performed on the covariance matrix of the retained coefficients. The largest eigenvalues and corresponding eigenvectors were used until their cumulative values exceeded 99%. We find that the four factors F_1, F_2, F_3 and F_4 only span a two-dimensional space because of the high correlations among the first three factors.

We also show in Table 6.5, the average absolute error difference between the PQS and the MOS scores assigned to the encoded images. Table 6.5 can be examined in several ways. First, we rank the importance of the factors taken in groups. The most important single factor is F_5. For two factors, we would use F_4 and F_5; for three F_2, F_4 and F_5; for four F_1, F_2, F_4 and F_5.

The importance of the factor F_5 by itself stands out in the examination of Table 6.5, and suggests that it could be used as a quality metric by itself. Note, however, that a study of the relationship between the MOS and the PQS, when F_5 alone is used as a metric, indicates that such a metric can be used to assess pictures of high quality, whilst its performance is much worse than the PQS, given in (6.30), for evaluating pictures of lower quality.

6.1.6.3   Other distortion factors

One such distortion factor would quantify the jagged distortion of smooth edge contours which is introduced by some image coding techniques. It is possible to define such a distortion factor by analyzing locally the discrepancies between the direction of the original image edge and that of the corresponding edge in the coded image. We have found that such a factor has a slight effect on one of the images in our test set, principally at high quality, but does not affect the overall results presented. We also observed that a Moiré type of disturbance occurs in the scarf area of the test image "Barbara" at specific quality levels. This effect occurs in such isolated instances by which the overall approaches and results are not affected. However, this additional study has pointed out some of the limitations of our results, and will be elaborated next.

6.1.7   Discussion

We consider, first, some of the limitations and promises of the methodology presented here for applications of recent interest. Second, we discuss the components of an overall strategy for systematic advances in perceptually based image coding.

6.1.7.1   Limitations in applications

The set of distortion factors that we have defined spans a three-dimensional space. Thus, image impairments are characterized by a three-dimensional vector. Multiple regression analysis, a statistical regression technique, has allowed the reduction of this vector to a single number, the PQS, which has a good correlation with the MOS.

The performance of the PQS can be further improved by using a piecewise linear model or by two separate measures, one for low to medium quality images and the other for medium to high quality. We have performed a complete analysis for the lower quality

Table 6.5: Correlation coefficient between PQS and MOS

Factors					ρ	ρ^*	$E\{|error|\}$
1					0.5665	0.5582	0.8347
	2				0.5373	0.5282	0.8510
		3			0.5714	0.5633	0.8269
			4		0.7426	0.7384	0.6551
				5	0.8985	0.8971*	0.4268
1	2				0.5547	0.5461	0.8418
1		3			0.5729	0.5564	0.8289
1			4		0.7499	0.7417	0.6362
1				5	0.9001	0.8972	0.4206
	2	3			0.5779	0.5616	0.8222
	2		4		0.7539	0.7459	0.6336
	2			5	0.8991	0.8962	0.4236
		3	4		0.7600	0.7522	0.6302
		3		5	0.9015	0.8986	0.4144
			4	5	0.9180	0.9156*	0.3773
1	2	3			0.5716	0.5550	0.8265
1	2		4		0.7516	0.7435	0.6352
1	2			5	0.8996	0.8967	0.4220
1		3	4		0.7600	0.7522	0.6302
1		3		5	0.9015	0.8986	0.4144
1			4	5	0.9271	0.9239	0.3536
	2	3	4		0.7600	0.7522	0.6302
	2	3		5	0.9015	0.8986	0.4144
	2		4	5	0.9302	0.9271*	0.3460
		3	4	5	0.9279	0.9247	0.3551
1	2	3	4		0.7600	0.7522	0.6302
1	2	3		5	0.9015	0.8986	0.4144
1	2		4	5	0.9285	0.9253*	0.3501
1		3	4	5	0.9279	0.9247	0.3551
	2	3	4	5	0.9279	0.9247	0.3551
1	2	3	4	5	0.9279	0.9247	0.3551

range, and find that the contributions of the distortion factors are substantially different from what we reported for a fit over the entire quality range. The predictive value of this limited range PQS is also improved. These observations are related to the choice of subjective quality assessment scales and performance ranges to be discussed next.

6.1.7.2 Visual assessment scales and methods

The visual assessment methods of the ITU-R BT. Rec. 500-10 [ITU00b] were developed for use in entertainment broadcast television, i.e., for pictures of fairly high quality, and targeted at non-expert viewers for images and video with low information content. Image coding applications now have a much wider range. The subjective 5-point impairment scale for images or video of the ITU-R BT. Rec. 500-10 is too broad and not precise enough for many current applications. It is not for coder design.

For instance, for the video conferencing which mainly contains head and shoulder images, acceptable image quality is much lower than for broadcast television and still image coding. Thus all such encoded images and video will cluster at the low end of the impairment scale. The same comment applies to high quality image coding, where the preservation of critical details is important. Thus, alternative quality scales [All83, vDMW95], and a different subjective assessment method, such as using anchor images, are needed [AN93].

Recent work considers each image impairment as a component of a vector, and analyzes this multi-dimensional subjective space to determine the orthogonality and relative weights of such impairments in subjective space [MK96, KM96a, KM96b]. The introduction of adaptive coding techniques, where the bit-rate constraints dynamically modify some of the quantization parameters, such as in the MPEG-2, has also led to the use of a variable and continuous rating of quality [Pro95].

6.1.7.3 Human vision models and image quality metrics

In our work, we have made use of a simple global, or single channel, model of visual perception. Much progress has been made in the development of multi-channel models of visual perception [WLB95, TH94, ZH89] and in the study of the visual masking thresholds for each of the channels [Wat93b].

Such threshold models have been used by Daly [Dal93] in the determination of a Visual Distortion Predictor (VDP) that computes the probability of detection of a visible error. These perceptual models have application to the evaluation of quality in the processing of very high quality images, where errors are small and close to threshold. We have compared the PQS with the VDP for high quality image coding. The distortion images predicted by both methods were quite similar for pictures of high quality [AA96].

The promise of metrics based strictly on perceptual models is that they would apply to all types of image impairments, and not only to image coding. The limitation of such an approach, besides its restricted applicability to impairments at the visual threshold, is in understanding or controlling image display and viewing conditions to fit the models. Note that successful limited use of properties of visual perception for image processing or encoding applications have been reported through the years [AFC95, HA84, Net77, Lim78, G⁺88, W⁺83, Wat93a].

6.1.7.4 Specializing PQS for a specific coding method

We have proposed a methodology for a "general" PQS which will give useful results for different types of coders and images. When the coding method is fixed and the test image data and the coder are fixed, we can tune the picture quality scale to that coder and obtain then more accurate results [Miy88]. Such a specialization is useful for coder design or adaptive coding parameter adjustment [AFMN93].

6.1.7.5 PQS in color picture coding

The PQS methodology has been applied to color picture coding. For color, the importance of a perceptually uniform color space leads us to consider color differences as defined in the Munsell Renotation System which is psychometrically uniform and metric. Then, we can utilize the color differences instead of $e[m, n]$ for a PQS applicable to coded color images [MY88, KGMA95].

6.1.8 Applications of PQS

There are a number of applications for a perceptually relevant distortion metric such as the PQS. Coding techniques can now more confidently be compared by checking their PQS values for the same bit-rate, or bit-rates for the same image quality. For instance, for wavelet coders, alternatives in the choice of wavelets, quantization strategies, and error free coding schemes have been compared [LAE96, LAE95]. This comparison makes possible a systematic choice of parameters, on the basis of a meaningful measure. The same study shows that the SNR does not help in making such choices.

Optimization of quantization parameters in coders based on the PQS has also been examined [AFMN93]. For adaptive coders, some of the factors, $\{F_i\}$, can be estimated locally by a decoder within the encoder. Hence, the objective picture quality metric PQS that measures the degradation of quality in the coded picture can then be reduced by adjustments of the encoder parameters [AFMN93].

6.1.9 Conclusions

We have proposed a methodology for devising a quality metric for image coding and applied it to the development of an objective picture quality scale, i.e., the PQS, for achromatic still images. This PQS metric was developed over the entire range of image quality defined by the impairment scale of the ITU-R BT. Rec. 500. The PQS is defined by taking into account known image impairments due to coding, and by weighting their quantitative perceptual importance. To do so, we use some of the properties of visual perception relevant to global image impairments, such as random errors, and emphasize the perceptual importance of structured and localized errors. The resulting PQS closely approximates the mean opinion score, except at the low end of the image quality range. We have also interpreted the PQS system as composed of a linear combination of three essential factors of distortion: the "amount of error", the "location of error' and the "structure of error". We have also discussed some of the extensions and applications of such an objective picture quality metric.

Systematic studies of the objective evaluation of images, as well as their subjective assessment, are difficult and only now becoming active areas of research [JJS93]. However, the rapid increase in the range and use of electronic imaging and coding and their increasing economic importance justifies renewed attention and specialization of perceptually relevant image quality metrics as a critical missing component for systematic design and for providing the quality of service needed in professional applications.

6.2   Application of PQS to a Variety of Electronic Images

PQS is applicable to any basic discussion of coding errors and their relationship to degradation of image quality. It will be useful for a wide variety of purposes. However, if we consider the overall quality of the five image categories listed below, we will see that, as their intended purposes are different, we must consider not only the PQS but also other factors as well. The five image categories are:

1. Conventional TV images (with spatial pixel resolution of 480×720 at 30 frames/s or 576×720 at 25 frames/s)
2. HDTV (with spatial pixel resolution of 1080×1920 at 30 Frames/s or 720×1280 at 60 Frames/s) [HPN97]
3. Extra high quality images
4. Cellular phone type
5. Personal computer (PC), computer graphics (CG)

Details are explained in this section and in Sections 3, 4 and 5.

It is only reasonable to compare the image quality of conventional TV and the center portion of an HDTV image, because the difference between them is almost entirely the width of the frames. The purposes of the other image categories are quite different, so that, for instance, comparing a cellular phone with the other types will not be reasonable.

Essential problems in the evaluations of displayed images by electronic devices such as CRT, LCD, PDP and FED are considered in detail.

6.2.1 Categories of Image Evaluation

In general, image evaluation techniques (as shown in Table 6.6) fall into one of three categories.

1. Assess faithfulness of the images to the original signal, weighted with visual perception characteristics. It is based on a close comparison of the deterioration of an image compared to the original. It is used to design good coders and to evaluate the performance of an image communication system. The PQS described in Section 6.1 is a representative of this category.

2. Determine overall quality, representing a total impression for an entire program, rather than testing still or short moving scenes as described in the ITU-R BT. Rec. 500-10 [ITU00a].

3. An absolute evaluation (without reference images) on the quality of an image. There will be many cases in which the quality of an image is better than its original image signal. For instance, the original image signal might contain a flicker disturbance, which is successfully removed in the processed image. In other cases, the original image might be remade and modified to improve the quality.

Category 3 which involves enhancing or remaking the original image is quite different from a picture quality evaluation in which "goodness" is defined as faithfulness to the original images. Thus, our discussion does not cover Category 3. Category 2 is designed for mass evaluation of the images. The ITU-R BT. Rec. 500-10 seems to describe the evaluation procedure not only for Category 1 but also for Category 2. However, as explained in Section 6.1, it is not adequate to use the ITU-R BT. Rec.500-10 when discussing parameters of an image coder related to image quality, because the ITU-R BT. Rec.500-10 is not directly describing the relationship between the deterioration of image quality and related coding characteristics.

6.2.1.1 Picture spatial resolution and viewing distance

In discussing the relation between spatial resolution and image quality, we must refer to the relation between the spatial resolution and displayed frame size, which is one of the most important factors in determining overall image quality.

Strictly maintain the relationship between the designed viewing distance and the size of the displayed frame

In the following discussions, we consider practical cases in which an image (not the originally designed size) is evaluated:

Table 6.6: Selection of test methods

Selection of test methods

Assessment problem	Method used	Description
Measure the quality of systems relative to a reference	Double-stimulus continuous quality-scale (DSCQS) method[1]	Rec. ITU-R BT.500, § 5
Measure the robustness of systems (i.e. failure characteristics)	Double-stimulus impairment scale (DSIS) method[1]	Rec. ITU-R BT.500, § 4
Quantify the quality of systems (when no reference is available)	Ratio-scaling method[2] or categorical scaling (under study)	Report ITU-R BT.1082
Compare the quality of alternative systems (when no reference is available)	Method of direct comparison, ratio-scaling method[2] or categorical scaling (under study)	Report ITU-R BT.1082
Identify factors on which systems are perceived to differ and measure their perceptual influence	Method under study	Report ITU-R BT.1082
Establish the point at which an impairment becomes visible	Threshold estimation by forced-choice method or method of adjustment (under study)	Report ITU-R BT.1082
Determine whether systems are perceived to differ	Forced-choice method (under study)	Report ITU-R BT.1082
Measure the quality of stereoscopic image coding	Double stimulus continuous quality-scale (DSCQS) method[3]	Rec. ITU-R BT.500, § 5
Measure the fidelity between two impaired video sequences	Simultaneous double stimulus for continuous evaluation (SDSCE) method	Rec. ITU-R BT.500, § 6.4
Compare different error resilience tools	Simultaneous double stimulus for continuous evaluation (SDSCE) method	Rec. ITU-R BT.500, § 6.4

[1] Some studies on contextual effects were carried out for the DSCQS and the DSIS methods. It was found that the results of the DSIS method are biased to a certain degree by contextual effects. More details are given in Appendix 3 to Annex 1.

[2] Some studies suggest that this method is more stable when a full range of quality is available.

[3] Due to the possibility of high fatigue when evaluating stereoscopic images, the overall duration of a test session should be shortened to be less than 30 min.

1. To display a large size image on a small size display, resulting in high spatial resolution images

2. To display a small size image on a large size, resulting in low spatial resolution images

Both of above cases are unfair. We must not admire or be disappointed with a quality of a displayed image without considering such cases.

The discrepancy between the original design and a practical usage in Case 2 above will be easily understood, however the discrepancy in Case 1 is sometimes misunderstood. For instance, if an HDTV image is displayed on a small size display, the viewing angle between adjacent pixels is smaller than originally designed, and appears to be sharper than the designed condition. Another example will be to display a conventional TV image on a small screen, for instance, a 15 inch display.

Other problems which we must consider regarding the viewing distance are as follows.

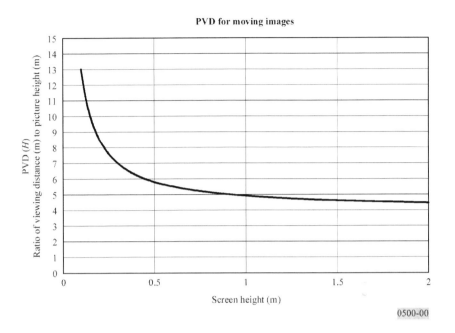

Figure 6.5: PVD (preferable viewing distance) for moving images (from ITU-R Rec. BT.500-10, © 2000 ITU-R).

6.2.1.2 Constancy of viewing distance

According to Section 2.1.1.1 of the ITU-R BT. Rec. 500-10, for displays smaller than 30 inches, the Preferred Viewing Distance (PVD) does not change proportionally to the display size; the observers maintain a rather constant viewing angle (Figure 6.5). The PVD is 1.8 m for a 12 inch display, 2.1 m for a 20 inch display, and 2.7 m for a 29 inch display. This is the reason why human viewers have the tendency to maintain a constant viewing distance. In the case of 1125 lines TV, the PVD is the same as conventional TV if the height of frame is the same as that of a conventional TV.

The degree of constancy of viewing distance is shown and explained in Figure 6.5 and the absolute viewing distance is calculated and shown by D in Table 6.7 using PVD values as shown in Section 2.1.1.13-3 of the ITU-R BT. Rec. 500-10. The constancy of viewing distance seems to be dependent upon the size of the frame. For larger size displays, viewing distance is from $4H$ to $5H$. In the case of smaller displays, the PVD will depend upon the observers' perception of the total environment, which is settled by a reasonable and adequate psychological distance between the observers. In the case of large size displays, the balance of involvement and resolution will be important. A sense of pressure will prevent a person getting too close to a display.

6.2.1.3   Viewing angle between adjacent pixels

Let us calculate the viewing angle between adjacent pixels for each defined viewing distance.

Using parameters listed in the ITU BT. Rec. 601-5 (Table 6.9) and the ITU-R BT. 709-3 (Table 6.10), calculated results are shown in Table 6.7 and Table 6.8.

Table 6.7: Calculated viewing distance (D) and viewing angle (θ) of pixel spacing (Δ) of each system (in the case of Standard Definition TV with 525 lines/480 active lines and 60 fields/s)

screen diagonal	screen height (H)	$\Delta = H/480$	PVD	D	θ
4/3 (inch)	(m)	(m)	(H)	(m)	deg.min.' sec."
12	0.18	0.000375000	9	1.62	00' 47.75"
15	0.23	0.000479167	8	1.84	00' 53.71"
20	0.30	0.000625000	7	2.10	01' 01.39"
29	0.45	0.000937500	6	2.70	01' 11.62"
60	0.91	0.001895833	5	4.55	01' 25.94"
>100	1.53	0.003187500	4	6.12	01' 47.43"
>100	1.53	0.003187500	3	4.59	02' 23.24"

Table 6.8: Comparison of viewing distance (D), viewing angle (θ) of pixel spacing (Δ) of each system (in the case of High Definition TV with 60 fields/s and 1035 active lines), and calculated viewing distance

screen diagonal	screen height (H)	$\Delta = H/1035$	PVD	D	θ
16/9 (inch)	(m)	(m)	(H)	(m)	deg.min.' sec."
15	0.18	0.000173913	9	1.62	00' 22.14"
18	0.23	0.000222222	8	1.84	00' 24.91"
24	0.30	0.000289855	7	2.10	00' 28.47"
36	0.45	0.000434783	6	2.70	00' 33.21"
73	0.91	0.000879227	5	4.55	00' 39.86"
>120	1.53	0.001478261	4	6.12	00' 49.82"
>120	1.53	0.001478261	3	4.59	01' 06.43"

6.2.2   Linearization of the Scale

From the calculated results, it is clear that the spatial resolution of the SDTV is decided essentially based on the visual acuity: around a visual acuity of 1.0 or a visual angle of one minute, except for the case of small sized displays (less than about 20 inches) and

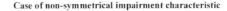

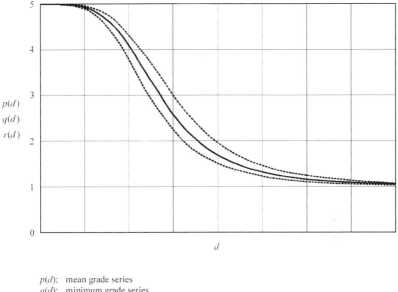

Figure 6.6: Image quality versus distortion curve (from Fig.18 in ITU-R Rec. BT.500-10, © 2000 ITU-R).

CG displays. In the case of the HDTV, it is clear that the PVD is not closer than that of the SDTV. A viewing distance of $2H$ was first considered and used in the development of the HDTV. Although the visual quality for still images was good at this viewing distance, it was not acceptable for rapidly moving images because the images moved too fast. Therefore, the viewing distance was changed to $4H$ for the HDTV, and high resolution performance was utilized to improve further resolution for still images. (However, the resolution is not sufficient, for instance, if we consider that the vernier acuity is about 30 times higher than the visual acuity of 1 minute. Details are discussed in Subsections 6.3.1 and 6.3.2.) In the case of CG display, the visual angle is several times higher because the CG display is a monitor which requires high resolution to display fine images such as font characteristics.

The relationships between image qualities and distortions are discussed in the ITU-R BT. Rec. 500-10 and the image quality versus distortion curves are shown in Figure 6.6.

Regarding linearity of the curve, "log d" rather than direct "d" should be used where d is the distortion or coding error, because the perceptual properties are almost proportional to "log d" regardless of the gamma value. The modified Figure 6.6 is shown in Figure 6.7.

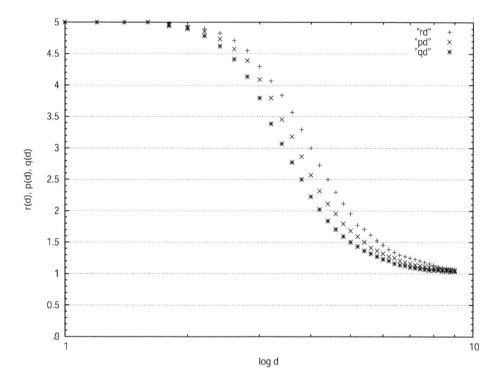

Figure 6.7: Modified Figure 6.6. The scale of horizontal axis is "log d" instead of "d."

It can be clearly seen that the curve in Figure 6.7 is greatly improved, becoming almost a straight line between PQS of 4.5 and 1.5. The linear characteristics of the curve which show the relation between the PQS and the distortion are indispensable in the discussion of the PQS independent of any PQS value. By this linearity, we can discuss the overall degree of improvement and deterioration of images even in the case of plural signal processing and coding. Section 6.1 is described on the basis of these facts.

6.2.3   Importance of Center Area of Image in Quality Evaluation

As described in the beginning of Section 6.2, the center area of an image is most important to picture quality assessment. The reason why we evaluate the center part is that the contents and quality of the center area are more important than the other parts of the image. However, picture quality evaluation and comparison which are based on the center part of an image alone are not always appropriate. A comparison of the image quality of a group with that of another group is meaningless, because the main purpose of each class of images may not be the same. HDTV aims at viewer's involvement rather than at fine display resolution. In contrast, CG aims at high resolution, for instance, detailed font shapes should be displayed.

6.2.4   Other Conditions

Viewing conditions and the order in which the test sequence is displayed are also important.

If we consider all of the above factors simultaneously, the evaluation becomes very complicated. Nonetheless, to carry out the overall evaluation is indispensable. The ITU-R BT. Rec. 500-10 describes in detail how to carry out picture quality assessment tests. As described above, this procedure is primarily for a macro evaluation of images, in other words, ITU-R BT. Rec. 500-10 recommends evaluating the overall picture quality of a transmitted image or to provide a comparative measure for sales of TV sets. It is not intended for technical design of a video coder.

6.3   Various Categories of Image Systems

6.3.1   Standard TV Images with Frame Size of about 500×640 Pixels

Specifications of NTSC, SECAM, PAL and component codings in this group are described in the ITU-R BT. Rec. 601-5. Discussions on the PQS for these standard TV images have already been presented in Section 6.1.

6.3.2   HDTV and Super HDTV

The HDTV (whose resolution is about 1000 x 1600 pixels) is designed as a wide screen TV. Its design philosophy is explained as follows. If we cut out an area the same size as conventional TV frame out of the center part of a wide HDTV frame, the picture quality of the images is about the same, because the HDTV's absolute viewing distance is the same as that of the conventional TV (Figure 6.5) [I.U82].

An HDTV is a wide screen TV with its resolution four times as large as that of a conventional TV, covering additional areas which surround the center part of the image.

A Super HDTV (2000 × 4000) is also a wide screen TV. As explained in Subsection 6.2.1.3, the viewing distance for the HDTV was changed to $4H$ to improve the perception of rapidly moving images, although it was first designed to be about $2H$. The fact

Table 6.9: Encoding parameters of digital television for studios (from ITU-R BT. Rec. 601-5)

Encoding parameter values for the 4:2:2 of the family

Parameters	525-line, 60 field/s systems	625-line, 50 field/s systems
1. Coded signals	*Y, R-Y, B-Y*	
2. Number of samples per total line: - luminancd signal (*Y*) - each colour-difference signal (*R-Y, B-Y*)	858 429	864 432
3. Sampling structure	Orthogonal, line, field and picture repetitive. *R-Y* and *B-Y* samples co-sited with odd(1st, 3rd, 5th, etc.) *Y* samples in each line.	
4. Sampling frequency: - luminancd signal - each colour-difference signal	13.5 MHz 6.75 MHz	
5. Form of coding	Uniformly quantized PCM, 8bits per sample, for the luminance signal and each colour-difference signal	
6. Number of samples per digital active line: - luminance signal - each colour-difference signal	720 360	
7. Correspondence between video signal levels and quantization levels: - luminance signal - each colour-difference signal	220 quantization levels with the black level corresponding to level 16 and the peak white level corresponding to level 235 224 quantization levels in the centre part of the quantization scale with zero signal corresponding to level 128	

that the viewing distance was changed from $2H$ to $4H$ alleviated the rapid dizziness and improved the preferability. However, the sense of viewer involvement was decreased as a result, and we need to double the size of display frame, because the frame size is very important to delivering a strong sense of involvement of the viewer. From Figure 6.8, it is reasonable to use a viewing distance of $4\,m$ to $5\,m$ for a Super HDTV with 2000 $\times$ 4000 pixel resolution and a display of $H \approx 2\,m$, $W \approx 4\,m$, where H and W are the height and the width of the image, respectively. In this case, a viewing distance of $2\,m$ to $3\,m$ (i.e., $2H$) will also be fine, provided that the camera is stable and video contains relatively slow motion.

As explained above, a special merit of the HDTV and the Super HDTV is that it is "wide," which brings about a sense of involvement. The PQS is also applicable and successful in picture quality assessment when taking into account of coding errors.

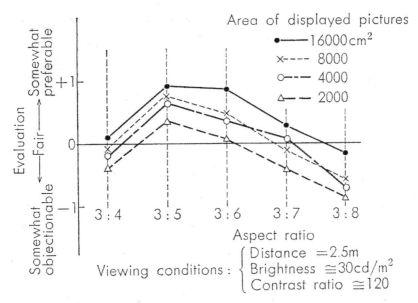

Results of subjective evaluation of preference
as to picture size and aspect ratio.

Figure 6.8: Picture quality versus size of screen (from NHK Monograph, No.32, Fig.4)

6.3.3 Extra High Quality Images

The information that creates deep emotions in the mind of the observer (viewer) is important. What characteristics are most important to transmit in order to evoke the high order sensation? Up to now, the following six factors or characteristics have been identified [MIS+98]:

1. gradient of 4096

2. black level reproduction of lower than 0.5 cd/m^2 down to 0.002 cd/m^2

3. common mode noise

4. step response

Table 6.10: Parameter values for the HDTV standards for production and international program exchange (from ITU-R BT. Rec. 709-3, © 2000 ITU-R)

Picture characteristics

Item	Parameter	Value	
		1125/60/2:1	1250/50/2:1
2.1	**Aspect ratio**	16:9	
2.2	**Samples per active line**	1920	
2.3	**Sampling lattice**	Orthogonal	
2.4	**Active lines per picture**	1035	1152

5. halation

6. cross modulation among R, G and B

The other unknown factors are not clear yet, but the common sense pertaining to our knowledge about image quality seems to be very different. For instance,

1. Up to 100 times precise reproduction seems to be necessary compared to traditional Landolt Ring visual acuity. This acuity is keener than the Vernier Acuity.

2. A deep impression of eyes of a portrait is obtained by a coding technique (A) which generates aliasing disturbance in case where the pixel density is not sufficient, and is better than the other coding technique (B) which does not generate aliasing by a smoothing process; namely, coding (A) is better than coding (B).

The reasons for the above two observations are as follows. There is a view that our ancestors had a very keen sense of visual perception, which was more than a visual acuity 1.0 utilizing all of the body's sense to enhance visual acuity. Primitive humans seemed to have had visual perception more than 100 times higher than Landolt Ring visual acuity, in order not to be attacked and fallen prey. This old keen sense seems to be related to feeling deep emotions. Therefore, if we can muster high-order information to stimulate the primitive center of the brain, deep feeling will be aroused. Research on this topic and related issues is currently ongoing.

6.3.4   Cellular Phone Type

Cellular phones have low channel capacity, where a high compression rate coding is adopted for video transmission. Therefore, comprehension and understanding of displayed image information are important for the viewing distance which varies between the distance of distinct vision and that is constrained by the length of the arm.

6.3.5 Personal Computer and Display for CG

Fine resolution is important in LCDs for which the absolute viewing distance is about 20 inches (the *distance of distinct vision*) independent of the size of the display. In this category, "monitoring of image" is rather the more important criterion than the conventional use. In this case, any discussion on picture quality will not be appropriate. What is more important here is to have more numbers of pixels and quantization levels in order to display the shape of the font clearly.

6.4 Study at ITU

In the ITU, the ITU-T SG9 and the ITU-R WP6Q are in charge of objective picture quality assessment.

The SG9 studies towards standardization of superior schemes for three assessment models of FR (Full Reference), RR (Reduced Reference) and NR (No Reference), respectively, which are specified in Recommendation J.143 [ITU00e] described below. FR recommendation J.144 [ITU01] has already been approved, and it is expected to be revised in the future according to R&D results. Recommendations for RR and NR are planned to be drafted according to the study results of Video Quality Experts Group (VQEG) [VQE].

The WP6Q has published a report on objective picture quality assessment, the ITU-R BT. Rec. 2020-1: "Objective quality assessment technology in a digital environment" [ITU00d]. For the practical operation of digital transmission where quick and short-time judgment is important, it recommends that RR and NR schemes be adopted, and the detail discussion for coder design in Section 6.1 is omitted.

The next subsection explains the recommendations related to picture quality assessment drafted by the SG9.

6.4.1 SG9 Recommendations for Quality Assessment

The ITU-T SG9 has 21 questions (1/9-22/9, Q.3 has been merged to Q.2). Q.4/9 (Rapporteur: Ms. Alina Karwowska-Lamparska, Poland) and Q.21/9 (Rapporteur: Mr. Arthur Webster, U.S.A.) are in charge of quality assessment. In addition to J.143 (User requirements for objective picture quality) and J.144 (Method for objective picture quality assessment with full reference), the SG9 has approved following recommendations:

- J.133: Measurement of MPEG-2 transport streams in networks [ITU02a]
- J.146: Loop latency issues in contribution circuits for conversational TV programs [ITU02b]

- J.147: Objective picture quality measurement method by use of in-service test signals [ITU02c]
- J.148: Requirements for an objective perceptual multimedia quality model [ITU03]

Each recommendation is explained below.

6.4.2   J.143

Rec. J.143: "User requirements for objective perceptual video quality measurements in digital cable television" [ITU00e] describes the user requirements for objective video quality assessment, especially perceptual quality measurements. It defines three frameworks for the measurements:

1. Full Reference (FR): Compares the processed picture with the original picture directly
2. No Reference (NR): Measures the quality of the processed picture without any information about the original picture
3. Reduced Reference (RR): Measures the quality of the processed picture with some reference data extracted from the original picture

These concepts are shown in Figure 6.9.

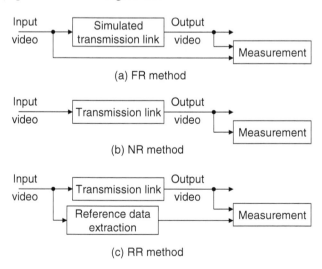

Figure 6.9: Concepts of FR, NR, and RR methods.

6.4.2.1 FR scheme

FR compares the input to and the output from a processing system. To establish a method that correlates well with subjective picture quality, the VQEG [VQE] has been studying the scheme [Roh00, HMM99]. In the VQEG study Phase I, no picture quality scale was found to be superior to the PSNR [Roh00]. [1]

Phase II study limited the compressed bitrate range from 768kbps to 5Mbps without transmission errors, and four proposed FR schemes were found to be superior to the PSNR.

The four methods are recommended with two other methods in J.144: "Objective perceptual video quality measurement techniques for digital cable television in the presence of a full reference" [ITU01], which is described later.

Future direction of FR study: While the target of FR studies has been television quality according to the ITU-R BT. Rec. 500 [ITU00c], the study of multimedia quality is planned by the VQEG and the FR study for multimedia will be continued. When mobile phones and PDAs are used as target platforms, the conditions for compressed bitrates, viewing distance, and displays are expected to be very different from the conventional FR studies. This may lead to a new recommendation for FR-multimedia quality assessment.

6.4.2.2 NR scheme

The NR scheme has been studied for quality assessment using only processed pictures, which tries to model the Single Stimulus Continuous Quality Evaluation (SS-CQE) [ITU00c] for subjective assessment. However, it has difficult issues such as distinction between video freeze failure and normal still pictures, which may cause false alarms. A practical and sufficiently accurate NR method has not been established.

Future direction of NR study: VQEG plans to study NR schemes for television and multimedia (RRNR-TV and RRNR-MM). As the testing conditions such as the display and the viewing distance are different for television and multimedia, separate measurement models will be considered. SSCQE test results will be used as reference for performance evaluation.

6.4.2.3 RR scheme

The RR scheme extracts image features from the original picture, and evaluates the output picture quality using the image features and the output picture itself [S.O97,

---

[1] $PSNR = 10 \log_{10}(255^2/MSE)$ where MSE is represented as follows using $b(x,y)$ (the original picture [8 bit]) and $b_p(x,y)$ (the processed picture): $MSE = (1/N) \sum_{(x,y)} \{b(x,y) - b_p(x,y)\}^2$ The summation is over the whole sequence. N is the total number of pixels.

J.B98]. It is midway between the FR and the NR. The assessment is expected to be more precise than the NR, and unlike the FR, the original picture itself is not necessary. Therefore, the RR is appropriate for applications such as operational monitoring for TV transmission.

Some image features for this use are presented in an ANSI standard [ANS96]. However, no specific measurement method has been established, and VQEG plans to continue studying it.

Relationship between RR and PQS: The target of the VQEG RR model is a high correlation with perceptual visual quality. Since the model for the FR is formulated only recently, difficulty is expected for establishing an RR model considering the limited amount of reference data (image features from the original picture). However, strictly specifying experimental conditions may lead to a certain level of high correlation with perceptual quality, as the VQEG attempted while establishing FR models.

6.4.3   J.144

The FR methods are recommended in J.144: "Objective perceptual video quality measurement techniques for digital cable television in the presence of a full reference" [ITU01].

Originally it contained eight methods studied in the VQEG Phase I. After the VQEG study Phase II, it has been revised to recommend the four methods which are superior to the PSNR under the condition that degradation is induced by coding noise at 768kbps to 5Mbps without transmission errors.

6.4.4   J.133

J.133: "Measurement of MPEG-2 transport streams in networks" recommends methods for measurement of MPEG-2 transport streams in television transmission networks, especially PCR (Program Clock Reference) jitters that are very important. It defines four parameters to be measured to identify various PCR jitters (Fig.6.10):

1. PCR_FO (PCR frequency offset)

2. PCR_DR (PCR drift rate)

3. PCR_AC (PCR accuracy)

4. PCR_OJ (PCR overall jitter)

While the FR/RR/NR is for baseband video, this recommendation is for MPEG-2 transport streams, and the method is also specified in a DVB technical report (ETSI TR 101 290: "Measurement guidelines for DVB systems") [ETS01].

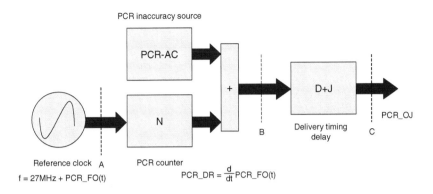

Figure 6.10: Concept of ITU-T Rec.J.133. PCR: Program Clock Reference; PCR_FO: PCR Frequency Offset; PCR_DR: PCR Drift Rate; PCR_AC: PCR Accuracy; PCR_OJ: PCR Overall Jitter; D: a constant representing the mean delay through the communications network; J: jitter in the network delay and its mean value over all time is defined to be zero; N: an idealized PCR count. © 2002 ITU-T.

6.4.5 J.146

J.146: "Loop latency issues in contribution circuits for conversational TV programs" recommends the perceptually permissible transmission delay for TV programs that contain conversation between remote places.

It provides the guideline that 1.5 to 2 seconds are the maximum allowable delay time from subjective evaluation experiments with different delay times.

6.4.6 J.147

J.147: "Objective picture quality measurement method by use of in-service test signals" recommends a method for estimating picture quality by the false detection rate of invisible markers embedded into the video before transmission (Figure 6.11). A method for embedding markers which uses spread spectrum and orthogonal transform [ORMS01] is introduced.

This method evaluates baseband video signals just as the FR/RR/NR. It can monitor the quality using only received signals just as the NR, but it uses invisible markers, so that it is considered as another framework, leading to the separate recommendation.

Specific methods for embedding and detecting markers can be revised according to future R&D results.

6.4.7 J.148

J.148: "Requirements for an objective perceptual multimedia quality model" specifies user requirements for evaluating perceptual multimedia quality. It identifies the impor-

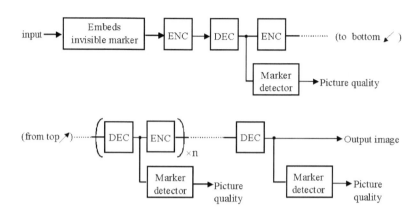

Figure 6.11: Concept of ITU-T Rec.J.147. (ENC: Encoder; DEC: Decoder.) © 2002 ITU-T.

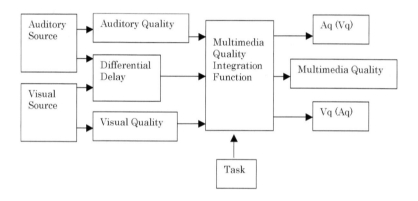

Figure 6.12: Concept of ITU-T Rec.J.148. (VQ: Video Quality; AQ: Audio Quality.)

tance of audio quality in addition to video in multimedia, and it considers multimedia quality also defined by cross-effect of audio and video including synchronization of audio and video. It presents a framework for an integrated evaluation model (Figure 6.12).

6.5 Conclusion

Evaluation of picture quality is a very difficult task which is compounded by too many factors whose impacts on visual picture quality as perceived by human viewers are difficult to explain clearly. We have attempted to explain these factors as systematically and essentially as we possibly can. We believe that our descriptions are useful for both beginners and the whole engineering community to develop new image engineering methods and technologies.

References

[AA96] N. Avadhanam and V. R. Algazi. Prediction and measurement of high quality in still image coding. In *Proceedings of SPIE Very High Resolution and Quality Imaging Conference, EI 96*, 2663:100–109, San Jose, California, 1996.

[AFC95] V. R. Algazi, G. E. Ford, and H. Chen. Linear filtering of images based on properties of vision. *IEEE Transactions on Image Processing*, 4(10):1460–1464, October 1995.

[AFMN93] V. R. Algazi, G. E. Ford, M. Mow, and A. Najmi. Design of subband coders for high quality based on perceptual criteria. In *Proceedings of SPIE, Applications of Digital Image Processing XVI*, 2028:40–49, SPIE. San Diego, California, 1993.

[All83] J. Allnatt. *Transmitted-picture Assessment*. New York: John Wiley and Sons, 1983.

[AN93] A. J. Ahumada, Jr. and C. H. Null. Image quality: A multidimensional problem. In A. B. Watson, Ed., *Digital Image and Human Vision*. Cambridge, MA: The MIT Press, 1993.

[ANS96] ANSI T1.801.03-1996. *Digital transport of one-way video signals–parameters for objective performance assessment*, February 1996.

[AV98] N. Avadhanam and V.R.Algazi. Anchored subjective quality assessment scale for still image coding. *submitted to Signal Processing*, February 1998.

[Bed79] R. A. Bednarek. On evaluation impaired television pictures by subjective measurements. *IEEE Transactions on Broadcasting*, BC-25(2):41–46, 1979.

[C. 65] C. H. Graham et al. *Vision and Visual Perception*. New York: John Wiley and Sons, 1965.

[CCI82] CCIR. Rec. 567-1. Transmission performance of television circuits designed for use in international connections, pl-38. In Recommendations and reports of the CCIR and ITU, Geneva, 1982.

[CKL66] F. W. Campbell, J. J. Kulikowsky, and J. Z. Levinson. The effect of orientation on the visual resolution of gratings. *J. Physiol.*, 187:427–436, 1966.

[CP77] S. Chatterjee and B. Price. *Regression Analysis by Example*. New York: John Wiley and Sons, 1977.

[Dal93] S. Daly. The visible difference predictor: An algorithm for the assessment of image quality. In A. B. Watson, Ed., *Digital Image and Human Vision*. Cambridge, MA: The MIT Press, 1993.

[Dau88] I. Daubechies. Orthonormal bases of compactly supported wavelets. *Commun. on Pure and Appl. Math.*, 41:909–996, November 1988.

[ETS01] ETSI TR 101 290. *Digital Video Broadcasting (DVB); Measurement guidelines for DVB systems,*, V.1.2.1, May 2001.

[Fuj73] T. Fujio. Two-dimensional processing of TV signal. Technical Report Vol. 25, NHK Technical Research Laboratory, March 1973.

[G+88] B. Girod *et al.* A subjective evaluation of noise-shaping quantization for adaptive intra-/interframe dpcm coding of color television signals. *IEEE Transactions on Communications*, COM-36(3):332–346, March 1988.

[HA84] T. A. Hentea and V. R. Algazi. Perceptual models and the filtering of high-contrast achromatic images. *IEEE Transactions on Systems, Man and Cybernetics*, SMC-14(2):230–246, March 1984.

[HM87] Y. Horita and M. Miyahara. Image coding and quality estimation in uniform perceptual space. IECE Technical Report IE87-115, IECE, January 1987.

[HMM99] T. Hamada, S. Miyaji, and S. Matsumoto. Picture quality assessment system by three-layered bottom-up noise weighting considering human visual perception. *SMPTE Journal*, 108(1):20–26, January 1999.

[HPN97] B. G. Haskell, A. Puri, and A. N. Netravali. *Digital Video: An Introduction to MPEG-2*. New York: Chapman & Hall, 1997.

[Ind] Independent JPEG Group. Independent JPEG group's free JPEG software. Home archive is ftp.uu.net/graphics/jpeg.

[ITE87] ITE. *ITE Television System Chart (Version II). Digital Standard Picture*. ITE, December 1987.

[ITU00a] ITU. Rec. ITU-R BT.500-10, ITU Rec. BT.601-5, ITU-R BT.709-3 (details are described in the text.), 2000.

[ITU00b] ITU-R Recommendation BT.500-10. *Methodology for the subjective assessment of the quality of television pictures*, 2000. (Question ITU-R 211/11).

[ITU00c] ITU-R Recommendation BT.500-10. *Methodology for the subjective assessment of the quality of television pictures*, May 2000.

[ITU00d] ITU-R Report BT.2020-1. *Objective quality assessment technology in a digital environment*, September 2000.

[ITU00e] ITU-T Recommendation J.143. *User requirements for objective perceptual video quality measurements in digital cable television*, May 2000.

[ITU01] ITU-T Recommendation J.144. *Objective perceptual video quality measurement techniques for digital cable television in the presence of a full reference*, March 2001.

[ITU02a] ITU-T Recommendation J.133. *Measurement of MPEG-2 transport streams in networks*, July 2002.

[ITU02b] ITU-T Recommendation J.146. *Loop latency issues in contribution circuits for conversational TV programmes*, July 2002.

[ITU02c] ITU-T Recommendation J.147. *Objective picture quality measurement method by use of in-service test signals*, July 2002.

[ITU03] ITU-T Recommendation J.148. *Requirements for an objective perceptual multimedia quality model*, May 2003.

[I.U82] I.Uyama. II fundamental requirements for high-definition television systems, II-1 large-screen effects. *NHK Monograph*, 32:14–20, June 1982.

[Jai89] A. K. Jain. *Fundamentals of Digital Image Processing*. Englewood Cliffs, NJ: Prentice Hall, 1989.

[J.B98] J.Baïna, et al. Quality of MPEG2 signal on a simulated digital terrestrial television. *ieee*, BR-44(4):381–391, December 1998.

[JJS93] N. Jayant, J. Johnston, and R. Safranek. Signal compression based on models of human perception. *Proceedings of the IEEE*, 81(10):1385–1422, October 1993.

[JM86] B. L. Jones and P. M. McManus. Graphics scaling of qualitative terms. *SMPTE Journal*, 95:1166–1171, November 1986.

[Joh80] J. D. Johnston. A filter family designed for use in quadrature mirror filters. In *Proceedings of IEEE ICASSP 1980*, 291–294. IEEE, 1980.

[K+85] M. Kunt *et al.* Second-generation image-coding techniques. *Proceedings of the IEEE*, 73:549–574, April 1985.

[Ken75] M. G. Kendall. *Multivariate Analysis*. London: Charles Griffin, 1975.

[KGMA95] K. Kotani, Q. Gan, M. Miyahara, and V. R. Algazi. Objective picture quality scale for color image coding. In *Proceedings of International Conference on Image Processing, ICIP 95*, 3:133–136, Washington D. C., October 1995.

[KK95] S. A. Karunasekera and N. G. Kingsbury. A distortion measure for blocking artifacts in images based on human visual sensitivity. *IEEE Transactions on Image Processing*, 4(6):713–724, June 1995.

[KM96a] V. Kayargadde and J.-B. Martens. Perceptual characterization of images degraded by blur and noise: experiments. *Journal of the Optical Society of America A (Optics, Image Science and Vision)*, 13(6):1166–1177, June 1996.

[KM96b] V. Kayargadde and J.-B. Martens. Perceptual characterization of images degraded by blur and noise: model. *Journal of the Optical Society of America A (Optics, Image Science and Vision)*, 13(6):1178–1188, June 1996.

[LA65] N. W. Lewis and J. A. Allnatt. Subjective quality of television pictures with multiple impairments. *Electronic Letters*, 1:187–188, July 1965.

[LAE95] J. Lu, V. R. Algazi, and R. R. Estes, Jr. A comparison of wavelet image coders using a picture quality scale. In *Proceedings of SPIE Wavelet Applications Conference*, Orlando, Florida, April 1995. SPIE.

[LAE96] J. Lu, V. R. Algazi, and R. R. Estes, Jr. A comparative study of wavelet image coders. *Optical Engineering*, 9:2605–2619, September 1996.

[LB82] F. X. J. Lukas and Z. L. Budrikis. Picture quality prediction based on a visual model. *IEEE Transactions on Communications*, COM-30(8):1679–1692, July 1982.

[Lim78] J. O. Limb. On the design of quantizer for dpcm coder: A functional relationship between visibility, probability, and masking. *IEEE Transactions on Communications*, COM-26(5):573–578, May 1978.

[Lim79] J. O. Limb. Distortion criteria of the human viewer. *IEEE Transactions on Systems, Man, and Cybernetics*, SMC-9(12):778–793, December 1979.

[MIS+98] M. Miyahara, T. Ino, H. Shirai, S. Taniho, and R. Algazi. Important factors to convey high order sensation. *IEEE Transactions on Communications*, E81-B(11):1966–1973, November 1998.

[Miy88] M. Miyahara. Quality assessments for visual service. *IEEE Communications Magazine*, 26(10):51–60, October 1988.

[MK96] J. B. Martens and V. Kayargadde. Image quality prediction in a multidimensional perceptual space. In *Proceedings of 3rd IEEE International Conference on Image Processing, Lausanne*, 1:877–880, 1996.

[MKA98] M. Miyahara, K. Kotani, and V. R. Algazi. Objective picture quality scale (pqs) for image coding. *IEEE Transactions on Communications*, 46(9):1215–1226, September 1998.

[MY88] M. Miyahara and Y. Yoshida. Mathematical transformation of (R,G,B) color data to munsell (H,V,C) color data. In *SPIE Visual Communications and Image Processing III*, 1001–1118, November 1988.

[Net77] A. N. Netravali. Interpolative picture coding using a subjective criterion. *IEEE Transactions on Communications*, COM-25(5):503–508, May 1977.

[O+87] A. Oosterlick *et al*. *Image Coding Using the Human Visual System*, chap. 5. Belgium: Katholieke Universiteit, 1987.

[ORMS01] O.Sugimoto, R.Kawada, M.Wada, and S.Matsumoto. A method for estimating PSNR of coded pictures using embedded invisible markers without reference pictures. In *ispacs*, 42–46, Nashville, Tennessee, USA, November 2001.

[Pro95] Proc. MOSAIC Workshop:. Advanced methods for the evaluation of television picture quality. Proceedings of the mosaic workshop, Institute for Perception Research, Eindhoven, Netherlands, 1995.

[PW84] G. C. Phillips and H. R. Wilson. Orientation bandwidth of spatial mechanisms measured by masking. *J. Opt. Soc. Am. A*, 1:226–231, 1984.

[Roh00] A. M. Rohaly et al. Video Quality Experts Group: Current results and future directions. In *Proc. SPIE*, 4067:742–753, Perth, Australia, 2000.

[Sak77] D. J. Sakrison. On the role of the observer and a distortion measure in image transmission. *IEEE Transactions on Communications*, COM-25(11), November 1977.

[S.O97] S.Olsson, et al. Objective methods for assessment of video quality: state of the art. *ieee*, BR-43(4):487–495, December 1997.

[TH94] P. C. Teo and D. J. Heeger. Perceptual image distortion. In *International Conference on Image Processing*, 2:982–986. ICIP, 1994.

[vDMW95] A. M. van Dijk, J. B. Martens, and A. B. Watson. Quality assessment of coded images using numerical category scaling. In *Proc. SPIE Advanced Image and Video Communications and Storage Technologies, Amsterdam*, 2451:90–101, 1995.

[VQE] VQEG. *The Video Quality Experts Group Web Site*. http://www.its.bldrdoc.gov/vqeg/.

[W+83] R. Wilson *et al*. Anisotropic non-stationary image estimation and its applications: Part ii - Predictive image coding. *IEEE Transactions on Communications*, COM-31(3):398–406, March 1983.

[Wat93a] A. B. Watson. DCT quantization matrices visually optimized for individual images. In *Proc. SPIE Human Vision, Visual Processing, and Digital Display*, 1913:202–216, 1993.

[Wat93b] A. B. Watson, Ed. *Digital Image and Human Vision*. Cambridge, MA: The MIT Press, 1993.

[WLB95] S. J. P. Westen, R. L. Lagendijk, and J. Biemond. Perceptual image quality based on a multiple channel HVS model. In *International Conference on Acoustics, Speech, and Signal Processing*, 4:2351–2354. ICASSP, 1995.

[YH68] M. Yasuda and K. Hiwatashi. A model of retinal neural networks and its spatio-temporal characteristics. *Japanese journal of medical electronics and biological engineering*, 6, no.1:53–62, 1968.

[ZH89] C. Zetzsche and G. Hauske. Multiple channel model for the prediction of subjective image quality. In *SPIE Proceedings*, 1077:209–216. SPIE, 1989.

Chapter 7

Structural Similarity Based Image Quality Assessment

Zhou Wang[†], Alan C. Bovik[‡] and Hamid R. Sheikh[§]

†   *University of Texas at Arlington, U.S.A.*
‡   *University of Texas at Austin, U.S.A.*
§   *Texas Instruments, Inc., U.S.A.*

It is widely believed that the statistical properties of the natural visual environment play a fundamental role in the evolution, development and adaptation of the human visual system (HVS). An important observation about natural image signals is that they are highly structured. By "structured signal", we mean that the signal samples exhibit strong dependencies amongst themselves, especially when they are spatially proximate. These dependencies carry important information about the structure of the objects in the visual scene. The principal hypothesis of *structural similarity* based image quality assessment is that the HVS is highly adapted to extract structural information from the visual field, and therefore a measurement of structural similarity (or distortion) should provide a good approximation to perceived image quality.

In this chapter, structural similarity is presented as an alternative design philosophy for objective image quality assessment methods. This is different from and complementary to the typical HVS-based approaches, which usually calculate signal difference between the distorted and the reference images, and attempt to quantify the difference "perceptually" by incorporating known HVS properties.

7.1   Structural Similarity and Image Quality

In *full-reference* image quality assessment methods, the quality of a test image is evaluated by comparing it with a reference image that is assumed to have perfect quality. The goal of image quality assessment research is to design methods that quantify the strength of the perceptual similarity (or difference) between the test and the reference images. Researchers have taken a number of approaches to this end.

Figure 7.1: Comparison of 8bits/pixel *Einstein* images with different types of distortions. (a) original image, MSE = 0, MSSIM = 1; (b) contrast stretched image, MSE = 144, MSSIM = 0.9133; (c) mean shifted image, MSE = 144, MSSIM = 0.9884; (d) JPEG compressed image, MSE = 142, MSSIM = 0.6624; (e) blurred image, MSE = 144, MSSIM = 0.6940; (f) salt-pepper impulsive noise contaminated image, MSE = 144, MSSIM = 0.8317.

The first approach, which is called the error sensitivity approach, considers the test image signal as the sum of the reference image signal and an error signal. Assuming that the loss of perceptual quality is directly related to the visibility of the error signal, most HVS-based image quality assessment models attempt to weigh and combine different aspects of the error signal according to their respective visual sensitivities, which are usually determined by psychophysical measurements. One problem with this approach is that larger visible differences may not necessarily imply lower perceptual quality. An example is shown in Figure 7.1, where the original *Einstein* image is altered with different distortions: contrast stretch, mean shift, JPEG compression, blurring, and impulsive salt-pepper noise contamination. Each type of distortion was adjusted to yield the same mean squared error (MSE) relative to the original image, except for the JPEG compressed image, which has a slightly smaller MSE. Despite their nearly identical MSE, the images can be seen to have significantly different perceptual qualities. It is important to note that although the difference between the contrast stretched image (Figure 7.1(b)) and the reference image (Figure 7.1(a)) is easily discerned, the contrast stretched image has good perceptual quality.

The second approach is based on the conjecture that the purpose of the entire visual observation process is to efficiently extract and make use of the information represented in natural scenes, whose statistical properties are believed to play a fundamental role in the evolution, development and adaptation of the HVS (e.g., [SO01]). One distinct example of the second approach is the *structural similarity* based image quality assessment method [WBSS04], which is motivated from the observation that natural image signals are highly "structured," meaning that the signal samples have strong dependencies amongst themselves, especially when they are spatially proximate. These dependencies carry important information about the structure of the objects in the visual scene. The principal premise of the structural similarity approach is that the major goal of visual observation is to extract such information, for which the HVS is highly adapted. Therefore, a measurement of structural information change or structural similarity (or distortion) should provide a good approximation to perceived image quality. Let us again take the contrast stretched image in Figure 7.1(b) as an example. Although its visible difference from the reference image is significant, it preserves almost all of the important information that reflects the structure of the objects represented in the image. In fact, the reference image can almost be fully recovered via a simple point-wise inverse linear luminance transform. Consequently, a high quality score should be assigned. On the other hand, some structural information in the original image is severely distorted and permanently lost in the JPEG compressed and the blurred images, and therefore they should be assigned lower quality scores.

The natural question that follows is then: What constitutes the important information that reflects the structure of objects represented in an image? This is the key issue that will define the specific implementation of the image quality assessment algorithm. While it is difficult to directly provide a relatively small set of features that sufficiently describe the structural information in an image, it is worthwhile to consider its opposite: what is the information in an image that is not important for representing the structure of the objects? A simple answer comes from the perspective of image formation. Recall that the luminance of the surface of an object being observed is the product of the illumination and the reflectance, but the structures of the objects in the scene are independent of the illumination. Consequently, we wish to separate out the influence of illumination from the information that is more important for representing object structures. Intuitively, the major impact of illumination change is the variation of the average luminance and contrast in the image. Since luminance and contrast can vary across a scene, they are preferably measured locally. This leads to a localized image similarity measure that separates (and perhaps removes) the influence of luminance and contrast variation from the remaining attributes of the local image region.

The first instantiation of the structural similarity-based method was made in [Wan01, WB02] and promising results on simple tests were achieved. This method was further generalized and improved in [WBSS04, WSB03]. It was also adapted for video quality assessment in [WLB04]. In Sections 7.2 and 7.3 of this chapter, we will mainly have a

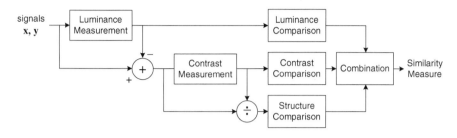

Figure 7.2: Diagram of the SSIM system.

close look at the Structural SIMilarity (SSIM) index introduced in [WBSS04].

7.2   The Structural SIMilarity (SSIM) Index

The system diagram of the SSIM image quality assessment system is shown in Figure 7.2. Suppose that $\mathbf{x}$ and $\mathbf{y}$ are two non-negative image signals, which have been aligned with each other (e.g., spatial patches extracted from each image). The purpose of the system is to provide a similarity measure between them. The similarity measure can serve as a quantitative measurement of the quality of one signal if we consider the other to have perfect quality. Here $\mathbf{x}$ and $\mathbf{y}$ can be either continuous signals with a finite support region, or discrete signals represented as $\mathbf{x} = \{\, x_i \,|\, i = 1, 2, \cdots, N \,\}$ and $\mathbf{y} = \{\, y_i \,|\, i = 1, 2, \cdots, N \,\}$, respectively, where i is the sample index and N is the number of signal samples (pixels).

The system separates the task of similarity measurement into three comparisons: luminance, contrast and structure. First, the luminance of each signal is compared. Assuming discrete signals, this is estimated as the mean intensity:

$$\mu_x = \bar{x} = \frac{1}{N} \sum_{i=1}^{N} x_i \,. \tag{7.1}$$

The luminance comparison function $l(\mathbf{x}, \mathbf{y})$ is then a function of μ_x and μ_y:

$$l(\mathbf{x}, \mathbf{y}) = l(\mu_x, \mu_y) \,. \tag{7.2}$$

Second, we remove the mean intensity from the signal. In discrete form, the resulting signal $\mathbf{x} - \mu_x$ corresponds to the projection of vector $\mathbf{x}$ onto the hyperplane of

$$\sum_{i=1}^{N} x_i = 0 \,. \tag{7.3}$$

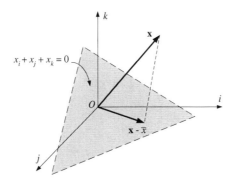

Figure 7.3: Projection onto the hyperplane of $\sum x_i = 0$. Note: this is an illustration in 3-D space. In practice, the number of dimensions is equal to the number of pixels.

as illustrated in Figure 7.3. We use the standard deviation (the square root of variance) as an estimate of the signal contrast. An unbiased estimate in discrete form is given by

$$\sigma_x = \left(\frac{1}{N-1} \sum_{i=1}^{N} (x_i - \mu_x)^2 \right)^{1/2}.$$ (7.4)

The contrast comparison $c(\mathbf{x}, \mathbf{y})$ is then the comparison of σ_x and σ_y:

$$c(\mathbf{x}, \mathbf{y}) = c(\sigma_x, \sigma_y).$$ (7.5)

Third, the signal is normalized (divided) by its own standard deviation, so that the two signals being compared have unit standard deviation. The structure comparison $s(\mathbf{x}, \mathbf{y})$ is conducted on these normalized signals:

$$s(\mathbf{x}, \mathbf{y}) = s\left(\frac{\mathbf{x} - \mu_x}{\sigma_x}, \frac{\mathbf{y} - \mu_y}{\sigma_y} \right).$$ (7.6)

Finally, the three components are combined to yield an overall similarity measure:

$$S(\mathbf{x}, \mathbf{y}) = f(l(\mathbf{x}, \mathbf{y}), c(\mathbf{x}, \mathbf{y}), s(\mathbf{x}, \mathbf{y})).$$ (7.7)

An important point is that the three components are relatively independent. For example, the change of luminance and/or contrast has little impact on the structures of images.

In order to complete the definition of the similarity measure in Eq. (7.7), we need to define the three functions $l(\mathbf{x}, \mathbf{y})$, $c(\mathbf{x}, \mathbf{y})$, $s(\mathbf{x}, \mathbf{y})$, as well as the combination function $f(\cdot)$. We also would like the similarity measure to satisfy the following conditions:

1. Symmetry: $S(\mathbf{x}, \mathbf{y}) = S(\mathbf{y}, \mathbf{x})$. Since our purpose is to quantify the similarity

between two signals, exchanging the order of the input signals should not affect the resulting similarity measurement.

2. Boundedness: $S(\mathbf{x}, \mathbf{y}) \leq 1$. Boundedness is a useful property for a similarity metric since an upper bound can serve as an indication of how close the two signals are to being perfectly identical. This is in contrast with most signal-to-noise ratio type of measurements, which are typically unbounded.

3. Unique maximum: $S(\mathbf{x}, \mathbf{y}) = 1$ if and only if $\mathbf{x} = \mathbf{y}$ (in discrete representations, $x_i = y_i$ for all $i = 1, 2, \cdots, N$). In other words, the similarity measure should quantify any variations that may exist between the input signals. The perfect score is achieved *only* when the signals being compared are exactly the same.

For luminance comparison, we define

$$l(\mathbf{x}, \mathbf{y}) = \frac{2\,\mu_x\,\mu_y + C_1}{\mu_x^2 + \mu_y^2 + C_1}. \tag{7.8}$$

where the constant C_1 is included to avoid instability when $\mu_x^2 + \mu_y^2$ is very close to zero. Specifically, we choose

$$C_1 = (K_1\,L)^2, \tag{7.9}$$

where L is the dynamic range of the pixel values (255 for 8-bit grayscale images), and $K_1 \ll 1$ is a small constant. Similar considerations also apply to contrast comparison and structure comparison as described later. Equation (7.8) is easily seen to obey the three properties listed above.

Equation (7.8) is also connected with Weber's law, which has been widely used to model light adaptation (also called luminance masking) in the HVS. According to Weber's law, the magnitude of a just-noticeable luminance change ΔI is approximately proportional to the background luminance I for a wide range of luminance values. In other words, the HVS is sensitive to the *relative* luminance change, and not the absolute luminance change. Letting R represent the ratio of luminance change relative to background luminance, we rewrite the luminance of the distorted signal as $\mu_y = (1 + R)\mu_x$. Substituting this into (7.8) gives

$$l(\mathbf{x}, \mathbf{y}) = \frac{2(1 + R)}{1 + (1 + R)^2 + C_1/\mu_x^2}. \tag{7.10}$$

If we assume C_1 is small enough (relative to μ_x^2) to be ignored, then $l(\mathbf{x}, \mathbf{y})$ is a function only of R instead of $\Delta I = \mu_y - \mu_x$. In this sense, it is qualitatively consistent with Weber's law. In addition, it provides a quantitative measurement for the cases when the luminance change is much more than the visibility threshold, which is out of the application scope of Weber's law.

The contrast comparison function takes a similar form:

$$c(\mathbf{x}, \mathbf{y}) = \frac{2\,\sigma_x\,\sigma_y + C_2}{\sigma_x^2 + \sigma_y^2 + C_2}, \tag{7.11}$$

where C_2 is a non-negative constant

$$C_2 = (K_2\,L)^2, \tag{7.12}$$

and K_2 satisfies $K_2 \ll 1$. This definition again satisfies the three properties listed above. An important feature of this function is that with the same amount of contrast change $\Delta\sigma = \sigma_y - \sigma_x$, this measure is less sensitive to the case of high base contrast σ_x than low base contrast. This is related to the contrast masking feature of the HVS.

Structure comparison is conducted after luminance subtraction and contrast normalization. Specifically, we associate the direction of the two unit vectors $(\mathbf{x} - \mu_x)/\sigma_x$ and $(\mathbf{y} - \mu_y)/\sigma_y$, each lying in the hyperplane (Figure 7.3) defined by Equation (7.3), with the structures of the two images. The correlation (inner product) between them is a simple and effective measure to quantify the structural similarity. Notice that the correlation between $(\mathbf{x} - \mu_x)/\sigma_x$ and $(\mathbf{y} - \mu_y)/\sigma_y$ is equivalent to the correlation coefficient between $\mathbf{x}$ and $\mathbf{y}$. Thus, we define the structure comparison function as follows:

$$s(\mathbf{x}, \mathbf{y}) = \frac{\sigma_{xy} + C_3}{\sigma_x\,\sigma_y + C_3}. \tag{7.13}$$

As in the luminance and contrast measures, a small constant has been introduced in both denominator and numerator. In discrete form, σ_{xy} can be estimated as:

$$\sigma_{xy} = \frac{1}{N-1} \sum_{i=1}^{N} (x_i - \mu_x)(y_i - \mu_y). \tag{7.14}$$

Geometrically, the correlation coefficient corresponds to the cosine of the angle between the vectors $\mathbf{x} - \mu_x$ and $\mathbf{y} - \mu_y$. Note also that $s(\mathbf{x}, \mathbf{y})$ can take on negative values.

Finally, we combine the three comparisons of Equations (7.8), (7.11) and (7.13) and name the resulting similarity measure the Structural SIMilarity (SSIM) index between signals $\mathbf{x}$ and $\mathbf{y}$:

$$\text{SSIM}(\mathbf{x}, \mathbf{y}) = [l(\mathbf{x}, \mathbf{y})]^\alpha \cdot [c(\mathbf{x}, \mathbf{y})]^\beta \cdot [s(\mathbf{x}, \mathbf{y})]^\gamma, \tag{7.15}$$

where $\alpha > 0$, $\beta > 0$ and $\gamma > 0$ are parameters used to adjust the relative importance of the three components. It is easy to verity that this definition satisfies the three conditions given above. In particular, we set $\alpha = \beta = \gamma = 1$ and $C_3 = C_2/2$. This results in a specific form of the SSIM index:

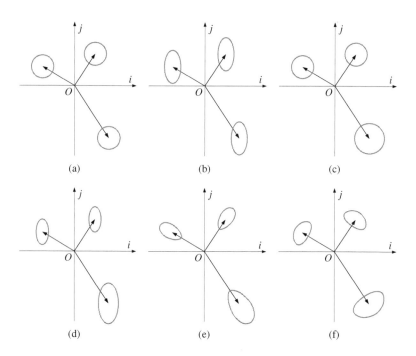

Figure 7.4: Equal-distortion contours for different quality measurement systems. (a) Minkowski error measurement systems; (b) component-weighted Minkowski error measurement systems; (c) magnitude-weighted Minkowski error measurement systems; (d) magnitude and component-weighted Minkowski error measurement systems; (e) SSIM measurement system (a combination of Eqs. (7.11) and (7.13)) with more emphasis on $s(\mathbf{x}, \mathbf{y})$; (f) SSIM measurement system (a combination of Eqs. (7.11) and (7.13)) with more emphasis on $c(\mathbf{x}, \mathbf{y})$. Each image is represented as a vector, whose entries are image components. Note: this is an illustration in 2-D space. In practice, the number of dimensions should be equal to the number of image components used for comparison (e.g., the number of pixels or transform coefficients).

$$\mathrm{SSIM}(\mathbf{x}, \mathbf{y}) = \frac{(2\,\mu_x\,\mu_y + C_1)\,(2\,\sigma_{xy} + C_2)}{(\mu_x^2 + \mu_y^2 + C_1)\,(\sigma_x^2 + \sigma_y^2 + C_2)}\,. \tag{7.16}$$

The SSIM index may be better understood geometrically in a vector space of signal components as in Figure 7.4. These signal components can be either image pixel intensities or other extracted features such as transformed linear coefficients. Figure 7.4 shows equal-distortion contours drawn around three different example reference vectors, each of which could, for example, represent the local content of one reference image. For the purpose of illustration, we show only a two-dimensional space, but in general the dimensionality should match that of the signal components being compared. Each contour represents a set of test signals with equal distortion relative to the respective reference signal. Figure 7.4(a) shows the results for a simple Minkowski metric. Each contour has the same size and shape (a circle here, as we are assuming an exponent of 2). That is, perceptual distance corresponds to Euclidean distance. Figure 7.4(b) shows a Minkowski metric in which different signal components are weighted differently. This could be, for example, weighting according to the contrast sensitivity function (CSF), as is common in many quality assessment models. Here the contours are ellipses, but still are all the same size. More advanced quality measurement models may incorporate contrast masking behaviors, which has the effect of rescaling the equal-distortion contours according to the signal magnitude, as shown in Figure 7.4(c). This may be viewed as a simple type of *adaptive* distortion measure: it depends not just on the difference between the signals, but also on the signals themselves. Figure 7.4(d) shows a combination of contrast masking (magnitude weighting) followed by component weighting. In comparison of the vectors $\mathbf{x} - \mu_x$ and $\mathbf{y} - \mu_y$, the SSIM index corresponds to the comparison of two independent quantities: the vector lengths, and their angles. Thus, the contours will be aligned with the axes of a polar coordinate system. Figures 7.4(e) and 7.4(f) show two examples of this, computed with different exponents. Again, this may be viewed as an *adaptive* distortion measure, but unlike the other models being compared, both the size and the shape of the contours are adapted to the underlying signal.

7.3 Image Quality Assessment Based on the SSIM Index

For image quality assessment, it is useful to apply the SSIM index locally rather than globally. First, image statistical features are usually highly spatially non-stationary. Second, image distortions, which may or may not depend on the local image statistics, may also be space-variant. Third, at typical viewing distances, only a local area in the image can be perceived with high resolution by the human observer at one time instance (because of the foveation feature of the HVS, e.g., [GB95]). Fourth, localized quality measurement can provide a spatially varying quality map of the image, which delivers more information about the quality degradation of the image and may be useful in some applications.

In [Wan01, WB02], the local statistics μ_x, σ_x and σ_{xy} (Eqs. (7.1),(7.4) and (7.14)) are computed within a local 8×8 square window. The window moves pixel-by-pixel from the top-left corner to the bottom-right corner of the image. At each step, the local statistics and SSIM index are calculated within the local window. One problem with this method is that the resulting SSIM index map often exhibits undesirable "blocking" artifacts as exemplified by Figure 7.5(a). Such kind of "artifacts" is not desirable because it is created from the choice of the quality measurement system (local square window), but not from image distortions. In [WBSS04], a circular-symmetric Gaussian weighting function $\mathbf{w} = \{ w_i \, | \, i = 1, 2, \cdots, N \}$ with unit sum ($\sum_{i=1}^{N} w_i = 1$) is adopted. The estimates of local statistics μ_x, σ_x and σ_{xy} are then modified accordingly as

$$\mu_x = \sum_{i=1}^{N} w_i \, x_i . \tag{7.17}$$

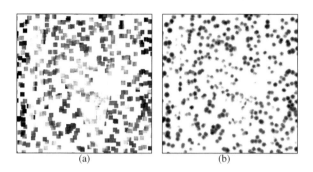

(a)	(b)

Figure 7.5: SSIM index maps of the impulse noise contaminated *Einstein* image (Figure 7.1(f)). Brightness indicates the magnitude of the local SSIM index value. (a) square windowing approach; (b) smoothed windowing approach.

$$\sigma_x = \left(\sum_{i=1}^{N} w_i \left(x_i - \mu_x \right)^2 \right)^{1/2}. \tag{7.18}$$

$$\sigma_{xy} = \sum_{i=1}^{N} w_i \left(x_i - \mu_x \right)\left(y_i - \mu_y \right). \tag{7.19}$$

With such a windowing approach, the quality maps exhibit a locally isotropic property, as shown in Figure 7.5(b).

In practice, one usually requires a single overall quality measure of the entire image. We use a mean SSIM (MSSIM) index to evaluate the overall image quality:

$$\text{MSSIM} = \sum_{j=1}^{M} W_j \cdot \text{SSIM}_j, \tag{7.20}$$

where M is the number of samples in the quality map, SSIM_j is the SSIM index value at the j-th sample, and W_j is the weight given to the j-th sample and

$$\sum_{j=1}^{M} W_j = 1. \tag{7.21}$$

If all the samples in the quality map are equally weighted, then $W_j = 1/M$ for all j's. Depending on the application, it is also possible to assign spatially varying weights to different samples in the SSIM index map. For example, region-of-interest image processing systems may give different weights to different segmented regions in the image. For another example, it has been observed that different image textures attract human fixations with varying degrees (e.g., [PS00, RCB03]), and therefore a fixation probability model can be used to define the weighting model. Further, since the visual resolution decreases gradually as a function of the distance from the fixation point (e.g., [WB01]), a smoothly varying foveated weighting model can also be employed to define the weights. For the experiments described in this chapter, however, we use uniform weighting. A MATLAB implementation of the SSIM index algorithm is available online at [WBSS03].

Many image quality assessment algorithms have been shown to behave consistently when applied to distorted images created from the same original image, using the same type of distortions (e.g., JPEG compression). However, the effectiveness of these models degrades significantly when applied to a set of images originating from different reference images, and/or including a variety of different types of distortions. Thus, cross-image and cross-distortion tests are critical in evaluating the effectiveness of an image quality metric. It is impossible to show a thorough set of such examples, but the images in Figure 7.1 provide an encouraging starting point for testing the cross-distortion capability of the quality assessment algorithms. The MSE and MSSIM measurement results are given in the figure caption. Obviously, the MSE performs very poorly in this case. The MSSIM values exhibit much better consistency with the qualitative visual appearance.

For a more thorough test, we apply the SSIM index algorithm to the LIVE database of JPEG and JPEG2000 compressed images that were evaluated by a number of subjects for perceptual quality [SWBC03]. The database was created with 29 high-resolution 24 bits/pixel RGB colour images (typically 768×512 or similar size) compressed at a range of quality levels using either JPEG or JPEG2000, producing a total of 175 JPEG images and 169 JPEG2000 images. The bit rates were in the range of 0.150 to 3.336 and 0.028 to 3.150 bits/pixel for JPEG and JPEG2000 images, respectively, and were chosen non-uniformly such that the resulting distribution of subjective quality scores was approximately uniform over the entire range. Subjects viewed the images from comfortable viewing distances and were asked to provide their perception of quality on a continuous linear scale that was divided into five equal regions marked with adjectives "Bad," "Poor," "Fair," "Good" and "Excellent." Each JPEG and JPEG2000 compressed image was viewed by $13 \sim 20$ subjects and 25 subjects, respectively. The subjects were mostly male college students. Raw scores for each subject were normalized by the mean and variance of scores for that subject (i.e., raw values were converted to Z-scores) and then scaled and shifted by the mean and variance of the entire subject pool to fill the range from 1 to 100. Mean opinion scores (MOSs) were then computed for each image, after removing outliers (most subjects had no outliers). The image database, together with the subjective score and standard deviation for each image, has been made available on the Internet at [SWBC03].

The luminance component of each JPEG and JPEG2000 compressed image is averaged over a local 2×2 window and downsampled by a factor of two before the MSSIM value is calculated. Our experiments with the current dataset show that the use of the other colour components does not significantly change the performance of the model, though this should not be considered generally true for colour image quality assessment. Note that no specific training procedure is employed before applying the SSIM algorithm, because the SSIM index is intended for general-purpose image quality assessment, as opposed to specific application types (e.g., image compression) only.

Figure 7.6 shows two sample JPEG and JPEG2000 images from the database, together with their SSIM index maps and absolute error maps. By closer inspection of corresponding spatial locations in the SSIM index and the absolute error maps, we observe that the SSIM index is generally more consistent with perceived quality measurement. In particular, note that at low bit rates, the coarse quantization in JPEG and JPEG2000 algorithms often results in smooth representations of fine-detail regions in the image (e.g., the tiles in Figure 7.6(c) and the trees in Figure 7.6(d)). Compared with other types of regions, these regions may not be worse in terms of pointwise difference measures such as the absolute error. However, since the structural information of the image details is nearly completely lost, they exhibit poorer visual quality. Comparing Figure 7.6(e) with Figure 7.6(g), and Figure 7.6(f) with 7.6(h)), we can see that the SSIM index is better in capturing such poor quality regions.

The scatter plots of the MOS versus the PSNR and the MSSIM image quality prediction are shown in Figure 7.7, where each sample point represents one test image. It can be observed that the MSSIM supplies better prediction capability of the subjective scores than the PSNR. In order to provide quantitative comparisons on the performance of the SSIM index measure, we use the logistic function adopted in the video quality experts group (VQEG) Phase I FR-TV test to provide a non-linear mapping between the objective/subjective scores [VQE00]. The fitted curves are shown in Figure 7.7. After fitting, a set of quantitative measures are computed, which include the Pearson correlation coefficient (CC), the mean absolute error (MAE), the root mean squared error (RMS), the outlier ratio (OR, defined as the proportion of predictions that are outside the range of two times of the standard error in the subjective test), and the Spearman rank-order correlation coefficient (SROCC). Readers can refer to Appendix A [VQE00, VQE03] for details about how these measures are calculated. It can be seen that the MSSIM outperforms the PSNR in all these comparisons by clear margins (Table 7.1).

7.4   Discussions

This chapter discusses the motivation, the general idea, and a specific SSIM index algorithm of the structural similarity-based image quality assessment method. It is worthwhile to look into the relationship between this method and the traditional error sensitivity based image quality assessment algorithms.

On the one hand, we consider "structural similarity" as a substantially different design principle for image quality assessment and would like to emphasize two distinct features of this method in comparison with the error sensitivity-based models. First, in

Figure 7.6: Sample JPEG and JPEG2000 compressed images and quality maps (cropped from 768×512 to 240×160 for visibility). (a) and (b) are the original *Buildings* and *Stream* images, respectively. (c) JPEG compressed *Buildings* image, 0.2673 bits/pixel; (d) JPEG2000 compressed *Stream* image, 0.1896 bits/pixel; (e) and (f) show SSIM maps of the compressed images, where brightness indicates the magnitude of the local SSIM index (squared for visibility). (g) and (h) show absolute error maps of the compressed images, where brighter point indicates smaller error (for easier comparison with the SSIM map).

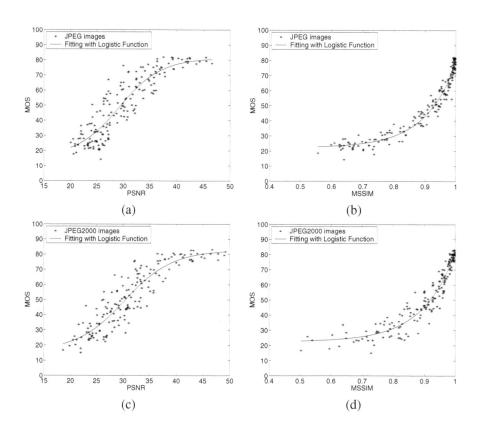

(a)   (b)

(c)   (d)

Figure 7.7: Scatter plots of subjective mean opinion score (MOS) versus model prediction. Each sample point represents one test image. (a) PSNR prediction for JPEG images; (b) MSSIM prediction for JPEG images; (c) PSNR prediction for JPEG2000 images; (d) MSSIM prediction for JPEG2000 images.

Table 7.1: Performance comparison of PSNR and MSSIM using the JPEG and JPEG2000 image databases. CC: correlation coefficient; MAE: mean absolute error; RMS: root mean squared error; OR: outlier ratio; SROCC: Spearman rank-order correlation coefficient.

Database	JPEG		JPEG2000	
Model	PSNR	MSSIM	PSNR	MSSIM
CC	0.904	0.978	0.910	0.958
MAE	6.769	3.324	6.300	4.352
RMS	8.637	4.176	8.062	5.540
OR	0.200	0.006	0.095	0.024
SROCC	0.893	0.973	0.906	0.955

terms of the nature of the distortions that a quality measure attempts to capture, it is targeted at *perceived structural information variation*, instead of *perceived error*. Second, in terms of the construction of the quality assessment system, it is a *top-down* approach that mimics the hypothesized functionality of the overall HVS, as opposed to a *bottom-up* approach that simulates the function of relevant early-stage components in the HVS.

On the other hand, we also view the structural similarity-based methods as being complementary to, rather than opposed to, the typical error sensitivity based approaches. Notice that error sensitivity based methods often involve signal decompositions based on linear transforms such as the wavelet transforms (e.g., [Lub93, Dal93, TH94, WYSV97]). Such signal decompositions can be thought of as specific descriptive representations of the signal "structures." In this sense, the error between transformed wavelet coefficients implicitly suggests that the structural change between the image signals be compared. From a different angle, the SSIM indexing method as described in the previous sections might be converted into an equivalent "error" measure in a specific coordinate system, only that such a coordinate system is locally adaptive, non-linear, and input-dependent. It needs to be mentioned that certain divisive-normalization based masking models (e.g., [MNE$^+$00, EGM03]) exhibit input-dependent behavior in measuring signal distortions, which leads to a departure from the distortion contours shown in Figures 7.4(a)-(d), although precise alignment with the axes of a polar coordinate system as in Figures 7.4(e) and 7.4(f) is not observed. Even though not clear at this moment, we think it is possible that the two types of approaches may eventually converge into similar solutions.

The SSIM indexing algorithm is quite encouraging not only because it achieves good quality prediction accuracy in the current tests, but also because of its simple formulation and low complexity implementation. This is in contrast with many complicated HVS-based quality assessment systems. Its simplicity makes it much more tractable in the context of algorithm and parameter optimizations (e.g., the derivative of the SSIM index with respect to the image can be explicitly computed [WS04]) for the development of perceptually-optimized image processing and coding systems.

Finally, we would like to point out that the SSIM indexing approach is only a particular implementation of the philosophy of structural similarity, from an image formation point of view. Under the same philosophy, other approaches may emerge that could lead to algorithms significantly different from the SSIM index. Creative investigation of the concepts of structural information and structural distortion is likely to drive the success of these innovations.

References

[Dal93]   S. Daly. The visible differences predictor: An algorithm for the assessment of image fidelity. In A. B. Watson, Ed., *Digital Images and Human Vision*, 179–206. Cambridge, MA: The MIT Press, 1993.

[EGM03] I. Epifanio, J. Gutirrez, and J. Malo. Linear transform for simultaneous diagonalization of covariance and perceptual metric matrix in image coding. *Pattern Recognition*, 36(8):1679–1923, Aug. 2003.

[GB95] W. S. Geisler and M. S. Banks. Visual Performance. In M. Bass, Ed., *Handbook of Optics*, 25.1–25.55. New York: McGraw–Hill, June 1995.

[Lub93] J. Lubin. The use of psychophysical data and models in the analysis of display system performance. In A. B. Watson, Ed., *Digital Images and Human Vision*, 163–178. Cambridge, MA: The MIT Press, 1993.

[MNE⁺00] J. Malo, R. Navarro, I. Epifanio, F. Ferri, and J. M. Artifas. Non-linear invertible representation for joint statistical and perceptual feature decorrelation. *Lecture Notes on Computer Science*, 1876:658–667, 2000.

[PS00] C. M. Privitera and L. W. Stark. Algorithms for defining visual regions-of-interest: comparison with eye fixations. *IEEE Trans. Pattern Analysis and Machine Intelligence*, 22(9):970–982, Sep. 2000.

[RCB03] U. Rajashekar, L. K. Cormack, and A. C. Bovik. Image features that draw fixations. In *Proc. IEEE Inter. Conf. Image Processing*, 3:313–316, Sep. 2003.

[SO01] E. P. Simoncelli and B. A. Olshausen. Natural image statistics and neural representation. *Annu. Rev. Neurosci.*, 24:1193–1216, May 2001.

[SWBC03] H. R. Sheikh, Z. Wang, A. C. Bovik, and L. K. Cormack. *Image and video quality assessment research at LIVE*. University of Texas at Austin, 2003. (accessible at: *http://live.ece.utexas.edu/research/quality/*).

[TH94] P. C. Teo and D. J. Heeger. Perceptual image distortion. In *Human Vision, Visual Processing, and Digital Display V, Proc. SPIE*, 2179:127–141, 1994.

[VQE00] VQEG. *Final report from the video quality experts group on the validation of objective models of video quality assessment.* VQEG, Mar. 2000. (accessible at: *http://www.vqeg.org/*).

[VQE03] VQEG. *Final report from the video quality experts group on the validation of objective models of video quality assessment, Phase II.* VQEG, Aug. 2003. (accessible at: *ftp://ftp.its.bldrdoc.gov/dist/ituvidq/frtv2_final_report/*).

[Wan01] Z. Wang. *Rate scalable foveated image and video communications.* Ph.D. thesis, Dept. of ECE, The University of Texas at Austin, Texas, 2001.

[WB01] Z. Wang and A. C. Bovik. Embedded foveation image coding. *IEEE Trans. Image Processing*, 10(10):1397–1410, Oct. 2001.

[WB02] Z. Wang and A. C. Bovik. A universal image quality index. *IEEE Signal Proc. Lett.*, 9(3):81–84, March 2002.

[WBSS03] Z. Wang, A. C. Bovik, H. R. Sheikh, and E. P. Simoncelli. *The SSIM Index for Image Quality Assessment*. New York University and University of Texas at Austin, 2003. (accessible at: *http://www.cns.nyu.edu/~lcv/ssim/*).

[WBSS04] Z. Wang, A. C. Bovik, H. R. Sheikh, and E. P. Simoncelli. Image quality assessment: From error measurement to structural similarity. *IEEE Trans. Image Processing*, 13(4):600–612, Apr. 2004.

[WLB04] Z. Wang, L. Lu, and A. C. Bovik. Video quality assessment based on structural distortion measurement. *Signal Processing: Image Communication*, 19(2):121–132, Feb. 2004. Special issue on video quality metrics.

[WS04]   Z. Wang and E. P. Simoncelli. Stimulus synthesis for efficient evaluation and refinement of perceptual image quality metrics. In B. E. Rogawitz and T. N. Pappas, Eds., *Human Vision and Electronic Imaging IX, Proc. SPIE*, 5292:99–108, Jan. 2004.

[WSB03] Z. Wang, E. P. Simoncelli, and A. C. Bovik. Multiscale structural similarity for image quality assessment. In *Proc. IEEE Asilomar Conf. Signals, Systems & Computers*, 1398–1402, Nov. 2003.

[WYSV97] A. B. Watson, G. Y. Yang, J. A. Solomon, and J. Villasenor. Visibility of wavelet quantization noise. *IEEE Trans. Image Processing*, 6(8):1164–1175, Aug. 1997.

Chapter 8

Vision Model Based Digital Video Impairment Metrics

Zhenghua Yu[†] and Hong Ren Wu[‡]

† *National Information Communication Technology Australia (NICTA)*
‡ *Royal Melbourne Institute of Technology, Australia*

8.1 Introduction

A proliferation of applications of digital image and video compression technology has been witnessed in the past two decades, from digital camera, digital television, video conferencing, to mobile multimedia, to name a few. Whilst digital image and video compression explores statistical and psychological redundancies in raw image and video, in a need to achieve high compression ratios, such compression is usually lossy, which will inevitably lead to noticeable coding artifacts in decoded image and video at medium to low bit rates. Coding artifacts have been well classified in literature [YW98], among which blocking, ringing and blurring artifacts and temporal fluctuations and granular noise are typical artifacts.

Measuring digital image and video quality/distortion has long been an important, albeit challenging, task in the field. Raw error measures, most notably the Peak Signal to Noise Ratio (PSNR) or its MSE (Mean Squared Error) equivalent, are widely used. Nonetheless, it is also well-acknowledged that the PSNR does not necessarily reflect the perceived quality [Gir93]. The second major type of measures is based on vision models which achieves a more accurate measurement through emulating the mechanisms of visual quality/distortion perception in the human visual system (HVS).

Whilst a great number of vision model based metrics concentrate on the overall image/video quality, it is of theoretical and practical importance to measure individual coding artifacts.

- Theoretically, most vision models are devised based on experiments for a single purpose such as frequency sensitivity with simple stimuli. It is less known to what degree the visual characteristics measured by such experiments can be applied to accurately predict more complex artifacts in natural image and video [KK95]. Research on vision model based impairment metrics of different artifacts for digital image and video can validate vision models from a different perspective. Meanwhile, it is likely that better understanding of the visual characteristics regarding individual artifacts can lead to better understanding of the human visual system and feed into more accurate vision models.

- From a practical point of view, different image/video coding systems may exhibit different types and degrees of artifacts. Therefore impairment metrics can measure coding system performance from various angles and potentially could point out ways to improve the system. One example is post-processing of image and video where blocking impairment metrics have been used to design deblocking algorithms [LB02].

The research on quantitative measurement of video quality, to the best of our knowledge, was pioneered by Lukas and Budrikis in 1982 [LB82]. The history of digital image and video impairment metrics research is relatively short (less than a decade). However, several well established or promising impairment metrics have been proposed, which are briefly reviewed below.

Karunasekera and Kingsbury are among the pioneers in vision model based impairment metrics research, as demonstrated by their work on a vision model based blocking artifacts metric for still images [KK95]. The metric proposed was for measurement of vertical blocking artifacts only, but it could be easily extended for horizontal blocking artifacts, given that human sensitivity to horizontal and vertical orientations was found to be similar. Both the original and the processed images are inputs to the metric. Firstly the difference between the original image and the processed images is calculated, creating a difference image. Then vertical edges in the difference image are detected through a combination of high-pass filtering along the rows and low-pass filtering along the columns. After that, $N \times 1$ pixel blocks are selected as the base regions to compute the visibility of vertical edges due to masking. Two masking effects are considered: activity masking and brightness masking, which are about the masking of the edge due to background activity and brightness. Here, activity and brightness are all calculated from a $N \times N$ block in the original image. Activity is modeled as the weighted energy in frequency domain of the block. Brightness is taken as the average pixel value of the $N \times N$ block. Then a nonlinear transformation in the form of exponential function is applied, in order to simulate the nonlinear amplitude sensitivity characteristics of the human visual system. Finally, the value of perceived vertical blocking artifacts is calculated as the mean of the absolute nonlinearly transformed masked errors.

Specially designed subjective tests were conducted to parameterize the metric in [KK95], which involved using special test patterns to turn on/off blocks in the metric. While this approach can parameterize the metric, it is quite time consuming and expensive for a single lab to conduct large scale subjective tests. On the other hand, experimental results of the human visual system in literature can not be directly utilized.

A set of features related to individual artifacts was defined in ANSI T1.801.03-1996, including blocking, blurring, jerkiness, etc. and the sum of these features was used to derive the metric of overall quality [WJP+93, Sec96]. The blocking feature alone could serve as a blocking metric described as follows. Given an input video sequence, a spatial-temporal region of the sequence is selected for computing. Horizontal and vertical edge enhancement filters are applied to the region and the filter responses can be denoted as $H(i, j, t)$ and $V(i, j, t)$, respectively, where i and j are the spatial coordinates and t the temporal coordinate. These responses can be converted into polar coordinate (R, θ) where R is the strength of the H and V responses and θ is the angle coordinate. If plotted in (R, θ) histogram, blocking artifacts will add more pixels along the $m\frac{\pi}{2}$ axis, therefore, high concentration of R energy along the HV axis. Consequently the HV to non-HV edge energy difference is defined as a blocking feature. The blocking features are calculated independently for both the original and the processed sequences. If blocking artifacts exist, the blocking feature of the processed sequence will be greater than that of the original sequence. Therefore the relative blocking feature in the original and processed sequences can indicate the degree of blocking artifacts. However, the blocking feature was designed as part of an overall quality metric and, therefore, no formula was provided to map the relative blocking feature into a normalized range as in other dedicated blocking metrics. Since the calculation on the processed sequence is independent of the original sequence, only the blocking feature of the original sequence is needed to be transmitted to the receiver for quality assessment, instead of the entire original sequence. Therefore, this is a reduced-reference metric.

Wu and Yuen [WY97] proposed a generalized blocking impairment metric (GBIM) for digital images that only requires the processed image. It uses weighted mean squared difference along block boundaries as the blockiness measure, taking into account the effects of spatial masking and the luminance level on the visibility of block distortions. The GBIM's significance is twofold. It is the first blocking artifacts measure without a reference (no-reference assessment), and it is simple and works reasonably well. However GBIM's HVS model is too simple and does not extend well to video.

Vision model based digital video impairment metrics were firstly studied by Yu, Wu and Chen in [YWC00a]. The work has led to both the Perceptual Blocking Distortion Metric (PBDM) [YWWC02] and the Perceptual Ringing Distortion Metric (PRDM) [YWC00b, ZWY+03], which are based on the observation that blocking and ringing artifacts dominate certain regions of the processed sequence. Both metrics have adopted the vision model proposed by Teo and Heeger [TH94a] to compute distortions at every pixel level. Algorithms have been developed to segment the blocking and ringing dominant regions. Perceptual distortions in the blocking/ringing dominant regions are summed up to form blocking and ringing measures.

The metric proposed in [WBE00] is a no-reference blocking artifacts metric. The key idea is to model the blocky image as a non-blocky image interfered with by a pure blocky signal. Blocking artifacts are measured by detecting and estimating the power of the blocky signal. As the first step, inter-pixel differences are calculated for each pair of pixels. Then the differences are organized as 1-D signal and cut into segments of size N. These segments go through the FFT and the power spectrum of the segments is estimated afterwards. The power spectrum curve is smooth-filtered with a median filter. Peak powers at N/8, 2N/8, 3N/8 and 4N/8 are summed to obtain the blocking measure in one direction (horizontal or vertical). Horizontal and vertical blockinesses are calculated independently and the overall blockiness measure is the average of horizontal and vertical measures.

An alternative way to blind measure blocking artifacts is through DCT domain, by modeling the artifacts as 2-D step functions in shifted blocks [BL01]. A major advantage of this approach is the reduced computational complexity, due to the sparseness of the DCT coefficients. The metric has been applied in reduction of blocking artifacts [LB02].

Although in these blind measurement methods, examples have been given to integrate with some vision models, it may be difficult to integrate with more complete and advanced vision models while preserving the "no-reference" feature. For example, excitatory-inhibitory contrast gain control models require both the original and the processed sequences. Meanwhile performance evaluation of the metrics with subjective tests has yet to be reported. Besides, these metrics are based on prior-knowledge of the coding algorithm, that is, block edge artifacts exist at spatially fixed block boundaries. This assumption is questionable in the case of low bit rate video coding, where blocking artifacts could propagate into reconstructed frames and appear anywhere in the frame, not just block boundaries, due to motion compensation and non-zero motion vectors. The blocking energy may not concentrate as greatly as in the image coding case. Consequently it may be difficult to extend the metric to digital video.

Digital image and video impairment metrics could be classified from several angles, depending on:

- the type of targeted impairment, for example, blocking metric, ringing metric. Most impairment metrics proposed so far have been blocking metrics, due to the importance of blocking artifacts and the well understood and modeled cause of blocking artifacts.

- the type of media — image or video.

- the adoption of vision model. Impairment metrics can be classified as HVS based or non-HVS based.

- the use of reference image/video. They can be classified as full reference, reduced reference or no reference impairment metrics (Chapter 11). Full reference metrics

Table 8.1: Classification of Impairment Metrics

	impairment	vision model	reference	image/video
Karunasekera et al [KK95]	blocking	Y	full	image
ANSI T1.801.03 [Sec96]	blocking	N	reduced	both [1]
GBIM [WY97]	blocking	Y	no	image
PBDM [YWC00a, YWWC02]	blocking	Y	full	video
PRDM [YWC00b, ZWY+03]	ringing	Y	full	video
Wang et al [WBE00]	blocking	Y	no	image
Bovik et al [BL01]	blocking	Y	no	image

require the original image/video frame while no original image/video is required for no reference ones. Reduced reference is a third category where the amount of reference information required from the video input to perform meaningful impairment measurement is less than the entire video frame.

The metrics described above are classified in Table 8.1. One observation can be made from the table: although digital video has been an essential enabler in digital television and video conferencing, few impairment metrics are designed for video. It is partially because a video metric is more complex to model and more time consuming to compute.

This chapter addresses the topic of vision model based digital video impairment metrics, based on the work of the PBDM and the PRDM. Section 8.2 will review the formation of vision model for the objective measurement of coding artifacts. The PBDM and its experimental results are presented in Section 8.3, while the PRDM is covered in Section 8.4. Section 8.5 concludes this chapter.

8.2 Vision Modeling for Impairment Measurement

The perceptual blocking distortion metric (or PBDM) [YWC00a, YWWC02] and the perceptual ringing distortion metric (or PRDM) [YWC00b, ZWY+03] have been formulated upon research outcomes of the human visual system (HVS). Although vision models have been discussed in Chapter 2, it is imperative to review the formation of the vision model used in these impairment metrics, for completeness of this chapter.

---

[1] Although ANSI T1.801.03 caters for video, the blocking feature is only based on spatial information, therefore the metric is largely an image impairment metric.

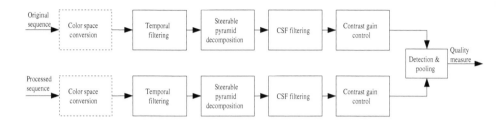

Figure 8.1: Block diagram of the vision model

Modeling human vision has long been a challenging research topic [Wan95, Win99a], as the human visual system is extremely complex and difficult to understand. The vision model adopted in the PBDM and the PRDM is largely based on the work of Teo and Heeger and van den Branden Lambrecht [TH94a, van96a]. The model is focused on the pattern sensitivity aspect of human vision and simulates several key elements of the HVS: color perception, spatio-temporal contrast sensitivity and contrast gain control. A block diagram of the model is shown in Figure 8.1. It takes two video sequences as the input, one being the original and the other being the processed, and computes the quality degradation between these two sequences to generate the output which is the measured visual quality. Every stage of the model is described in more detail below.

8.2.1   Color Space Conversion

Digital image and video can be represented in several color spaces, with the RGB and the YC_bC_r being the most commonly used. The vision model adopted in the PBDM and the PRDM works in the YC_bC_r color space, and more specifically, on the luminance (Y) channel only, for reasons explained shortly. Signal conversion from (R, G, B) to (Y, C_b, C_r) is an optional stage in the vision model, applicable when the input video sequences are in the RGB color space.

Natural color can be decomposed into three primary colors: red (R), green (G) and blue (B). Psychological and physiological experiments on human perception of natural color have led to opponent-color models [HJ57, PW96]. A typical model is the pattern-color separable opponent colors space [PW96], whose three principal color components are Black-White (B-W), Red-Green (R-G) and Blue-Yellow (B-Y). Meanwhile it has been noted that among the visual pathways, human vision has the highest acuity in the black-white pathway. Although the color conversion matrix in [PW96] is not exactly the same as the matrix normally used with the YC_bC_r color space [ITU95], the black-white pathway and the Y channel possess similar characteristics in carrying the luminance component, therefore the Y channel could be a substitute of the black-white pathway. The Color Moving Pictures Quality Metric (CMPQM) adopts the opponent color space, however a restrained number of filters are used in the R-G and B-Y pathways than in the B-W pathway, given these two pathways are less sensitive than the B-W pathway [van96b]. Further research on the influence of the choice of color space on the performance of vision models has concluded that it is possible for the vision model to work on the luminance (Y) component only without a dramatic degradation in prediction accuracy [Win00]. The PBDM and the PRDM follow this approach and use only the luminance component, to take the advantage of reduced computational complexity.

8.2.2 Temporal Filtering

Input video sequences are filtered in temporal domain to simulate the spatio-temporal contrast sensitivity mechanism of human vision.

It is well known that the visual system processes information by contrast rather than absolute light level. Measuring contrast sensitivity with regard to a set of stimuli (e.g. with varying spatial/temporal frequencies) has been adopted as a method to characterize the human visual system. Multi-channel vision models have been proposed to explain those contrast sensitivity experiments [Wat83, WR84]. Among the models proposed, spatio-temporal separable models have been widely accepted [TH94a, van96a]. In signal processing technology, usually the models are implemented as spatio-temporally separable filter banks, which have obtained a reasonably good approximation to the HVS [van96a, Win99b].

It is generally believed that two temporal mechanisms exist, one being transient and the other sustained [HS92]. Consequently, two temporal filters have been employed in [van96a, Win99b]. IIR digital filters have been chosen in order to reduce the number of taps and hence the delay, where the sustained mechanism is low-pass and the transient one is band-pass with a peak frequency around 8 Hz [van96a]. Temporal domain representation of the IIR filter could be depicted as follows

$$y[n] = b[1] * x[n] + b[2] * x[n-1] + ... + b[n_b + 1] * x[n - n_b]$$
$$-a[2] * y[n-1] - ... - a[n_a + 1] * y[n - n_a]$$

(8.1)

where $x[n]$ means a pixel in input frame n and $y[n]$ is the pixel at the same spatial location in output frame n. The input-output description of this filtering operation in the z-transform domain is a rational transfer function [OS89]

$$Y[z] = \frac{b[1] + b[2]z^{-1} + ... + b[n_b + 1]z^{-n_b}}{1 + a[2]z^{-1} + ... + a[n_a + 1]z^{-n_a}} X(z)$$

(8.2)

References [van96a] and [Win99b] adopted filter coefficients designed in [Lin96] as shown in Table 8.2. As an example, the corresponding frequency responses of the 30 Hz filters are shown in Figure 8.2.

Table 8.2: Temporal Filter Coefficients

Filter	a[2] ... a[n_a+1]	b[1] b[2] ... b[n_b+1]
low pass 30Hz	-0.080000	0.800000, 0.120000
band pass 30Hz	0.264560, 0.532900	0.205877, -0.079503, -0.182592, 0.105776
low pass 25Hz	-0.080000	0.859813, 0.060187
band pass 25Hz	0.599392, 0.348100	0.222765, -0.210861, -0.099928, 0.182667

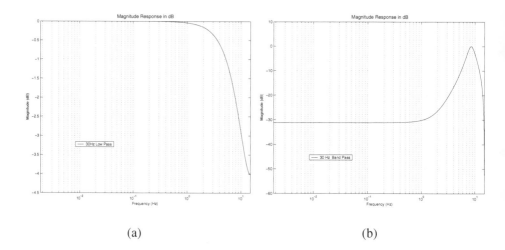

(a)   (b)

Figure 8.2: Frequency response of the temporal filters. (a) 30Hz Low Pass (b) 30Hz Band Pass

However, most visual detail information is carried in the sustained channel. This is partially the reason why some early single channel models work well under certain simple test conditions. Objective and subjective tests conducted with the metric in [Win00] have confirmed that a majority of distortions exist in the sustained channel. Therefore in the PBDM [YWWC02] only the sustained temporal channel is used, with the benefit of reduced computational complexity.

8.2.3   Spatial Filtering

Compared with the temporal mechanism, more research efforts have been put into modeling the spatial mechanism of the HVS. For the achromatic visual pathway, experiments have revealed that there exist several spatial frequency and orientation tuned channels

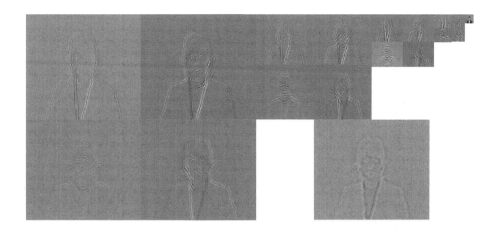

Figure 8.3: Example: *Claire* image after the steerable pyramid transform.

with frequency bandwidths of one to two octaves and orientation bandwidths between 30 and 60 degrees [Val82, PW84], suggesting 4 to 7 spatial frequency bands.

A steerable pyramid transform (SPT) vision model introduced by Teo and Heeger [SF95, TH94a, TH94b] is adopted in the PBDM. The transform decomposes input sequences into six frequency levels including an isotropic low-pass (LP) level, four bandpass (BP) levels with four orientations each centered around 0°, 45°, 90° and 135°, and an isotropic high-pass (HP) level. Figure 8.3 illustrates an example of *Claire* image after the steerable pyramid transform. The isotropic low-pass level is shown at the top-right corner, the isotropic high-pass level is shown at the bottom-right corner, and the other levels have four orientations each. All subband coefficients are normalized to the range of 0 to 255 for display purposes.

In vision research, the sensitivity in the detection of spatial patterns at various spatial frequencies is characterized through spatial Contrast Sensitivity Functions (CSFs). When applied in quality metrics, CSFs can be simulated by digital filtering with the respective filter response. Since the spatial filter bank already decomposes the sequences into several frequency levels, CSF filtering can be implemented by multiplying every subband with a proper CSF coefficient [Win99b]. Overall, this constitutes a coarse, nonetheless very fast, approximation of the desired CSF filter response.

8.2.4   Contrast Gain Control

Another essential effect in vision modeling is pattern masking. When a target pattern and a masking pattern are mixed, the perception of the target pattern will differ depending on the characteristics of the masking pattern. A number of prominent contrast gain control models of pattern masking have been proposed by Foley [Fol94], Teo and

Heeger [TH94a], and Watson and Solomon [WS97]. The latter two are based on multi-channel models, and they are "image-driven" [WS97], which explains their popularity in video quality metrics [van96a, Win99b].

The contrast gain control stage in a multi-channel vision model follows the spatio-temporal decomposition and the CSF filtering. In the PBDM and the PRDM, Teo and Heeger's contrast gain control model is used [TH94a], which consists of an excitatory-inhibitory stage and a normalization stage. In the excitatory visual pathway, the input coefficient values are locally squared. In the inhibitory visual pathway, pooling is performed over the squared subbands. The ratio of the excitatory energy versus the inhibitory energy is calculated in the normalization stage. Let $\mathbf{X}[j, f, \theta, k, l]$ be a coefficient after the steerable pyramid decomposition and the CSF filtering (cf. Subsection 8.2.3), where $(j, f, \theta, , k, l)$ represent the frame number, the spatial frequency, the orientation, horizontal and vertical spatial coordinates, respectively [TH94a]. Considering masking over orientation subbands at the same spatial frequency level, the squared and normalized output (i.e., the perceptual response) $\tilde{\mathbf{X}}_R[j, f, \theta, k, l]$ is then computed as:

$$\tilde{\mathbf{X}}_R[j, f, \theta, k, l] = \sum_{i=1}^{4} w_i \frac{(\mathbf{X}[j, f, \theta, k, l])^2}{\sum_{\phi}(\mathbf{X}[j, f, \phi, k, l])^2 + \gamma_i^2}, \qquad (8.3)$$

where w_i is the overall scaling constant for a contrast mechanism, γ_i^2 is the corresponding saturation constant, and $\phi = 0^\circ, 45^\circ, 90^\circ, 135^\circ$. Four contrast mechanisms (i.e., with four pairs of w_i and γ_i) are summed up to compute the output of one input coefficient $\mathbf{X}[j, f, \theta, k, l]$ in order to overcome the rapid saturation of each mechanism [TH94b]. As an example, coefficients adopted in [YWWC02] are shown in Table 8.3. For a particular mechanism, the dynamic range of the acceptable input is limited by the value of γ_i. If the values of input coefficients $\mathbf{X}[j, f, \phi, k, l]$ are too small compared with γ_i, $\sum_{\phi}(\mathbf{X}[j, f, \phi, k, l])^2 + \gamma_i^2$ roughly equals to γ_i^2 which is constant for a given subband. As a result, the output of one contrast mechanism $w_i \frac{(\mathbf{X}[j,f,\theta,k,l])^2}{\sum_{\phi}(\mathbf{X}[j,f,\phi,k,l])^2+\gamma_i^2}$ is determined by the input coefficient $\mathbf{X}[j, f, \theta, k, l]$ only. When the values of input coefficients $\mathbf{X}[j, f, \phi, k, l]$ move closer to γ_i, the output of the contrast mechanism will be jointly affected by $\mathbf{X}[j, f, \phi, k, l]$ and γ_i, which is desirable. However, a value of $\mathbf{X}[j, f, \theta, k, l]$ much larger than γ_i will saturate the output of the contrast gain control mechanism, that is, $w_i \frac{(\mathbf{X}[j,f,\theta,k,l])^2}{\sum_{\phi}(\mathbf{X}[j,f,\phi,k,l])^2+\gamma_i^2}$ will roughly equal to w_i. In this case, it is likely that both the original and the processed sequences will produce output coefficients with the same value of w_i, and therefore perceptual differences between the original and the processed will become indistinguishable. To overcome this problem, four contrast mechanisms (i.e., with four pairs of w_i and γ_i) are used to cover a wide range of input coefficient values. In the example given by Table 8.3, γ_i spans from 4.4817 to 54.5982, so input coefficients from 0 to few times of 54.5982 are within the desired range. Contrast sensitivity weights, c_i for ($i = 1, 2, \ldots, 6$), are also shown in Table 8.3, for the low-pass, four bandpass and the high-pass frequency bands, respectively.

Table 8.3: Metric Coefficients

c_i (CSF)	0.41, 1.25, 1.2, 0.4097, 0.083, 0.001
w_i	1, 1, 1, 1
γ_i	4.4817, 7.3891, 12.1825, 54.5982

When implementing the contrast gain control model together with the steerable pyramid transform, the output of the steerable pyramid will be the input to the contrast gain control model. The same set of w_i and γ_i coefficients are used for all bands. There is no problem for other frequency bands except for the low-pass band. The DC energy will accumulate in the low-pass band so that the pixel values in the band may frequently exceed the value of 1000 which is much larger than 54.5982 and hence outside of the desired dynamic range of the contrast gain control model. Therefore, there is an optional stage of mean value subtraction [YWWC02] to prevent the accumulation of the DC energy into the low-pass band.

Another implementation consideration is related to the high-pass and low-pass bands after the steerable pyramid transform. The contrast gain control equation (8.3) has the summation over all the orientations (θ) at the same frequency level. The high-pass and low-pass bands are isotropic, i.e., there is only one orientation. These two bands are still squared and normalized using (8.3), although there is no summation over orientation bands.

8.2.5 Detection and Pooling

Up to the contrast gain control stage, the original sequence and the processed sequence are handled independently. The detection and pooling stage then integrates data from the different bands of both sequences according to a summation rule. This stage simulates the integration process of the visual cortex. Since the steerable pyramid transform decomposes the input sequence into different subbands with different resolutions, the sensor outputs of the original sequence $\tilde{X}_{R_o}[j, f, \theta, k, l]$ and the sensor outputs of the processed sequence $\tilde{X}_{R_p}[j, f, \theta, k, l]$ need to be normalized by the decimation factor of each band:

$$X_{R_o}[j, f, \theta, k, l] = \frac{\tilde{X}_{R_o}[j, f, \theta, k, l]}{X_f} \tag{8.4}$$

$$X_{R_p}[j, f, \theta, k, l] = \frac{\tilde{X}_{R_p}[j, f, \theta, k, l]}{X_f} \tag{8.5}$$

where X_f is the decimation factor of band f which could be 1, 4, 16, 64, 256, etc. The pooling is implemented as a squared error norm of the difference between the normalized sensor outputs of the original and the processed sequences using:

$$\Delta \mathbf{X}_R = \frac{\sum_{j,f,\theta,k,l} \left| \mathbf{X}_{R_o}[j, f, \theta, k, l] - \mathbf{X}_{R_p}[j, f, \theta, k, l] \right|^2}{N},$$
(8.6)

where $\Delta \mathbf{X}_R$ is a measure of perceptual distortion, N is the number of frames, and the other notations follow (8.3).

In this stage, the perceptual distortion map of each frame can also be generated after summing over spatial frequency and orientation subbands. This distortion map is of the same resolution as that of the original frame. Each pixel value represents the perceptual distortion at that spatial location. As will be shown in the following sections, the pixel-level distortion map is essential to the measurement of impairments in the PBDM and the PRDM.

8.2.6   Model Parameterization

These are several parameters in the vision model described in the above subsections, such as the CSF coefficients. Model parameterization is an important step and there are at least two approaches: parameterization with vision research experiments and parameterization with video quality experiments.

8.2.6.1   Parameterization by vision research experiments

Typical psychovisual experiments use simple synthetic patterns as stimuli. Sine waves and Gabor patterns are among the typical test patterns. An example test pattern is shown in Figure 8.4. The experiments are usually designed to expose one particular aspect of the HVS, such as the CSF, or a specific type of masking, etc.

One approach to parameterization of a comprehensive vision model is to parameterize each model component individually with corresponding human visual experiments, i.e., CSF coefficients are determined by CSF experiments, masking coefficients are arrived at from masking experiments, etc.. The MPQM and the NVFM are parameterized using this approach [van96a]. Sometimes, specially designed subjective tests were conducted to parameterize the metric [KK95], where special test patterns were used to turn on or off functional blocks in the metric.

Although this approach has been effective in deriving model parameters, there are some issues to be discussed below [YWWC02]. In most psychovisual experiments, only a small group of assessors are employed; therefore, the data are usually sparse and it is often hard to compare results obtained by different researchers under different condi-

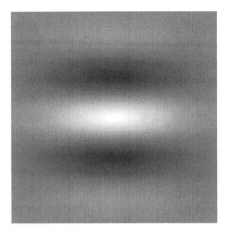

Figure 8.4: Example Gabor patch stimulus

tions. These problems have motivated the Modelfest effort to provide a reliable and general data set for common psychovisual stimuli [Car99]. Besides, it is important to note that most experiments in vision research are threshold experiments, while in applications such as video quality assessment and video coding, natural video scenes over a wide quality range are involved, where the stimuli may well exceed the perceptual threshold. There is also a significant difference between the tasks of subjects in the subjective tests. In vision research, assessors typically are asked for qualitative judgements (yes/no), while in video quality assessment, they need to rate the stimuli quantitatively using quality scales. The above considerations have prompted the development of a second parameterization approach, using video quality experiments.

8.2.6.2 Parameterization by video quality experiments

In this approach, the basic architecture of the HVS as revealed by vision research experiments is adopted. However, the vision model is optimized for video quality assessment using natural video sequences [YWWC02]. The comprehensive vision model described in the above subsections is able to measure objective video quality, given an original video and a processed video sequence. Meanwhile, the subjective quality of the same video sequences can also be assessed through subjective tests. With a collection of video sequences, how accurate the objective model is in predicting subjective quality can be measured through the correlation of subjective and objective data. As the subjective data are fixed for the given video sequences, the correlation is a function of the vision model parameters only. The basic problem is to find a set of parameters that produce the highest correlation between the subjective and the objective data, which can be solved through an optimization process.

With this approach, the available video quality subjective test data sets, such as the ANSI T1A1 [ANS93] and the Video Quality Experts Group (VQEG) [VQE00] tests, can all be used. These tests involved several well-known test laboratories with sufficient number of subjects, producing reasonably reliable test data. The VQEG Phase I test included both 50 Hz and 60 Hz tests, where in the 60 Hz tests, ten scenes and sixteen Hypothetical Reference Circuits (HRCs) were used [VQE00]. A subset of the VQEG Phase I 60 Hz subjective test data has been used to parameterize the PBDM [YWWC02].

The optimization approach in [YWWC02] is summarized here. The task is to obtain six CSF coefficients (c_i) for six frequency levels, and four pairs of w_i and γ_i. Given a pair of original and processed sequences, for any set of c_i, w_i and γ_i coefficients, the objective Video Quality Rating (VQR) can be obtained as described below. Firstly, the sequences are passed through the steerable pyramid decomposition without temporal filtering, and the subband coefficients of frames 30 and 120 are used for VQR calculation. The VQR of a frame is then obtained by applying CSF filtering, contrast gain control and pooling on the stored subband coefficients of that frame. For the same pair of original and processed sequences, the subjective Mean Opinion Score (MOS) can be taken from the VQEG subjective data.

Given a collection of original and processed sequences (seven HRCs and ten scenes were used in [YWWC02]), a collection of objective and subjective scores can be obtained. The agreement between the rank orders of the corresponding objective and subjective scores can be assessed with the Spearman rank order correlation [Sac84] (see also Appendix A). Given a collection of bivariate pairs (MOS_i, VQR_i), the Spearman correlation coefficient r_s is defined as:

$$r_s = 1 - \frac{6 \sum_l D_l^2}{N(N^2 - 1)}, \tag{8.7}$$

where D_l is the difference between the ranked pairs, l is the index of the data sample and N is the number of data samples.

The model optimization problem can be formulated as follows. Let c_i ($i = 1, 2, ..., 6$), w_i ($i = 1, 2, 3, 4$), and γ_i ($i = 1, 2, 3, 4$) represent the model coefficients in the objective model, and $\mathbf{C}, \mathbf{W}, \mathbf{\Gamma}$ be the vector representation of the corresponding set of coefficients. As an example, $\mathbf{W} = [w_1, w_2, w_3, w_4]^T$. Given a collection of VQR and MOS pairs, the problem is to find the optimal $\mathbf{C}_{opt}$, $\mathbf{W}_{opt}$, and $\mathbf{\Gamma}_{opt}$ that maximize the Spearman rank order correlation $r_s(\mathbf{C}, \mathbf{W}, \mathbf{\Gamma})$. Here the Spearman rank order correlation is used as the cost function to be maximized.

A search algorithm was proposed in [YWWC02], which was adapted to the architecture of the PC cluster computing facility used. The optimization algorithm can be described by the following pseudo-code.

Algorithm 1. *Model optimization*

(1) *initialize* $\mathbf{C}$ *and* $\mathbf{\Gamma}$ *to those of the NVFM [van96a],* $\mathbf{W}$ *to* $[1, 1, 1, 1]^T$ *and* $current\_r_s = 0$

(2) **repeat**

(3) *let* $previous\_r_s = current\_r_s$

(4) *fix* $\mathbf{C}, \mathbf{W}$ *and search for optimal* $\mathbf{\Gamma}$ *using Algorithm 2 (see below)*

(5) *fix* $\mathbf{C}, \mathbf{\Gamma}$ *to the optimal sets from (4) and search for optimal* $\mathbf{W}$ *using Algorithm 2*

(6) *fix* $\mathbf{\Gamma}, \mathbf{W}$ *to the optimal sets from (4) and (5), search for optimal* $\mathbf{C}$ *using Algorithm 2, and record the highest* r_s *as* $current\_r_s$

(7) **until** $current\_r_s - previous\_r_s < T$

(8) *output the* $\mathbf{C}, \mathbf{W}, \mathbf{\Gamma}$ *associated with the highest* r_s *as* $\mathbf{C}_{opt}, \mathbf{W}_{opt}, \mathbf{\Gamma}_{opt}$

As part of *Algorithm 1*, a search algorithm has been developed to find an optimal set of coefficients while the other two are fixed. For each coefficient being optimized and at each iteration, the optimal coefficient is selected from up to four candidate data points. As an example, let c_i be a coefficient to be determined. The candidate coefficient set can be denoted as $\{c_{ij} : j = 1, 2, 3, 4\}$, where c_{ij} is the jth candidate for coefficient c_i, and $c_{i1} < c_{i2} < c_{i3} < c_{i4}$. The candidate set is constructed around a seed coefficient[1] with a known search step size. A coarse search (i.e., with a large search step size $\Delta c_{ij,j+1} = \|c_{ij} - c_{ij+1}\|$) is conducted in the first iteration. If a locally maximal coefficient vector $\tilde{\mathbf{C}}$ has been found, a smaller search step will be used to search in the vicinity of $\tilde{\mathbf{C}}$ in the next iteration unless the improvement in correlation is below the threshold T. The locally maximal coefficient vector $\tilde{\mathbf{C}}$ is defined as the coefficient vector in the candidate set that meets the following two conditions:

a) it yields the highest correlation, and

b) none of its component coefficients $\tilde{c}_{ij}$ is at the boundary of the candidate set (i.e., $j \neq 1$ and $j \neq 4$ for coefficient c_i).

After one iteration, human interaction is required to review the search step size, to decrease the step size when needed, and to reduce the number of candidates for a particular coefficient if the previous iteration demonstrates that varying this coefficient has less influence on the correlation than other coefficients. Sometimes condition (a) can be met, but some component coefficients cannot satisfy condition (b). In this case, only these component coefficients will be varied in the next iteration, while the other component coefficients are fixed at the previous optimal value.

---

[1]The initial seed coefficient is obtained from *Algorithm 1* and passed on to *Algorithm 2*, e.g., the initial $\mathbf{\Gamma}$ in Step (4) of *Algorithm 1* is a seed coefficient vector.

The search algorithm can be summarized as follows.

Algorithm 2. *Search for an optimal coefficient vector among a candidate set.*

(1)   *initialize the seed coefficient vector* **S** *(passed from Algorithm 1), search step size* Δc *and set* $current\_r_s = 0$.

(2)   **repeat**

(3)   *let* $previous\_r_s = current\_r_s$

(4)   **repeat**

(5)   *construct the candidate coefficient set* $\mathcal{C}$ *in the vicinity of* **S** *with step size* Δc

(6)   *compute the corresponding* r_s *for every vector* **C** *in* $\mathcal{C}$

(7)   *find the vector* $\tilde{\mathbf{C}}$ *which yields the highest* r_s *(denoted as* $\tilde{r}_s$)

(8)   *let* $\mathbf{S} = \tilde{\mathbf{C}}$

(9)   **until** $\tilde{\mathbf{C}}$ *is a locally maximal coefficient vector*

(10)   *let* $current\_r_s = \tilde{r}_s$ *and* $\Delta c = \Delta c/2$

(11)   **until** $current\_r_s - previous\_r_s < T$

(12)   *output* $\tilde{\mathbf{C}}$ *as the optimal coefficient vector*

T was set to 0.0001 in the experiments. There were three to four iterations with *Algorithm* 2 before the optimization converged, and two iterations with *Algorithm* 1. Although the coefficients obtained are not globally optimized, a high correlation can still be achieved with this method. The optimized coefficients are shown in Table 8.3 [YWWC02]. Among the variable coefficients, w_i has little influence on the correlation. As a result, all w_i coefficients were fixed at 1.

8.3   Perceptual Blocking Distortion Metric

Over the development of image/video codecs and quality/impairment metrics, it has been noticed that different types of distortion are predominant in different regions of digitally compressed frames [Bal98, YWWC02]. Blocking artifacts, defined as discontinuities across block boundaries (see Chapter 3) [YW98], is a major type of distortion in images and videos coded with the block based DCT algorithm, and is more noticeable in smooth regions. In contrast, ringing artifacts mostly occur around high contrast edges of objects. The PBDM has been developed based on this observation. A region

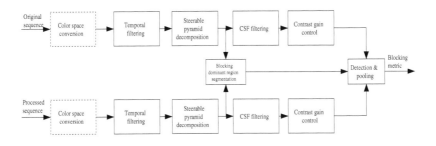

Figure 8.5: Block diagram of the perceptual blocking distortion metric

segmentation algorithm has been devised to separate blocking dominant regions in the processed frames. It is assumed that only distortions in the blocking dominant regions contribute to the human perception of blocking artifacts. The vision model, as described in Section 8.2, is adopted to measure perceptual distortions introduced to each pixel. Objective measurement of blocking artifacts can be obtained by calculating the cumulative perceptual distortions in the blocking dominant regions.

A block diagram of the PBDM is shown in Figure 8.5. Main components of the PBDM include temporal filtering, steerable pyramid decomposition, blocking region segmentation, CSF filtering, contrast gain control, detection and pooling.

The PBDM is heavily based on the vision model discussed in Section 8.2. The pixel-level distortion calculation chain remains unchanged, which consists of temporal filtering, steerable pyramid decomposition, CSF filtering and contrast gain control. Both the original and the processed sequences will go through the distortion calculation chain and the perceptual distortion of every pixel can be obtained. Compared with the quality metric model shown in Figure 8.1, the addition of a blocking dominant region segmentation stage is a major difference. Besides, the detection and pooling stage has been modified to sum up distortions in the blocking dominant regions. These two stages, new or modified in the PBDM, are described in detail as follows.

8.3.1 Blocking Dominant Region Segmentation

Blocking artifacts can be characterized as artificial horizontal or vertical edges in the processed images. There is an intimate relationship between edges and regions, therefore region segmentation can be solved through edge finding methods [YGvV97]. Gradient based and zero-crossing based methods have been traditional and effective methods for edge finding. Similarity to these well-known methods can be drawn with the edge finding algorithm in the PBDM. However, because of the difference in the filter bank used and the specific task to locate block edges, the algorithm has to be designed differently.

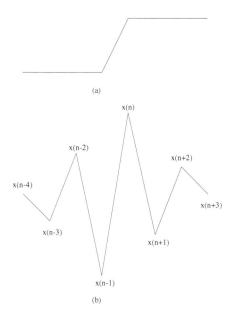

Figure 8.6: Gray level of part of one scan line before and after decomposition as an example. (a) Waveform in the pixel domain; (b) waveform in the DWT domain.

The spatial-temporal decomposition filter bank, described in Section 8.2, highlights different frequency components in different subbands. The characteristics of the spatial high-pass subband are very similar to those of the Laplacian filter, including, transforming a sharp transition of pixel values into a pair of local maximum and minimum in the coefficient domain. A typical waveform is shown in Figure 8.6. Therefore, detection of edges in a line or a row could be accomplished by detecting local maximum and minimum pairs in the high-pass subband of the processed sequence. However, an edge in a line or a row alone does not necessarily constitute a blocking artifact. Blocking artifacts are characteristics of a block of pixels, or a small region of pixels. To become visible blocking artifacts, the edges and, consequently, the local maximum and minimum pairs need to be persistent horizontally or vertically across several adjacent lines or rows. Such persistency is checked in the detection of block edges.

Before describing the block edge detection algorithm [YWWC02], it is necessary to review the notation. Images are two-dimensional data, as shown in Figure 8.7. As an example, a vertical block edge means that there are artificial and unified pixel value differences between pixels to the left hand side and to the right hand side of the block edge. When examining subband coefficients in a row around a vertical block edge, such as coefficients $x[m, n-4]$ to $x[m, n+3]$, most likely these coefficients will have values similar to those illustrated by the waveform in Figure 8.6(b). Normally the transition between $x[m, n-1]$ and $x[m, n]$ in such a one-dimensional waveform is called an "edge,"

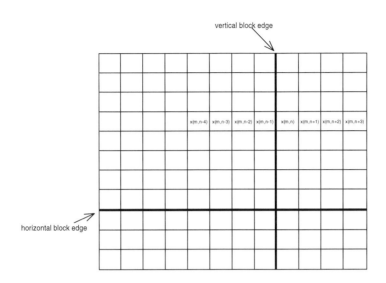

Figure 8.7: Notation of horizontal and vertical block edges

but to avoid confusion with "block edge," the pair $x[m, n-1]$ and $x[m, n]$ will be called an "edge point" instead in this chapter. The persistency of edges in adjacent lines to constitute blocking artifacts implies that a vertical block edge is characterized as several vertically adjacent edge points. Similar notations are applied to horizontal block edges.

The blocking dominant region segmentation algorithm consists of seven steps as detailed in the following subsections. The algorithm works on the high-pass band after the steerable pyramid transform. Figure 8.8 is used as an example to demonstrate how the algorithm works. It shows one frame of the *Claire* sequence before the decomposition (a), the high-pass subband of that frame after the decomposition (b), a magnified part of the images (c and d), and images after every stage of the segmentation algorithm.

8.3.1.1 Vertical and horizontal block edge detection

Vertical and horizontal block edge detections are conducted independently in both directions. In one direction, the detection is conducted in two stages. Vertical block edge detection is described below. First, all horizontally adjacent pixels are examined to locate edge points. A vertical block edge is then detected if there is a significant number of adjacent edge points along a common column.

At the first stage, edge points are detected as adjacent maximum-minimum pairs. The index pair $(k, l - 1)$ denotes the coordinate where an edge is to be located. If a vertical block edge is to be detected, the edge point detection will first be executed horizontally. At the beginning, the edge data point counter is reset to zero. If an adjacent maximum-minimum pair is detected, it is counted as an edge point. Mathematically, let $\mathbf{X}[k, l]$ represent the coefficient value at position (k, l). In the kth row, the notation $\mathbf{X}[k, l]$ can be rewritten as $\mathbf{X}_k[l]$. A typical 1-D waveform is illustrated in Figure 8.6b. If $\mathbf{X}_k[l - 1]$ and $\mathbf{X}_k[l]$ are a pair of local maximum and minimum, then $\mathbf{X}_k[l - 1]$ is marked as an edge point, the sign of $d_k[l - 1, l] = \mathbf{X}_k[l] - \mathbf{X}_k[l - 1]$ is stored, and the counter of edge data points gets an increment. The same procedure is applied to the other five data points which lie vertically below $\mathbf{X}[k, l - 1]$.

At the second stage, characteristics of six consecutive data points are considered because of the most commonly used 8×8 block size. A block edge is located if a certain amount of edge points exists within these six points, and that the edge points are persistent. The top row and the bottom row adjacent to the six rows under investigation may not show up as typical rows with edge points, because they are also affected by the neighboring blocks after the decomposition. Therefore these two rows are not included in the detection process.

Mathematically, the absolute sums $\hat{d}_i = \sum_{j=k}^{k+5} |d_j[i, i + 1]|$, $i \in [l - 3, l + 1]$ are calculated. If the following conditions are met, a vertical block edge is found:

- The edge data point counter is greater than four.

- The number of sign changes of $d_j[l - 1, l]$ between two vertically adjacent edge data points $[j, l - 1]$ and $[j, l]$ is less than two.

- The following inequalities are satisfied:

$$
\begin{aligned}
\hat{d}_{l-1} &> T_{B_1}, \\
\hat{d}_{l-1} &> \hat{d}_l + T_{B_2}, \\
\hat{d}_{l-1} &> \hat{d}_{l-2} + T_{B_2}, \\
\hat{d}_{l-2} &> T_{B_1}/2, \\
\hat{d}_l &> T_{B_1}/2, \\
\hat{d}_{l-3} &< \min(\hat{d}_{l-2}/2, T_{B_1}/2 + T_{B_2}), \text{ and} \\
\hat{d}_{l+1} &< \min(\hat{d}_l/2, T_{B_1}/2 + T_{B_2}),
\end{aligned}
$$

where T_{B_1} and T_{B_2} are positive thresholds determined by experiments.

If a vertical block edge is found, all the six data points plus the other two vertically neighboring points at two ends are marked as vertical edge points. A similar procedure is applied to horizontal block edges. Both the reference and the processed sequences are subject to edge detection, and edge indicator maps are generated for both sequences. Taking the *Claire* sequence as an example, the detected horizontal and vertical block edges in part of the processed and the reference frames are shown in Figure 8.8e-f.

8.3.1.2   Removal of edges coexisting in original and processed sequences

Edges detected in the original sequence are natural edges in the video content instead of blocking artifacts. Therefore only additional edges in the processed sequence need to be retained for blocking region segmentation. The edge maps of the original and the processed sequences are compared at this step. If one data point is marked as a vertical edge point in the original sequence, then the data point at the corresponding spatial location in the processed sequence and the other four horizontally neighboring points will be removed from the vertical edge map. Horizontal edge points are processed in a similar manner. Figure 8.8g illustrates the edges in the example, a frame from *Claire*, after this stage.

8.3.1.3   Removal of short isolated edges in processed sequence

Isolated short edges will not become visible blocking artifacts. Therefore, if the length of an edge detected in the processed sequence is shorter than a threshold (e.g., the size of a block — 8 pixels — in the PBDM [YWWC02]), and no pixel in the edge appears as the crossing of a horizontal edge and a vertical edge, then the edge is considered as a short isolated edge and should be removed from the edge indicator map. The algorithm scans the edge indicator map from top to bottom and left to right, and removes all short isolated edges. The edges in the example, a frame from *Claire*, as a result of this stage are shown in Figure 8.8h.

8.3.1.4   Adjacent edge removal

Two detected adjacent edges represent only one real edge. Therefore, one of them should be removed from the edge map. The algorithm searches for any parallel adjacent horizontal or vertical edges and keeps only one edge (the left or the top) of the adjacent pairs in the edge map. Figure 8.8(i) shows the edges in the example, a frame of *Claire*, after this step.

8.3.1.5   Generation of blocking region map

Blocking artifacts are characteristics of a region of coefficients. Therefore, it is most likely that eight rows or columns surrounding an edge will show up as blocking artifacts. This assumption is the basis of the blocking region generation algorithm. In the case of vertical edges, any coefficient detected as a vertical edge point, together with eight adjacent coefficients on the left and the right sides of the coefficient are classified as in the blocking region. Similar processing is applied to horizontal edges. The generated blocking region map for the example, a frame from *Claire*, is shown in Figure 8.8j.

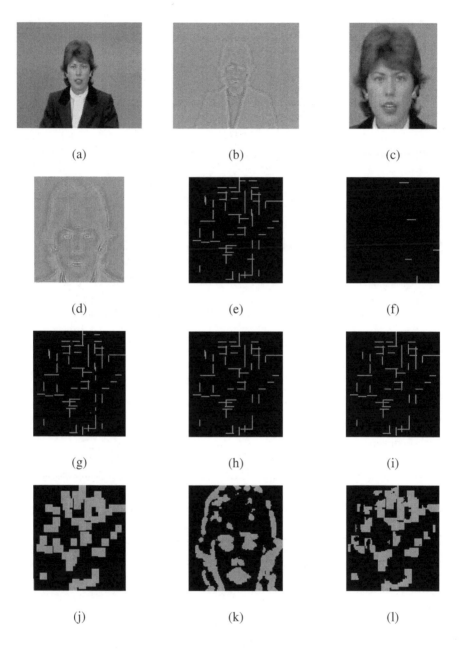

Figure 8.8: One frame of the *Claire* sequence. (a) Processed frame; (b) High-pass subband after the SPT decomposition; (c) Magnified part of the processed frame; (d) Magnified part of the high-pass subband (e) Horizontal and vertical edges in the processed sequence; (f) Horizontal and vertical edges in the original sequence; (g) After the removal of edges coexisting in the original and processed sequences; (h) After the removal of short edges; (i) After the removal of adjacent edges; (j) Blocking region map; (k) Ringing region map; (l) After the removal of ringing regions.

8.3.1.6 Ringing region detection

Strong ringing artifacts will occur along high-contrast edges of objects after encoding and decoding. Since the steerable pyramid decomposes images into several spatial frequency-orientation bands, these high contrast object edges will induce strong oscillations in the high-pass band of the original sequence. Such oscillations usually appear as large value difference between adjacent coefficients. Based on this property, ringing dominant regions can be determined. The detection algorithm is performed on the original sequence. Given a coefficient $\mathbf{X}[k, l]$, the inter-coefficient differences of its surrounding 5×5 block are calculated. If the following condition is met:

$$
\begin{aligned}
\sum_{j=-2}^{2} \sum_{i=-2}^{1} (\mathbf{X}[k+j, l+i] - \mathbf{X}[k+j, l+i+1])^2 + \\
\sum_{j=-2}^{1} \sum_{i=-2}^{2} (\mathbf{X}[k+j, l+i] - \mathbf{X}[k+j+1, l+i])^2 > T_{B_3}
\end{aligned}
\tag{8.8}
$$

it indicates large inter-coefficient differences likely caused by ringing artifacts and, consequently, the 3×3 block surrounding $\mathbf{X}[k, l]$ is detected as in a ringing region. The detected ringing regions in the frame of *Claire* are shown in Figure 8.8k.

8.3.1.7 Exclusion of ringing regions from blocking region map

The major distortions in the ringing dominant regions are the reconstruction errors of edges, which appear as ringing artifacts. Even though additional block edges may still be detected in the regions, they will not appear as visible blocking artifacts since they are usually masked by strong ringing artifacts. Therefore, ringing dominated regions are removed from the blocking region map.

After these seven steps, the final blocking region map is obtained, which represents the regions where blocking artifacts are dominant. Using the *Claire* sequence as an example, the magnified part of the blocking region map is illustrated in Figure 8.8(l), which is consistent with subjective observation.

Three thresholds for blocking region segmentation, T_{B_1}, T_{B_2} and T_{B_3}, were parameterized experimentally. Two video sequences, *Claire* (CIF) and *Car Phone* (QCIF) were used for this purpose. The thresholds were adjusted so that blocking regions segmented by the algorithm described above coincided with the subjective segmentation. The resulting values are $T_{B_1} = 8$, $T_{B_2} = 1$ and $T_{B_3} = 300$.

8.3.2 Summation of Distortions in Blocking Dominant Regions

In this stage, the summation of the differences between outputs from the original and the processed sequences is carried out over spatial frequency and orientation subbands according to the blocking region map, and then averaged by the number of frames N, in order to obtain the blocking distortion d_B. Mathematically, this step can be expressed as

$$d_B = \frac{\sum_{j,f,\theta,k,l} \left| \mathbf{X}_{R_o}[j,f,\theta,k,l] - \mathbf{X}_{R_p}[j,f,\theta,k,l] \right|^2}{N}, \forall (k,l) \in \mathcal{B} \qquad (8.9)$$

where $\mathcal{B}$ denotes the set of coefficient coordinates in the blocking dominant region, and the other notations follow (8.6). After the summation, d_B needs to be converted to the Objective Blocking Rating (OBR) on a scale between 1 and 5, corresponding to the five-grade impairment scale commonly used in subjective tests [ITU98]. This OBR is defined as the PBDM and formulated as:

$$PBDM = \begin{cases} 5 - d_B^{0.6} & \text{if } d_B < 4^{1/0.6}, \\ 1 & \text{otherwise.} \end{cases} \qquad (8.10)$$

The saturation threshold $4^{1/0.6}$ is a precaution in case d_B is a very large. The exponent 0.6 was derived through experiments to best fit the MOS with the PBDM.

8.3.3   Performance Evaluation of the PBDM

Both subjective and objective tests were conducted to evaluate the performance of the PBDM, where evaluation metrics used included correlations and prediction errors between the subjective and the objective data [YWWC02].

The test video sequences were selected from the ANSI T1A1 data set, including *disgal, smity1, 5row1, inspec* and *ftball* [ANS95], and the VQEG data set, i.e., *Src13 Balloon-pops, Src14 NewYork 2, Src15 Mobile & Calendar, Src16 Betes pas betes* and *Src18 Autumn leaves* [VQE99]. The sequences are interlaced with a frame rate of 30 frames per second and at a spatial resolution of 720×480 pixels. The lengths of the ANSI and the VQEG sequences are 360 and 260 frames, respectively.

The notation of the HRC (Hypothetical Reference Circuit) refers to the codec system set with different bit rates, resolutions, and methods of encoding. Target bit rates were selected to cover a full range of blocking impairments.

The ANSI sequences were encoded by a software MPEG-2 encoder [MPE94] at five different bit rates (five HRCs)

- HRC31, MPEG-2 at 1.4 Mbps
- HRC32 MPEG-2 at 2 Mbps
- HRC33, MPEG-2 at 3 Mbps
- HRC35 MPEG-2 at 768 kbps
- HRC36, MPEG-2 at 5 Mbps

The VQEG sequences were coded at two bit rates:

- HRC9, MPEG-2 at 3 Mbps, full resolution
- HRC14, MPEG-2 at 2 Mbps, 3/4 horizontal resolution

The characteristic of the VQEG MPEG-2 encoder is different from that of the encoder used for the ANSI sequences in that it is effective in minimizing blocking artifacts, even at low bit rates. Therefore, blurring artifacts are dominant in most of the VQEG sequences, whilst blocking artifacts are less obvious and only visible in some sequences, and ringing artifacts are even less noticeable.

The Double-Stimulus Impairment Scale variant II (DSIS-II) method as standardized in ITU-R BT. 500 [ITU98] was adopted as the subjective test method as described in Chapter 4. The DSIS-II method presents two sequences to the participants, first the original, then the processed, and then the sequences are repeated in sequel. The assessors were asked to vote on the degree of blocking distortions only. A five-grade impairment scale was used: (5) imperceptible; (4) perceptible, but not annoying; (3) slightly annoying; (2) annoying; and (1) very annoying. The panel consisted of five expert viewers. Before the test, instructions were given to the subjects, followed by a training session.

The display used was a SONY BVM-20F1E color video monitor with an active display diagonal of 20 inches. The sequences were played back using a SONY D-1 digital video cassette recorder and converted to component analog video by an Abekas A26 D1 to analog converter. The viewing distance was five times the screen height. The test session was randomized so that no two consecutive trials presented the same video sequence or HRC.

Test data were analyzed following [ITU98] and [VQE00]. First, the MOS and the associated confidence intervals were calculated. Screening of subjective data was carried out and no subject had to be discarded as a result.

The outputs of the PBDM (i.e., the OBR) were calculated for all test sequences. The performance of the OBR as a prediction of the MOS was evaluated using three metrics: the Spearman rank order correlation, the Pearson correlation with logistic fit, and the average absolute error between the MOS and the PBDM ($E|error|$) [MKA98]. Table 8.4 presents the evaluation results. The 95% confidence bounds of the Pearson-logistic metric were calculated using the method described in [SC89]. The average standard deviations of the subjective data are also shown in the table. Figure 8.9 illustrates the scatter plots of the PBDM and the PSNR versus the MOS, respectively. As shown experimentally, the PBDM achieves a very good agreement with the MOS, as reflected by the high correlations and low prediction errors. The performance of the PSNR as a blocking metric is also included in the table as well as in the figure, which reveals that the PSNR does not provide an accurate measurement of subjective blocking distortions.

Figure 8.10 shows two sample images, representative of two test sequences. The PSNR indicates similar quality for the two images. However, more severe blocking

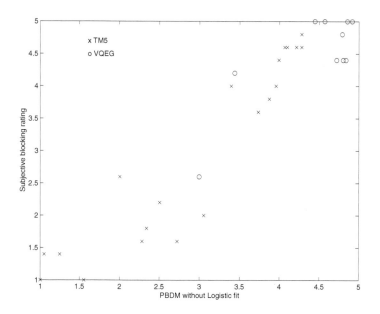

(a) PBDM without the logistic fit

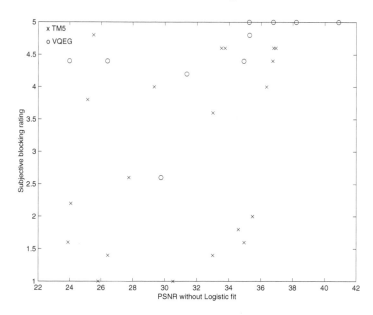

(b) PSNR (in dB) without the logistic fit

Figure 8.9: Scatter plots of objective ratings versus mean subjective ratings.

Table 8.4: Blocking Metric Performance

Metric	Pearson-logistic metric			Spearman	$E\lvert error\rvert$	Average standard deviation in the subjective test data
	Correlation	Upper bound	Lower bound			
PBDM	0.961	0.982	0.918	0.910	0.296	0.455
PSNR	0.489	0.726	0.149	0.517	1.038	

distortions are discerned in the *inspec* sequence, which is accurately predicted by the PBDM.

8.4 Perceptual Ringing Distortion Measure

Another major type of distortion in hybrid MC/DPCM/DCT coded videos is the ringing artifacts, which are fundamentally related to the Gibb's phenomenon when quantization of individual coefficients results in high-frequency irregularities of the reconstructed block. Mosquito artifacts also could be regarded as the temporal version of ringing artifacts, where high frequency fluctuations in areas around high contrast edges appear due to coarsely quantized higher frequency AC coefficients. More detailed discussions can be found in Chapter 3.

An HVS based objective measurement of ringing artifacts for digital video was first proposed in [YWC00b], and was based on the Teo-Heeger vision model, as described in Section 8.2. The work was extended in [ZWY+03] to normalize the objective ringing rating to the scale of 0 to 5, which is the range commonly adopted in subjective tests [ITU98]. Meanwhile in [ZWY+03], the performance of the objective ringing metric also was verified through subjective tests. These two pieces of work are collectively referred to as the Perceptual Ringing Distortion Metric (PRDM) in this chapter.

Ringing artifacts are most evident along high contrast edges, if either side of the edge is in the area of generally smooth texture [YW98]. Based on this property, the ringing dominant regions are detected in the PRDM. The same vision model as described in Section 8.2 is used to calculate the distortion at every pixel location, and the distortions in the ringing regions are summed up to form an objective measure of ringing artifacts.

The PRDM consists of the following five stages: spatio-temporal decomposition, ringing region segmentation, CSF filtering, contrast gain control, and detection and pooling. The block diagram of the metric is illustrated in Figure 8.11. The metric takes the original and the processed video sequences as the inputs. In the current met-

(a) *MOS*: 3.6; PBDM: 3.79; PSNR: 32.99dB.

(b) *MOS*: 1.4; PBDM: 1.22; PSNR: 33.00dB.

Figure 8.10: Performance comparison of different metrics for (a) *5row1* and (b) *inspec* sequences

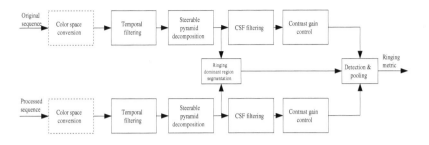

Figure 8.11: Block diagram of the perceptual ringing distortion metric

ric, the calculation is only applied to the luminance component of the video. Both the original and the processed sequences are decomposed by the spatio-temporal filter-bank, as discussed in Section 8.2. However, since most of the distortions exist in the sustained temporal channel, only the temporal low-pass channel is used for the spatial decomposition and distortion calculation, thus reducing the computational complexity. Following the decomposition, ringing dominant regions are segmented and an associated ringing region map is generated. CSF filtering is achieved by multiplying each spatial frequency subband with a weighting coefficient. The Teo-Heeger contrast gain control model [TH94a] is then implemented to account for pattern masking. In the last stage, differences between the outputs from the original and the processed sequences are summed up over spatial frequency and orientation channels, and a JND (Just Noticeable Distortion) map is generated afterwards. According to the ringing region map, distortions above the JNDs in the ringing regions are summed up and scaled to the range of 0 to 5 as the PRDM rating.

Among the components of the PRDM, only ringing region segmentation and detection and pooling are tailored for ringing artifacts, and hence described in detail below.

8.4.1   Ringing Region Segmentation

The spatio-temporal multi-channel mechanism of the vision model decomposes the stimuli into several temporal, spatial frequency and orientation tuned channels, while the spatial high-pass channel preserves and highlights texture and edge information. Sharp edges appear as strong rippling and different types of textures demonstrate different degrees of local pixel value variations. Ringing artifacts are most apparent in this spatial high-pass channel as strong rippling around object edges.

There are at least two methods to detect object edges. One is to use edge detection algorithms directly. However, if the area of the object is too small or the edge exists in the area of high complexity texture, distortions around the edge will not show up as ringing artifacts. It is usually hard to distinguish these situations by an automated method. For this reason, an alternative approach has been adopted in the PRDM, which is based on region segmentation and growing. Firstly, all pixels are classified into three types depending on the variance of its neighboring pixels: smooth texture, complex texture, and unknown regions. A region growing algorithm is then used to grow the smooth texture and the complex texture regions in order to classify the unknown pixels into either type of regions. The boundaries between the smooth texture and the complex texture regions are detected as regions of ringing artifacts.

Based on this idea, a boundary detection algorithm is developed. The ringing dominant region detection is performed on the highest frequency band of the original sequence and consists of the following steps.

8.4.1.1 Modified variance computation

The variance of each 5×5 coefficient block is computed and saved as the variance value of the central location. In the modified variance computation, the absolute value is used instead of the squared value to reduce the computational complexity. Mathematically, the mean, $\mu[k, l]$, is defined as

$$\mu[k, l] = \frac{1}{25} \sum_{j=-2}^{2} \sum_{i=-2}^{2} \mathbf{X}[j + k, i + l] \tag{8.11}$$

and the variance, $\sigma_{abs}[k, l]$,

$$\sigma_{abs}[k, l] = \sum_{j=-2}^{2} \sum_{i=-2}^{2} |\mathbf{X}[j + k, i + l] - \mu[k, l]| \tag{8.12}$$

where $\mathbf{X}[k, l]$ is the coefficient value at coordinate (k, l) of the high-pass band and $\sigma_{abs}[k, l]$ is the associated variance.

The modified variance is subject to 3×3 block averaging to remove all possible noises.

$$\sigma_m[k, l] = \frac{1}{9} \sum_{i=-1}^{1} \sum_{j=-1}^{1} \sigma_{abs}[i + k, j + l] \tag{8.13}$$

8.4.1.2 Smooth and complex region detection

First of all, a threshold process is performed on all coefficients to detect the extremely smooth or complex areas. If the modified variance, σ_m, of a coefficient is less than a low threshold (T_{R_1}), it is classified as in the smooth texture region $(\mathcal{R}_1)$. If σ_m is higher than a high threshold (T_{R_2}), it is grouped as in the complex region ($\mathcal{R}_2$). A complex region includes high contrast edges and complex textures. Other coefficients with modified variances between T_{R_1} and T_{R_2} are of unknown type and are classified as in region $\mathcal{R}_3$.

After the threshold process, elements in $\mathcal{R}_3$ are basically boundary coefficients and need to be classified into either $\mathcal{R}_1$ or $\mathcal{R}_2$ so that the exact boundary between the smooth texture region ($\mathcal{R}_1$) and the complex region ($\mathcal{R}_2$) can be detected. A region growing algorithm, as described by the pseudo code in Algorithm 3, is designed for this purpose in which every element is classified based on its value and the property of neighboring elements. Here, the neighborhood of one element means the nearest eight elements.

Both the smooth texture and the complex regions can be grown with this algorithm. The thresholds are determined by experiments which give Threshold 1 (T_{R_1}) = 25, Threshold 2 (T_{R_2}) = 90 and Threshold 3 (T_{R_3}) = 60.

Algorithm 3. *Smooth Region Growing*

```
pre_num=-1;
cur_num=0;
while (cur_num>pre_num)
{
     pre_num=cur_num;
     for (every coefficient at (k,l) of the coefficient array)
        if ((modified variance at (k,l) > Threshold 1) &&
           (modified variance at (k,l) < Threshold 3) &&
           (coefficient at (k,l) belongs to Region 3) &&
           (any neighboring element belongs
             to Region 1))
           {
             mark element (k,l) in Region 1;
             cur_num++;
           }
}
```

8.4.1.3 Boundary labeling and distortion calculation

The boundaries between $\mathcal{R}_1$ and $\mathcal{R}_2$ are labeled as the high contrast object boundaries. Then the 9×9 blocks around the boundary elements are marked as in the ringing region. In some circumstances $\mathcal{R}_3$ still exists after region growing. The boundary of $\mathcal{R}_3$ will not be marked because no sharp edge exists in this type of boundary and ringing artifacts therein will not be obvious.

8.4.2   Detection and Pooling

As discussed previously, following the vision model, the JND map can be generated where each coefficient value represents the visible distortion at that location. Combining the JND map and the ringing region map, the objective ringing artifact measure d_R can be formed by summing up the differences of all ringing region elements over all the spatial frequency and orientation channels between the original and processed sequences using Minkowski summation. Then d_R is converted to a scale of 0 to 5 as the PRDM with the following formula:

$$PRDM = \begin{cases} 5 - d_R^{0.75} & \text{if } d_R < 5^{4/3}, \\ 0 & \text{otherwise.} \end{cases} \tag{8.14}$$

The formula is acquired through the best fit of the objective distortion to the MOS using the nonlinear least-square regression method [ZWY$^+$03].

8.4.3   Performance Evaluation of the PRDM

The performance of the PRDM was evaluated by subjective tests, using MPEG-2 coded video sequences covering a range of degrees of ringing artifacts [ZWY$^+$03].

Two test sequences, Mobile & Calendar (MC) and Table Tennis (TT), were selected as they could exhibit various degrees of ringing artifacts after the MPEG-2 encoding. The two original sequences were MPEG-2 encoded at ten different bit rates, namely, 1.5, 1.75, 2.0, 2.25, 2.5, 2.75, 3.0, 4.0, 5.0 and 7.5 Mbps.

The subjective test method used was derived from the standard DSIS-II method [ITU98], with the major difference being the grading scale. A continuous scale between zero and five was employed, to avoid quantization errors associated with the original five grade scale scheme. In the test, sequences from the two sources were presented in an interleaved order. The sequences with the same scene and different coding bit rates were presented in a pseudo-random order.

The panel consisted of five expert viewers and two non-expert viewers. The assessors were required to vote on the degree of ringing distortions only, and were trained before the subjective test to differentiate ringing artifacts from other distortions. This was achieved using four processed sequences — Sailboat (1.0 Mbps and 3.0 Mbps) and Tempete (1.0 Mbps and 2.0 Mbps).

In the data analysis stage, the mean opinion scores (MOS) and the standard deviations were calculated. Screening of the observers was performed following ITU-R BT.500 [ITU98] and no subject was discarded after screening.

Three evaluation metrics were used to compare the output of the PRDM software model with the MOS, i.e., Pearson correlation for prediction accuracy, Spearman rank-

Table 8.5: Ringing Metric Performance

Metric	Pearson	Spearman	E|error|	Average standard deviation in the subjective test data
PRDM(both)	0.9397	0.9805	0.2651	0.4402
PSNR(both)	0.7674	0.7489	0.5358	
PRDM(MC)	0.9642	1	0.2457	0.4219
PSNR(MC)	0.9341	1	0.6659	
PRDM(TT)	0.9945	1	0.2845	0.4649
PSNR(TT)	0.9709	1	0.4056	

order correlation for prediction monotonicity [VQE00], and the average absolute error between the MOS and the PRDM (E$|error|$) [MKA98].

The performance of the PSNR as a ringing metric was also evaluated. Since the raw results of the PSNR did not fall into the range of 0 to 5, they were passed through a logistic fit to scale into the range, by assuming a monotonic nonlinear relationship.

Table 8.5 presents the evaluation results of the PRDM and the PSNR [ZWY$^+$03]. Figure 8.12 also illustrates a scatter plot of the PRDM versus the MOS and the PSNR versus the MOS [ZWY$^+$03]. The experimental results show that the PRDM is quite accurate in measurement of the ringing artifacts and has a performance superior to that of the PSNR.

8.5 Conclusion

Solid foundations have been laid and encouraging progress has been made in vision model based objective measurement of digital image and video coding artifacts. Several metrics have proven their accuracy in measuring certain artifacts, e.g., blocking and ringing. However, this field is still new and far from mature. To conclude this chapter, the authors would like to point out several important and challenging research topics and open issues:

- Better understanding of the human visual system and consequently applying the knowledge to impairment metrics. So far the understanding and application of the HVS is still limited and several important phenomena, most notably temporal masking, have yet to be modeled.

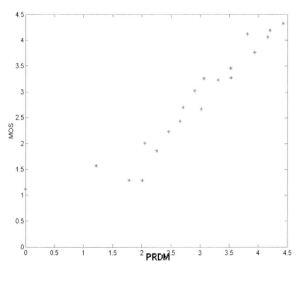

(a) *PRDM*

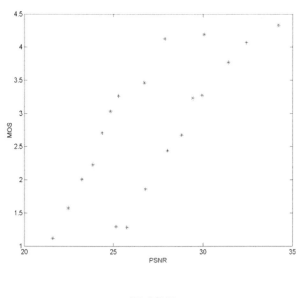

(b) *PSNR*

Figure 8.12: Scatter plots of objective rating versus mean subjective ratings on ringing artifacts

- To extend the impairment metrics to other distortions, including blurring, temporal jerkiness, etc.

- Better and more accurate vision model based no-reference and reduced-reference metrics.

- The application of the impairment metrics in other areas, e.g., to improve image and video coding systems.

References

[ANS93]   ANSI Accredited Standards Working Group T1A1 contribution number T1A1.5/94-118R1. *Subjective Test Plan (Tenth and Final Draft)*. Alliance for Telecommunications Industry Solutions, 1200 G Street, NW, Suite 500, Washington DC, Oct. 1993.

[ANS95]   ANSI T1.801.01-1995. *ANSI T1.801.01-1995, American National Standard for Telecommunication — Digital Transport of Video Teleconferencing/Video Telephony Signals - Video Test Scenes for Subjective and Objective Performance Assessment*. American National Standards Institute, 1995.

[Bal98]   I. Balasingham et al. Performance evaluation of different filter banks in the jpeg-2000 baseline system. In *Proc. IEEE Int'l. Conf. Image Processing*, 2:569–573, 1998.

[BL01]   A. C. Bovik and S. Liu. Dct-domain blind measurement of blocking artifacts in dct-coded images. In *Proc. IEEE ICASSP*, 3:1725–1728, 2001.

[Car99]   T. Carney et. al. The development of an image/threshold database for designing and testing human vision models. In *Human Vision, Visual Processing, and Digital Display IX, Proc. SPIE*, 3644:542–551, 1999.

[Fol94]   J. M. Foley. Human luminance pattern-vision mechanisms: Masking experiments require a new model. *J. Opt. Soc. Am.*, 11:1710–1719, 1994.

[Gir93]   B. Girod. What's wrong with mean-squared error. In A. B. Watson, Ed., *Digital images and human vision*, 207–220. Cambridge, MA: The MIT Press, 1993.

[HJ57]   L. M. Hurvich and D. Jameson. An opponent-process theory of color vision. *Psych. Rev.*, 64(2):384–404, 1957.

[HS92]   R. F. Hess and R. J. Snowden. Temporal properties of human visual filters: Number, shapes and spatial covariation. *Vision Res.*, 32(1):47–59, 1992.

[ITU95]   ITU-R. *BT.601-5, Studio Encoding Parameters of Digital Television for Standard 4:3 and Wide-Screen 16:9 Aspect Ratios*. ITU, Geneva, Switzerland, 1995.

[ITU98]   ITU-R. 500-9. *ITU-R BT. 500-9, Methodology for the Subjective Assessment of the Quality of Television Pictures*. ITU, Geneva, Switzerland, 1998.

[KK95]   S. A. Karunasekera and N. G. Kingsbury. A distortion measure for blocking artifacts in images based on human visual sensitivity. *IEEE Trans. Image Processing*, 4(6):713–724, Jun. 1995.

[LB82]   F. J. Lukas and Z. L. Budrikis. Picture quality prediction based on a visual model. *IEEE Trans. Commun.*, COM-30:1679–1692, Jul. 1982.

[LB02]   S. Liu and A. C. Bovik. Efficient dct-domain blind measurement and reduction of blocking artifacts. *IEEE Trans. Circ. and Syst. for Video Technology*, 12(12):1139 –1149, Dec. 2002.

[Lin96]   P. Lindh. *Perceptual Image Sequence Quality Metric Using Pixel Domain Filtering*. Linkoping Institute of Technology, Linkoping, Sweden, April 1996. Masters Thesis.

[MKA98]   M. Miyahara, K. Kotani, and V. R. Algazi. Objective picture quality scale (PQS) for image coding. *IEEE Trans. Commun.*, 46(9):1215–1226, 1998.

[MPE94]   MPEG Software Simulation Group. *mpeg2encode/mpeg2decode version 1.1*, Jun. 1994. (accessable at: *ftp.netcom.com*).

[OS89]   A. V. Oppenheim and R. W. Schafer. *Discrete-Time Signal Processing*. Englewood Cliffs, NJ: Prentice–Hall, 1989.

[PW84]   G. C. Philips and H. R. Wilson. Orientation bandwidth of spatial mechanisms measured by masking. *J. Opt. Soc. Am. A*, 1(2):226–232, 1984.

[PW96]   A. B. Poirson and B. A. Wandell. Pattern-color separable pathways predict sensitivity to simple colored patterns. *Vision Res.*, 36(4):515–526, 1996.

[Sac84]   L. Sachs. *Applied Statistics*. Berlin and Heidelberg: Springer–Verlag, 1984.

[SC89]   G. W. Snedecor and W. G. Cochran. *Statistical Methods*. Ames, IA: The Iowa State University Press, 1989.

[Sec96]   Secretariat, Alliance for Telecommunications Industry Solutions. *T1.801.03-1996, American National Standard for Telecommunications - Digital Transport of One-Way Video Signals - Parameters for Objective Performance Assessment*. ANSI, 1996.

[SF95]   E. P. Simoncelli and W. T. Freeman. The steerable pyramid: A flexible architecture for multi-scale derivative computation. In *Proc. IEEE 2nd Int'l Conf. On Image Processing*, 175–184, Washington DC, Oct. 1995.

[TH94a]   P. T. Teo and D. J. Heeger. Perceptual image distortion. In *Proc. First IEEE International Conference on Image Processing*, 2:982–986, Nov. 1994.

[TH94b]   P. T. Teo and D. J. Heeger. Perceptual image distortion. In *Proc. SPIE*, 2179:127–141, 1994.

[Val82]   R. L. D. Valois et. al. Spatial frequency selectivity of cells in macaque visual cortex. *Vision Res.*, 22(5):545–559, 1982.

[van96a]   C. van den Branden Lambrecht. *Perceptual Models and Architectures for Video Coding Applications*. Ph.D. thesis, Swiss Federal Institute of Technology, Lausanne, Switzerland, 1996.

[van96b]   C. J. van den Branden Lambrecht. Color moving pictures quality metric. In *Proc. ICIP*, 1:885–888, Lausanne, Switzerland, 1996.

[VQE99]   VQEG. *VQEG Subjective Test Plan Ver. 3*. VQEG, Jul. 1999. (accessable at: *http://www.vqeg.org/*).

[VQE00]   VQEG. *Final report from the video quality experts group on the validation of objective models of video quality assessment.* VQEG, Mar. 2000. (accessible at: *http://www.vqeg.org/*).

[Wan95]   B. A. Wandell. *Foundations of Vision.* Sunderland, MA: Sinauer Associates, Inc., 1995.

[Wat83]   A. B. Watson. Detection and recognition of simple spatial forms. In A. C. Slade, Ed., *Physical and Biological Processing of Images*, 100–114. Berlin: Springer–Verlag, 1983.

[WBE00]   Z. Wang, A. C. Bovik, and B. L. Evan. Blind measurement of blocking artifacts in images. In *Proc. IEEE ICIP*, 3:981–984, 2000.

[Win99a]   S. Winkler. Issues in vision modeling for perceptual video quality assessment. *Signal Processing*, 78(2):231–252, 1999.

[Win99b]   S. Winkler. A perceptual distortion metric for digital color video. In *Human Vision and Electronic Imaging, Proc. SPIE*, 3644:175–184. SPIE, 1999.

[Win00]   S. Winkler. Quality metric design: A closer look. In *Proc. SPIE Human Vision and Electronic Imaging Conference*, 3959:37–44, Jan. 2000.

[WJP$^+$93] A. A. Webster, C. T. Jones, M. H. Pinson, S. D. Voran, , and S. Wolf. An objective video quality assessment system based on human perception. In *Proc. SPIE Human Vision, Visual Processing, and Digital Display IV*, 1913:885–888, San Jose, CA, Feb. 1993.

[WR84]   H. R. Wilson and D. Regan. Spatial-frequency adaptation and grating discrimination: Predictions of a line-element model. *J. Opt. Soc. Am.*, 1:1091–1096, 1984.

[WS97]   A. B. Watson and J. A. Solomon. A model of visual contrast gain control and pattern masking. *J. Opt. Soc. Am. A*, 14:2379–2391, 1997.

[WY97]   H. R. Wu and M. Yuen. A generalized block-edge impairment metric (GBIM) for video coding. *IEEE Signal Processing Letters*, 4(11):317–320, Nov. 1997.

[YGvV97]   I. T. Young, J. J. Gerbrands, and L. J. van Vilet. Image processing fundamentals. In V. K.Madisetti and D. B.Williams, Eds., *The Digital Signal Processing Handbook*. Boca Raton, FL: CRC Press, 1997.

[YW98]   M. Yuen and H. R. Wu. A survey of hybrid mc/dpcm/dct video coding distortions. *Signal Processing*, 70:247–278, 1998.

[YWC00a]   Z. Yu, H. R. Wu, and T. Chen. Perceptual blocking distortion measure for digital video. In *Proceedings of the SPIE Visual Communication and Image Processing 2000*, 4067:768–779, June 2000.

[YWC00b]   Z. Yu, H. R. Wu, and T. Chen. A perceptual measure of ringing artifact for hybrid mc/dpcm/dct coded video. In *Proceedings of the IASTED International Conference on Signal and Image Processing (SIP 2000)*, 94–99, Nov. 2000.

[YWWC02] Z. Yu, H. R. Wu, S. Winkler, and T. Chen. Objective assessment of blocking artifacts for digital video with a vision model. *Proc. IEEE*, 90(1):154–169, Jan. 2002. (An invited paper in Special Issue on Translating Human Vision Research into Engineering Technology).

[ZWY$^+$03] Z. Zhe, H. R. Wu, Z. Yu, T. Ferguson, and D. Tan. Performance evaluation of a perceptual ringing distortion metric for digital video. In *Proceedings of ICASSP*, III:613–616, April 2003.

Chapter 9

Computational Models for Just-Noticeable Difference

Weisi Lin
Institute for Infocomm Research, Singapore

9.1 Introduction

Digital images are acquired, synthesized, enhanced, watermarked, compressed, transmitted, stored, reconstructed, authenticated, displayed, or printed before being presented to the human visual system (HVS). In various image processing tasks, pixels are processed for visual quality improvement, compact signal representation or efficient data protection. It is well known that the HVS cannot sense all changes in an image due to its underlying physiological and psychological mechanisms. Therefore it is advantageous to incorporate knowledge of the HVS visibility thresholds into image processing algorithms and systems, since the HVS is the ultimate receiver of the majority of processed images and video. With such knowledge, the scarce system resources (computing power, bandwidth, memory space, display/printing resolution, and so on) can be allocated to achieve the maximum perceptual significance, accessory information (e.g., for watermarking, authentication, and error protection) can be concealed in the regions with the least HVS sensitivity to the incurred changes, and visual quality of processed images can be evaluated for better alignment with the human perception. Incorporating the HVS visibility thresholds appropriately can play an important role in shaping and optimizing many image processing algorithms.

Just-noticeable difference (JND)[1] refers to the visibility threshold below which any change cannot be detected by the HVS [JJS93]. Its determination in general is a complex

---

[1] JND represents just-noticeable distortion in some literature. However, the term of just-noticeable difference is general since the change incurred in a processed image does not necessarily lead to visual distortion; an example is image sharpening (i.e., sharpened edges usually enhance the perceived quality in spite of the introduction of objective pixel errors).

and challenging task, because it is related to the HVS characteristics, as well as some recognition process in the human brain, and is adaptive to the contents of the visual signal under consideration. Other affecting factors include viewing conditions (such as viewing distance, ambient lighting, the context of preceding display, the pixel position in the image), and even viewers' preference and experience (e.g., trained observers are able to spot certain changes more easily).

9.1.1   Single-Stimulus JND Tests

In psychophysical studies, the relationship between an external physical stimulus (sensation) and its psychological representation in the mind (perception) has been investigated. As aforementioned, JND is the minimum sensory difference in the image that is perceived by the HVS. Various computational models for JND (e.g., [AP92, Wat93, HM98, CL95, CB99, RPG99]) are mainly based upon the effect of the basic stimuli described below.

A test image x_t for the psychophysical experiments on JND can be expressed as:

$$x_t = x_o + t \tag{9.1}$$

where x_o represents the original image, and t a visual stimulus.

The concept of JND can be traced back to the discovery of Weber-Fechner law (in 1800's). If x_o is a uniform image and t is a luminance variation, Weber-Fechner law states that the just noticeable t increases with background luminance x_o, and the law holds for any x_o above 10 cd/m^2.

When x_0 is still a uniform image but t is with changing contrast (e.g., a sine-wave), the JND threshold also depends upon spatial and temporal frequencies of t (known as the contrast sensitivity function (CSF) [Bar99]). Spatial frequencies reflect the signal variation in image domain, while temporal frequencies relate to the change of visual contents (e.g., motion or scene change) over the time. The psychophysical experiments with sine-wave gratings and other patterns (e.g., DCT basis functions) as t show that the CSF takes on an approximate paraboloid shape in the spatio-temporal space for both chromatic and achromatic channels. That is, the contrast threshold decreases towards both lower and higher ends of spatial/temporal frequencies. Note that the spatial CSF also adapts to background luminance x_o, and its interpretation depends on the viewing distance because spatial frequencies are in units of cycles per degree (*cpd*) of visual angle (i.e., if an observer moves close to an image, a feature of a fixed size in the image takes up more degrees of visual angle).

Contrast masking occurs when both x_o and t are not uniform images (e.g., both of them are sine-wave gratings or DCT basis functions). The visibility of t (the maskee) in the presence of x_o (the masker) has been investigated with a same orientation and with different orientations. The masked contrast threshold increases with the masking contrast when the latter is not too low[1]. In general, more significant masking effect exhibits when the masker and the maskee are closer in frequency, orientation and location.

9.1.2 JND Tests with Real-World Images

Subjective tests can be also conducted to determine the JND in a real-world distorted image $\mathbf{x}_d$ against the original image $\mathbf{x}_o$, and the test image $\mathbf{x}_t$ for such experiments can be expressed as

$$\mathbf{x}_t = \mathbf{x}_o + h(\mathbf{x}_d - \mathbf{x}_o) \tag{9.2}$$

where h is a weighting factor $(0 < h < 1)$ to be adjusted in the tests. $\mathbf{x}_d$ is a processed (e.g., decompressed) version of $\mathbf{x}_o$. Typical coding artifacts that may appear in $\mathbf{x}_d$ include blockiness, blurring, ringing, luminance fluctuations [YW98], etc. In the tests, observers compare $\mathbf{x}_o$ and $\mathbf{x}_t$ with an increasing h, and a JND can be determined with an h value when 75 % of observers are able to distinguish $\mathbf{x}_t$ from $\mathbf{x}_o$ [Wat00]. If the distorted image $\mathbf{x}_t$ corresponding to a JND is denoted as $\mathbf{x}_1$, a 2-JND difference can be decided when $\mathbf{x}_o$ is substituted with $\mathbf{x}_1$ in Formula (9.2), i.e., $\mathbf{x}_1$ is regarded as the "original" image. The differences with 3-JND, 5-JND, and so on, can be determined with a similar process.

The method can be used to evaluate the HVS tolerance for the distortion in the particular groups of images/video, or for the common artifacts (like blockiness, ringing, blurring, mosquito noise, etc.). The tests have not been primarily designed to facilitate mathematical JND modeling since the JND determined in this way is highly contextual to the contents of the image under test. However, such JND maps can be utilized to calibrate a perceptual distortion metric that has been developed (e.g., [Cor97]). The calibration turns the non-linear problem of perceptual distortion measurement into a linear one. As a result, a calibrated metric is expected to predict perceptual visual distortion in the units of JNDs.

9.1.3 Applications of JND Models

Knowledge on JND no doubt can help in designing, shaping and optimizing many image processing algorithms and systems. For visual quality/distortion prediction, a metric can be defined or fine-tuned according to JND [Wat93, Cor97, LDX03] for better matching the HVS perception. JND has been used to determine not only the noticeable visual distortion (as in the majority of existing relevant metrics) but also the possibly noticeable visual quality enhancement (against the original image) [LDX03]. A JND-based

---

[1]With very low masking contrast, the masked contrast threshold stays constant and may even exhibit the opposite effect, *facilitation* [LF80] (i.e., the maskee that is invisible by itself can be detected due to the presence of the masker).

perceptual metric can be also adopted beyond the quality evaluation purpose (e.g., for image synthesis [RPG99]).

The JND profile facilitates perceptual compression for image and video. Since the quantization process is the major cause of coding errors, proper quantization steps can be adaptively chosen according to the JND for a given bandwidth [Wat93, CL95, SJ89, HK02]. For motion estimation, the JND information helps in deciding suitable estimation mode [MGE$^+$01] and efficient search process [YLL$^+$03b]. The JND determination can bring about new insight in many other manipulations for an image and video coding process, like inter-frame replenishment [CB99], bit allocation, object-based coding, and filtering of motion estimated residues [YLL$^+$03b] or DCT coefficients [Saf94].

In many practical applications, certain accessory data have to be embedded inside visual signal itself (e.g., for watermarking, authentication, and error protection). With the JND indication, it is possible to insert such data in an image with minimum visual difference [WPD99, Zen99].

New applications can be found in many other visual signal processing tasks, since a JND profile provides localized information on perceptual sensitivity to the changes introduced by various processes. JND-based algorithm building and/or process parameter determination are likely to result in pro-HVS performance under certain resource constraints (e.g., computing power, bandwidth, memory, resolution for display or printing).

9.1.4   Objectives and Organization of the Following Sections

After the above introduction of the psychophysical experiments related to JND and the possible applications of a JND model, the rest of this chapter is devoted to presenting the computational JND models suitable for practical use in different tasks. These models are mainly based upon the psychophysical evidence briefed in Section 9.1.1. According to the operating domains, we can divide JND models into two basic categories, namely, *subband-based* models and *pixel-based* models. The former category has been relatively well investigated with DCT decomposition [AP92, Wat93, HM98, TS96, TV98], because of the popularity of DCT-based coders for compression. However, the latter category [CL95, CB99, YLL$^+$03a] is more convenient to be used in some situations (e.g., in motion estimation and residue manipulations [YLL$^+$03a], quality/distortion evaluation with decoded signal [LDX03], and other manipulations [CB99, RPG99]).

In Section 9.2, the models with DCT subbands will be introduced with a general and easily-adopted formulation. As expected, more elaborate models have been developed for DCT subbands with different aspects of HVS characteristics. There has been some initial research work in modelling with wavelet subbands, and the interested readers can refer to [Zen99, WYSV97] for the topic. Pixel-based models are to be presented as Section 9.3, under a general framework for the existing methods. In this section,

Table 9.1: Major Notations.

$\mathbf{x}$	an image or a frame of video
W	horizontal dimension of an image
H	vertical dimension of an image
(i, j)	pixel position index in an image
$\mathbf{x}[i, j]$	a luminance pixel in an image or a frame of video
N	dimension of a DCT block (square block)
n	position index of a DCT block in an image
(k, l)	subband index in a DCT block
$\mathbf{X}[n, k, l]$	DCT coefficient
$s(n, k, l)$	JND for a DCT subband
ϕ_u	DCT normalizing coefficient
(x, y)	pixel position index within an image block
$P(i, j)$ or $P(n, x, y)$	JND for an image pixel
$P_v(i, j)$	JND for a pixel in a frame of video
G	the maximum number of grey levels for an image

the methodology is also discussed for the conversion between subband-based and pixel-based JNDs. Section 9.4 introduces methods of comparing and evaluating various JND models, while Section 9.5 concludes the chapter.

For the convenience of the readers, the major notations used in the following sections are listed in Table 9.1.

9.2 JND with DCT Subbands

Discrete cosine transform (DCT) has been used in most of the prevalent compression standards [Gha03] (i.e., JPEG, H.261/3, MPEG-1/2/4, and the emerging H.264). Therefore, a JND model with DCT subbands is very useful, especially in perceptual visual coding and watermarking.

Hereinafter in this chapter, x denotes an image or a frame of video with size of $W \times H$, n denotes the position of an $N \times N$ DCT block in x (i.e., $n=0,1,\ldots,W \cdot H/N^2 - 1$), (k, l) denotes a DCT subband (i.e., $k, l = 0, 1,\ldots, N$-1), and $\mathbf{X}[n, k, l]$ denotes a corresponding DCT coefficient. The major factors affecting JND are CSF, luminance adaptation, intra-band masking, and inter-band masking. The JND can be formulated as the product of the base threshold due to the spatial CSF and a number of elevation parameters due to other effects. That is, the DCT-based JND is expressed as:

$$s(n, k, l) = t_{s-csf}(n, k, l) \prod_{\wp} \alpha_{\wp}(n, k, l) \qquad (9.3)$$

where $t_{s-csf}(n, k, l)$ is the base threshold due to the spatial CSF, and $\alpha_\wp(n, k, l)$ is the elevation parameter due to the effect $\wp$ ($\wp$ represents *lum* for luminance adaptation, *intra* for intra-band masking, and *inter* for inter-band masking). In this section, the widely-adopted formula [AP92] is to be introduced for incorporating the spatial CSF (i.e., for $t_{s-csf}(n, k, l)$). As for determining the other factors ($\alpha_\wp(n, k, l)$), the most practical solutions developed so far will be presented.

9.2.1   Formulation for Base Threshold

9.2.1.1   Spatial CSF equations

The spatial CSF describes the effect of spatial frequency towards the HVS sensitivity. The relation of the contrast sensitivity S with the absolute threshold T is defined as:

$$S = \frac{L}{T} \tag{9.4}$$

where L is the background luminance.

S is a function of L and spatial frequency f. The model devised in [AP92] is based upon Van Nes and Bouman's measurements [vNB67] with monochromatic light. Further analysis of the experimental data [vNB67] is given in [OT86]. For each L level, S approximately follows a parabola curve (with downward concavity) against f [Bar99, OT86]. The spatial CSF with a different L is represented by a different parabola: the vertex and the steepness of each parabola depend on L. For each L level, the vertex corresponds to the maximum sensitivity S_{max} (or T_{min}) at spatial frequency f_p (typically varying from 1 to 10 *cpd*). With the increase in L, the responding parabola moves upwards on the $S - f$ plane, i.e., S increases at a specific f and f_p (the vertex of the parabola) also increases.

The experimental data in [vNB67] can be fitted to the parabola equation in logarithmic scale as in [AP92]:

$$log\,S = log\,S_{max} - K(log\,f - log\,f_p)^2 \tag{9.5}$$

or alternatively,

$$log\,T = log\,T_{min} + K(log\,f - log\,f_p)^2 \tag{9.6}$$

where the positive constant K determines the steepness of the parabola. The actual spatial CSF curve varies with L, so T_{min}, K and f_p can be empirically determined [AP92] using the data in [PPMP91] towards the DCT basis functions:

$$T_{min} = \begin{cases} 0.142(\frac{L}{13.45})^{0.649} & \text{if } L \leq 13.45 \; cd/m^2 \\ \frac{L}{94.7} & \text{otherwise} \end{cases} \tag{9.7}$$

$$f_p = 6.78(\frac{L}{300})^{\beta_f} \tag{9.8}$$

$$K = 3.125(\frac{L}{300})^{\beta_K} \tag{9.9}$$

where

$$\beta_f = \begin{cases} 0.182, & \text{if } L \leq 300 \; cd/m^2 \\ 0, & \text{otherwise} \end{cases} \tag{9.10}$$

$$\beta_K = \begin{cases} 0.0706, & \text{if } L \leq 300 \; cd/m^2 \\ 0, & \text{otherwise} \end{cases} \tag{9.11}$$

9.2.1.2   Base threshold

When Equation (9.6) is used for the (k,l)-th DCT subband, some modification is needed. First of all, the basis of Equation (9.6) is with the experimental setting of a single signal stimulus at a time, so the spatial summation effect [AP92] has to be accounted for if changes from more than one subband (as in the real-world processed images) are present concurrently. This is because an invisible change in a subband may become visible if presented together with changes from other subbands. A simple way to avoid the over-estimation of T is to introduce a coefficient b ($0 < b < 1.0$) into the first term at the right-hand side of Equation (9.6) to cause a decrease of T_{min}, as in Equation (9.12) below.

Secondly, there also exists the oblique effect [AP92, PAW93] since T in Equation (9.6) is obtained from the experimental data of only vertical spatial frequencies (i.e., $k=0$). For an arbitrary subband ($k \neq 0$ and $l \neq 0$), the threshold is actually higher than the one given by Equation (9.6). Therefore another coefficient r ($r=0.7$ in [AP92]) is hence introduced for the absolute threshold for the (k,l)-th DCT subband for the n-th block:

$$logT^0(n,k,l) = log\frac{b \cdot T_{min}(n)}{r + (1-r)cos^2\theta(k,l)} + K(n) \cdot (logf(k,l) - logf_p(n))^2 \tag{9.12}$$

where

$$f(k,l) = \frac{1}{2N}\sqrt{\frac{k^2}{\omega_x^2} + \frac{l^2}{\omega_y^2}} \tag{9.13}$$

$$\theta(k,l) = arcsin\frac{2 \cdot f(k,0) \cdot f(0,l)}{f^2(k,l)} \tag{9.14}$$

with ω_ρ ($\rho = x, y$) being the horizontal and vertical visual angles of a pixel. When k=0 or l=0, $\theta(k,l)$ =0 and there is no oblique effect since $[r + (1 - r)cos^2\theta(k,l)] = 1$; when $i=j$, $\theta(k,l) = 90°$ if $\omega_x = \omega_y$, and the oblique effect takes the maximum because $[r + (1 - r)cos^2\theta(k,l)] = r$.

ω_ρ is calculated based on viewer distance d and the display width of a pixel, Λ_x or Λ_y, on the monitor:

$$\omega_\rho = 2 \cdot arctan(\frac{\Lambda_\rho}{2d}) \tag{9.15}$$

Equation (9.12) can be extended to chrominance components [PAW93], with the associated parameters to be determined for a color channel.

Equation (9.12) is not applicable to $T^o(n,0,0)$, and a conservative estimation of $T^o(n,0,0)$ is given [AP92, PPMP91] as:

$$T^o(n,0,0) = min\{T^o(n,1,0), T^o(n,0,1)\} \tag{9.16}$$

Because $T^o(n,k,l)$ is expressed in luminance measure, it can be converted linearly to a grey level value, and then scaled to the DCT domain to arrive at the base threshold due to the spatial CSF:

$$t_{s-csf}(n,k,l) = \frac{G}{\phi_k\phi_l(L_{max} - L_{min})}T^o(n,k,l) \tag{9.17}$$

where L_{max} and L_{min} are the maximum and the minimum display luminance values corresponding to the grey levels 255 and 0, respectively; G is the maximum number of grey levels (G=256 for 8-bit image representation); and

$$\phi_u = \begin{cases} \sqrt{\frac{1}{N}}, & \text{if } u = 0 \\ \sqrt{\frac{2}{N}}, & \text{otherwise} \end{cases} \tag{9.18}$$

being the DCT normalizing coefficient.

The local background luminance for Formulae (9.7)–(9.11) with the n-th block can be obtained based upon the DC components within a small neighborhood [HK02]:

$$L(n) = L_{min} + \frac{L_{max} - L_{min}}{G}\left(\frac{\sum_{m \in F(n,0,0)} \mathbf{X}[m,0,0]}{N \cdot N_F}\right) \qquad (9.19)$$

where $F(n,0,0)$ denotes the foveal region (covering the visual angle v) centered at location n; N_F denotes the number of DC coefficients in the foveal region $F(n,0,0)$ and can be approximately calculated as:

$$N_F = \left(\lfloor \frac{2d \cdot tan(\frac{v}{2})}{N \cdot \Lambda_x} \rfloor\right) \cdot \left(\lfloor \frac{2d \cdot tan(\frac{v}{2})}{N \cdot \Lambda_y} \rfloor\right) \qquad (9.20)$$

where the symbol $\lfloor \ \rfloor$ denotes rounding to the nearest smallest integer, and v takes approximately 1°. If $d = 24$ inches, $\Lambda_x = \Lambda_y = 1/80$ inch, and $N=8$, $N_F \approx 4\text{x}4 = 16$.

9.2.2 Luminance Adaptation Considerations

As can be seen from the above-presented formulae, $t_{s-csf}(n,k,l)$ increases with background luminance $L(n)$, i.e., Weber-Fechner law is observed. Since the luminance variation over different blocks in a digital image merely spans a very small luminance range in the spatial CSF experiments [Bar99, vNB67, OT86], a single spatial CSF parabola may be used for all blocks in the image. Therefore, a simplified base-threshold can be derived by using the average grey level in the image L_0 as $L(n)$ for all blocks in the formulae of Section 9.2.1 and the resultant $t_{s-csf}(k,l)$ [1] is then compensated for luminance adaptation [Wat93]:

$$\alpha_{lum}^{(1)}(n) = \left(\frac{\mathbf{X}[n,0,0]}{\mathbf{X}_L}\right)^\epsilon \qquad (9.21)$$

where $\mathbf{X}[n,0,0]$ is the DC coefficient of the n-th block under consideration, $\mathbf{X}_L$ is the DC coefficient corresponding to L_0, and $\epsilon = 0.649$.

As a further simplification, the average gray level used in base-threshold calculation can be set to a constant (e.g., 128 for 8-bit image representation), because digital images have been converted to a limited range of grey levels (just 256 levels for 8-bit image representation) and adjusted for the average gray level (so the average gray level is not very different from one image to another); consequently, $t_{s-csf}(k,l)$ is image-independent and can be therefore computed off-line for time-critical applications.

However, the full base-threshold model in Section 9.2.1 and the above-stated simplified schemes tend to underestimate the visibility threshold in low grey-level (i.e., dark) regions, since luminance adaptation in real-world digital images is more complex than what is described by Weber-Fechner law. First of all, apart from the influence of local luminance $L(n)$, the visibility threshold is also affected by the global luminance L_0. When L_0 is fixed, the contrast threshold exhibits a parabola curve against local

---

[1] t_{s-csf} does not depend upon n in this case.

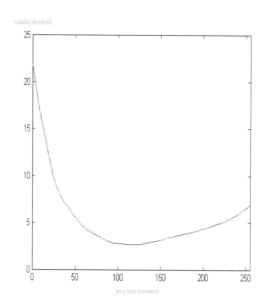

Figure 9.1: Luminance Adaptation for Digital Images (From [CL95],©1995 IEEE).

luminance $L(n)$ [Nad00]. For digital images, much stronger dependence on $L(n)$ is exhibited than on L_0, because as above mentioned, L_0 for digital images is not very different from one image to another. Other factors affecting luminance adaptation in digital images include the ambient illumination falling on the display and the γ-correction of the display tube that partially compensates for Weber-Fechner law's effect [NH88]. As a result, more realistic luminance adaptation for real-world digital images takes a quasi-parabolic curve [CL95, SJ89, NH88], as approximately illustrated in Figure 9.1 (G=256).

Since $L(n)$ can be estimated by the DC component $\mathbf{X}[n, 0, 0]$ of a DCT block, luminance adaptation can be adjusted [ZLX03, Zha04] as follows, in consideration of the quasi-parabolic curve in Figure 9.1:

$$\alpha_{lum}^{(2)}(n) = \begin{cases} \kappa_1(1 - \frac{2 \cdot \mathbf{X}[n,0,0]}{G \cdot N})^3, & \text{if } \mathbf{X}[n, 0, 0] \leq \frac{G \cdot N}{2} \\ \kappa_2(\frac{2 \cdot \mathbf{X}[n,0,0]}{G \cdot N} - 1)^2, & \text{otherwise} \end{cases} \tag{9.22}$$

where κ_1 and κ_2 determine the maximum value of $\alpha_{lum}^{(2)}(n)$ when $\mathbf{X}[n, 0, 0] = 0$ (dark local luminance) and the value of $\alpha_{lum}^{(2)}(n)$ when $\mathbf{X}[n, 0, 0] = G \cdot N$ (bright local luminance), respectively. Equation (9.22) adapts better to the HVS characteristics for luminance adaptation in low and high luminance regions, in comparison with the full base-threshold model in Section 9.2.1 and the simplified schemes with Equation (9.21).

9.2.3 Contrast Masking

9.2.3.1 Intra-band masking

Intra-band masking refers to the unnoticeable error tolerance due to the signal component in the subband itself. When $\mathbf{X}[n, k, l]$ is changed to $\widehat{\mathbf{X}}[n, k, l]$ (due to compression, watermarking, etc.), the subband error is:

$$\triangle \mathbf{X}[n, k, l] = |\mathbf{X}[n, k, l] - \widehat{\mathbf{X}}[n, k, l]| \qquad (9.23)$$

A stronger $\mathbf{X}[n, k, l]$ is usually able to mask a larger $\triangle \mathbf{X}[n, k, l]$, and therefore the intra-band effect can be accounted for with [Wat93, HK02]:

$$\alpha_{intra}(n, k, l) = max(1, |\frac{\mathbf{X}[n, k, l]}{t_{s-csf}(n, i, j) \cdot \alpha_{lum}(n)}|^{\zeta}) \qquad (9.24)$$

where ζ lies between 0 and 1.

9.2.3.2 Inter-band masking

Inter-band masking is more complicated than intra-band masking, because it involves multiple subband components. The inter-band masking effect can be estimated by considering the strength of the masking subband, and the difference (for both value and orientation) between the masking and the masked subbands [HM98]. The process can be simplified by grouping $\{\mathbf{X}[n, k, l]\}$ in Cortex space [TS96], and the masking is evaluated only within the same Cortex band (typically spanning several DCT subbands).

For further implementation efficiency, inter-band masking can be evaluated at the DCT-block level [TV98]. Let $R_L(n)$, $R_M(n)$ and $R_H(n)$ represent the sums of the absolute DCT coefficients in the low-frequency (LF), medium-frequency (MF) and high-frequency (HF) groups, respectively, within a block, as shown in Figure 9.2. The MF and HF energy for the block is:

$$E_{mh}(n) = R_M(n) + R_H(n) \qquad (9.25)$$

the relative LF strength is:

$$\widetilde{E}_d(n) = \frac{\overline{R_L}(n)}{\overline{R_M}(n)} \qquad (9.26)$$

and the relative LF and MF strength is:

$$\widetilde{E}_{dm}(n) = \frac{\overline{R_L}(n) + \overline{R_M}(n)}{\overline{R_H}(n)} \qquad (9.27)$$

where $\overline{R_L}(n)$, $\overline{R_M}(n)$ and $\overline{R_H}(n)$ denote the means of $R_L(n)$, $R_M(n)$ and $R_H(n)$, respectively.

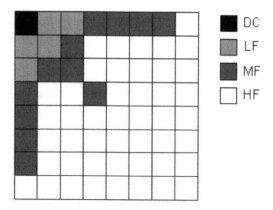

Figure 9.2: Coefficient Grouping for DCT Block Classification (N=8) (From [TV98], ©1998 IEEE).

Based upon the similar methodology in [TV98], each DCT block is assigned to one of the three classes with descending order of the HVS sensitivity, namely, *Low-Masking*, *Medium-Masking* and *High-Masking* classes, according to $E_{mh}(n)$:

- $E_{mh}(n) \leq \mu_1$: the block is assigned to *Low-Masking* class.

- $\mu_1 \leq E_{mh}(n) \leq \mu_2$: if condition (9.28) or (9.29) is met, the block is assigned to *Medium-Masking* class; otherwise it is assigned to *Low-Masking* class.

- $\mu_2 \leq E_{mh}(n) \leq \mu_3$: if condition (9.28) or (9.29) is met, the block is assigned to *Medium-Masking* class; otherwise it is assigned to *High-Masking* class.

- $E_{mh}(n) > \mu_3$: if condition (9.28) or (9.29) is met for $\varphi_\tau = \tau \cdot \varphi$ and $\chi_\tau = \tau \cdot \chi$ (where $\tau < 1$), the block is assigned to *Medium-Masking* class; otherwise it is assigned to *High-Masking* class.

The condition used above for determining the *Medium-Masking* class is defined as:

$$\widetilde{E}_{dm}(n) \geq Q \tag{9.28}$$

$$max\{\widetilde{E}_d(n), \widetilde{E}_{dm}(n)\} \geq \varphi \quad and \quad min\{\widetilde{E}_d(n), \widetilde{E}_{dm}(n)\} \geq \chi \tag{9.29}$$

where $\varphi > \chi$.

The possible block classification according to $E_{mh}(n)$ is illustrated in Figure 9.3. The elevation with inter-band masking can be then determined as:

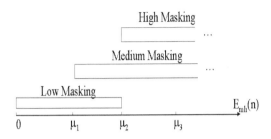

Figure 9.3: Block Classification for Inter-band Masking.

$$\alpha_{inter}(n) = \begin{cases} 1, & \text{for } \textit{Low-Masking} \text{ blocks} \\ \delta_1, & \text{for } \textit{Medium-Masking} \text{ blocks, and } R_L(n) + R_M(n) \leq R_0 \\ \delta_2, & \text{for } \textit{Medium-Masking} \text{ blocks, and } R_L(n) + R_M(n) > R_0 \\ (1 + \frac{E_{mh}(n)-\mu_2}{2\cdot\mu_3-\mu_2})\delta_2, & \text{otherwise} \end{cases}$$

(9.30)

where $\delta_2 > \delta_1 > 1$.

A reference set of the model parameters is: $\mu_1 = 125, \mu_2 = 290, \mu_3 = 900, \varphi = 7, \chi = 5, \tau = 0.1, Q = 16, R_0 = 400, \delta_1 = 1.125$ and $\delta_2 = 1.25$.

9.2.4 Other Factors

The spatial CSF, luminance adaptation, intra-band contrast masking and inter-band contrast masking are the major factors in modelling the DCT-based JND. Depending upon the applications, however, other factors may need to be also considered for more accurate JND generation in images or video. Examples include color masking and temporal masking.

The masking effect to color channel θ by color channel ϱ can be evaluated as $e_{\theta\varrho} \cdot u_\varrho^2$ [AK01], where u_ϱ is the contrast in the masking channel and $e_{\theta\varrho}$ is the associated weighting parameter. The overall color masking is then obtained via accumulating the effects from all channels.

In video, more details are masked by faster object movement or scene change that leads to the occurrence of higher temporal frequencies. The temporal masking effect can be simplified as the inverse of the response of a low-pass temporal filter [WHMI01].

Further research is needed to sufficiently model color masking and temporal masking, as well as to convincingly demonstrate the application in various situations.

9.3   JND with Pixels

Pixel-wise JND estimators have been developed [CL95, CC96, CB99, RPG99, YLL$^+$03a, Zha04] and applied for the situations where manipulations are more conveniently performed in image domain, e.g., motion estimation [YLL$^+$03b], quality evaluation with decoded images [LDX03] and other visual processing [CB99, RPG99]. Pixel-wise JND can be derived from pixel domain [CL95, CB99, YLL$^+$03a] or from subbands [RPG99, Zha04]. In the former case, luminance adaptation and texture masking are the major factors being considered, while in the latter case, spatial CSF can be also incorporated for more accurate estimation. It will be also demonstrated that JND thresholds in one domain can be derived from those in a different domain for a same image.

9.3.1   JND Estimation from Pixel Domain

Let $\mathbf{x}[i,j]$ denote the luminance intensity of a pixel at $[i,j]$ in a frame of visual signal, and $0 \le i \le W - 1, 0 \le j \le H - 1$. In the case of images, the pixel-wise JND, $P(i,j)$, is to be estimated with visual information within an image [CL95, CB99, YLL$^+$03a]. In the case of video, the associated pixel-wise JND, $P_v(i,j)$ is then obtained by integrating temporal (interframe) masking effect [CC96] with $P(i,j)$. The formulae in this section are expressed for luminance components; the same set of formulae may be used for chrominance components with the parameters adjusted [YLL$^+$03a]. Further studies are required for the more accurate estimation in chrominance channels.

9.3.1.1   Spatial JNDs

Two major factors have been considered for the spatial JND in pixel domain: luminance adaptation and texture masking. An approximate curve on visibility threshold versus background luminance has been illustrated in Figure 9.1 for digital images. For texture masking, the spatial activities in the neighborhood reduce the visibility of changes, and therefore textured regions can conceal more error than smooth or edge areas.

Luminance adaptation and texture masking usually exist within a same image region, and this allows bigger pixel changes without being noticed, when compared with the case of one source of masking alone. A nonlinear additivity model can be hence used to determine the visibility threshold for the overall masking effect [YLL$^+$03a]:

$$P(i,j) = T^L(i,j) + T^t(i,j) - C^{Lt}(i,j) \cdot \min\{T^L(i,j), T^t(i,j)\} \qquad (9.31)$$

where $T^L(i,j)$ and $T^t(i,j)$ are the visibility thresholds for luminance adaptation and texture masking, respectively; $C^{Lt}(i,j)$ accounts for the overlapping effect in masking, and $0 < C^{Lt}(i,j) \leq 1$.

$T^L(i,j)$ can be determined by a piecewise approximation of Figure 9.1 [CL95]:

$$T^L(i,j) = \begin{cases} 17(1 - \sqrt{\frac{L(i,j)}{127}}) + 3, & \text{if } L(i,j) \leq 127 \\ \frac{3}{128}(L(i,j) - 127) + 3 & \text{otherwise} \end{cases} \qquad (9.32)$$

where $L(i,j)$ is the average background luminance at (i,j) within a small (e.g., $N{\times}N$) neighborhood.

For more accurate estimation for texture masking, smooth, edge and non-edge regions have to be distinguished. The masking effect in smooth regions can be largely neglected; distortion around edges is easier to notice than that in textured regions because edge structure attracts more attention from a typical HVS: 1) edge is directly related to important visual contents, such as object boundaries, surface crease and reflectance change; 2) edge structure is perceptually simpler than a textured one and observers have good prior knowledge on what an edge should look like. Therefore, texture masking can be estimated as:

$$T^t(i,j) = \beta \cdot g(i,j) \cdot e(i,j) \qquad (9.33)$$

where β is a control parameter; $g(i,j)$ denotes the weighted average of gradients around (i,j) [CL95]; $e(i,j)$ is an edge-related weight of the pixel at (i,j), and its corresponding matrix e is computed as [YLL$^+$03a]:

$$\mathbf{e} = \bar{\mathbf{b}} * \mathbf{h} \qquad (9.34)$$

where $\bar{\mathbf{b}}$ is based on edge (e.g., detected by Canny detector) of the original image, with possible element values of 0.1, 0.3 and 1.0 for smooth, edge and texture pixels, respectively; h is a $m \times m$ Gaussian low pass filter with standard deviation σ (e.g., $m = 7$ and $\sigma = 0.8$).

9.3.1.2 Simplified estimators

When $C^{Lt}(i,j) \equiv 1$ (i.e., no overlapping is considered), Equation (9.31) becomes the JND estimator devised in [CL95]:

$$P(i,j)_{simplified-I} = \max\{T^L(i,j), T^t(i,j)\} \qquad (9.35)$$

That is, only the major effect between the two contributing factors is selected.

When $T^L(i,j)$ is considered as the dominant masking factor, i.e., $\min\{T^L(i,j),$

$T^t(i,j)\} \equiv T^t(i,j)$, Equation (9.31) becomes:

$$P(i,j)_{simplified-II} = T^L(i,j) + C'(i,j) \cdot T^t(i,j) \qquad (9.36)$$

where $C'(i,j) = 1 - C^{Lt}(i,j)$. If $C'(i,j) = \frac{p}{T^L_{(i,j)}}$, where p takes a value between 0.5 and 1, Equation (9.36) is equivalent to the JND estimator proposed in [CB99]; since $T^L(i,j)$ is much greater than 1 for most pixels, $C'(i,j)$ is much smaller than 1 so $P(i,j)_{simplified-II}$ is determined largely by $T^L(i,j)$.

9.3.1.3 Temporal masking effect

Simple temporal masking effect can be added to scale the spatial JND for video. Usually bigger inter-frame difference (caused by motion or content changes) leads to larger temporal masking. The JND for video can be therefore denoted as:

$$P_v(i,j) = f(a(i,j)) \cdot P(i,j) \qquad (9.37)$$

where $f(a)$ is the empirical function to reflect the increase in masking effect with the increase in interframe changes, as denoted in Figure 9.4; and $a(i,j)$ represents the average inter-frame luminance difference between the current frame $\mathbf{x}$ and the previous frame $\mathbf{x}^p$ [CC96]:

$$a(i,j) = \frac{\mathbf{x}[i,j] - \mathbf{x}^p[i,j]}{2} + \frac{L_0 - L_0^p}{2} \qquad (9.38)$$

where L_0 and L_0^p are the average luminance of $\mathbf{x}$ and $\mathbf{x}^p$, respectively.

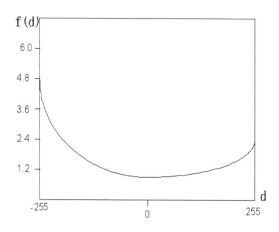

Figure 9.4: Empirical Function on Temporal Masking Effect (From [CC96],©1996 IEEE).

9.3.2 Conversion between Subband- and Pixel-Based JNDs

9.3.2.1 Subband summation to pixel domain

Pixel-wise JNDs can be obtained via summation of subband JNDs (for both the DCT domain [Zha04] and a non-DCT domain [RPG99]). Compared with the methods presented in Section 9.3.1, this approach may incur higher computational expense because of the need for image decomposition; however, it is capable of directly incorporating spatial CSF for higher estimation accuracy. The derivation from DCT subbands [Zha04] is to be introduced as follows, due to the fact that DCT is commonly in use (especially in compression related systems) and therefore DCT decomposition leads to less extra hardware or software requirement than other types of image decomposition.

With the same notations in Section 9.2, $\mathbf{X}[n, k, l]$ and $s(n, k, l)$ denote the DCT coefficient and the associated JND threshold, for the (k, l)-th DCT subband of the n-th image block; the JND for each pixel can be obtained by summation of the influence from all DCT subbands within the same block.

The contribution from $s(n, i, j)$ is evaluated as:

$$t'(n, i, j) = \begin{cases} sign(\mathbf{X}[n, k, l]) \cdot s(n, k, l) & \text{if } |\mathbf{X}[n, k, l]| \geq s(n, k, l) \\ 0 & \text{otherwise} \end{cases} \tag{9.39}$$

where $sign(\mathbf{X}[n, k, l])$ denotes the sign of $\mathbf{X}[n, k, l]$. The truncation involved above avoids the overestimation in smooth regions in which the magnitude of $\mathbf{X}[n, k, l]$ is small; the use of $sign(\mathbf{X}[n, k, l])$ avoids the possible introduction of artificial patterns over blocks or discontinuity on boundaries of neighboring blocks.

The compound effect for an image pixel at position (x,y) of n-th block is obtained by summing (inverse-DCT transforming) the DCT-based thresholds in the same block:

$$t_p(n, x, y) = \sum_{u=0}^{N-1} \sum_{v=0}^{N-1} \phi_u \phi_v cos(\frac{u(2x+1)\pi}{2N}) cos(\frac{v(2y+1)\pi}{2N}) \cdot t'(n, u, v) \tag{9.40}$$

where x and y are the pixel position indices in a block and vary from 0 to N-1.

The JND threshold at a pixel can be determined as:

$$P(n, x, y) = max(|t_p(n, x, y)|, t_l(n)) \tag{9.41}$$

where $t_l(n)$ accounts for the local background luminance adaptation and can be directly obtained from the DC-subband JND value, $s(n, 0, 0)$, for the block:

$$t_l(n) = s(n, 0, 0)/N \tag{9.42}$$

Table 9.2: Exemplar Weighting Parameters $\{z(k,l)\}$ for Pixel-domain Decomposition (N=4) (From [CL95],©1995 IEEE).

0.0211	0.0197	0.0273	0.0445
0.0211	0.0261	0.0365	0.0581
0.0396	0.0464	0.0626	0.0954
0.0865	0.0990	0.1286	0.1865

9.3.2.2   Pixel domain decomposition into subbands

If the pixel-wise JND, $P(i,j)$, is available, the corresponding subband JND can be approximated by assigning higher JND energy to higher spatial frequency subbands (in line with the spatial CSF). More specifically, the JND of the (k,l)-th subband can be estimated as [CL95]:

$$s(n,k,l) = \sqrt{E_P(n) \cdot z(k,l)} \tag{9.43}$$

where the pixel-wise JND energy in the n-th NxN image neighborhood is calculated as:

$$E_P(n) = \sum_{\zeta=0}^{N-1} \sum_{\eta=0}^{N-1} P^2(\zeta + (n - \lfloor \frac{n \cdot N}{W} \rfloor \frac{W}{N}) \cdot N, \eta + \lfloor \frac{n \cdot N}{W} \rfloor \cdot N) \tag{9.44}$$

where the symbol $\lfloor \ \rfloor$ denotes rounding to the nearest smallest integer.

The weighting parameter, $z(k,l)$, is determined according to the spatial CSF, and

$$\sum_{u=0}^{N-1} \sum_{v=0}^{N-1} z(u,v) = 1 \tag{9.45}$$

$z(k,l)$ is inversely proportional to the spatial sensitivity in the $[k,l]$-th subband. In the example calculated in [CL95] for N=4, $\{z(k,l)\}$ take the values in Table 9.2, when the image size is W=H=256, and the viewing distance d is six times the image height.

9.4   JND Model Evaluation

The accuracy of a JND model can be evaluated for its effectiveness in shaping noise in images or video. In image domain, JND-guided noise injection can be made via [CL95]:

$$\widehat{\mathbf{x}}[i,j] = \mathbf{x}[i,j] + q \cdot S^{random}(i,j) \cdot P(i,j) \tag{9.46}$$

where $\mathbf{x}[i,j]$ is the original image, q is a parameter to control the noise level, and $S^{random}[i,j]$ takes either +1 or -1 randomly, regarding i and j, to avoid introduction of fixed pattern of changes.

In DCT subbands, JND-guided noise injection can be made into DCT coefficients:

$$\widehat{\mathbf{X}}[n, k, l] = \mathbf{X}[n, k, l] + q \cdot S^{random}(n, k, l) \cdot s(n, k, l) \qquad (9.47)$$

where $S^{random}(n, k, l)$ takes +1 or -1 randomly regarding n, k and l. The noise-injected image can then be yielded from $\widehat{\mathbf{X}}[n, k, l]$ with IDCT.

In Equations (9.46) and (9.47) , q (> 0) regulates the total noise energy to be injected. If $q < 1$ and there is not overestimation in the JND model adopted, the noise injection is perceptually lossless. Perceptual visual quality of the resultant noise-injected images can be compared and evaluated with subjective viewing tests [ITU00]. The resultant mean opinion score (MOS) is regarded as a fair indicator of perceptual quality for each image if a sufficient number of observers are involved.

Under a same level of total error energy (e.g., a same MSE or PSNR), the better perceptual quality the noise-injected image/video has, the more accurate the JND model is; alternatively, with a same level of perceptual visual quality, a more accurate JND model is able to shape more noise (i.e., resulting in lower MSE or PSNR) in an image.

Figure 9.5 shows the noise-injected versions of the *Cameraman* image with the full DCT-based model introduced in Sections 9.2.1~9.2.3 and the pixel-based model introduced in Section 9.3.1.1, with a same MSE level.

9.5 Conclusions

This chapter presents the computational JND (just-noticeable difference) models in both subbands and pixels, under an integrated formulation. It gives a systematic introduction in the field to date, as well as a practical user's guide for the related techniques. For JND estimation in DCT subbands, the most practical existing algorithms have been rationalized and detailed, for incorporating CSF (contrast sensitivity function), luminance adaptation, intra-band masking and inter-band masking. For JND estimation from image pixels, different existing models have been presented under a general framework. The JND conversion between subbands and image domain has been also discussed. Subjective visual evaluation of images/video after JND-guided noise injection provides a way of benchmarking various JND models. As previously mentioned, more careful research is needed for modelling color components and the temporal effect.

The most direct application of the subband- and pixel- based JND derivations is in visual quality/distortion metrics. However, JND models can also facilitate noise shaping in many other applications of visual processing with different requirements. In various image/video processing tasks, be they compression, data hiding or error protection, ef-

(a)DCT based JND Model

(b)Pixel based JND Model

Figure 9.5: Noise Injected Images with JND Models (MSE=112).

fectively shaping the inevitable noise or accessory data onto perceptually less significant subbands/regions can lead to resource (computational power, bitrate, memory, etc.) saving or/and performance (e.g., resultant visual quality) enhancement.

References

[AK01]   A. J. Ahumada and W. K. Krebs. Masking in color images. *Proc. of SPIE Human Vision and Electronic Imaging VI*, 4299:187–194, January 2001.

[AP92]   A. J. Ahumada and H. A. Peterson. Luminance-model-based DCT quantization for color image compression. *SPIE Proc. Human Vision, Visual Processing, and Digital Display III*, 365–374, 1992.

[Bar99]   P. Barten. Contrast Sensitivity of the Human Eye and Its Effects on Image Quality. *SPIE Press*, 1999.

[CB99]   Y. J. Chiu and T. Berger. A Software-only Videocodec Using Pixelwise Conditional Differential Replenishment and Perceptual Enhancement. *IEEE Trans. Circuits Syst. Video Technology*, 9(3):438–450, April 1999.

[CC96]   C. H. Chou and C. W. Chen. A Perceptually Optimized 3-D Subband Image Codec for Video Communication over Wireless Channels. *IEEE Trans. Circuits Syst. Video Technology*, 6(2):143–156, 1996.

[CL95]   C. H. Chou and Y. C. Li. A perceptually tuned subband image coder based on the measure of Just-Noticeable-Distortion Profile. *IEEE Trans. Circuits Syst. Video Technology*, 5(6):467–476, December 1995.

[Cor97]   S. Corporation. Sarnoff JND vision model. *Contribution to IEEE G-2.1.6 Compression and Processing Subcommittee*, August 1997.

[Gha03]   M. Ghanbari. Visibility of wavelet quantization noise. *Standard Codecs: Image Compression to Advanced Video Coding*, August 2003.

[HK02]   I. Hontsch and L. J. Karam. Adaptive image coding with perceptual distortion control. *IEEE Trans. On Image Processing*, 11(3):214–222, March 2002.

[HM98]   P. J. Hahn and V. J. Mathews. An analytical model of the perceptual threshold function for multichannel image compression. *Proc. IEEE Int'l Conf. Image Processing (ICIP)*, 3:404–408, 1998.

[ITU00]   ITU. Methodology for the Subjective Assessment of the Quality of Television Pictures. *ITU-R Recommendation 500-10*, 2000. Geneva, Switzerland.

[JJS93]   N. Jayant, J. Johnston, and R. Safranek. Signal compression based on models of human perception. *Proc. IEEE*, 81:1385–1422, October 1993.

[LDX03]   W. Lin, L. Dong, and P. Xue. Discriminative Analysis of Pixel Difference Towards Picture Quality Prediction. *Proc. IEEE Int'l Conf. Image Processing (ICIP)*, 3:193–196, September 2003.

[LF80]   G. E. Legge and J. M. Foley. Contrast masking in human vision. *Journal of the Optical Society of America*, 70:1458–1471, 1980.

[MGE$^+$01]   J. Malo, J. Gutierrez, I. Epifanio, F. Ferri, and J. M. Artigas. Perceptual feedback in multigrid motion estimation using an improved DCT quantization. *IEEE Trans. Image Processing*, 10(10):1411–1427, October 2001.

[Nad00]   M. J. Nadenau. *Integration of Human Color Vision Models into High Quality Image Compression*. PhD thesis, EPFL, Lausanne, 2000.

[NH88]   A. N. Netravali and B. G. Haskell. Digital Pictures: Representation and Compression. *Proc. SPIE Human Vision, Visual Processing, and Digital Display VI*, 1988.

[OT86]   L. A. Olzak and J. P. Thomas. Seeing spatial patterns, (vol.) 1. In *Handbook of Perception and Human Performance*. New York: John Wiley and Sons, 1986.

[PAW93]   H. A. Peterson, A. J. Ahumuda, and A. B. Watson. An improved detection model for DCT coefficient quantization. *Proc. SPIE Human Vision, Visual Processing, and Digital Display VI*, 1913:191–201, 1993.

[PPMP91] H. A. Peterson, H. Peng, J. H. Morgan, and W. B. Pennebaker. Quantization of color image components in the DCT domain. *Proc. SPIE Human Vision, Visual Processing, and Digital Display II*, 1453:210–222, 1991.

[RPG99]   M. Ramasubramanian, S. N. Pattanaik, and D. P. Greenberg. A perceptually based physical error metric for realistic image synthesis. *Computer Graphics (SIGGRAPH'99 Conference Proceedings)*, 73–82, August 1999.

[Saf94]   R. J. Safranek. A JPEG compliant encoder utilizing perceptually based quantization. *Proc. SPIE Human Vision, Visual Proc., and Digital Display V*, 2179:117–126, February 1994.

[SJ89]   R. J. Safranek and J. D. Johnston. A perceptually tuned sub-band image coder with image dependent quantization and post-quantization data compression. *Proc. IEEE Int'l Conf. Acoustics, Speech, and Signal Processing (ICASSP)*, 1945–1948, 1989.

[TS96]   T. D. Tran and R. Safranek. A locally adaptive perceptual masking threshold model for image coding. *Proc. IEEE Int'l Conf. Acoustics, Speech, and Signal Processing (ICASSP)*, 4:1883–1886, 1996.

[TV98]   H. Y. Tong and A. N. Venetsanopoulos. A perceptual model for JPEG applications based on block classification, texture masking, and luminance masking. *Proc. IEEE Int'l Conf. Image Processing (ICIP)*, 3:428–432, 1998.

[vNB67]   F. L. van Nes and M. A. Bouman. Spatial modulation transfer in the human eye. *Journal of the Optical Society of America*, 57:401–406, March 1967.

[Wat93]   A. B. Watson. DCTune: A technique for visual optimization of DCT quantization matrices for individual images. *Society for Information Display Digest of Technical Papers XXIV*, 946–949, 1993.

[Wat00]   A. B. Watson. Proposal: Measurement of a JND Scale for Video Quality. Submission to the IEEE G-2.1.6 Subcommittee on Video Compression Measurements Meeting on August 7th, 2000.

[WHMI01] A. B. Watson, J. Hu, and J. F. McGowan III. DVQ: A digital video quality metric based on human vision. *Journal of Electronic Imaging*, 10(1):20–29, January 2001.

[WPD99]   R. B. Wolfgang, C. I. Podilchuk, and E. J. Delp. Perceptual Watermarks for Digital Images and Video. *Proc IEEE*, 87(7):1108–1126, July 1999.

[WYSV97] A. B. Watson, G. Y. Yang, J. A. Solomon, and J. Villasenor. Visibility of wavelet quantization noise. *IEEE Trans. Image Processing*, 6(8):1164–1175, August 1997.

[YLL+03a]   X. Yang, W. Lin, Z. Lu, E. P. Ong, and S. Yao. Just-noticeable-distortion profile with nonlinear additivity model for perceptual masking in color images. *Proc. IEEE Int'l Conf. Acoustics, Speech, and Signal Processing (ICASSP)*, 3:609–612, 2003.

[YLL+03b]   X. K. Yang, W. Lin, Z. K. Lu, E. P. Ong, and S. S. Yao. Perceptually-adaptive Hybrid Video Encoding Based On Just-noticeable-distortion Profile. *SPIE 2003 Conference on Video Communications and Image Processing (VCIP)*, 5150:1448–1459, 2003.

[YW98]   M. Yuen and H. R. Wu. A survey of MC/DPCM/DCT video coding distortions. *Signal Processing*, 70(3):247–278, 1998.

[Zen99]   W. Zeng. Visual Optimization in Digital Image Watermarking. *Proc. ACM Mulitimedia Workshop on Multimedia and Security*, 1999.

[Zha04]   X. H. Zhang. Just-noticeable distortion estimation for images. Master's thesis, School of Electrical and Electronic Engineering, Nanyang Technological University, Singapore, February 2004.

[ZLX03]   X. H. Zhang, W. S. Lin, and P. Xue. A new DCT-based just-noticeable distortion estimator for images. *Proc. Joint Conference of 4th Int'l Conf. Information, Comm. and Signal Proc. and 4th IEEE Pacific-rim Conf. Multimedia (ICICS-PCM)*, 2003. Singapore.

Chapter 10

No-Reference Quality Metric for Degraded and Enhanced Video

Jorge E. Caviedes[†] and Franco Oberti[‡]
†   *Intel Corporation, U.S.A.*
‡   *Philips Research, The Netherlands*

10.1   Introduction

The goal of objective image quality (OIQ) metrics is to accurately predict standard subjective quality ratings (e.g., according to ITU-R Recommendation 500 [ITU00]). Most of the work on OIQ in the last five years has been carried out by the Video Quality Experts Group (VQEG) under the aegis of the ITU-T and the ITU-R, and driven by the broadcast industry [VQE] (see also Chapter 11). Thus, VQEG work is focused on fidelity metrics, i.e., measures of quality loss along the broadcast section of the video chain.

In contrast with the fidelity-oriented approaches to quality, which originated in the broadcast field and are primarily concerned only with the degradation effects of transmission, in the age of high resolution digital imaging systems, quality measures must address degradation as well as enhancement. Modern image acquisition, storage, and reproduction systems include components such as resolution enhancement, re-formatting, coding and transcoding. For this reason, in order to assess objective quality, quality metrics must be able to use any reference along the video chain, or even a virtual reference. This modality is known as *no-reference* to differentiate itself from the *full-reference* modality, which uses pixel-per-pixel comparison of processed and original images in order to assess quality. There are presently no perceptual models of distortion or enhancement that may apply to the no-reference case. However, a promising approach consists of identifying the minimum set of features which influence quality in most situations and examine their ability to predict perceived quality. We believe that this is a significant step towards the development of content-independent, no-reference objective quality models as it sheds light on the key quality factors and how they influence image quality.

In this chapter, we present work on a no-reference objective quality metric (NROQM) that has resulted from extensive research on impairment metrics, image feature metrics, and subjective image quality in several projects in Philips Research, and participation in the ITU Video Quality Experts Group [VQE]. The NROQM is aimed at requirements including video algorithm development, embedded monitoring and control of image quality, and evaluation of different types of display systems. The NROQM is built from metrics for desirable and non-desirable image features (sharpness, contrast, noise, clipping, ringing, and blocking artifacts), and accounts for their individual and combined contributions to perceived image quality. This set of features is the minimum required to estimate perceived image quality, with the ability to measure improvement and degradation. The set of relevant image features is selected based on their perceptual impact on quality, and whether they can be accurately detected and quantified. The assumed point of measurement is at the end of the video chain, i.e., after all possible treatments have taken place.

Although predicting subjective ratings without a strict comparison of the processed image against the original version seems to be an academic challenge, its applications have tremendous importance in the market. The most important application of OIQ metrics is the implementation of flexible control systems able to deliver high quality to the consumer of video products and services on a continuous, real-time basis. Reliable metrics enable transparent and competitive ratings of quality of service (QoS) and quality of products, which will benefit the customers as well as the providers.

This chapter is organized as follows. Section 10.2 presents a brief summary of the state-of-the-art. Section 10.3 discusses the OIQ metric design process and its components. Section 10.4 explains the NROQM model. Section 10.5 presents the results of performance tests of the NROQM, and Section 10.6 deals with conclusions and future research.

10.2   State-of-the-Art for No-Reference Metrics

The main product in the market is the Digital Video Quality Analyzer, DVQ, from Rohde & Schwarz [ROH]. This device measures quality of encoded video based on a blocking artifact metric corrected for motion and spatial masking (conditions which may hide the impairment). The DVQ does not take into account other MPEG artifacts besides blockiness, does not work on the decoded image (its input is the MPEG transport stream), and is not able to measure picture improvements such as sharpness enhancement or noise reduction.

Similar to the DVQ product is the Picture Appraisal Rating (PAR) from Snell & Wilcox [SNE]. The PAR also takes the MPEG stream as the input, and estimates the

Peak Signal to Noise Ratio (PSNR) from available quantization data. Improving over the PAR, at Philips Research we have developed a no-reference PSNR estimation metric that works on the decoded video image [TCC02].

Other devices such as the PQA200 from Tektronix [Inc97] and the VP200 from Pixelmetrix [PIX] are full-reference metrics, which require the processed video and the original as inputs, and calculate quality scores based on the pixel-per-pixel differences on the perfectly aligned images. Tektronix also offers the PQM300, which uses a no-reference model based on blocking artifact metrics; and Pixelmetrix also offers a no-reference product that works on the MPEG stream.

Regarding full-reference models, a second round of tests has been carried out by the VQEG [VQE03]. Based on the results of this evaluation, four models will be recommended by the ITU-R SG-6 [ITU03]. These models are: British Telecom, Yonsei University/Radio Research Laboratory/SK Telecom, CPqD and NTIA/ITS. Description of these models can be found in [ITU03].

No-reference models are expected to be tested by the VQEG in 2004-2005. We anticipate that most models will use image features. A model proposed by Genista Corp. [GEN] includes features such as blockiness, blur, and jerkiness as fidelity indicators.

10.3 Quality Metric Components and Design

The components of the NROQM are measures of image features which can be computed directly without using the original image. Among many possible features we have chosen those which have a strong influence on perceived quality. For instance, MPEG artifacts affect quality negatively so that the more the artifacts the lower the quality; while a desirable feature, such as sharpness, clearly affects quality in a positive way. Measuring quality without a reference means estimating quality from the values of the image features. There are three principles which allow measuring quality in a no-reference manner:

1. Degraded video shows impairments due to transmission (e.g., noise), compression (e.g., MPEG artifacts) and image processing (e.g., clipping due to finite precision arithmetic). These impairments have a monotonic although not continuous effect on quality, and the best quality is that of the impairment-free image.

2. Enhanced video shows improved attributes such as sharpness and contrast, plus reduced artifacts.

3. Original video is assumed to have zero impairments and no significant enhancement. This loosely defines the *virtual reference*.[1]

---

[1]The label no-reference really means free of choice rather than absence of reference.

Thus, the combined assessment of desirable and non-desirable features is expected to be a reasonable indicator of overall image quality as perceived on the average by human subjects. A no-reference metric represents a virtual instrument which takes as its input video images and produces a quality score which indicates on an arbitrary scale the extent of degradation or enhancement.

Figure 10.1 shows the framework for the development of a no-reference quality metric. The design process starts by computing the selected features on a set of video sequences called the *training set*. That same training set is also used in a subjective test to obtain the subjective scores. Next, an objective model based on the image features is built in order to predict subjective quality from objective scores. The set is chosen so that the data points, i.e., quality scores, are spread out to cover the range of interest (e.g., from noticeably degraded to noticeably enhanced video), and with enough points in between to provide the best resolution in the scale (e.g., the smallest noticeable difference).

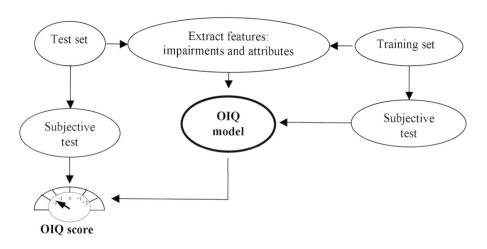

Figure 10.1: Framework for the development of no-reference objective quality metric.

Although one can compute more than a hundred features of an image, we have found that a small set of them can be used effectively to account for perceived quality. Even in these conditions, a complete test set (i.e., needed for a full ANOVA design) is too large and thus the test set must be chosen based on relevance and partitioned designs.

Each feature metric must have high precision and performance which should correctly quantify the attribute or impairment, i.e., it must have a minimum of false positive detections, and a maximum of true positive detections. The performance of each feature metric can be assessed on video strongly and only influenced by that feature, to the extent possible (sometimes it is impossible to isolate the effects of a single feature). In the following subsections we describe the feature metrics and discuss their influence on perceived quality.

10.3.1 Blocking Artifacts

Among of the best-known compression impairments are the blocking artifacts (see Figure 10.3 and Chapter 3). Blocking artifacts are the result of coarse quantization of DCT coefficients (of 8×8-pixel blocks) in MPEG coding. The amount and visibility of blocking artifacts increases with increased compression, i.e., lower bitrate. Perceived quality is strongly affected by blocking artifacts.

Blocking artifacts can be measured as the number of 8-pixel edges found on the 8×8 grid overlaying the image. To differentiate block edges from natural edges it is assumed that natural edges are strong (i.e., very steep) transitions while block edges are weak and regularly spaced.

One method to measure blocks is by the size of the discontinuity for pixels n and $n + 1$ where n is a multiple of 8 (the grid size). If the discontinuity exists and is lower than a threshold, e.g., 20, then it is on a block edge. If the edge is found in 8 consecutive lines (from line i to $i + 8$), then it is a blocking artifact. In actual images, the transitions are gradual, and the edge is at the center of such transition (usually detected as the point where the second derivative is zero).

Another simple method is to calculate the extrapolated discontinuity between 8×8 blocks. Let $B[m, n, t] = [\mathbf{x}[m \times 8 + i, n \times 8 + j, t]], (i, j) \in [0 \dots 7]$ be the luminance block of the sequence situated at column m, row n, and time t. The value of the artifact between the two blocks $B[m, n, t]$ and $B[m + 1, n, t]$ is the discontinuity at the frontier evaluated for the eight lines with the extrapolated values of the neighboring pixels, as illustrated by Figure 10.2. For each row j of the blocks, extrapolated pixel values $(E_l)_j$ and $(E_r)_j$ are computed according to a first order extrapolator:

$$
\begin{aligned}
(E_l)_j &= \tfrac{3}{2}\mathbf{x}[m \times 8 + 7, n \times 8 + j, t] &-& \tfrac{1}{2}\mathbf{x}[m \times 8 + 6, n \times 8 + j, t] \\
(E_r)_j &= \tfrac{3}{2}\mathbf{x}[(m+1) \times 8 + 0, n \times 8 + j, t] &-& \tfrac{1}{2}\mathbf{x}[(m+1) \times 8 + 1, n \times 8 + j, t]
\end{aligned}
\tag{10.1}
$$

And the vertical artifact value is the mean of the eight discontinuities:

$$
D_v = \frac{1}{8}\sum_{j=0}^{7}(\Delta A_v)_j = \frac{1}{8}\sum_{j=0}^{7}\left|(E_r)_j - (E_l)_j\right|
\tag{10.2}
$$

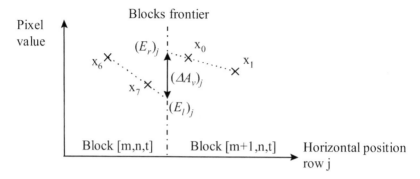

Figure 10.2: Quantization of blocking artifact intensities.

Horizontal artifacts are calculated in a similar fashion. For other blocking artifact metrics, see [GPLL99, KK95, WL01, WY96, YWC00].

10.3.2 Ringing Artifacts

The second best known MPEG artifact is called *ringing*. Ringing is a shimmering effect around high contrast edges. Depending on the orientation it manifests itself as edge doubling. See Figure 10.3 for a magnified view of ringing and blocking artifacts. Ringing is not necessarily correlated with blocking as the amount of ringing depends on the amount and strength of edges in the image, while blocking depends on the presence of uniform or smoothly changing regions.

An algorithm to detect and to measure ringing includes the following steps:

1. Detect strong edges using a high threshold for the edge transitions.

2. Detect low activity regions, or adjacent to strong edges where most local variances are very low.

3. Detect ringing pixels in the low activity regions as pixels where the local variance is large compared with the other variances. For example, if the local variance for most pixels in a low activity region is less than or equal to 3, the local variance for a ringing pixel must be at least four times that value. The sum of all ringing pixels on the image is the ringing value.

An alternative method to measure ringing artifacts is called the visible ringing measure (VRM) [Ogu99, YHTN01]. The VRM is based on the average local variance calculated on a small-size window in the vicinity of major edges. Those regions, which exclude the edges, are detected through morphological operations.

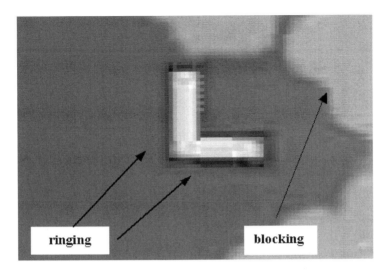

Figure 10.3: Magnified view of ringing and blocking artifacts.

10.3.3 Clipping

Clipping is a truncation in the number of bits of the image values (luminance and chrominance components) imposed by the arithmetic precision of the process being used. It results in abrupt cutting of peak values at the top and bottom of the dynamic range, which leads to aliasing artifacts caused by the high frequencies created at those discontinuities.

The sharpness enhancement technique known as peaking can cause clipping [dH00]. Peaking works by adding positive and negative overshoots to the edges. However, if the extreme values are beyond the limits of the dynamic range, saturation occurs and the pixels are clipped (i.e., pixels take maximum/minimum values of 255 or 0 for 8 bit precision).

The simplest clipping measurement is a function of the number of clipped pixels found in the image. The clipping metric is defined as 0.0 when no pixels are clipped and 1.0 when 1% or more of the pixels are clipped. A fixed margin can be applied to the image on the left, right, top and bottom to avoid counting any blanking or black bars, as well as to speed up the measurement.

A clipping measurement algorithm includes:

1. Test every pixel on the image except the margin on top, bottom, left and right.
2. If the pixel is 0 or Max (e.g., 255 if the precision is 8 bits) increase the count.
3. Calculate the percentage of clipped pixels for the tested image.

4. If 0%, clipping is 0. If 1% or more, clipping is 1.0. Other values are simply the percentage.

10.3.4   Noise

Noise is a random variation in the spatial or temporal dimension, which appears in video images as a result of random processes linked to transmission and generation techniques. It is most noticeable in smooth regions or regions with smooth transitions. It gives the subjective impression that the image is not clean, or that something unintended is superimposed on the image.

In some cases, small amounts of high-frequency noise add to the "naturalness" of textures (in contrast with a plastic or synthetic appearance) and have been found to increase perceived quality. Most noise, however, obscures details and reduces quality of the visual information.

To measure noise, most algorithms assume that any image contains small areas of constant brightness (i.e., no details are present). Hence, whatever variation in these areas is nothing but noise. A typical noise measurement algorithm would consist of the following steps:

1. Divide the image into small blocks.
2. Measure the intensity variations for every block.
3. Assuming that the intensity of the noise is much smaller in magnitude than the signal, the block with the least variation (or the average of the blocks with the smallest variation) should correspond to a constant brightness region.
4. Use a set of high-pass filters or a band-pass filter to filter out the DC component. The sum of the outputs of the filters, clipped using perceptual thresholds proposed in [WYSV97], is used to compute the variance or noise.

10.3.5   Contrast

In simple terms, contrast is related to the difference between the luminance of pixels of interest and the background, and largely depends on the dynamic range of the luminance signal. Contrast sensitivity is the ability to distinguish objects from the background. The perception of contrast depends on several factors including a mental reference image of the object in question, overall luminance (although not in all cases), background, and color. For work on contrast perception issues, see, for example, Fairchild [Fai99] and Spehar et al [SDZ96].

A very basic algorithm to measure contrast includes the following steps:

1. Compute the luminance histogram of the image, excluding a fixed margin on the left, right, top and bottom.
2. Separate the upper and lower parts of the histogram that contain each a certain percentage of the total energy.
3. Calculate the difference between the luminance of the upper and lower parts of the histogram and normalize by the average luminance.

This metric is relevant because for a given image, contrast increases are associated with improved quality. Thus it is useful as a first approach. Many interactions with other features can be expected. However, this rudimentary definition of a contrast metric is content dependent and will probably require further research to reduce or eliminate that dependency.

10.3.6 Sharpness

Image sharpness is the informal, subjective evaluation of the clarity of detail and contours of an image. Objectively, it can be measured by the definition of edges in the spatial domain, or by the characteristics of the high frequencies in the transform domain. Perceived sharpness is highly dependent on content, and also on spatial resolution, contrast, and noise as reported by Johnson et al [JF00]. For a no-reference sharpness metric we require reduced dependency on content, i.e., minimum baseline, and with the same dynamic range for any image.

We have created a new sharpness metric based on the local edge kurtosis [CG02]. This metric takes into account spatial and local frequency information and uses the weighted kurtosis of 8×8 blocks that contain the edge pixels.

The algorithm consists of the following steps:

1. Create the edge image using an edge detection method such as Canny [Can86].
2. Assign edge pixels to 8×8 blocks (e.g., the MPEG grid blocks can be used).
3. Compute 2-D kurtosis for each block and weight by the number of edge pixels.
4. Sharpness is the 2-D kurtosis averaged over all blocks.

This metric by itself shows high correlation with perceived sharpness, i.e., quality changes are affected only by sharpness. Therefore it is used as a perceptual sharpness metric. Figure 10.4 shows the edge detection and block assignment procedures for a frame of the doll sequence.

10.4 No-Reference Overall Quality Metric

The components of the NROQM are impairment and attribute metrics implemented as described in previous sections. The contrast, noise and clipping metrics were devel-

Figure 10.4: Edge detection and block assignment for the doll sequence.

oped for a first version of the quality metric used in a video chain optimization system described in [vZA01]. The blockiness and ringing metrics used here were developed in Philips Research, and proposed for a quality metric based on MPEG impairments [CDGR00, CJ01]. Features are computed for each video in a set of training sequences. The NROQM model accounts for the individual and combined effects of the features on perceived quality.

10.4.1   Building and Training the NROQM

In this subsection we give a description of the method used to build and train the NROQM model. As indicated in Figure 10.1, in Section 10.3, we use a set of video sequences (i.e., the training set), their subjective scores, and the measured values of a set of features for each video sequence to build and train the model. The set of sequences used constitutes a significant but still limited test set. Subjective scores have been obtained using a standardized subjective evaluation.

Instead of using a brute force approach, e.g., taking the six feature measurements plus the subjective scores for all sequences in the training set and running them through a regression program, we chose instead to apply an incremental, heuristic, knowledge-driven approach. The reason for this choice is that a brute force approach is only applicable to a combinatorial, exhaustive set. In our case, working by subsets allows selecting the effects and interactions best represented, and identifying and avoiding weaknesses in the feature metrics.

Computing the video features and incrementally assessing their (individual and combined) effects on quality is a safe way to verify data integrity and robustness of model components. The advantage of our approach also stems from the fact that any terms or functions included can be checked against plausible theories or hypotheses linked to perceptual effects on quality.

The training method thus consists of the following steps:

1. Compute the six feature measurements for the entire test set.

2. Partition the test set into (i) coded (include coded-enhanced), (ii) noisy (include noise-enhanced), and (iii) original and original-enhanced sequences.

3. Further refine the coded and noisy sets into sets that exhibit variations predominantly in one feature, e.g., blocking, ringing, or noise.

4. Fit the values of the feature of interest to the subjective data in each set of Step 3 through a perceptual function. A perceptual function is just a function that maximizes correlation between feature values and subjective scores. At this step, one must check the dynamic range and spread of feature values in order to identify weaknesses of the feature metrics or the training set, and improve them or introduce temporary workarounds.

5. Expand each set by including sequences whose quality is influenced by more than one feature and find out whether they are additive, and if there are interactions such as masking or facilitation. Refine the fitting objective to subjective data.

6. Consider the set of (original) enhanced video sequences, and fit objective to subjective data. This step includes first taking into account the effect of sharpness, and then the combined effects of contrast and noise (if any).

7. Fit objective to subjective data for coded-enhanced and noise-enhanced sequences. This takes care of interactions not considered before, and scaling factors.

8. Consider the entire set and merge the fitting functions found so far. Review overall fitness of the model to all subjective data. Redo or iterate on previous steps as needed.

The flowchart shown in Figure 10.5 shows the generic application of the method just described, from left to right and top to bottom. Level 1 corresponds to the calculation of the main features, blocking, ringing, clipping, noise, contrast and sharpness. In levels 2 and 3 we build the components of the overall metric following the method described before. The components built in levels 2 and 3 are associated with the perceptual effect of the basic feature measures (i.e., B, R, C, N, CN, and SH) and their interactions (as well as necessary scaling factors). In level four we arrive at the formula:

$$NROQM\,(B,R,C,N,CN,SH) =$$

$$-\left(1 + CN + \frac{SH}{a}\right)\frac{B^b}{c} \qquad (a)$$

$$-\left(1 + CN + \frac{SH}{a}\right)^{\frac{B}{1+B}}\frac{R^b}{c} \qquad (b)$$

$$-(1 + CN)(1 + dC)^e \qquad (c) \qquad\qquad (10.3)$$

$$-f\,(N)\,(SH)\,(CN) \qquad (d)$$

$$1mm] - \frac{\left(\frac{N}{CN}\right)^g}{(1 + dC)^h} \qquad (e)$$

$$-(i + jCN)^k \qquad (f)$$

$$+SH \qquad (g)$$

Term (10.3a) contains the blocking contribution. Exponent b is linked to the perceptual effect. The perceptual effect maps the feature measurement onto the subjective score scale. This must be done using a function that is monotonic within the range of interest. In this case, the range of interest for blockiness is $B \in [0, 1000)$. The term $(1 + CN + SH/a)$ is linked to the facilitation effect of contrast and sharpness on blockiness (i.e., contrast and sharpness increase the visibility of blockiness). Constants a and c are scaling factors.

Term (10.3b) contains the ringing contribution and is similar to (10.3a) except that it includes $B/(1 + B)$ linked to the masking effect of blocking on ringing, i.e., when blocking is low, ringing becomes more noticeable. It also makes sure that if blocking is zero, any ringing detected is not taken into account, as it should not be associated with MPEG coding.

Term (10.3c) contains the clipping contribution. The perceptual effect is included using the power of e. The facilitation effect of increased contrast, which makes clipping more likely, is accounted by including $(1 + CN)$. Constant d is a scaling factor.

Terms (10.3d) and (10.3e) contain the contribution of noise. The facilitation effect of sharpness and contrast are included as $(SH)(CN)$ in (10.3d), and the masking effect of clipping (itself affected by contrast), are included in (10.3e) (strong clipping would tend to reduce the noise effect). Exponent g is linked to the perceptual effect of noise on quality; h is designed to fit the data showing the masking effect, and d is a scaling factor.

Term (10.3f) is an ad-hoc term intended to reduce the contrast baseline. The current measure of contrast is content dependent and it can therefore shift the scores noticeably when there is a change of scene.

Term (10.3g) is the contribution of sharpness. It is the only positive term in the formula (a positive contrast term cannot be used yet because our contrast metric is strongly

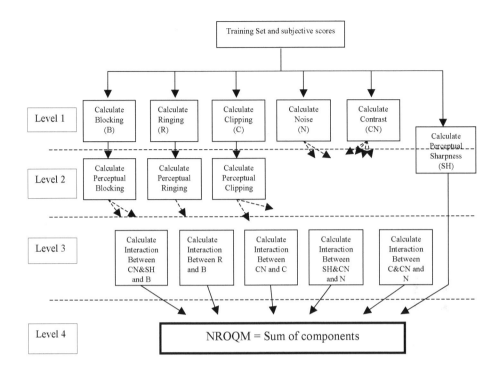

Figure 10.5: NROQM algorithm flowchart.

content dependent). Since our sharpness metric is itself very well correlated with the perceptual effect, there is no need for a perceptual function. The optimized values of the parameters are shown in Table 10.1.

Table 10.1: Parameter values for the NROQM model

Param	a	b	c	d	e	f	g	h	i	j	k
Value	5	0.5	40	2.5	1.5	2	0.5	2	0	4	0.5

10.5 Performance of the Quality Metric

The quality metric NROQM has been trained to obtain a high correlation with the subjective scores of the training set. The training set consists of 65 sequences (including

Table 10.2: Training set treatments and descriptions

Description	Processing
Source (3)	Original, no processing
Sharpness (3×4)	Gains of 0.25, 0.5, 0.75, 1.0
Coded (3×3)	1Mb/s, 3Mb/s, 5Mb/s
Coded-postprocessed (3)	1Mb/s-DFD, 3Mb/s-DFD, 5Mb/s-DFD different source
Noisy (3×3)	30dB, 35dB, 40dB
Noisy + sharpness enhanced (14 comb)	30dB-0.25(1)/0.5(2)/0.75(2), 35dB-0.25(2)/0.50(2)/0.75(1), 40dB-0.25(1)/0.5(2)/0.75(1)
Coded + Sharpness enhanced (3×2×2)	3Mb/s-0.25, 3Mb/s-0.5, 5Mb/s-0.25, 5Mb/s-0.5
Noisy + swan noise reduction (3, one source each)	30dB-swan, 35dB-swan, 40dB-swan

a mix of coding, sharpness enhancement, and noise reduction treatments) derived from three original scenes (*Doll, Dolphins, Lawnmowerman*). See more details in Table 10.2.

To obtain the subjective scores of the training set, a comparative, categorical subjective evaluation modality was chosen, which uses two monitors. A group of 20 non-expert viewers were instructed to evaluate each processed sequence vs. its original and to score on a seven point scale (much worse, worse, slightly worse, same, slightly better, better, much better). The experimental design has been patterned following recommendation ITU-R BT.-500-10 [ITU00]. A plot of subjective vs. objective scores for the training set is shown in Figure 10.6. The overall correlation with subjective scores (obtained following ITU-Rec 500) was 0.85 (while the PSNR correlation was 0.399), and the average correlation per scene was 0.93 (0.93 for doll, 0.94 for dolphins, and 0.92 for *Lawnmowerman*).

10.5.1   Testing NROQM

As explained in Section 10.3 (Figure 10.1), after training the model, tests on an independent set of video sequences must follow. The test material should be different from that previously used, and the subjective testing must be reliable and repeatable. The first of several planned tests was carried out during the 2002 Philips Concern Research Exhibition in Eindhoven.

A constant value could be added to (10.3) to make the graphs overlap, but it would not change the correlation results, thus it has not been included. Some of the systematic

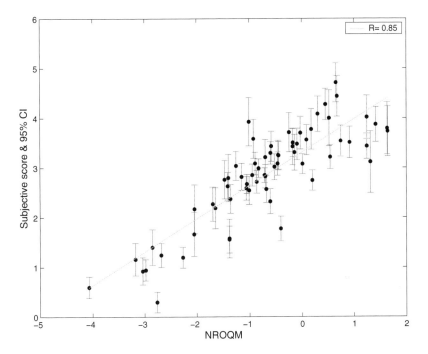

Figure 10.6: Performance of NROQM on the training set. CI: Confidence Interval.

errors found can be traced back to specific problems with the model or its components. For example, for post-proccessed sequences, the NROQM overestimates quality due to underestimation of the blockiness (remaining after deblocking) by the blockiness metric. This problem was known from the early phases of training, and can be solved by refining the blockiness metric.

The test was intended as a first assessment of the performance of the NROQM against subjective scores on a small set of sequences including degradation and enhancement treatments. It was not a full-blown experiment to validate the NROQM or its resolution, but rather to get a feel for its performance and to obtain early information on areas which need improvement. Two original sequences were used, *Fade* and *Boat*. *Fade* is an indoor scene of a woman, using a slow pan-zoom. *Boat* is an outdoor shot of a boat traveling down a river, under overcast sky, and including detailed scenery.

From each source, 6 sequences were generated according to the treatments shown in Table 10.3. These treatments (two cases of noise plus noise-reduction, three cases of coding plus sharpness enhancement) insured a variety of quality levels generated by a different mix of image features.

Table 10.3: Treatments applied to each source video for the performance test

Sequence	Source material	Video Enhanced Algorithm
1	Clean	None
2	Clean+Noise (28dB)	Noise reduction 1 (INFINEON)
3	Clean+Noise (28dB)	Noise reduction 2 (SWAN-DNR)
4	MPEG 3Mbs	Peaking setting No. 1
5	MPEG 3Mbs	Peaking setting No. 2
6	MPEG 3Mbs	Peaking setting No. 3

The test consisted of a 2 alternative forced choice (2AFC) on 21 possible pairs for each scene. Between 33 and 87 subjects were used per comparison. The subjective ranking can easily be obtained from the test, and when the percentages (of one choice over the other) are used, a psychometric technique described in [Eng00] can be applied to obtain subjective scores. Although it is not an ITU recommended test, it is a simpler and very reliable way to obtain reference subjective data when enough subjects are available. The results showed correlations of 0.848 and 0.943 for *Fade* and *Boat*, respectively.

10.5.2   Test with Expert Viewers

A test using expert viewers has also been carried out on a small scale. The test set consists of 15 sequences, which include 5 sharpness-enhanced versions (gains from 0.25 to 2.0) of each of three originals (high definition, progressive video output, named *Bicycle*, *Trailer* and *Birds*).

This test set is beyond the capabilities of the NROQM because the training set did not include extreme over enhancement leading mainly to clipping values not found in the training set, the subjective scores used for training were not for expert viewers, and the (expected) increased precision of the expert test. Nevertheless, the test was useful to learn about the NROQM strengths and weaknesses.

We found that subjective scores drop for excessive sharpness, and although the sharpness metric sees enhancement, the NROQM picks up the degradation thanks to the clipping component. The maximum NROQM values are the same as the subjective results, except for trailer sequence, where the NROQM detects degradation not seen by the experts (this points to the need to train the NROQM for the broader range of clipping values).

The following points summarize the results of the expert viewer test:

1. High clipping values, not seen before by the NROQM (the maximum clipping in

the training set was 0.24, while here the maximum is 1), pull the scores for *Bicycle* and *Trailer* too far down and may cause the NROQM to miss the maximum for *Trailer* (clipping of 0.6).

2. The perceptual effect of sharpness in the presence of high contrast (maximum contrast in the training set is 0.52, while maximum contrast in this set is 0.88) was not represented in the training set, and thus the NROQM does not fully reflect enhancement under these conditions.

3. Although there is no MPEG artifact, the blocking and ringing metrics detect important values. This is because the enhancement algorithm works in a block-wise manner, and because over-sharpening causes ringing. However, we have not studied the perceptual effect of these impairments or the ability of the metric to account for them.

4. The degradation of *Birds* equence (at 2.0 sharpness gain) is not reflected by the NROQM, i.e., the objective score is higher than that of the 1.5 sharpness gain, while the subjective score is lower. This is because visible distortions, other than clipping, brought about by over sharpening, are not currently accounted for in the NROQM model.

10.6   Conclusions and Future Research

An approach to objective quality metrics has been presented in this chapter, which does not require full access to the original source. The quality model has been designed to meet the broad requirements of the consumer video industry, including measuring improvement and degradation, supporting algorithm development, and embedded, real-time monitoring and control.

The results so far are encouraging. Incremental and heuristic training of the NROQM model on a variety of video sequences has proven feasible. The first validation test showed a good correlation between subjective and objective scores and did not reveal any serious drawbacks or unknown weaknesses. However, a formal mathematical modeling approach can be pursued if an exhaustive test set is used and the drawbacks of the feature metrics are minimized.

Further validation tests and improvement of the feature metrics and the model should be pursued in areas including blockiness metric for post-processed video, content independent contrast metric, automatic calibration of a virtual reference upon scene cut (if absolute content independency is not possible), and optimization of metric resolution (i.e., the smallest subjective quality difference that the metric can detect, defined according to the VQEG approved method [ATI01]). Although further tune-up of the constants used in the NROQM formula is still possible, e.g., to deal with the out-of-range values for clipping and contrast found in the expert test, the general model appears to include the terms necessary to model image quality.

In supplemental work, we have also found that metrics for motion artifacts can further improve the overall performance of the NROQM at the expense of more complexity, i.e., one more term and its relevant interactions. Temporal impairment metric candidates are:

- Energy difference in successive frames (in low motion cases it indicates artifacts);
- Judder (e.g., unnatural modulation of smooth motion, caused by uncompensated frame rate conversion);
- IBP drift (a difference between I frame and previous one, caused by motion estimation errors which increase in the B and P frames between two I frames or GOP);
- Motion field smoothness (standard deviation of the motion vectors);
- Mosquito noise (noise around stationary edges, calculated as sum or absolute frame differences among pixels in a small neighborhood around an edge pixel).

Future validation tests we expect to carry out include VQEG validation of no-reference models and others. Since the VQEG test sequences include only degraded video (main requirement for the broadcast industry), we must also pursue additional testing that includes enhanced video as well. Expert viewer tests are also recommended. We have carried out preliminary expert tests aimed at scoring perceived features (sharpness, contrast, noise, and coding artifacts) and to evaluate sharpness enhancement. Although the results are useful to informally validate and find specific areas of improvement of the NROQM, further research is necessary to define the role of expert tests in training and validation, e.g., as a substitute or as a complement for naive viewer tests.

Finally, our approach to dealing with individual impairments and enhancements, and the combined effects of impairments and enhancements offers a divide-and-conquer method that can be further exploited. This method entails dividing the training sequences into impaired, enhanced, and impaired-enhanced sets. A new category that poses future challenges to the model is the over-enhanced video. Particularly, if over-enhancing does not increase impairments that cannot be directly measured by the model. The best example is over-sharpening which results in the so-called *plastic face* distortion.

References

[ATI01]   Methodological framework for specifying accuracy and cross-calibration of video quality metrics, September 2001. Technical Report No. T1.TR.PP.72.2001, ATIS T1A1.1.

[Can86]   J. Canny. A computational approach to edge detection. *IEEE Trans. Pattern Anal. Mach. Intell.*, 8(6):679–698, November 1986.

[CDGR00] J. E. Caviedes, A. Drouot, A. Gesnot, and L. Rouvellou. Impairment metrics for digital video and their role in objective quality assessment. In *Proc. VCIP2000*, 791–800, Perth, Australia, June 2000.

[CG02]   J. E. Caviedes and S. Gurbuz. No-reference sharpness metric based on local edge kurtosis. In *Proc. ICIP2002*, III: 53–56, Rochester, NY, September 2002.

[CJ01]   J. E. Caviedes and J. Jung. No-reference metric for a video quality control loop. In *Proc. SCI2001*, 290–295, Orlando, FL, July 2001.

[dH00]   G. de Haan. *Video processing for multimedia systems*. Eindhoven: University Press Facilities, 2000.

[Eng00]   P. G. Engeldrum. *Psychometric Scaling: A Toolkit for Imaging Systems Development*. Winchester, MA: Imcotek Press, 2000.

[Fai99]   M. D. Fairchild. A victory for equivalent background – on average. In *Proc. IS&T/SID 7th Color Imaging Conference*, 87–92, Scottsdale, AR, 1999.

[GEN]   Genista Corp. homepage, http://www.genista.com.

[GPLL99]   C. Glassman, A. Peregoudov, A. Logunov, and V. Lichakov. Video compression artifacts: predicting the perceptual ratings. In *Proc. IBC*, 560–564, Amsterdam, 1999.

[Inc97]   T. Inc. A guide to picture quality measurements for modern television systems. available from http://www.tektronix.com, Beaverton, OR, 1997.

[ITU00]   ITU-R. Methodology for the subjective assessment of the quality of television pictures, 2000. Recommendation ITU-R BT.500-10.

[ITU03]   Objective perceptual video quality measurement techniques for standard definition digital broadcast television in the presence of a full reference, 2003. Draft New Recommendation ITU-R BT.[Doc. 6/39].

[JF00]   G. M. Johnson and M. D. Fairchild. Sharpness rules. In *Proc. IS&T/SID 8th Color Imaging Conference*, 24–30, Scottsdale, AR, November 2000.

[KK95]   S. A. Karunasekera and N. G. Kingsbury. A distortion measure for blocking artifacts in images based on human visual sensitivity. *IEEE Trans. Image Processing*, 4(6):713–724, June 1995.

[Ogu99]   S. H. Oguz. *Morphological post-filtering of ringing and lost data concealment in generalized lapped orthogonal transform based image and video coding*. Ph.D. thesis, Univ. of Wisconsin, Madison, WI, 1999.

[PIX]   Pixelmetrix Corporation homepage, http://www.pixelmetrix.com.

[ROH]   Rohde & Schwarz homepage, http://www.rohde-schwarz.com.

[SDZ96]   B. Spehar, J. S. DeBonet, and Q. Zaidi. Brightness induction from uniform and complex surrounds: A general model. *Vision Research*, (36):1893–1906, July 1996.

[SNE]   Snell & Wilcox homepage, http://www.snellwilcox.com.

[TCC02]   D. S. Turaga, Y. Chen, and J. E. Caviedes. No-reference PSNR estimation for compressed pictures. In *Proc. ICIP2002*, III: 61–64, Rochester, NY, September 2002.

[VQE]   The Video Quality Experts Group, http://www.vqeg.org.

[VQE03]   VQEG. Final report from the Video Quality Experts Group on the validation of objective models of video quality assessment, phase II, 2003. Accessable at http://www.vqeg.org.

[vZA01]   K. van Zon and W. Ali. Automated video chain optimization. *IEEE Trans. on Consumer Electronics*, 47(3):593–603, August 2001.

[WL01]   A. Worner and J. Lauterjung. A real time single ended algorithm for objective quality monitoring of compressed video signals. In *Proc. of SCI2001*, 13:329–334, Orlando, FL, July 2001.

[WY96]   H. R. Wu and M. Yuen. Quantitative quality metrics for video coding blocking artifacts. In *Proc. of Picture Coding Symposium 1*, 23–26, Melbourne, Australia, 1996.

[WYSV97]   A. Watson, G. Yang, J. Solomon, and J. Villasenor. Visibility of wavelet quantization noise. *IEEE Trans. Image Processing*, 6:1164–1175, August 1997.

[YHTN01]   S. Yang, Y. Hu, D. Tull, and T. Nguyen. Maximum likelihood parameter estimation for image ringing artifact removal. *IEEE Trans. on CVST*, 11(8):963–973, August 2001.

[YWC00]   Z. Yu, H. R. Wu, and T. Chen. Perceptual blocking distortion measure for digital video. In *Proc. VCIP2000*, 4067:768–779, Perth, Australia, June 2000.

Chapter 11

Video Quality Experts Group

Philip Corriveau
Intel Media and Acoustics Perception Lab, U.S.A.

In the hopes of addressing the need for a standardized objective method for picture quality assessment, the Video Quality Experts Group was established in 1997. For the first time on an international scale, two independent sectors of the International Telecommunications Union, T and R, will cooperate in conducting formal evaluations to compare the results of subjective assessments and objective methods.

The need for setting official standards on objective models for video quality evaluation has been recently recognized by both ITU-R and ITU-T. Consequently these standardization bodies have established questions of study related to this topic (ITU-R Q64-3/11, ITU-T Q11/12 and ITU-T Q22/9). In order to initiate an examination of objective methods in comparison to subjective methods, a group of experts was formed known as the Video Quality Experts Group (VQEG).

Since there are several classes of video material and applications, it may not be possible for VQEG to identify one unique model that is globally applicable. The hope is to narrow the selection of objective methods such that researchers, broadcasters and others have a choice when it comes to monitoring image quality. The desired outcome is to have a draft recommendation ready for both ITU-T and ITU-R in 1999.

11.1 Formation

In October 1997, a meeting was held at CSELT, in Turin, Italy to discuss the technical procedures needed to validate objective measures of video quality. Experts from ITU-T SG 12, SG 9 and ITU-R SG 11 took part in the meeting and contributed to the specification of a work plan for this activity. It was decided that this group would be responsible for carrying out and reporting the status and results of work related to video quality assessment.

VQEG formally proposed the validation of objective measurements of video quality and began the process of solicitation of submissions of objective models to be included in an ITU Verification Process leading to an ITU Recommendation.

The initial proposal stated that all objective models should be capable of receiving as input both processed sequences and corresponding source sequences. Based on this input, it must provide one unique figure of merit that estimates the subjective assessment value of the processed material. The objective model must be effective for evaluating the performance of block-based coding schemes (such as MPEG-2 [ISO94, RH96] and H.263 [ITU98b]) in bit rate ranges of 768 kbits/s to 36 Mbits/s and sequences with different amounts of spatial and temporal information.

Two ad hoc groups under VQEG were tasked with authoring both the subjective and objective test plans. The work is always split so that different members of the group are responsible for producing the documents that best fit their areas of expertise.

11.2   Goals

The desired outcome was to have a draft recommendation ready for both ITU-T and ITU-R in 1999. It was an aggressive test plan, but VQEG was confident in its ability to complete the task on time in a non-biased, scientifically grounded fashion, leading to a revolution in assessing and monitoring video image quality. The Video Quality Experts Group finally got results to the Standards bodies in 2003 after several years of hard work.

11.3   Phase I

An extensive testing process was finalized at the second meeting of VQEG held in May 1998 at the National Institute of Standards and Technology (NIST) facility in Gaithersburg, Maryland. A set of test sequences was selected by the Independent Lab and Selection Committee (ILSC) of VQEG. The sequences were kept confidential until the proponents submitted their final implementations of objective assessment methods to the ILSC (August 7, 1998). The final selection of test conditions (referred to as hypothetical reference circuits or HRCs) was approved by the VQEG members present at the meeting. Details concerning the processing of test sequences and the final distribution of these sequences to both proponents and testing facilities were finalized. The two testing phases (subjective and objective) were executed in parallel and a detailed timetable can be found in the report for the Gaithersburg VQEG meeting [VQE00a].

The subjective test was conducted by multiple laboratories who received the source and the processed video clips in random order on digital component video viewing tapes. Each facility conducted formal subjective assessments according to the subjective test plan developed and published by VQEG [VQE00c]. The subjective test plan followed the currently accepted testing procedures standardized in ITU-R Recommendation BT.500 [ITU98a]. The Double Stimulus Continuous Quality Scale (DSCQS) method was used for the subjective test. Due to the size of the test (20 test sequences × 16 HRCs) and the need to minimize contextual effects, the test design was broken down into four basic quadrants (50Hz/high quality, 50Hz/low quality, 60Hz/high quality and 60Hz/low quality). In addition, the need to repeat experiments in different laboratories necessitated the use of a large number of subjective testing facilities. The following eight laboratories conducted subjective tests:

- CRC (Canada)

- RAI (Italy)

- CCETT (France)

- CSELT (Italy)

- NHK (Japan)

- DCITA (Australia)

- Berkom (Germany)

- FUB (Italy)

In the case of objective testing, each proponent of an objective method received the video sequences on computer tapes, analyzed them, and reported the results-one number per HRC/source combination. (For details of the objective test procedure, see the objective test plan [VQE00b].) A randomly selected subset of these sequences (10%) was sent to several independent testing laboratories to confirm the results reported by the proponents. The following four laboratories performed this verification:

- CRC (Canada)

- IRT (Germany)

- FUB (Italy)

- NIST (USA)

The proponents of objective methods of video quality participating in this test are listed below (the numbers in brackets were assigned to the proponents for analysis purposes and are referred to later in the Results and Discussion section):

- **[P0]** Peak Signal to Noise Ratio (PSNR)

- **[P1]** CPqD (Brazil)

- **[P2]** Tektronix / Sarnoff (USA)

- **[P3]** NHK (Japan Broadcasting Corporation)/ Mitsubishi Electric Corp. (Japan)

- **[P4]** KDD (Japan)

- **[P5]** Swiss Federal Institute of Technology (EPFL) (Switzerland)

- **[P6]** TAPESTRIES (European Union)

- **[P7]** NASA (USA)

- **[P8]** KPN Research (The Netherlands) / Swisscom CIT (Switzerland)

- **[P9]** NTIA/ITS (USA)

11.3.1   The Subjective Test Plan

The Subjective Test Plan is necessary to ensure methodological consistencies among laboratories. Its components adhere to recommendations established by the international standards bodies. Before acceptance, the plan was reviewed by proponents and testing facilities to ensure a common set of testing parameters. The plan defines these parameters in detail and discusses the selection, length and presentation of test material, along with viewing conditions. These and other details had been previously agreed upon and reflect the standards defined in ITU-R Recommendation BT.500-9. The screening process for selection of participants, as well as the methods of data collection and analysis are further clarified in the Subjective Test Plan. For additional information concerning these and other details, along with a list of participating laboratories, please refer to the VQEG Subjective Test Plan (15).

11.3.2   The Objective Test Plan

The first goal of the VQEG is to find an objective measurement system that most accurately predicts subjective test scores. Implementation details are secondary. A number of candidate objective measurement methods are being considered by the VQEG. So far, approximately ten organizations have expressed interest in participating in the first validation test. This first test will be primarily for MPEG-2 systems at bit rates from 2 Mb/s to 36 Mb/s. Some H.263 coded video at 768 kb/s and 1.5 Mb/s will also be included as well as a limited number of error conditions. Future tests are planned to validate assessment methods for video at much lower bit rates and with more error conditions included. All tests will compare the results of objective measures with subjective scores taken on the same video sequences.

11.3.3 Comparison Metrics

A number of attributes characterize the performance of an objective video quality model as an estimator of subjective video image quality. Due to the nature of subjective assessments, modifications to the metrics were made to account for inherent differences in mean ratings caused by fluctuations within the viewer population and differences between multiple testing facilities. VQEG established a specific set of metrics to measure these attributes.

It was determined that an objective model Video Quality Rating (VQR) should be correlated with the subjective Difference Mean Opinion Score (DMOS) in a predictable and reliable fashion. VQEG agreed that the relationship between the VQR and the DMOS can be non-linear as long as it was stable and the data set's error variance was determined to be usefully predictive.

Other evaluation metrics chosen dealt with the prediction accuracy of the model, prediction monotonicity of a model, prediction consistency of a model and an analysis of variance. Details on the entire objective test plan and the comparison metrics are provided in the VQEG Objective Video Quality Test Plan [VQE00b].

11.3.4 Results

Depending on the metric that is used, there are eight or nine models (out of a total of ten) whose performance is statistically indistinguishable. Note that this group of models includes PSNR. PSNR is a measure that was not originally included in the test plans but it was agreed at the third VQEG meeting held in September 1999 in The Netherlands to include it as a reference objective model. It was also discussed and determined at this meeting that three of the proponent models did not generate proper values due to software or other technical problems. Please refer to reference [VQE99] for explanations of their performance.

Based on the analyses presented in its report [VQE99], VQEG is not presently prepared to propose one or more models for inclusion in ITU Recommendations on objective picture quality measurement. Despite the fact that VQEG is not in a position to validate any models, the test was a great success. One of the most important achievements of the VQEG effort is the collection of an important new data set. Prior to the current effort, model developers have had a very limited set of subjectively-rated video data with which to work. Once the VQEG data set is released, future work is expected to substantially improve the objective measures of video quality.

With the finalization of this first major effort conducted by VQEG, several conclusions stand out:

- no objective measurement system in the test is able to replace subjective testing

- no one objective model outperforms the others in all cases

- the analysis does not indicate that a method can be proposed for ITU Recommendation at this time

- a great leap forward has been made in the state of the art for objective methods of video quality assessment

- the data set produced by this test is uniquely valuable and can be utilized to improve current and future objective video quality measurement methods

11.4 Phase II

In 2001-2003, VQEG performed a second validation test, FR-TV Phase II, the goal being to obtain more discriminating results than those obtained in Phase I [VQE03]. The Phase II test contains a more precise area of interest, focused on secondary distribution of digitally encoded television quality video. The Phase II test contains two experiments, one for 525-line video and one for 625-line video. Each experiment spans a wide range of quality, so that the evaluation criteria are to determine statistical differences in model performance. The Phase II test contains a broad coverage of typical content (spatial detail, motion complexity, color, etc.) and typical video processing conditions to assess the ability of models to perform reliably over a very broad set of video content (generalizability). To address the concern that standardization bodies would prefer to recommend a complete system, models submitted to the Phase II test were required to supply their own video calibration (e.g., spatial registration, temporal registration, gain and level offset).

The FR-TV Phase II test utilized three independent labs. Two labs, Communications Research Center (CRC, Canada) and Verizon (USA), performed the 525 test and the third lab, Fondazione Ugo Bordoni (FUB, Italy), performed the 625 test. Of the initial ten proponents that expressed interest in participating, eight began the testing process and six completed the test. The six proponents of the FR-TV Phase II are:

- NASA (USA, Proponent A)

- British Telecom (UK, Proponent D)

- Yonsei University / Radio Research Laboratory (Korea, Proponent E)

- CPqD (Brazil, Proponent F)

- Chiba University (Japan, Proponent G)

- NTIA (USA, Proponent H)

11.4.1 Results

The results of the two tests (525 and 625) were similar but not identical. According to the formula for comparing correlations in "VQEG1 Final Report" (June, 2000, p. 29), correlations must differ by 0.35 to be different in the 525 data (with 66 subjects) and must differ by 0.55 to be different in the 625 data (with 27 subjects). By this criterion, all six Video Quality Metrics (VQMs) in the 525 test perform equally well, and all VQMs in the 625 test also perform equally well. Using the supplementary statistical analyses, the top two VQMs in the 525 test and the top four in the 625 test perform equally well and also better than the others in their respective tests.

The Pearson correlation coefficients for the six models ranged from 0.94 to 0.681. It should not be inferred that VQEG considers the Pearson correlation coefficient to be the best statistic. Nevertheless, the ranking of the models based upon any of the seven metrics is similar but not identical.

Using the F test, finer discrimination between models can be achieved. From the F statistic, values of F smaller than approximately 1.07 indicate that a model is not statistically different from the null (theoretically perfect) model. No models are in this category. Models D and H performed statistically better than the other models in the 525 test and are statistically equivalent to each other.

For the 625 data, the same test shows that no model is statistically equal to the null (theoretically perfect) model but four models are statistically equivalent to each other and are statistically better than the others. These models are A, E, F, and H. PSNR was calculated by BT, Yonsei and NTIA. The results from Yonsei were analyzed by six of the seven metrics used for proponents' models. For both the 525 and 625 data sets, the PSNR model fits significantly worse than the best models. It is very likely that the same conclusions would hold for PSNR calculated by other proponents.

VQEG believes that some models in this test perform well enough to be included in normative sections of Recommendations [VQE03].

In October 2003, the International Telecommunications Union — Radiocommunications Working Party Six — prepared and released for approval a Draft Standard [Wor03]. In this recommendation, the group specified four methods for estimating the perceived video quality of full referenced video systems. The four systems tested in 2002 were included in the draft recommendation, at this point the four included are: British Telecom, Yonsei University/Radio Research Laboratory/SK Telecom, CPqD and NTIA/ITS. For now, these methods are disclosed in the recommendation.

It is stated that the intention of the study group is to eventually recommend only one normative full reference method. They also state that for consideration to be standardized, other methods must be validated by open independent bodies like VQEG [Wor03].

The test environments in different facilities around the world vary from extremely complex to very simple. One of the most advanced video quality test facilities is the Communications Research Centre (CRC) / Industry Canada in Ottawa, Canada. Figure 4.4 shows a typical room layout at CRC for a dual monitor evaluation. This facility meets and exceeds the standards defined in the ITU-R Recommendation BT 500.7. Other facilities like those at NTIA/ITS and Intel Corporation use soundproof chambers that are designed to meet industry standards, but accommodate much smaller displays than that at the CRC.

11.5   Continuing Work and Directions

The Video Quality Experts Group is continuing with work that will lead to standards in not only the area of full reference evaluation but in reduced reference and multimedia. The international group has such diverse skills and resources that it will continue to help evaluate image quality issues facing industry for years to come.

11.6   Summary

The areas of subjective assessment and objective assessment have been around for years and exploded in the 1980s and 1990s to address new television technologies and changes in distribution and availability of media content. The importance of using end users to evaluate image quality will never decrease and the need for developing black box solutions that model human perception will be a required industry that will always exist.

Hopefully these last few pages have given the reader enough to understand subjective testing such that he/she can conduct a test and that he/she has an understanding of the use and place of both subjective and objective testing methods. There are a lot of issues that need to be solved and everyone is encouraged to explore, develop new methods and publish so that knowledge can continue to be shared.

References

[ISO94]   ISO/IEC 13818. Information technology–generic coding of moving pictures and associated audio information. Technical report, ISO/IEC, November 1994.

[ITU98a]   ITU. ITU-R BT. 500-9, methodology for the subjective assessment of the quality of television pictures. *ITU-R BT*, 1998.

[ITU98b]   ITU-T Video Coding Experts Group. Video coding for low bit rate communication. ITU-T recommendation H.263, January 1998.

[RH96]   K. R. Rao and J. J. Hwang. *Techniques and standards for image/video/audio coding.* Upper Saddle River, NJ: Prentice Hall, 1996.

[VQE99]   VQEG. *Final Report from the Video Quality Experts Group on the Validation of Objective Models of Video Quality Assessment.* VQEG, December 1999. (Accessible at *ftp://ftp.crc.ca/crc/vqeg*).

[VQE00a]   VQEG. *VQEG meeting report for the Gaithersburg meeting.* VQEG, January 2000. (Accessible at anonymous ftp site: *ftp://ftp.its.bldrdoc.gov/dist/ituvidq/vqeg2min.rtf*).

[VQE00b]   VQEG. *VQEG Objective Test Plan.* VQEG, January 2000. (Accessible at anonymous ftp site: *ftp://ftp.its.bldrdoc.gov/dist/ituvidq/obj_test_plan_final.rtf*).

[VQE00c]   VQEG. *VQEG Subjective Test Plan.* VQEG, January 2000. (Accessible at anonymous ftp site: *ftp://ftp.its.bldrdoc.gov/dist/ituvidq/subj_test_plan_final.rtf*).

[VQE03]   VQEG. *Final Report from the Video Quality Experts Group on the Validation of Objective Models of Video Quality Assessment, Phase II.* VQEG, August 2003. (Accessible at anonymous ftp site: *ftp://ftp.its.bldrdoc.gov/dist/ituvidq/*).

[Wor03]   Working Group 6Q. *Objective perceptual video quality measurement techniques for standard definition digital broadcast television in the presence of a full reference.* ITU-R, October 2003. Working Group 6Q DRAFT NEW RECOMMENDATION ITU-R BT.[DOC. 6/39].

Part III

Perceptual Coding and Processing of Digital Pictures

Chapter 12

HVS Based Perceptual Video Encoders

Albert Pica, Michael Isnardi and Jeffrey Lubin
Sarnoff Corporation Inc., U.S.A.

12.1 Introduction

The primary difference between perceptual based encoders and more traditional ones is the metrics used in controlling their performance. Typically, a mean square error (MSE) based metric is use in the encoder's rate control loop. Although MSE is easy to compute, and therefore widely used, it is not a very accurate predictor of encoder-induced artifact visibility. Figure 12.1 shows how a perceptual-based visual distortion model (VDM) has much higher correlation with DSIS[1] ratings across different contents compared to traditional MSE. In this example, the VDM used is Sarnoff's Visual Discrimination Model–JNDmetrix^{TM}.

Because of the much higher correlation to subjective ratings, models of the human visual system (HVS) are being employed in video encoding systems to more accurately estimate the visibility of video artifacts introduced by the encoder. However, systems that use HVS based distortion models (actually fidelity models) are typically computationally expensive and can introduce significant temporal delay to the encoder. We now discuss the principles behind HVS models and examine different approaches to using these models.

12.1.1 Outline

The makeup of this chapter is selected to give an overview of: 1) why use perceptual based approaches, 2) several possible architectures and their applications, and 3) future

---

[1]DSIS (Double Stimulus Impairment Scale) is a commonly used fidelity rating scale used to determine the visibility of encoder distortion artifacts.

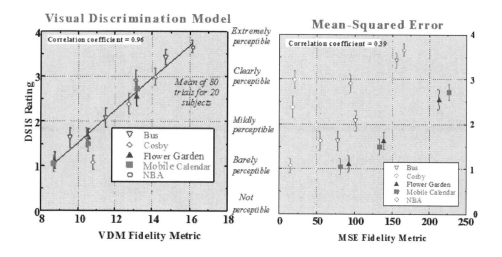

Figure 12.1: VDM vs. DSIS rating and MSE vs. DSIS.

directions in this area. Specifically, the following topics are covered in this chapter:

- Noise Visibility and Visual Masking – sensitivities of the human visual system to structured and unstructured noise, and their application to video encoding.

- Architectures that support perceptual based encoding – description of encoder architectures that specifically take advantage of human vision model. Architectures include macroblock, picture level control and look-ahead processing for a MPEG-2 encoder.

- Standards-specific features – origins of blockiness in standards, and in-loop filtering as a means to remove block visibility.

- Salience/masking pre-processing – determining in advance of encoding key areas to be preserved and the application of region based filtering to reduce bitrate requirements for a frame.

- Application to multi-channel encoding – using statistical variations in multiple video program streams to improve the efficiency of a transmission channel.

- Future Challenges – the potential of using advanced image interpolation techniques to compression.

12.2   Noise Visibility and Visual Masking

All lossy video coding systems add noise, or distortion, to the image, and the visibility of the noise is a key feature of the compression algorithm. Much research has been done

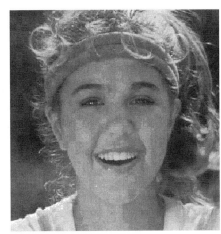

(a) MSE=27.1   (b) MSE=21.3

Figure 12.2: Visibility of noise - (a) uncorrelated (Gaussian) vs. (b) correlated noise (Quantized DCT Blocks).

on where to introduce image distortions so that they are least visible. The characteristics of the noise, as well as the ability of the content to visually mask the noise, are exploited in advanced video coding systems.

As demonstrated in Figure 12.2, there is a significant perceptual difference in the visibility of noise, depending on its structure, or self-correlation. The distortion in the image containing structured noise (Figure 12.2(b), in this case blockiness) is much more visible than the distortion in the image containing unstructured noise (Figure 12.2(a), in this case Gaussian white noise). In addition, note that the blockiness in Figure 12.2(b) is highly visible in the more uniform areas, such as the face, and much less visible in the busy areas, such as the hair. This demonstrates the ability of the content itself to mask certain distortion.

Most video compression systems are transform based, which means that image detail is transformed into a frequency or space-frequency representation that has desirable properties, such as energy compaction and better matching to HVS noise sensitivity. The result of the transform is a set of coefficients that represent the amplitudes of elemental basis functions; the image detail can be considered a weighted sum of these basis functions. The collection of basis functions is called a *basis set*.

The most widely used basis set for video compression is the DCT, or Discrete Cosine Transform [RY90, PM93]. There are many chip sets and processing cores that make the DCT and inverse DCT (needed in decoding) inexpensive to implement. Usually, it is performed on an 8×8 block of pixels, and its basis set consists of 64 two-dimensional

(a)   (b)

Figure 12.3: (a) JPEG (block-based DCT) vs. (b) JPEG2000 [TM02](full-image wavelet) still image compression at the same normalized bit rate (0.25 bits/pixel).

cosine functions as shown in Figure 12.13. As discussed in more detail later, finely quantized DCT coefficients, upon inverse transformation back to pixels, have the desirable property of introducing noise at edges of objects, where the edge itself masks most of the distortion. However, coarsely quantized DCT coefficients introduce blockiness that is highly visible in near-uniform areas of the scene. Blockiness is a key distortion that needs to be minimized in block based compression approaches.

Alternative basis sets that do not induce structured noise can have a significant perceptual advantage over the DCT, but to realize this advantage, distortion measures must reflect the actual visibility of the distortion. One alternative basis set that is becoming more widely accepted is the DWT, or Discrete Wavelet Transform [TM02, SCE01, CSE00, RJ02, CP02]. For maximal efficiency, the DWT is often applied to large blocks, up to and including the full image size. In Figure 12.3, a comparison between 8×8 block-based DCT and a full-image DWT is shown. As can be seen, the appearance of the compression artifacts are quite different between the two transform methods.

12.3 Architectures for Perceptual Based Coding

For a particular scene, the bit rate can be used to predict quality but the variation from scene to scene can be quite significant. As shown in Figure 12.4, the subjective rating (DMOS[1]) of compressed content improves (the impairment difference becomes lower) with bit rate, as expected. However, there is a large variation in the shape of this rate-distortion curve as a function of scene content. Some content improves continuously as bit rate increases, while other content exhibits rapid quality saturation above a certain bit rate.

---

[1]DMOS, or Difference Mean Opinion Score, is a common subjective rating metric.

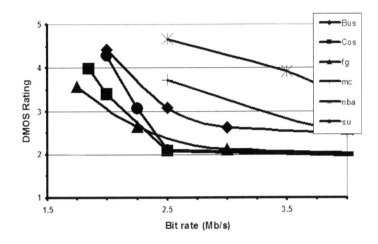

Figure 12.4: Rate-Distortion Curves (Bit rate vs. DMOS). Note the large variation in the shape of the curve as a function of scene content

The major problem is that these rate-distortion curves are not known in advance. As we will see, they can either be estimated (using, for example, a look-ahead technique) or can be computed by multiple coding/decoding passes, that can be computationally expensive and also add significant encoding delay.

Figure 12.5 shows a prototypical MPEG-2 [Gha03] encoder in block diagram form. As shown in this figure, perceptual based metrics can be introduced either inside or outside the main coding loop.

The main advantage of introducing perceptual coding metrics inside the main coding loop is its low latency. The current frame can be analyzed for its visual maskability, which can in turn control the quantization. The reconstructed frame can be compared to the original using visual discrimination model (VDM) (see loop with vertical hatch processing box labelled Q, Q^{-1}, IDCT, VDM Analysis). If visible differences are detected by the VDM model, the architecture can support a second pass or multiple passes to recode the frame until pixel differences are below a preset visible threshold. A drawback of this approach is that this perceptual-based encoding functionality must be designed into the encoder; it is difficult or even impossible to retrofit existing encoding implementations with this feature.

In order to retrofit pre-existing encoder implementations with perceptual-based rate control, some form of outside-the-loop mechanism is required in the vertical hatch. As also shown in Figure 12.5, the reconstructed video is spatio-temporally aligned with the

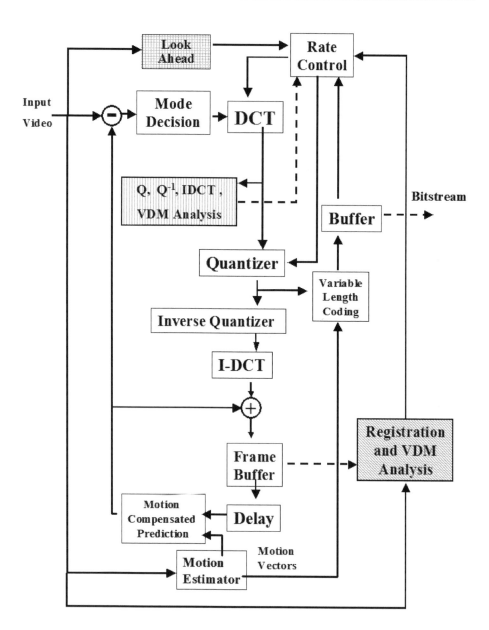

Figure 12.5: Prototypical block diagram of an MPEG-2 encoder with possible VDM components noted by diagonal, vertical and cross hatch boxes. Blocks not shaded are implied by the MPEG-2 standard.

corresponding input frame, and compared with the vertical hatch using a VDM metric (see loop using diagonal hatch processing box labelled VDM Analysis). This outer feedback loop, however, exhibits significant latency, and may not work well for rapidly changing content or across scene cuts. It has been demonstrated to work well for setting overall GOP level targets for IPB frames. To improve performance, a look-ahead, feed forward visual maskability metric can be used as an input to the rate control. This will give advance warning of significant changes in scene content.

12.3.1 Masking Calculations

A key aspect to all vision distortion models is the estimate of visual masking. One common feature of the masking calculations is that they calculate the summation of energies across different frequency bands, and tend to indicate high maskability along high contrast edges and other visually salient features that are broadband in frequency. However, visually salient regions may require special treatment to ensure that artifacts remain invisible. In our experience, maskability calculations for video optimization applications should include both a computation of local spatio-temporal masking energy, and a salient region rejection operation that explicitly discards regions of highest energy.

One way to achieve this is as follows. Spatio-temporal masking energy is computed as the sum of energies across local spatio-temporal frequency and/or orientation bands, but with highly salient features first removed from the summation. In one embodiment, illustrated in Figure 12.6, the computation of local spatio-temporal masking energy is performed by first subjecting each frame of luminance input to a Laplacian pyramid operation that decomposes the input into different spatial frequency bands. These are then temporally low-pass filtered, down-sampled to the same spatial resolution, and then summed to produce an estimate of spatio-temporal energy at each local region in space and time. Each such energy value is then divided by the maximum luminance for the frame, to generate a measure of local contrast energy.

For salient feature rejection, a histogram of the local contrast energy in each frame is constructed, and then the regions with the highest percentiles in this histogram are set to zero. The percentile threshold in this operation is a free parameter, currently set to 75% in one embodiment.

Further spatio-temporal low-pass filtering is then applied to the output of the salience-rejection operation, to provide spatial and temporal summation of maskable energies, as suggested by masking models in the literature.

Other improvements, not indicated in the figure, include a luminance weighting of the final output. This is useful in that Weber's Law suggests that, other things being equal, areas of high luminance tend to support invisibility of additional fixed luminance increments better than low luminance areas.

Another common feature of masking models in the literature is that although they calculate the summation of energies across different frequency and orientation bands,

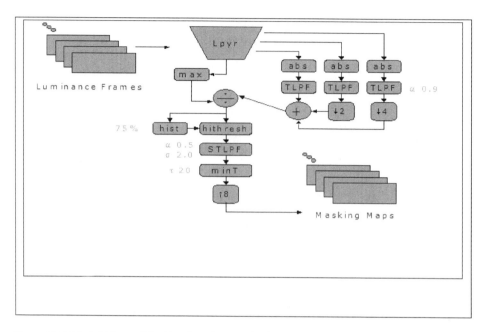

Figure 12.6: Maskability model with salient feature rejection. Lpyr = Laplacian Pyramid; abs = absolute value function; TLPF = Temporal Low Pass Filter. STLPF = Spatio-Temporal Low Pass Filter; Hist = histogram; hithresh = high threshold; minT extracts minimum in the temporal dimension.

few take into account the temporal dimension. For example, although some techniques exist to take advantage of spatio-temporal masking in specific situations, e.g., at scene cuts, these algorithms detect the feature of interest directly rather than as a result of a general purpose spatio-temporal masking model. On the other hand, a maskability calculation that sums spatio-temporal energies will correctly predict high maskability along scene cuts, as well as in other conditions of high incoherent motion energy.

As well, even more so than is the case for compression applications incorporating purely spatial masking models, one needs also to consider salience when constructing such a model in the spatio-temporal domain. In particular, high contrast moving edges are an especially salient feature, and need to be discarded from the maskability calculation, since areas of high salience will invite visual inspection and thus make detectability more likely.

In other words, for a useful spatio-temporal masking model, masking should be high when local motion energy accumulated across space and time is large, *except* when this high motion energy is associated with strong tracking signals, such as high contrast coherently moving edges.

In practice, one can implement these ideas based on a computation of local object motion. Some desired interactions include:

 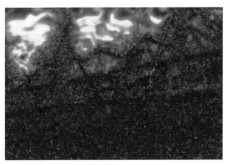

Figure 12.7: Masking map. Bright areas show areas of low masking. Perceptual based encoders encode the brighter areas with higher quality.

- Avoiding motion paths of salient objects. It is often difficult to "not watch" an interesting object in the scene. Therefore, do not quantize heavily on an intersecting path associated with this object.

- Quantize more heavily in areas in which there is a large amount of local (but incoherent) motion energy.

Masking maps can be used in perceptual coding. A masking map shows where slight changes in image detail would be visible. Figure 12.7 shows an original image, and a masking map produced from this image. Note that the bright, or "hot" areas are associated with the nearly uniform sky region. This means that special care must be taken to ensure that this region is coded with high fidelity, as any distortion, such as blockiness, will be readily noticed.

12.3.2 Perceptual Based Rate Control

Since quantization is a major factor contributing to the quality and coded bit rate of an image sequence, we now examine several ways of incorporating perceptual-based metrics into the rate control mechanism.

12.3.2.1 Macroblock level control

Setting the quantization level of individual macroblocks (16×16 blocks of luminance pixels) is the most difficult problem for any rate control scheme, including those that use perceptual models. Unfortunately, when applied to this level of granularity, most HVS models can only give relative estimates of visibility. However, this is still quite useful in making bit allocations for individual macroblocks. In Figure 12.5 the vertical hatch processing loop is used to control the macroblock quantization levels. Figure 12.8 demonstrates the results of this type of processing.

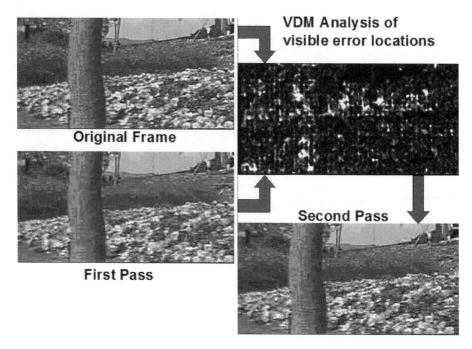

Figure 12.8: Demonstration of bit reallocation on the macroblock level using a visual discrimination model (VDM). This type of processing is shown in Figure 12.5 as the vertical hatch processing loop.

This algorithm is a variation on the MPEG Committee's TM5 rate control. During a first pass, the macroblock quantization is performed using a simple linear bit usage profile, the reconstructed field is then analyzed using a VDM that produces a distortion map as shown in Figure 12.8 (upper right quadrant). A second quantization is then performed using the distortion map to modify the original bit allocation. This process can be iterated until the distortion map meets a predetermined level of uniformity. The algorithm used is described in detail by Peterson and Lee [PL03]. To minimize the number of passes, a more sophisticated initial macroblock bit allocation should be performed (see [CZ97]). With a good initial start, a uniform distortion map can be achieved with only one additional pass.

12.3.2.2   Picture level control

Picture level control generally involves meeting a pre-assigned target bit rate (more precisely, a target bit count) for each picture type. Visual maskability of the source frame can give a starting point for the quantization profile to be used for the frame. After coding, the actual bit count can be measured, and the quantization profile can be scaled to reduce the difference. Figure 12.9 demonstrates the improvements that can be achieved

(a) Baseline Encoder-TM5   (b) DVC Encoder

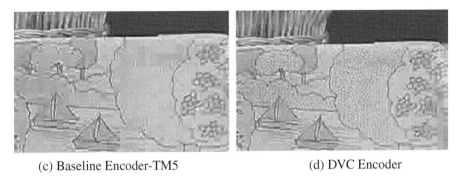

(c) Baseline Encoder-TM5   (d) DVC Encoder

Figure 12.9: Examples of vision based encoding — Optimization of IPB bit allocations based on visibility of artifacts. The DVC encoder incorporates the VDM enhancements discussed in this section.

with this approach. The diagonal hatch processing loop shown in Figure 12.5 was used to produce the images in Figure 12.9. For this type of processing, a frame's complexity measure (as defined in TM5) is monitored for each frame type (IPB). Typically, the remaining bits for the GOP are apportioned according to the ratio of frame type complexity measures. To improve this performance, the results of the VDM analysis are used as a modifier to the frame type target bit allocations. For example, define the target bits for picture type j as TBj where j=I,P,B, the picture quality for a picture type as PQj and the average picture quality across all picture types (for the GOP) as <PQ>. Then the new target bits are computed as:

$$TB'_j \equiv TB_j \times \left(1 - k \times \frac{PQ_j - <PQ>}{<PQ>}\right) \qquad (12.1)$$

where k is a reaction parameter that sets the rate of the compensation.. This approach typically takes one to two sub-GOPs (a sub-GOP is typically three frames long) after a scene change to settle into an optimal setting. For many scenes, this reallocation significantly improves image quality. (See [Lee03] for a detailed description of this approach.)

12.3.2.3 GOP level control

Group of picture (GOP) level rate control generally involves meeting a pre-assigned target bit rate (more precisely, a target bit count) for each GOP. In this case, a full HVS model can be used, it requires both the original and fully reconstructed images to be available. Because significant delay in the feedback loop can occur due to computational requirements of HVS model, its performance may be compromised. However, it can still be used to guide the relative bit allocations for the different picture types.

Bit allocations must be updated slowly or this control will introduce visible temporal variations called "breathing." In addition, this type of control usually takes several sub-GOPs before settling to optimum bit allocation levels.

An additional application of GOP level control is in the authoring of DVDs. In this process, the entire video is encoded using a VBR encoder. The bitstream is then decoded and analyzed using a visual distortion model at the frame level. The original video is then encoded a second time but the VDM output is used to modify the target bit allocations at the GOP level (to within buffer constraints). This automated authoring approach can significantly improve DVDs by giving them uniform visual quality even through long difficult to encode sequences.

12.3.3 Look Ahead

If the rate-distortion (R-D) curve of an image sequence was known prior to encoding, it would be straightforward to calculate the required rate in order to meet a specified distortion, or vice-versa. Unfortunately, R-D curves are not generally known prior to compression, and the curves themselves change as a function of scene content, coding parameters and viewing conditions[1].

One way to produce R-D curves for an image sequence is to measure the distortion for a given set of rates. One can pick, for instance, a group of pictures (GOP) and code it N times with different bit rates, or quantization scale settings. For each rate (R), a distortion measurement (D) is made. This produces a set of N points that can be connected to produce an R-D curve for that GOP, similar to those shown in Figure 12.4. Using this measured R-D curve, one can interpolate, if necessary, to find a rate that meets a given distortion target, and perform a final coding step at that rate.

While this approach works in theory, multipoint GOP-level R-D measurement is rarely used in practice because of the large amount of computation and latency involved. Instead, single-point R-D measurement is used in some real-time encoding systems. For instance, some real-time MPEG-2 encoders use a dual-pass architecture in which the

---

[1]Viewing conditions, such as viewing distance, ambient light and display settings, will affect perceptual-based distortion metrics, but will not affect objective distortion metrics, such as the PSNR.

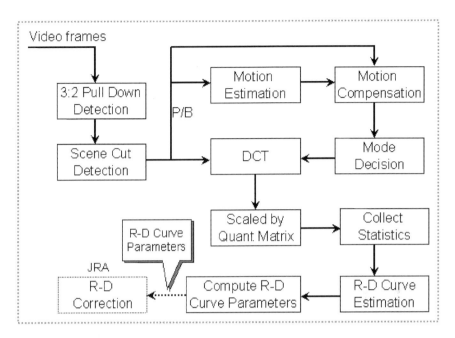

Figure 12.10: Block diagram of Sarnoff Look Ahead approach for R-D curve estimation. (JRA: Joint Rate Allocation.)

video is first encoded by a "look-ahead" encoder, whose output is processed to guide the second and final encoding step.

Recently, Sarnoff developed an approach to estimate a multi-point, picture-level R-D curve that is suited to real-time implementation. In this approach (Figure 12.10), the look-ahead module simulates the main motion-compensated transform coding loop of the actual encoder. In order to do this, the look-ahead module must know the coded picture type (I, P or B) of each incoming frame. The coded picture type can be readily calculated for fixed GOP structures, but must be predicted for scene-adaptive GOP structures[1].

For the most accurate R-D curve estimation for a given frame, the look-ahead module must make the same or similar coding decisions (e.g., macroblock coding type and motion estimation mode) as the actual encoder. Each block is transformed by an 8×8 DCT and the coefficients are scaled by the picture-level quant matrix (which must be the same or similar to the one used by the actual encoder).

The histogram of all the DCT coefficients, $\mathbf{X}[u, v]$ ($0 \leq u \leq N - 1$ and $0 \leq v \leq M - 1$), in the current frame, denoted by $H(X)$, is computed, where X is the value of

---

[1]This requires the monitoring of certain pre-processing functions (e.g., scene cut detection and 3:2 pulldown processing when working with mixed format material).

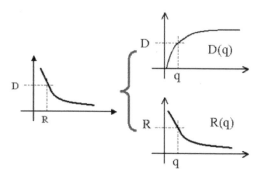

Figure 12.11: Composition of $R(D)$ from $R(q)$ and $D(q)$.

the DCT coefficient. For example, for $X = 10$, there are $H(10)$ DCT coefficients in the current video frame with value 10.

The rate-distortion (R-D) behavior of an encoder can be described by its rate-quantization and distortion-quantization curves (functions), denoted by $R(q)$ and $D(q)$, respectively, where q is the quantization scale parameter. The R-D function $D(R)$ is obtained by combining $R(q)$ and $D(q)$, as depicted by Figure 12.11. In Sarnoff's Look-Ahead approach, $R(q)$ and $D(q)$ are estimated as follows.

In [KHM01], it has been found that there is a linear relationship between the actual coding bit rate R (in bits per pixel) and the percentage of zeros among the quantized transform coefficients, denoted by ρ, in standard video coding systems, such as MPEG-2 encoding. In other words,

$$R(\rho) = \theta(1 - \rho) \qquad (12.2)$$

where θ is a constant for a given video frame and ρ ranges from 0 to 1. Note that ρ monotonically increases with the quantization scale q. Therefore, there is a one-to-one mapping between q and ρ, denoted by $\rho(q)$. Let $H(X)$ be the histogram of the DCT coefficients. If a uniform quantizer is used, we have

$$\rho(q) = \frac{1}{N \times M} \sum_{|X| \leq \alpha \times q} H(X) \qquad (12.3)$$

where $N \times M$ is the total number of DCT coefficients, and α is the quantization dead zone threshold, as shown in Figure 12.12. By default, the constant α is set to 1.25.

In order to estimate the slope θ, the following notations are introduced:

I-MB	INTRA MB in an I/P/B picture
P-MB	Non-INTRA MB in a P picture
B-MB	Non-INTRA MB in a B picture

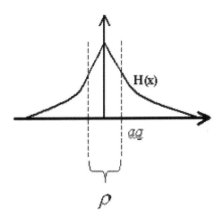

Figure 12.12: Calculation of $\rho(q)$.

Based on extensive simulations, we have found that the following assumptions are valid:

- all I-MBs have $\theta = 6.5$
- all P-MBs and B-MBs have $\theta = 7.5$

Therefore,

- An I picture has θ of 6.5, because all of its MBs are I-MBs
- For P and B pictures, the value of θ is estimated by a mixture formula:

$$\theta = (6.5 \times IntraN + 7.5 \times InterN)/(IntraN + InterN) \qquad (12.4)$$

where $IntraN$ and $InterN$ are the number of intra-coded and inter-coded macroblocks in the picture, respectively.

Now that the $\rho(q)$ function and the value of θ are known, the $R(q)$ function can be estimated as follows:

$$R(q) = R[\rho(q)] = \theta \cdot [1 - \rho(q)] \qquad (12.5)$$

The final step is to estimate $D(q)$. A simple objective distortion metric can be based on mean square error (MSE) as follows:

$$D(q) = \frac{1}{N \times M} \sum_{X \in \{\mathbf{X}[u,v],\, 0 \leq u \leq N-1,\, 0 \leq v \leq M-1\}} [X - X_q]^2 \times H(X) \qquad (12.6)$$

where X_q is the quantized version of the DCT value X.

However, Sarnoff has developed a perceptual-based metric called "Visual MSE" that is based on visual masking yet is amenable to real-time hardware implementations. Visual MSE, $D_i(q_s)$ is computed by dividing the computed MSE of each block by its

Visual Masking Ability, where $e_i(q_s)$ is the MSE of block i given a quantization scale q_s. V_i is the Visual Masking Ability of block i. $D_i(q_s)$ is given by the following formula:

$$D_i(q_s) = \frac{MSE\ of\ Block\ i\ given\ q_s}{Visual\ Masking\ Ability\ i} = \frac{e_i(q_s)}{V_i} \qquad (12.7)$$

where

$$V_i = \frac{Weighted\ AC\ Power}{DC\ Power} = \frac{\sum\sum_{k,l;k^2+l^2\neq 0} G_{k,l,i}^2 w_{k,l}}{G_{0,0,i}^2 + \varepsilon} + \beta, \qquad (12.8)$$

$G_{k,l,i}$ are the DCT coefficients of block i, with spatial frequency indices k,l and $w_{k,l}$ is a coefficient weighting matrix that is constant for all blocks. ε and β are empirically derived constants.

12.4   Standards-Specific Features

This section discusses features in video compression standards that are or can be used to reduce the visibility of coding artifacts, such as blockiness, ringing and mosquito noise.

12.4.1   Exploitation of Smaller Block Sizes in Advanced Coding Standards

12.4.1.1   The origin of blockiness

Block-based video compression standards [Gha03], such as MPEG-1/2/4 and H.261/3/4, can produce signal discontinuities at block boundaries. Signal discontinuities arise from basis functions that are reconstructed with possibly very different amplitudes compared to their original amplitudes. Since the basis functions span a single block, the differences in amplitude occur at the block boundaries.

The basis functions for the 8×8 DCT are shown in Figure 12.13.

12.4.1.2   Parameters that affect blockiness visibility

This in turn can result in noticeable blockiness or edge busyness, depending on the normalized bit rate, scene content and encoding parameters (e.g., GOP structure and quantization parameters). Figure 12.14 shows the visual effect of increasing the quantization scale factor for a single I-frame image using the MPEG-2 video compression standard. At lower levels of quantization, edge busyness is observed. At higher levels, blockiness is noticed. However, the visibility of blockiness for moving imagery depends on a number of important parameters, such as block size, viewing distance, local detail and temporal stationarity.

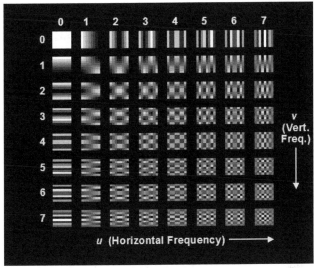

Figure 12.13: DCT basis functions.

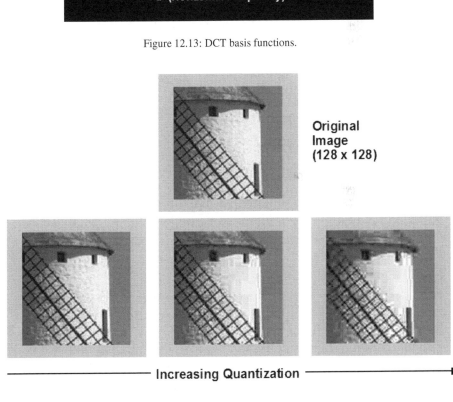

Figure 12.14: Impact of increasing quantization. (MPEG-2, I-frame.)

The first two parameters — block size and viewing distance — can be combined into a single metric: block size subtended on the retina. Because the HVS attenuates high spatial frequencies, blockiness visibility decreases with the angle subtended by the block boundaries. In practical terms, this means that:

- for a fixed block size, blockiness becomes less visible as viewing distance increases, and

- q for a fixed viewing distance, blockiness becomes less visible as block size decreases.

Figure 12.15 shows the effect of block size on blockiness visibility. At a viewing distance where the largest blocks are just visible, the smaller blocks should be nearly invisible.

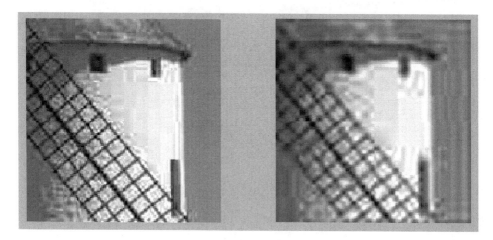

Figure 12.15: Effect of block size on visibility.

Local detail can mask blockiness to a certain degree. For the same level of signal discontinuity, blockiness is more visible in uniform areas than in highly textured areas of a scene. This is shown in Figure 12.16.

In video compression, the visibility of blockiness is also greatly affected by the temporal stationarity of the block edges. If the block edges are spatially aligned in time, their visibility will be much higher than if their positions were randomly moving. Temporally aligned block edges occur, for instance, in block-based compression systems that do not employ motion compensation. Many video editing formats prohibit motion compensation so that frame-accurate editing can be performed with minimal complexity. However, these systems must ensure that the Intraframe (I frame) video is compressed at a high enough bit rate so that blockiness is not visible. Intraframe-only coding in MPEG/H.26x and Motion JPEG and DV (both of which do not employ motion compensation) can produce visible, temporally aligned block edges as quantization is increased.

Figure 12.16: Visibility of blocks based on background content.

When motion compensation is used, only block edges for intra-coded macroblocks are temporally aligned. For efficient compression, most blocks should be motion compensated, which means that a prediction error is transmitted and added to a motion-compensated block at the decoder. The motion-compensated block may not be aligned to the block grid; in fact, this situation rarely occurs.[1] Most often, the motion-compensated block, whose position may be refined to half- or quarter-pixel resolution, will have block edges from the reference frame shifted spatially, and even attenuated due to the sub-pixel shifting. A predicted frame (P frame) that uses a heavily quantized I frame as a reference will shift the visible block edges in the I frame on a block-by-block basis, and the addition of the prediction error will reduce their visibility still further. Subsequent P or bi-directionally-predictive frames (B frames) use these "smeared" P frames as reference, propagating the smearing effect until the start of the next I frame, where the cycle repeats. The overall effect is to turn blockiness into an unstructured noise pattern on the image.

In advanced compression schemes, smaller block sizes are used primarily as a means to perform better motion compensation. However, a side benefit is a reduction in visible blockiness compared to schemes that use larger bocks. The recently approved H.264/MPEG-4 AVC video compression standard [Ric03] uses a 4×4 block size, com-

---

[1]This situation would only occur if all motion vector components were restricted to be a multiple of the block size. Such a restriction would defeat the whole purpose of motion compensation.

pared to MPEG's 8×8 block size, and is one of the reasons this standard enjoys an automatic reduction in blockiness visibility.

12.4.2   In-Loop Filtering

In some compression schemes, such as H.261, H.263 Profile 3 and H.264/MPEG-4 AVC, in-loop filtering is used to filter block edges in response to an edge-detection algorithm. In both H.261 and H.263 the deblocking filter is optional, whereas in H.264 and MPEG-4 AVC adaptive deblocking filter is specified. The de-blocking filter is present in both the encoder and decoder loops and helps to reduce the visibility of blockiness in motion-compensated frames. In addition to improved objective and subjective quality, in-loop filters generally eliminate the need for post-filtering in the decoder and help to lower cost and end-to-end delay.

In H.264/MPEG-4 AVC, the combination of smaller block size and in-loop deblocking filter causes the visibility of blockiness to be lower than previous standards, and is one of the reasons this compression scheme offers excellent compression efficiency. An excellent description and analysis of the H.264/MPEG-4 AVC de-blocking filter is given in [LJL^{+}03].

12.4.3   Perceptual-Based Scalable Coding Schemes

Scalable coding schemes produce multiple coded representations of a video sequence, usually separated into a base layer and one or more enhancement layers. In MPEG-2 video compression, several scalable modes are allowed, such as temporal, spatial and SNR scalabilities. However, these modes tend to produce large jumps in signal bandwidth and signal quality.

The MPEG-4 Visual standard introduced the concept of Fine Granularity Scalability[1] or FGS. In FGS, the bit planes of the DCT coefficients are sent as one or more enhancement layers. Any desired bit rate can be achieved by truncating the enhancement layer bitstream(s) at any point. This technology is finding use in streaming video applications and statistical multiplexing of pre-encoded content [PE02].

In [HWW03], the authors define a User Adaptive Video Quality Index (UAVQI) that incorporates a temporal quality decay constant that is specified by the user. An iterative technique is used to maximize UAVQI while minimizing the quality variation. For real-time operation, piecewise linear quality-rate curves are used, and the maximum number of iterations is fixed.

---

[1]This feature is documented in ISO/IEC 14496-2:2001/Amd.2:2002(E), "Information Technology — Coding of Audio-Visual Objects — Part 2: Visual, Amendment 2: Streaming Video Profile"

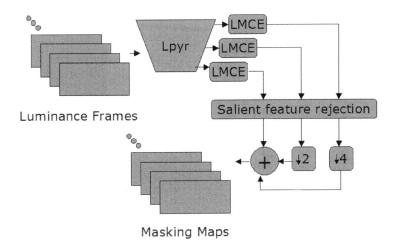

Luminance Frames

Masking Maps

LME: local motion contrast energy detectors

Figure 12.17: Pre-processor based maskability model with salient feature rejection.

In another work by the same authors [HW03], a model of foveation is used to increase the coded quality at the point of gaze, which is an input to the encoder. In the enhancement layer, each block's array of DCT coefficients is truncated at a point that depends on its distance from the point of gaze. An efficient rate control scheme that handles multiple foveation points is described.

12.5 Salience/Maskability Pre-Processing

Computation of visual salience and maskability can be used as a pre-processing step for any encoder system. As shown in Figure 12.17, a sequence of luminance frames is decomposed into spatio-temporally local representations. A salience rejection operation is then applied. As discussed previously, a relatively simple and effective rejector is an energy histogram, followed by a high percentile rejection operation in which the regions with the highest percentiles in this histogram are set to zero. The percentile threshold in this operation is a free parameter, currently set to 75% in one embodiment. Another useful rejector is based on the coherence of local motion with a spatio-temporal region. Local regions of motion energy all signaling the same or similar motion direction and speed are considered salient, since they tend to occur on coherently moving objects.

The efficacy of such pre-processing is shown in Figure 12.18. The images on the right are from Sarnoff's DVC encoder, where salience pre-processing was used to guide a spatially adpative pre-filter. This pre-filter automatically reduces the high-frequency content, and therefore coded bit rate, in areas where the high frequencies would not be noticed.

Baseline Encoder - TM5   DVC Encoder

Figure 12.18: The left images are from an MPEG-2 baseline encoder using TM5 rate control. The right images are from an MPEG-2 encoder (DVC Encoder) that uses saliency pre-processing to filter out high frequencies before they are encoded.

12.6 Application to Multi-Channel Encoding

An additional requirement for many video distribution systems is the need to multiplex multiple video programs into a single distribution channel. It is not uncommon to see from 8-12 video program streams being multiplexed into a 30 Mbit/sec transport stream. Most products available today (statistical multiplexers) use MSE as part of their joint rate allocation scheme. The look-ahead approach (see Section 12.3.3) that estimates the rate-distortion behavior for the incoming video stream is ideally suited for use in a multi-channel multiplexer (Figure 12.19). With a sufficiently large look-ahead buffer (approximately a GOP), joint rate allocation can specify an optimal (i.e., minimized visual distortion) target bit rate for each encoding stream. The function of the joint rate allocation (JRA) module is to determine bit rates that equalize distortions across video program streams while adhering to buffer constraints for the individual program streams and the transport stream. The JRA is well suited to linear programming techniques as well as approaches that are more heuristic.

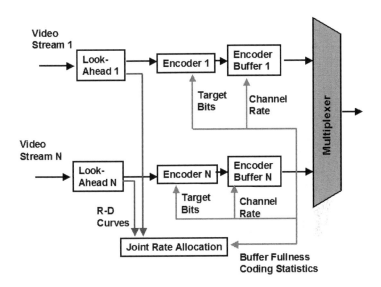

Figure 12.19: Schematic diagram of a multi-channel encoder.

References

[CP02]   S. Cho and W. Pearlman. A Full-Featured, Error-Resilient, Scalable Wavelet Vdeo Codec Based on the Set Partitioning in Herarchical Trees (SPIHT) Algorithm. *IEEE Transactions on Circuits and Systems for Video Technology*, 12(3):157–171, March 2002.

[CSE00]   C. Christopoulos, A. Skodras, and T. Ebrahimi. The JPEG2000 Still Image Coding System: An Overview. *IEEE Trans. Consumer Electronics*, 46(4):1103–1127, November 2000.

[CZ97]   T. Chiang and Y. Zhang. A New Rate Control Scheme using Quadratic Rate Distortion Model. *IEEE Transactions on Circuits and Systems for Video Technology*, 7(1):246–250, 1997.

[Gha03]   M. Ghanbari. *Standard Codecs: Image Compression to Advanced Video Coding*. U.K.: IEE, 2003.

[HW03]   C. Ho and J. Wu. Toward User Oriented Scalable Video by Using Foveated FGS Bitstreams. In *ICCE 2003 Conference Proceedings*, 46–47, June 2003.

[HWW03]   C. Ho, J. Wu, and S. Wang. A User Adaptive Perceptual Rate Control Scheme for FGS Videos. In *ICCE 2003 Conference Proceedings*, 42–43, June 2003.

[KHM01]   Y. K. Kim, Z. He, and S. K. Mitra. A Novel Linear Source Model and a Unified Rate Control Algorithm for H.263 / MPEG-2 / MPEG-4. In *Proceedings of International Conference on Acoustics, Speech, and Signal Processing*, 3:1777–1780, Salt Lake City, Utah, May 2001.

[Lee03]   J.-W. Lee. *Apparatus and Methods for Allocating Bits between Frames in a Coding System*. US Patent, 2003. Application 20020168007.

[LJL⁺03]   P. List, A. Joch, J. Lainema, G. Bjøntegaard, and M. Karczewicz. Adaptive De-blocking Filter. *IEEE Transactions on Circuits and Systems for Video Technology*, 13:614–619, July 2003.

[PE02]   F. Pereira and T. Ebrahimi(eds.). *The MPEG-4 Book*. Upper Saddle River, New Jersey: IMSC Press, 2002.

[PL03]   H. Peterson and J. Lee. *Apparatus and Method for Optimizing Encoding and Performing Automated Steerable Image Compression in an Image Coding System Using a Perceptual Metric*. US Patent, March 2003. 6,529,631 B1.

[PM93]   W. B. Pennebaker and J. Mitchell. *JPEG Still Image Data Compression Standard*. Norwell, MA: Van Nostrand Reinhold, 1993.

[Ric03]   I. Richardson. *H.264 and MPEG-4 Video Compression*. Hoboken, NJ: Wiley, 2003.

[RJ02]   M. Rabbani and R. Joshi. An Overview of the JPEG 2000 Still Image Compression Standard. *Signal Processing: Image Communication*, 17:3–48, 2002.

[RY90]   K. R. Rao and P. Yip. *Discrete Cosine Transform: Algorithms, Advantages, Applications*. New York: Academic Press, 1990.

[SCE01]   A. Skodras, C. Christopoulos, and T. Ebrahimi. The JPEG 2000 Still Image Compression Standard. *IEEE Signal Processing Magazine*, 18:36–58, September 2001.

[TM02]   D. Taubman and M. Marcellin. *JPEG2000: Image Compression Fundamentals, Standards and Practices*. Boston, MA: Kluwer, 2002.

Chapter 13

Perceptual Image Coding

Damian M. Tan and Hong Ren Wu
Royal Melbourne Institute of Technology, Australia

13.1 Introduction

Classical transform based coders performed excellently in reducing statistical redundancies. However, reduction of statistical redundancies does not always correspond to the reduction of psychovisual redundancies. The removal of visually redundant information has potential advantages. First, it ensures that only visually important information is encoded, thus maintaining visual quality. Second, discarding visually redundant information leads to better compression performance. Therefore, incorporating properties of the HVS to coding structures is seen as a beneficial step in enhancing the visual performance of image coders. Aspects of the HVS can be adapted into image coding in several ways. In [OS95], O'Rourke and Stevenson developed a set of non-separable wavelet filters for image coding, corresponding to the orientation sensitivity boundary of the HVS. The design of this filter set was undertaken based on studies of human vision sensitivities to sinusoidal gratings.

Quantization design is by far the most common method for adapting properties of HVS to coders. Perceptual quantization has been well studied and techniques have been developed to generate and/or regulate quantization matrices for luminance [NP77, Lim78, NLS89, Wat93, PAW93, WLB95, RW96, TPN96, FCH97, YZF99, HK02] and color [NR77, AP92, Wat94, DKMM99] DCT coders, wavelet coders [ZZQ97, RH00, Tau00] and sub-band coders [SJ89, HK00, Hem97, FCH97]. Applying vision modeling to quantization requires some degree of integration, especially when adaptive quantizers are involved. Alternatively, a simpler approach to adjusting visual significance within images can be made with coefficient weighting. Coefficient weighting in general, alters the magnitude of coefficients according to the perceived properties of images prior to quantization [TNH97, ML95, MH98]. However, some weighting schemes also require an inverse weighting process after de-quantization to complete image retrieval [HKS97, LJ97, LK98, RH99, LZFW00, ZDL00]. A less intrusive approach for embedding HVS properties to coding architectures is the application of perceptual pre-filtering for removal of visual redundancies. Efforts following this line of thinking have been reported, for spatial domain filtering, by Lan et al [LTK99] with sigma filters and Kopilovic et al [KS99] for anisotropic diffusion. Pre-filtering in the spectral domain has also been investigated by Lui et al [LW98] with sub-band dithering and Chen et al [CP84] with thresholding, while Nadenau [Nad00] looked at perceptual dithering for post-filtering. Aside from these mainstream approaches, Horowitz et al [HN97] described a matching-pursuit coder that used pruning to retain perceptually important information. Zeng [Zen99] on the other hand, presented segmentation as a way to isolate and code perceptually important regions within images while Wang and Bovik [WB01] devised a localized perceptual coding strategy based of the visual field of the fovea. Some of the techniques mentioned here are often combined together to deliver a more comprehensive perceptual coder.

In technical terms, there are strictly two classes of HVS based coders, rate constrained and quality driven. The rate constrained coders follow the traditional coding methodology in that given a target bit rate, the coder provides the visually optimum compressed image. By contrast, the quality driven coder encodes images to perceptually lossless or transparent quality at the lowest possible bitrate [SJ89, HK00]. Such an approach was inconceivable for conventional coders since it was impossible to quantify a consistent "perceived" quality with traditional distortion metrics such as the *Mean-Squared Error* (MSE) (13.42). In the following sections, discussions will be presented on a few popular contemporary perceptual coders which subscribe to the two classes of perceptual coding principles.

13.1.1   Watson's DCTune

DCTune [Wat93] as the name suggests, is an algorithm tailored for Discrete Cosine Transform (DCT) based image coding. It provides a method for generating quantization matrices for specific images and viewing conditions. The vision model used here takes into consideration luminance masking and contrast masking given by

$$m_k = max\left(t_k, |c_k|^{w_k}(t_k)^{1-w_k}\right), \tag{13.1}$$

where t_k and c_k are the luminance masking threshold and contrast masking threshold, respectively, of DCT block k. w_k is the masking exponent bounded by $0 < w_k < 1$. The perceptual distortion resulting from the quantization error, e_k, of DCT coefficients in the presence of masking is computed as

$$d_k = \frac{e_k}{m_k}. \tag{13.2}$$

Pooling this distortion with Minkowski's summation of power β,

$$D_P = \left(\sum_k |d_k|^\beta \right)^{\frac{1}{\beta}},$$ (13.3)

leads to the overall distortion of a DCT block. Adjusting the quantization error to correspond to the amount of masking within a block will ideally lead to visually lossless compression.

13.1.2 Safranek and Johnston's Subband Image Coder

Safranek and Johnston [SJ89] subband image coder is seen as a core template for subband vision based coders. It combines simple vision modeling with a coding structure that employs the 4×4 band *Generalized Quadrature Mirror Filter* (GQMF), DPCM and Huffman coding. The vision model exploits spectral and masking sensitivities of the HVS in GQMF domain. The spectral sensitivity, made in reference to GQMF subbands, was derived from threshold detection experiments. In the context of this coder, masking looks specifically at textured areas of images and hence it is modeled by the magnitude and variance of transform energy as

$$T_e(x, y) = \sum_{s=1}^{15} W_{MTF}(s) \cdot E(s, x, y) + W_{MTF}(0) \cdot \sigma(z)$$ (13.4)

where s, x and y are the subband, horizontal and vertical positions, respectively, E, the sum of subband energy within a 2×2 coefficient block, W_{MTF}, the subband weight derived from a modulation transfer function [Cor90] and $\sigma(z)$ the variance with $z \in \{(x, y), (x + 1, y), (x, y + 1), (x + 1, y + 1)\}$. The final perceptual threshold function,

$$P_t(s, x, y) = S_b(s) - 0.15 log_e(T_e(x, y)) - B_f,$$ (13.5)

comprised of texture masking, T_e, spectral sensitivity of GQMF subband, $S_b(s)$, and a brightness correction factor, B_f. Brightness correction was added to account for the changes in spectral sensitivities due to luminance variations. In the overall structure, the texture masking model regulates the DPCM coder and hence the bit rate to achieve constant quality coding.

13.1.3 Höntsch and Karam's APIC

Höntsch and Karam's Adaptive Perceptual Image Coder (*APIC*) [HK97, HK00] evolved from an earlier piece of work by Safranek and Johnston [SJ89], implementing a vision based adaptive quantization scheme with predictive (DPCM) coding. The quantization strategy in this coder was rather clever. It eliminates the need to transmit adaptive quantization step sizes by approximating the quantization level with coded and predicted coefficients, which were naturally available to both the encoder and the decoder. Based on the concept of *Just Noticeable Difference* (JND), the vision model adopted by this coder operates in the spectral domain of the GQMF and consists of contrast sensitivity, t_D, and masking sensitivity, a_C. The JND threshold has the expression

$$\hat{t}_{JND} = t_D(b, r, c) \cdot a_C(b, r, c). \tag{13.6}$$

Contrast sensitivity in this instance considers both luminance masking and luminance sensitivity. It was defined as

$$t_D(b, r, c) = t_B(b) \cdot a_D(b, \mu + \hat{i}_b(r, c)), \tag{13.7}$$

where b, r and c are the subband, vertical and horizontal positions respectively. $t_B(b)$, the base luminance sensitivity obtained experimentally as in [SJ89]. μ and $\hat{i}_b(r, c)$ represent the global mean and quantized/predicted transform coefficient respectively. a_D denotes the luminance adjustment function given by

$$a_D(b, m) = \frac{t_m(b)}{t_B(b)} \tag{13.8}$$

with t_m, the luminance masking threshold. Höntsch and Karam [HK00] found that luminance masking varied very little over different subbands. As a result, only one set of luminance adjustment function was used, specifically that which was tuned to the lowest frequency band, $a_D(0, m)$. Contrast masking combines both *intraband* (a_{C1}) and *interband* (a_{C2}) masking as follows,

$$a_C(b, r, c) = a_{C1} \cdot a_{C2}. \tag{13.9}$$

The intraband function adopts a normalized divisive inhibition model of the form

$$a_{C1}(b, r, c) = \begin{cases} 1 & , b = 0 \\ \max\left\{1, \left(\frac{|\hat{i}_b(r,c)|}{t_D(b,r,c)}\right)^{0.36}\right\} & , b \neq 0 \end{cases} . \tag{13.10}$$

Interband masking in this model accounts for both inter-frequency and inter-orientation masking. It was written as

$$a_{C2}(b,r,c) = \max \left\{ 1, \left(w_{MTF}(0) \cdot \sigma(\hat{i}_0(z)) + \sum_{n=1}^{b} w_{MTF}(n) \cdot \hat{i}_n^2(r,c) \right)^{0.035} \right\}$$

$$(13.11)$$

where w_{MTF} is the MTF weight and $\sigma(\hat{i}_0(z))$ the variance of neighboring coefficients with $z \in \{(r-1,c-1),(r-1,c),(r-1,c+1),(r,c-1),(r,c)\}$. The MTF adjustment reflects the sensitivity of an individual subband. The final quantization step size was taken to be

$$s(b,r,c) = 2\sqrt{3}\hat{t}_{JND}(b,r,c). \qquad (13.12)$$

As with its predecessor [SJ89], this coder is quality driven, thus it produces transparent quality images at specific viewing distances. However, the vision modeling in this coder is more extensive and accurate than its predecessor, particularly in the area of masking. This leads to superior performance as evident in the results [HK00].

13.1.4   Chou and Li's Perceptually Tuned Subband Image Coder

Chou and Li [CL95] presented a spatial domain HVS model for subband image coding. Threshold sensitivities from background luminance and spatial masking are used to estimate the JND and MND (*minimally noticeable difference*) profile to achieve transparent and near transparent coding, respectively. The MND profile is obtained by scaling the JND threshold which itself is defined as

$$t_{JND}(x,y) = \max\{t_M(x,y), t_C(x,y)\} \qquad (13.13)$$

with x and y the pixel coordinates within an image. t_M and t_C are the masking and background luminance sensitivity functions, respectively. These two functions were obtained heuristically with their definition given as follows,

$$t_M(x,y) = mg(x,y) \cdot (0.0001 \cdot bg(x,y) + 0.115) + (\lambda - bg(x,y)), \qquad (13.14)$$

$$t_C(x,y) = \begin{cases} T_0 \cdot (1 - (bg(x,y)/127)^{0.5}) + 3 & , bg(x,y) \leq 127 \\ \gamma \cdot (bg(x,y) - 127) + 3 & , bg(x,y) > 127 \end{cases}, \qquad (13.15)$$

where $bg(x,y)$ and $mg(x,y)$ are the mean background luminance and weighted average luminance difference around pixel (x,y), respectively. T_0 and γ, the gradient of the background luminance function for low and high luminance respectively and λ, the masking constant that determines the visibility threshold due to spatial masking. The parameters in (13.14) and (13.15) were obtained through subjective experiments and curve fitting. Once the JND/MND thresholds have been identified, they are mapped into the frequency domain by way of an MTF. The resulting spectral JND/MND profiles are then used to regulate the PCM/DCT coding of subband coefficients to achieve the desired image quality.

13.1.5 Taubman's EBCOT-CVIS

The EBCOT coder of Taubman [Tau00] looks at localized masking in the wavelet domain for the compression of images. This coder employs a *Rate-Distortion* (R-D) function to optimize compression gain. As such, vision modeling was embedded into the distortion function to estimate visual sensitivities to various levels of quantization error. The masking model was designed specifically for the bit-plane coding implementation and operates regardless of viewing conditions such as distances and resolutions. The distortion function has the form

$$D^n = w_b^2 \cdot \sum_k \frac{(\hat{s}_b^n[k] - s_b[k])^2}{\sigma_b^2 + |(V_b[k])^2|} \tag{13.16}$$

where $s[k]$ and $\hat{s}^n[k]$ are, respectively, the original and quantized coefficients of subband b of the transform. σ_b, the visibility floor provides the base masking threshold to avoid saturation, i.e., division by zero. w_b^2 is wavelet basis function and $V[j,k]$, the masker defined as

$$V_b[k] = \frac{\sum_{l \in \Theta_b[k]} |s_b[l]|^p}{\|\Theta_b[k]\|} \tag{13.17}$$

with $\Theta[k]$ the neighboring samples about $s[k]$ and p, the masking exponent bounded by the condition $0 < p < 1$. As $p \to 0$, the masking effects are minimized. The opposite is true when $p \to 1$. Following the structure of EBCOT, (13.16) only operates on blocks with typical dimensions of 16×16 coefficients. As a result, it only provides a coarse approximation of visual sensitivities of an area as opposed to a distortion estimation based on the individual pixel level. In practical terms, this means that small but visually important details within a block may be dampened if a large area of that block contains sufficient masking strength. This leads to the coarse-coding of small important details. Despite its shortcomings, this model has efficient implementation speed and superior visual performance to EBCOT with the MSE criterion. This performance however, can be improved upon with more advanced vision models.

13.1.6 Zeng et al.'s Point-Wise Extended Visual Masking

Zeng et al. [ZDL00] proposed a non-linear transducer model extended from [DZLL00] that combines self-masking and neighborhood masking. This model was developed around the framework of JPEG2000 [ISO00a], paying particular attention to the preservation of detailed edges within the masking function. It considered masking for both intra- and inter-frequency signals, though only the former has been implemented. Unlike the proposal in [Tau00], the masking function here is applied prior to quantization, adjusting the visual significance of individual coefficient. An adjusted coefficient is defined as

$$z_i = \frac{sign(x_i)|x_i|^\alpha}{1 + a \cdot \Sigma_k \frac{|\hat{x}_k|^\beta}{|\Phi|}} \tag{13.18}$$

where x_i is the ith transformed coefficient, α, the self-masking exponent and $\hat{x}_k$ the quantized neighboring coefficients of x_i. β and a are the neighborhood masking exponent and scaling factor, respectively, with Φ the number of adjacent pixels. The function $sign(x_i)$ gives the sign of x_i. Zeng et al. [ZDL00] prescribed the exponent values of $\alpha = 0.7$ and $\beta = 0.2$ with the rationale that a large value of α, relative to β, enhances self-masking and a small value for β dilutes the masking influence of a few large coefficients. This helps to distinguish edge coefficients from highly textured coefficients. The disadvantage of this approach lays at the decoding end where a reverse process is required to re-adjust the masked coefficients for the inverse frequency transform. The implication of having a "re-adjustment" function is two fold.

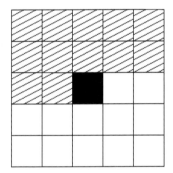

Figure 13.1: Shaded region: Causal neighborhood. Black region: Current coefficient (Non-Causal). Unshaded region: Non-Causal neighborhood. The decoder has access only to coefficients in the causal neighborhood.

First, the masking function could only operate on causal coefficients or coefficients which are available to the decoder as illustrated in Figure 13.1. Second, since only quantized coefficients are available to the decoder, if the encoder uses unquantized coefficients to obtain masked coefficients, then the re-adjustment processed can only be approximated at the decoder, resulting in some inaccuracies. On the other hand, if only quantized coefficients were to be used, then the masking adjustment will be coarse, due to quantization. This affects the accuracy of masking adjustments, especially at low bit-rates.

13.2   A   Perceptual   Distortion   Metric   Based   Image Coder

The perceptual coder presented in this section adopts the *Embedded Block Coding with Optimized Truncation* (EBCOT) coding structure.   EBCOT has many desirable features which include scalability and modularity [Tau00]. Modularity here refers to the distortion criterion which theoretically can take any form of error measure apart from the *Mean-Squared-Error* (MSE). Indeed, a distortion metric for quantifying perceptual quantization error was implemented in [Tau00] which delivered visual improvements over images optimized with the MSE metric. This metric, presented in Subsection 13.1.5, which shall be referred to as *CVIS*, only considers error masking localized within a frequency band, where in fact masking may also occur in other regions as well [Wan95]. Part of the reason for restricting vision modeling in CVIS is due to the limitations of the block-based architecture of the EBCOT coder. However, with minor modifications to the encoder, a more comprehensive model can be implemented, benefitting the overall visual performance of the EBCOT coder substantially. Though replacing the MSE with a perceptual distortion measure is by no means the only way of embedding HVS modeling into the EBCOT coder, this approach only involves model adaptation at the encoder, while the decoder remains unaltered. This maintenance of coder simplicity does come at a cost. As indicated in Subsection 13.1.5, applying perceptual measure over a block is less accurate than having individual treatments of coefficients. Nevertheless, Zeng et al. [ZDL00] have uncovered some shortcomings with perceptual enhancement of individual coefficients in the EBCOT environment as well (see Subsection 13.1.6). Overall, the best avenue here is to proceed with the simplicity of the perceptual distortion measure. It must the noted that the approach taken here is fully compliant to the EBCOT bitstream.

13.2.1   Coder Structure

A detailed description of the EBCOT coding structure is found in [Tau00]. Here, a general overview of the EBCOT coder is given pictorially in Figure 13.2. The EBCOT architecture is flexible and offers different coding parameters which affect its performance. The parameters of concern here are the decomposition structure and transform filter. For the proposed coding strategy, a five level dyadic wavelet decomposition with the bi-orthogonal 9/7 separable filter set [ABMD92] has been chosen. These filters are smooth and have relatively short lengths to facilitate computation speed. In addition, this filter set is symmetric, thus is linear phase.[1] The bi-orthogonal 9/7 filters produce 2-D spectral coefficients through a two stage filtering process. Each level of 2-D transformation generates four spectral bands, one isotropic and three orientated bands at approximately $0°$, $90°$ and $45°/135°$ orientations.

---

[1] non-linear phase filters requires phase re-alignment in order to reconstitute time-domain signals from spectral signals.

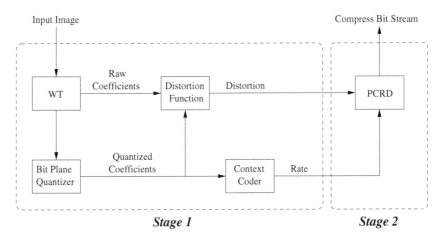

Stage 1   **Stage 2**

Figure 13.2: The EBCOT coding structure. In the first stage, the Wavelet Transform (WT) is applied to an image block. The resulting coefficients are bit-plane coded to produce the quantized coefficients. Both the bitrates and quantization errors (distortions) for these coefficients are determined by a context arithmetic coder and a distortion function, respectively. Given the rates and distortions, the final optimal bitstream, relative to the distortion function, is generated with the Post Compression Rate-Distortion (PCRD) optimizer.

13.2.2 Perceptual Image Distortion Metric

The present adaptation of vision modeling for distortion assessment has been motivated by its successful application in the evaluation of blocking distortions [YWWC02] and general quality [Win00] of digital video. The proposed *Perceptual Image Distortion Metric* (PIDM) adopts the *Contrast Gain Control* (CGC) model of Watson [WS97] with a three stage vision model following that of Yu et al. [YWWC02]. The CGC has three primary components: linear transform, masking response and detection. The first stage accounts for the human optical and cortical neural sensitivity, which is represented by a *contrast sensitivity function* (CSF) and a frequency transform, respectively. Stage two is concerned with the cortical responses, described by the masking function (13.20). Stage three then detects and pools the visible distortions. Figure 13.3 illustrates the structure of the PIDM. The PIDM replaces the MSE criterion in the error computation phase of the EBCOT coder. Hence, the PCRD optimization will control the bitrate in accordance to perceived distortions.

13.2.2.1 Frequency transform

According to neurological studies, it is understood that cortical neurons are sensitive to visual signals of specific frequencies' and/or orientations' bandwidth [DSKK58, DAT82]. As such, these neurons can be modeled by frequency and orientation selective linear transforms. Therefore, given an image, $\mathbf{x}$, the linear transform (LT) of $\mathbf{x}$, is defined as

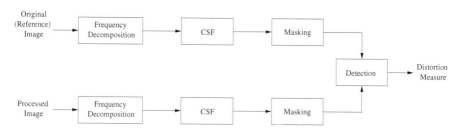

Figure 13.3: Perceptual Image Distortion Metric.

$$\mathbf{X} = LT(\mathbf{x}) \qquad\qquad (13.19)$$

where $\mathbf{X}$ is the spectral image of $\mathbf{x}$, maps real world images to frequency and orientation sensitive cortical neural images. Due to multiple layering and integration of neurons between the optic nerve and the visual cortex, it is difficult to ascertain the exact frequency and orientation responses of these neurons. However, for modeling purposes, a reasonable frequency decomposition which represents the HVS should have at least four equally spaced orientations and five frequency bands [van96]. Furthermore, it should ideally have minimal or no aliasing.[1] A transform that falls under this description is the *Steerable Pyramid Transform* (SPT) [SFAH92]. Despite having the desired properties for approximating neurons of the HVS, the SPT has limited applications in image compression. This is due primarily to the fact that the SPT produces an over-complete set of coefficients to minimize aliasing, i.e., the number of outputs exceeds the number of inputs. Transform kernels used for compression naturally avoid this since additional coefficients require additional resources to encode. In short, one can consider the requirements of transforms for the HVS modeling and compression as being mutually exclusive.

Clearly, the immediate solution to this is to maintain two separate transforms, one for coding and the other for distortion measure. Unfortunately, such an approach would increase the complexity and computation of the coder immensely, especially for a progressive bit-plane coder. Figure 13.4 illustrates the additional functionalities involved for maintaining two separate transforms as compared to that with one in Figure 13.2. To put the computational load into perspective, if one wishes to estimate all the distortions resulting from quantization, then an inverse wavelet transform and two SPT would be required for every bit-plane, operated upon by the coder, of every coefficient. Assuming that the computation costs for both the SPT and wavelet transform are equal, this would easily increase computation by three fold per bit-plane. Therefore, the realistic solution is to have one transform for both coding and distortion measure. Having the SPT as the coding transform is considered inefficient within the EBCOT framework at this point

---

[1]Aliasing in reference to sample aliasing.

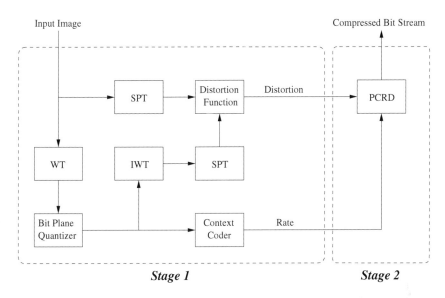

Figure 13.4: Alternative EBCOT structure with a perceptual distortion measure employing the steerable pyramid. This approach implements two different transforms, the wavelet transform (WT) for spatial decorrelation and the steerable pyramid transform (SPT) for vision modeling. The WT coefficients are quantized and arithmetically coded to produce the rate. The SPT emulates visual decomposition of the HVS and together with the distortion function determines the perceived distortion resulting from quantization error. Note that an inverse Wavelet Transform (IWT) projecting the quantized spectral coefficients back to the spatial domain is necessary before the SPT can be applied to the quantized coefficients.

in time, due to the overcompleteness of the SPT basis functions. The only practical alternative would be to apply wavelet transform with orthogonal or biorthogonal filters and accept some inaccuracies in approximating neural responses associated with sample aliasing. This is the approach taken by Taubman [Tau00] and Zeng et al. [ZDL00].

13.2.2.2   CSF

Contrast sensitivity can be estimated either in time-domain as in the case of Michelson's [Mic64] and Weber's [Bor57] contrast or in the frequency-domain with Peli's contrast [Pel90]. Although the PIDM estimates contrast in the frequency domain, the choice of transform filter here differs from that of Peli. To effect Peli's contrast would require the implementation of a Gabor filter. Needless to say, this would increase coder complexity and computation. A simple solution to this is to apply constrast sensitivity weights to individual frequency band in place of a CSF. A weighted coefficient, $\bar{c}_b$, of frequency band, b is generated as $\bar{c}_b = W_b \cdot c_b$ with W_b and c_b the CSF weight and transform coefficient, respectively. Winkler [Win00] reported that this approach saves time and has had little effect on the model accuracy. The coder described here will employ six CSF weights, one for each of five frequency bands, from the five level wavelet transform, and an additional weight for the isotropic low pass (DC) band.

13.2.2.3 Masking response

The masking function approximates cortical responses at the neuron level. It takes into account the masking phenomenon that occurs when visual stimulus excites both inhibitory and excitation fields of a high order neuron.[1] Masking occurs between neurons from similar (intra-) and different (inter) frequency, orientation and color channels. A function for describing masking that has gained popularity and acceptance is the divisive inhibition model of Legge and Foley [LF80] and its variations [FB94, Fol94]. This masking response function has the general form:

$$R = k \cdot \frac{E}{I + \sigma^q} \qquad (13.20)$$

where E and I are the excitation and inhibition functions, respectively, and k and σ are the scaling and saturation coefficients. The general definition of the excitation and inhibition functions of the two domains are given as follows:

$$E = X_l[\theta, m, n]^p \qquad (13.21)$$
$$I = \sum_\delta X_l[\theta_\alpha, m, n]^q \qquad (13.22)$$

where $X_l[\theta, m, n]$ is the transform coefficient at orientation θ, spatial frequency location (m, n) and resolution level l (see Figure 13.5). p and q are respectively, the excitation and inhibition exponents. The inhibition function pools the inhibiting factors from the different masking domains (δ): spatial, orientation, frequency and color.

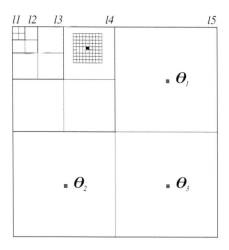

Figure 13.5: Mapping of orientation and spatial frequency locations.

---

[1]High order neurons are made in reference to neurons of the primary visual cortex which receives inputs from LGN neurons and neurons with lateral integration.

Legge and Foley's original model has been further refined by Teo and Heeger [TH94a, TH94b] and by Watson and Solomon [WS97]. Teo and Heeger's model has both exponents p and q set to 2, but requires at least four discrimination levels for an accurate account of masking. That is, there are four different sets of k_i and σ_i, producing different masking responses, R_i, each responsible for a different contrast range, $i = \{1, 2, 3, 4\}$. Watson and Solomon's model revealed that by setting the condition $p > q$ [WS97], only one level of discrimination is required to approximate masking. This reduces significantly, the computation load of Watson and Solomon's model compared to Teo and Heeger's model. Thus, the later is chosen for implementation here.

13.2.2.4 Detection

The final component of the PIDM is the detection function which measures the distortion between two neural images. An approach to this has been to take the square error (l_2 norm) [TH94a] defined as

$$D[i] = |R_a[i] - R_b[i]|^2 \qquad (13.23)$$

where R_a and R_b are the masking responses of the two images, a and b, at the ith coefficient. Pooling equation (13.23) for all coefficients, $i = \{1, 2, ..., N\}$, spanning all frequencies and orientations will provide the overall perceived distortion between the two images. Alternatively, Minkowski's summation,

$$D_M = \sqrt[\beta]{\sum_i^N |R_a[i] - R_b[i]|^\beta} \qquad (13.24)$$

can be employed to estimate visual distortion as in [WS97].

13.2.2.5 Overall model

The block coding principle in EBCOT assumes block independence. Therefore, each block is encoded based on the available information contained within itself. Accordingly, EBCOT by nature supports spatial masking since for any given point in a block, there are neighboring coefficients from which local masking can be approximated. However, to determine orientation and frequency masking, one would need to have coefficient transparency between blocks with different frequency and orientation bands. To do so would require alterations of the EBCOT coding structure. In the current masking model, only orientation and spatial masking have been considered. Frequency masking is somewhat more difficult to implement, requiring substantial modifications. In addition, there is also the issue of masking across scale. A coefficient at frequency level l is equivalent to a quarter of a coefficient at frequency level l-1. Although a frequency masking function would undoubtly improve masking estimation, the difficulty and uncertainty associated with its implementation led to its omission.

Watson's CGC model [WS97] called for a unified response function that merges masking in the various domains — spatial, frequency and orientation — into one inhibition factor. The approach taken here is to keep the masking functions from the different domains separate. The reason behind this is that separate masking functions allow greater flexibility in controlling the masking contributions from each domain. This would be extremely useful during model optimization. From (13.20), the neural masking response function is rewritten as:

$$R_{l,z}[m,n] = k_z \cdot \frac{E_z[m,n]}{I_z[m,n] + \sigma_z^q},$$ (13.25)

where $z \in \{\Theta, \Upsilon\}$, specifies the orientation and spatial masking domains, respectively. The excitation and inhibition functions for both the orientation and spatial masking functions are as follows:

$$E_{l,\Theta}[m,n] = X_l[\theta, m, n]^{p_\Theta}$$ (13.26)

$$E_{l,\Upsilon}[m,n] = X_l[\theta, m, n]^{p_\Upsilon}$$ (13.27)

$$I_{l,\Theta}[m,n] = \sum_{\alpha} X_l[\theta_\alpha, m, n]^q$$ (13.28)

$$I_{l,\Upsilon}[m,n] = \frac{8}{N} \sum_{u=m-l}^{m+l} \sum_{v=n-l}^{n+l} X_l[\theta, u, v]^q + \sigma_{var}^q.$$ (13.29)

$\sum_{\alpha} X_l[\theta_\alpha, m, n]^q$ represents the sum of transformed coefficients spanning all orientations, $\alpha = \{1, 2, 3\}$. $\sum_{u=m-l}^{m+l} \sum_{v=n-l}^{n+l} X_l[\theta, u, v]^q$ is the sum of neighboring coefficients about $X_l[\theta, m, n]$ (see Figure 13.5). The neighborhood, $N = (2l + 1)^2$, is the square area surrounding $X_l[\theta, m, n]$, the size of which is dependent on the frequency level of $X_l[\theta, m, n]$, $l = \{1, 2, 3, 4, 5\}$ (from lowest to highest). Thus, coefficients from the highest frequency level would have a larger neighborhood. This approach attempts to equalize the uneven spatial coverage between images of different frequency levels inherent in multi-resolution representations. The neighborhood variance, $\sigma_{var}^q = \frac{1}{N} \sum_{u=m-l}^{m+l} \sum_{v=n-l}^{n+l} (X_l[\theta, u, v] - \mu)^2$, with μ being the mean, has been added to the inhibition process to account for texture masking [SJ89]. Neighborhood variance has been introduced to improve the accuracy of distortion measures of texture masking. Large variances usually indicate high texture (i.e., grass, fur, etc.), resulting in the inhibition of neural signal responses, whereas low variances reflect relatively smooth regions of images, which in themselves offer very little texture masking. Exponents p_z and q are governed by the condition $p_z > q > 0$ according to [WS97]. Currently, q is set to 2. The last stage of the model utilizes the square error (13.30), re-written as:

$$D_{l,z}[m,n] = |R_{a,l,z}[m,n] - R_{b,l,z}[m,n]|^2$$ (13.30)

to determine perceptual distortion. Here, $R_{a,l,z}$ and $R_{b,l,z}$ represent the two neural responses from image a and b, with $z \in \{\Theta, \Upsilon\}$. The final distortion function encapsulating both spatial and orientation masking responses for block, b, at resolution level, l, is given by

$$D_l[b] = \sum_m^M \sum_n^N \left(G_\Theta \cdot D_{l,b,\Theta}[m,n] + G_\Upsilon \cdot D_{l,b,\Upsilon}[m,n] \right), \quad (13.31)$$

where G_Υ and G_Θ are the adjustable gains for spatial and orientation masking channel respectively.

13.2.3 EBCOT Adaptation

While the PCRD optimization offers a significant level of control as to the order in which bit-plane information is coded, it has one drawback with vision modeling. The problem with the vision based PCRD optimizer is that it is a dynamic R-D allocation system, but accumulates distortion incurred during bit-plane coding in a static way. The PCRD optimizer collects the rates and distortions for all sub-blocks and then generates the optimal compressed stream. For the current vision model, in the interest of visual accuracy, it is imperative that responses in (13.25) be derived relative to the actual states[1] of other coefficients that contribute to the masking process. Unfortunately, the states of these coefficients can only be approximated during the distortion calculation phase since the final bit-plane quantized/coded state of all coefficients are unknown until after the PCRD stage. Amending this problem with a dynamic PCRD environment would be impractical simply because the number of iterations required to generate the optimal compressed stream will be prohibitively high. The real solution here would be to find reasonable approximations for representing the states of all coefficients during distortion calculations. A reasonable approximation in this case would be to assume that all coefficients are quantized to the same bit-plane level.

R-D optimization measures the rate of change in distortion against the rate of change in bitrates to obtain best distortion reduction for least number of bits. In EBCOT, the change in visual distortions can be derived relative to the unquantized (original) or quantized coefficient. For the moment, let us just consider quantization at the whole bit-plane and ignore fractional bit-planes. Let $R_z[m,n]$ be the response of an unquantized coefficient, $\hat{R}_{z,i}[m,n]$ be the response of a coefficient quantized to the ith bit-plane and G_z the gain as defined in (13.31) with $z \in \{\Theta, \Upsilon\}$. Then in the first method, the perceptual distortion resulting from quantization to the ith bit-plane, which is relative to the original coefficient, is

$$D_i[m,n] = \sum_z G_z \cdot |\hat{R}_{z,i}[m,n] - R_z[m,n]|^2. \quad (13.32)$$

---

[1]The magnitude of coefficients resulting from quantization.

It is clear that as $i \rightarrow 0$, $R_{z,i} \rightarrow R_z$, so D_i also tends toward 0, i.e., perceptual distortion diminishes according to reduction in quantization step size. Comparatively, D_i exhibits more perceived distortions than D_{i+1}. Therefore, to determine the change in visual distortion between bit-planes i and $i+1$, one simply uses the measure

$$\Delta D_1[m,n] = D_i[m,n] - D_{i+1}[m,n]. \tag{13.33}$$

Hence, ΔD_1 is derived relative to the unquantized response, R_z. Alternatively, if only quantized coefficients were to be used, then the change in visual distortion between bit-planes i and $i+1$ can be measured by

$$\Delta D_2[m,n] = \sum_z G_z \cdot |\hat{R}_{z,i+1}[m,n] - \hat{R}_{z,i}[m,n]|^2. \tag{13.34}$$

In terms of visual accuracy, (13.34) would appear to be better since the distortion is taken directly from two visual responses, $\hat{R}_{z,i}$ and $\hat{R}_{z,i+1}$, of two bit-plane coded images. However, it is perhaps habitual that comparisons tend to be made between an original and a distorted signal which is why (13.33) was originally considered. As far as complexity and computational costs are concerned, (13.34) is simpler and more efficient than (13.33), therefore it is the method adopted in the proposed perceptual coder.

Spatial masking considers masking within a localized area or a neighborhood. Central to this is the size of this neighborhood. In their models, Zeng et al. [ZDL00] defined a neighborhood size of radius two while Taubman [Tau00] estimated local masking based on the coefficients of an entire block. The size of the latter was taken more for reasons of computational efficiency and simplicity rather than accuracy. In the dyadic wavelet decomposition, each successive resolution is a quarter the size of the previous. To ensure some form of uniformity for the masking area, the neighborhood sizes for spatial masking partially reflect the resolution level of the target coefficient. (13.29). In a five level hierarchical transform, assuming *L5* represents the highest frequency level and *L1* the lowest frequency level, then the radius of *L5* is five coefficients while at *L1*, the radius is one. However, at the lowest frequency level, coefficients have separate masking treatments. Figure 13.6 illustrates the size of neighborhoods for the other frequency levels.

A final necessary adjustment to the vision model for EBCOT adaptation is the alteration of the masking function for the lowest frequency, isotropic (DC) bands. Masking of DC coefficients is difficult to predict. These coefficients play a critical role in image formation and it is quite conceivable that due to this fact, little masking occurs among them. A point supported when the CGC model described previously was observed to have over estimated masking in the lowest frequency bands. In light of this, a modified self inhibiting masking response function has been used for DC coefficients. It takes the form:

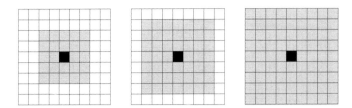

Figure 13.6: Neighborhood sizes (colored regions) for spatial masking relative to frequency level. *left*: *L2. center*: *L3. right*: *L4.*

$$R_{DC}[m,n] = k \cdot \frac{\bar{X}_{DC}[m,n]^p}{X_{DC}[m,n]^q + \sigma^q}, \tag{13.35}$$

where $\bar{X}_l[m,n]$ and $X_l[m,n]$ are the quantized and unquantized coefficients respectively. Effectively, (13.35) is roughly equivalent to (13.20) minus the spatial and orientation neighborhood masker. To simplify the problem, the parameters in (13.35) will follow those of the spatial masking function.

13.3 Model Calibration

The vision model presented in section 13.2.2 provides a generalized response of the HVS. As it is, the model lacks accuracy and would need some form of optimization before it meets the necessary requirements for perceptual coding purposes. Thus, the task at hand is not just an exercise in finding an optimal set of parameters, for the vision model in question, in evaluating the quality/distortion of images, but also to ensure that these parameters are optimized for visual coding as well.

Optimizing a vision model for coding purposes is difficult, to say the least, since there are no exact guidelines to follow. Within the scope of an image coder, vision model optimization is a tedious task when a considerable part of the process involves critical subjective assessment (see Figure 13.12). This is even more so, when one makes generous assumptions regarding the size of the search space which has to be optimized, i.e., when the possible ranges of all parameters are large. A practical solution to this problem is to break the actual optimization into two stages. In this ad hoc approach, the first stage will provide a coarse approximation of the parameters outside the scope the coder. The second stage will then refine the parameters within the scope of the coder.

Stage one of the optimization process fits the vision model to a set of subjective data. This is done by finding a parameter set which gives the best correlation between objective and subjective scores. There are two steps in accomplishing this task: subjective assessment and optimization of vision model. Section 13.3.1 looks at the selection of test images to be used for testing. Section 13.3.3 describes the methodology for acquiring subjective scores for the test images, while Subsections 13.3.9.1 and 13.3.9.2 deal with the optimization. Stage two optimizes the perceptual coder through manual fine-tuning of the model parameters. This is described in Subsection 13.3.9.3.

13.3.1   Test Material

The traditional approach to optimizing vision model parameters involves the measurement of the threshold response of human vision with synthetic test stimuli [WS97, van96, Win00]. These stimuli provide the means for gauging the frequency and orientation sensitivities of the HVS. Though it is understood that these simple test stimuli do not provide a realistic representation of natural images, their popularity as source materials in visual tests have perhaps been driven by their simplicity. Synthetic patterns like periodic waveforms have mathematical structures that are both easy to generate and manipulate. For the purpose of visual testing, this is useful since these patterns can be tailored to suit the needs of experiments. Recently however, there has been a move towards using natural images for the purpose of vision model optimization [Nad00, YWWC02] to validate the uses of these models in real world applications. For the current model parameterization, a set of twenty test images, each measuring 256×256 pixels, extracted from a set of natural images, listed in Table 13.1, will be used as the source material.

Table 13.1: Images used for subjective testing and optimisation [usc]

Image	Description	Dimensions
Barbara	A woman sitting in a room	720×576
Barbara2	A woman sitting on a chair	720×576
Black	Blackboard with toys on a table	720×576
Boat	Fishing boats in dry dock	720×576
Butterfly	A Monarch Butterfly	720×512
Goldhill	Houses along a stone road	720×576
Lena	Facial image of a woman	512×512
Pepper	Chilli and Capsicum	512×512
Sail	Windsurfers	720×512
Tulip	Garden tulips	720×512
Zelda	Facial image of a woman	720×576

13.3.2   Generation of Distorted Images

Distorted images are generated from the tests images listed in Table 13.1. To ensure that these distorted images cover all possible quality levels corresponding to that of the quality scale in Section 13.3.6, each test image is subjected to six different levels of distortions, producing six images of different qualities. Since the proposed coder implements a variable bit-plane quantization scheme, it is reasonable to assume that images distorted with a bit-plane masker should bear some resemblance to coded images. Bit-plane masks are applied to spectral coefficients obtained from a five level wavelet (Mallat) decomposition with the bi-orthogonal 9/7 filters [ABMD92]. A masked coefficient, $\hat{C}$, is defined as

$$\hat{C} = sign(C)(|C| \bullet M) \tag{13.36}$$

where $|C|$ is magnitude the spectral coefficient, $\bullet$, the boolean *AND* operator, $sign(C)$, the sign ($\pm$) of C and M, the masker as defined by

$$M = (\text{FFFF}) \ll (2l), \tag{13.37}$$

with $l \in \{1, 2, ..., 6\}$. FFFF is a hexadecimal number and $\ll$ represents the boolean left shift operator. The bit-plane mask is applied uniformly to all coefficients spanning all frequency levels and orientation bands. Once the bit-plane filtering has been applied, the inverse wavelet transform is used to project distorted images back to the time domain for subjective assessments.

13.3.3 Subjective Assessment

The ITU-R recommendation BT.500-10 [ITU00] provides two detailed methodologies for conducting subjective assessments of pictorial signals from the viewpoint of quality and impairment. The *double-stimulus impairment scale* (DSIS) method measures the distortions of processed signals. This method provides the means for evaluating the presence and strength of specific types of artifacts in images/videos. The *double-stimulus continuous quality-scale* (DSCQS) method on the other hand, measures the overall quality of processed images/videos.

Although the PIDM presented in Section 13.2.2 has been designed to approximate visual distortions between two images, the more appropriate subjective assessment method for parameterizing the PIDM is the DSCQS method. The reason behind this is that the DSCQS provides an overall subjective measure of material quality. The DSIS method however, was intended for the assessment of specific types of distortions. Yuen and Wu [YW98] have identified the existence of numerous types of artifacts in digital images which appear as a result of signal compression or processing operations. Though it is possible to conduct multiple DSIS tests to gauge all the possible types of distortions for a given set of images or videos, this approach is time consuming and there is also the problem of how to unify the results of various DSIS tests, on different types of artifacts into one magic number.

The DSCQS test methodology offers a general blueprint for subjective image assessment. However, to cater for specific requirements in subjective assessment, certain modifications may be necessary to obtain the desired results. As such, preliminary trial tests were set up to determine the most appropriate variation of the DSCQS test to be used as the final assessment structure. These trials were conducted with at least five participants and looked at the different methods of presenting and grading test materials. Therefore, for PDIM parameterization, the actual method adopted for subjective assessment is a variation of the DSCQS. The following subsections provide the arrangements and presentation of the test with analysis of the results as specified in [ITU00].

13.3.4   Arrangements and Apparatus

To ensure consistent presentation quality and to reduce analog distortions, the test materials are stored in digital format (D1) and presented with a grade A monitor. The apparatus used for conducting the DSCQS test are listed as follows:

- Abekas A65 Frame Store
- Abekas A26 D/A converter
- Sony BVM-20F1E Grade A monitor
- Sun Ultra 60 workstation

All test images are expanded to CCIR 601 format (YUV[1] 720×576 pixels) by superimposing each test image (256×256) over the center of a neutral grey frame (720×576). The expanded test images are kept in the A65 frame store. These images are converted to analog signals via the D/A converter and sent to the display monitor of subjective assessment. Figure 13.7 shows the connectivity of these devices. In this setup, the DSCQS test is automated since the presentations of test images are controlled by the workstation.

Two crucial elements for conducting proper subjective experiments are the environmental viewing conditions and the test conditions. The benchmark for laboratory viewing environment has been set in [ITU00]. For the current tests, these conditions are listed in table 13.2. As for the test conditions, they are listed as follows:

- Viewing distance - $3\times$ monitor height
- Maximum test duration per session - 25 minutes
- Maximum number of test subjects per session - 2

Prior to commence of tests, subjects are screened for visual acuity and instructed on the procedures of the test. As an added measure, these subjects were also trained to discriminate the different levels of image quality so that a level of consistency can be achieved with the subjective scores. This is reinforced at the beginning of each test session with stabilization images.

---

[1]The term YUV is commonly used in reference to component color of the digital $Y_d C_r C_b$ and/or analog $Y_t U_t V_t$ forms. Here, YUV represents digital component color.

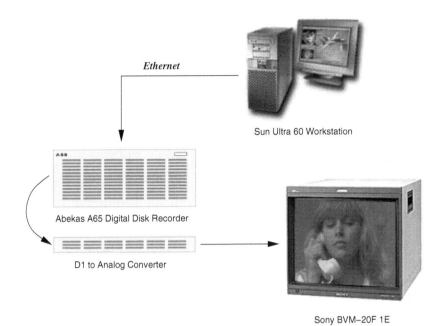

Ethernet

Sun Ultra 60 Workstation

Abekas A65 Digital Disk Recorder

D1 to Analog Converter

Sony BVM–20F 1E

Figure 13.7: Equipment Setup for the DSCQS Test.

Table 13.2: Laboratory viewing conditions for the variant DSCQS test

Ratio of luminance of inactive screen to peak luminance:	≤ 0.02
Ratio of the luminance of the screen, when displaying only black level in a completely dark room, to that corresponding to peak white:	≤ 0.01
Display brightness and contrast setup via PLUGE - Rec. ITU-R BT.814 [ITU94a] and ITU-R BT.815 [ITU94b].	
Maximum viewing angle:	$\pm 30^{o}$
Ratio of luminance of background behind picture monitor to peak luminance of picture:	≤ 0.15
Other room illumination:	very low

13.3.5   Presentation of Material

The DSCQS test organizes the presentation of materials into sets. Each test set contains a pair of images, one for reference and the other for assessment. According to [ITU00], the order of presentation of the reference and test images should be randomized. However, in the interest of keeping the test simple so as to avoid systemic errors in the grading phase, this order has been fixed with the presentation of the reference image followed by the test image. This structure is given in Figure 13.8.

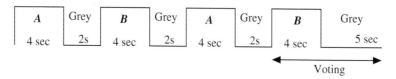

Figure 13.8: Sequence of presentation of images. *A* and *B* are the reference and test images respectively.

The test duration for each set of images lasts 27 seconds, segmented into two display repetitions (Figure 13.8). Although it has been suggested in [ITU00] that five repetitions may be appropriate for still image assessment, this has been deemed too lengthy by all participants of the preliminary trails. Each repetition consists of a presentation of the reference image followed by a mid-grey screen, then the test image and ends with another mid-grey screen. Subjects have nine seconds to register their scores for images which takes place at the start of the second presentation of the test image. While the order of presentation within an image set is fixed, the order between sets is randomized. This helps balance out the effects of fatigue during tests, and adaptation to images in the results. Each test session consists of 30 sets of test images plus a few stabilization sets and takes approximately 15 minutes to complete. The training session has 12 training sets and lasts around 6 minutes.

13.3.6   Grading Scale

Test images are assessed on a continuous scale, categorized into five levels: Excellent; Good; Fair; Poor and Bad. The grading scale measures 100 *mm* in length and is divided into the five respective segments. Scoring is applied at any point along the scale which takes any value between 0 and 100 (0 being the worst and 100 the best), with step size of 0.5 unit. Figure 13.9 illustrates two possible methods of collecting subjective scores, here referred to as the *absolute* and *difference* scale.

The absolute scale captures the quality of a test image relative to a reference. The difference scale, recommended in the standard DSCQS test [ITU00], calls for the grading of both the reference and test images with the final quality score obtained from the difference between the two. This scale proved to be somewhat inaccurate and unreliable due mainly to the fact that subjective scores of reference images were found to have varied at different stages of the trial test, sometimes considerably. For example, it has been observed that in numerous trial tests, an identical reference image presented on six different occasions during the test has attracted several different scores from the same assessor. Some of the variations in the scores may be attributed to fatigue. However, not all are since these variations manifest themselves at the beginning of the trial tests. Despite the fact that quality is measured based on the difference of two images, variations in subjective quality of the same reference image meant that there is no fixed point of reference quality. This is contradictory since a reference image should have same

Figure 13.9: Continuous quality grading scale. *Left*: Absolute scale. *Right*: Difference scale.

subjective appeal for the same pair of eyes regardless of how many times it is viewed or when it is viewed. In short, the reliability of the results is questionable in the absence of consistent subjective opinions of reference images. Another problem with the difference scale lies in the case where if the reference image is undistorted, the quality of the reference image could be graded in a category which may not reflect the true quality of the image itself. i.e., an original reference image should ideally be graded as excellent, but may be subjectively graded as anything other than excellent. In considering the above arguments, it is clear that the more appropriate grading scale to be used would be the absolute continuous quality scale. It must be noted that the minor alterations made to the DSCQS test is due primarily to the grading scale, not the method itself.

13.3.7 Results

Results of the subjective assessment are tabulated in Tables 13.3 – 13.4. Column one of the tables indicates the base image used to generate the distorted images in columns three to eight. It also lists the number of participants involved in evaluating that group of filtered images. The second column provides the statistics of interest, namely, the *mean opinion score* (MOS), μ, the standard deviation of the MOS, σ, plus the upper and lower bounds of the 99% confidence intervals, U and L respectively. The maximum and minimum standard deviations of the MOS are 15.9 and 3.7 respectively with an average of 9.95. Although the ITU-R BT.500-10 standard recommended the use of at least 15 subjects for all test material, not all of the images were tested with the minimum number of subjects. Nevertheless, in this particular instance, this issue has had little overall impact on the results.

The problem with subjective tests is that the results are subjective, hence any image may have different levels of appeal to different assessors. This problem is further com-

Table 13.3: DSCQS results A

Image	Statistics	Masking Level					
(No. of Subjects)		1	2	3	4	5	6
BarbaraA	μ	85.6	78.1	61.1	47.2	35.5	12.4
(19)	σ	7.3	8.9	13.7	13.9	11.8	8.1
	U	89.9	83.4	69.2	55.5	42.5	17.2
	L	81.3	72.9	53.0	39.0	28.5	7.6
Barbara2A	μ	88.8	67.9	59.8	38.8	30.0	18.1
(18)	σ	6.8	15.0	10.8	12.7	14.4	12.2
	U	92.8	76.8	66.2	46.3	38.5	25.3
	L	84.7	59.0	53.4	31.3	21.5	10.9
BlackA	μ	85.7	81.3	64.2	51.1	41.5	13.9
(19)	σ	5.1	6.3	11.5	13.3	14.4	8.9
	U	88.7	85.0	71.0	59.0	50.0	19.2
	L	82.7	77.6	57.4	43.2	33.0	8.6
BoatsA	μ	81.9	73.2	65.4	49.1	42.8	13.3
(18)	σ	8.8	13.5	13.9	12.1	11.3	8.7
	U	87.1	81.2	73.6	56.3	49.5	18.5
	L	76.7	65.3	57.2	42.0	36.2	8.2
ButterflyA	μ	88.5	82.3	64.0	50.0	39.2	11.0
(19)	σ	7.2	4.6	15.9	14.5	14.4	11.3
	U	93.4	85.4	74.6	59.6	48.8	18.6
	L	83.7	79.3	53.4	40.3	29.7	3.5
GoldhillA	μ	88.5	82.3	64.0	50.0	39.2	11.0
(15)	σ	7.2	4.6	15.9	14.5	14.4	11.3
	U	93.4	85.4	74.6	59.6	48.8	18.6
	L	83.7	79.3	53.4	40.3	29.7	3.5
LenaA	μ	91.3	84.9	72.5	53.2	47.7	29.5
(15)	σ	5.6	3.7	9.8	13.1	12.6	11.5
	U	95.0	87.4	79.0	62.0	56.2	37.2
	L	87.5	82.4	65.9	44.5	39.3	21.9
PepperA	μ	89.0	79.3	72.1	60.1	51.4	29.9
(15)	σ	7.0	13.2	12.4	9.8	13.2	15.0
	U	93.7	88.1	80.3	66.7	60.2	39.9
	L	84.4	70.5	63.8	53.6	42.6	20.0
SailA	μ	89.6	67.5	57.3	35.4	27.1	12.0
(15)	σ	6.0	10.8	12.3	13.0	14.8	11.0
	U	93.6	74.7	65.5	44.1	37.0	19.3
	L	85.6	60.3	49.1	26.8	17.3	4.7
TulipA	μ	90.8	83.0	69.8	46.3	39.0	15.3
(15)	σ	6.3	8.4	10.9	13.1	12.9	10.8
	U	95.0	88.6	77.0	55.1	47.5	22.4
	L	86.6	77.3	62.5	37.6	30.4	8.1

Table 13.4: DSCQS results B

Image	Statistics	Masking Level					
(No. of Subjects)		1	2	3	4	5	6
ZeldaA	μ	91.8	87.2	69.6	61.2	51.2	24.0
(14)	σ	5.2	4.5	8.8	8.7	11.7	8.7
	U	95.5	90.4	75.9	67.3	59.6	30.3
	L	88.2	83.9	63.3	55.0	42.9	17.8
BarbaraB	μ	87.6	79.5	65.6	46.8	40.3	17.1
(14)	σ	7.1	10.0	7.1	11.1	8.5	9.2
	U	92.6	86.6	70.7	54.8	46.4	23.7
	L	82.5	72.3	60.5	38.9	34.2	10.5
Barbara2B	μ	87.0	78.0	57.3	39.9	24.2	13.1
(14)	σ	6.1	8.8	9.9	13.0	7.5	6.8
	U	91.4	84.3	64.5	49.2	29.5	18.0
	L	82.6	71.7	50.2	30.6	18.8	8.3
BlackB	μ	87.2	78.6	68.0	58.5	49.7	21.5
(14)	σ	7.5	7.4	8.5	6.4	7.2	5.7
	U	92.6	83.9	74.1	63.1	54.9	25.6
	L	81.9	73.3	61.9	54.0	44.5	17.5
BoatsB	μ	84.9	74.5	63.5	45.4	35.0	15.0
(14)	σ	4.5	8.5	9.3	10.0	8.2	7.1
	U	88.1	80.6	70.2	52.5	40.9	20.1
	L	81.7	68.4	56.8	38.3	29.1	9.9
ButterflyB	μ	82.2	74.5	64.2	48.0	29.8	9.7
(15)	σ	11.4	13.9	11.1	13.2	14.9	8.8
	U	90.0	84.1	71.9	57.0	40.1	15.8
	L	74.3	64.9	56.5	38.9	19.5	3.6
GoldhillB	μ	89.5	82.0	63.8	47.1	40.7	12.6
(15)	σ	8.4	10.4	11.6	13.3	13.1	7.8
	U	95.3	89.2	71.8	56.3	49.7	17.9
	L	83.7	74.8	55.8	38.0	31.7	7.2
LenaB	μ	91.0	82.7	71.7	53.5	47.0	22.6
(15)	σ	6.0	5.0	7.7	7.4	6.0	13.7
	U	95.1	86.2	77.0	58.6	51.2	32.1
	L	86.9	79.2	66.4	48.4	42.9	13.2
SailB	μ	89.0	71.9	59.9	34.3	27.8	7.9
(15)	σ	8.7	11.8	11.9	10.1	11.6	6.3
	U	95.0	80.0	68.2	41.3	35.8	12.2
	L	83.0	63.7	51.7	27.4	19.8	3.5
GoldhillC	μ	90.6	77.4	66.4	50.6	39.6	21.1
(15)	σ	5.9	6.5	11.3	10.2	10.6	12.7
	U	94.7	81.9	74.2	57.7	46.9	29.8
	L	86.5	73.0	58.6	43.6	32.3	12.3

plicated by the fact that there is no clear line to distinguish the exact boundary between two quality levels. i.e., what is the difference between an image of "fair" quality and an image with "poor" quality? Therefore, though the deviations of the results were high, they were not unexpected. Irrespective of this, one may be inclined to accept the results in Tables 13.3 to 13.4 based solely on the merit that the maximum variation is only plus or minus one quality level.

It is evident from these results that high deviations seem to congregate around the medium quality levels. This suggests that subjects identify excellent and bad images more readily than good, fair and poor images. Although the distorted images were generated from the same set of filters, it is clear from the results that when any particular level of filtering is applied to two different images, the subjective quality response differs. This reinforces the fact that image content plays an important role in image quality and how it can be affected by coding structures.

13.3.8   Analysis

The standard deviation indicates the amount of variation in the results. Small deviations mean that a majority of subjects are in agreement with respect to the quality of images. It is understood that the primary factor which affects the level of variation in the results is dependent on the number of subjects. However, according to Tables 13.3 to 13.4, deviations seems to be source dependent as opposed to subject numbers. Images with less subjects performed equally well if not better than those with more subjects. A plausible explanation for this is that some images have regions which are structurally very different and hence exhibit different levels/types of regional distortions. When assessing these images, different subjects may concentrate on a specific region only and rate the quality of images accordingly, thus leading to substantial variations in results. Therefore, subjective tests designed to provide a unifying score for the overall quality of an entire image will undoubtedly lead to inaccuracies. A possible solution to this is to adopt a region based subjective assessment. This approach would require the identification and segmentation of image areas where the distortion levels and types may differ considerably.

13.3.9   Model Optimization

For a given set of static images, the purpose of the stage one model optimization is to fit the objective responses of the vision model, produced under a set of model parameters, to subjective responses from the test, obtained through methods specified in Section 13.3.7. Optimization is an iterative process and the term static images refers to the fact that the same images are used in each iteration of the optimization. Since only static images are used, there is no need to perform subjective assessment at each iteration.

Once one is in possession of both the subjective and objective scores, it is necessary to measure how accurate the vision model is in predicting the actual behavior of the HVS. To assess the correlation between the objective and subjective scores, one could employ the *Pearson's* correlation coefficient [JB87, BB99],

$$C_p = \frac{\sum_i^N (x - \mu_x) \cdot (y - \mu_y)}{[(\sum_i^N (x - \mu_x)^2) \cdot (\sum_i^N (y - \mu_y)^2)]^{\frac{1}{2}}}, \tag{13.38}$$

where x and y are the objective and subjective (MOS) scores, respectively, μ_x and μ_y are the means of x and y, respectively, and N, the number of test images. Alternatively, one may also use the *Spearman* rank order coefficient which has two forms of representation [BB99]

$$C_s = 1 - 6 \left(\frac{\sum_i^N R_i^2}{N(N^2 - 1)} \right) \tag{13.39}$$

and [JB87]

$$C_s = \frac{12 \sum_i^N \left(X_i - \frac{N+1}{2} \right) \left(Y_i - \frac{N+1}{2} \right)}{N(N^2 - 1)}, \tag{13.40}$$

where R represents the rank difference between the objective and subjective scores, X and Y, respectively of image i. Here, $X = \sum_k^M D_e(k)$, the sum of the neural distortion, D_e from (13.31), for each neuron, k. Both correlation coefficients are bounded by $-1 \leq C_z \leq 1$, $z \in \{p, s\}$. $C_z \rightarrow -1$ indicates maximum decorrelation. The opposite is true as $C_z \rightarrow 1$. Spearman's coefficient measures correlation based on rank order, ranked from highest to lowest quality or vice versa, whereas Pearson's correlation coefficient measures the actual differences between two sets of data. In this respect, Spearman's coefficient is more complicated to implement especially when data within a sample set has the same rank. In such a case, some form of arbitration will be required to break the deadlock. In the current optimization procedures, Pearson's correlation is used. Note also that for the Pearson's correlation to operate effectively, the objective scores were normalized to the subjective scale.

An essential component to any optimization problem is the *cost function*. The cost function is used to gauge the accuracy and control the direction of an optimization process. Optimization operations are designed to either find the maxima or minima of a cost function. In the current model optimization, for the former operation, one could use the correlation function (13.38). For the later operation, an error measure in the form of the squared difference, $\sum^N (x - y)^2$ [DS89], between the objective and subjective scores, x and y, could be used. Alternatively, one could also modify the correlation function (13.38) by inverting its response and bounding it between 0 and 1 as follows,

$$F_c = 1 - \frac{C_p + 1}{2}. \tag{13.41}$$

As $C_p \to 1$, $F_c \to 0$ and as $C_p \to -1$, $F_c \to 1$. Hence, F_c is bounded by $0 \leq F_c \leq 1$. Any optimization process that minimizes the cost function will maximize the correlation between objective and subjective responses. This is useful for optimization tools which minimizes the cost function. The steps for the model optimization are as follows:

- *Step 1*: For a given set of images (original and distorted), conduct the subjective test as described in Section 13.3.3.

- *Step 2*: For the same set of images, given a set of model parameters, calculate the objective responses.

- *Step 3*: Determine the correlation between the subjective and objective scores.

- *Step 4*: Based on past and current correlation, generate the next set of model parameters.

- go to *step 2*

Figure 13.10 outlines the model optimization process. The parameters produced by the decision process are dependent of the optimization method. Here, two approaches were employed, *full parametric* and *algorithmic*, in two separate phases.

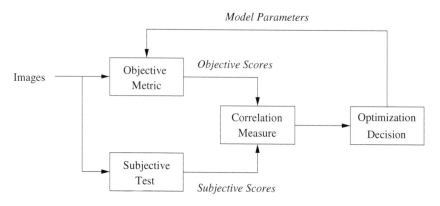

Figure 13.10: Stage one optimization. The objective metric follows that of Figure 13.3 while subjective test is described in Section 13.3.3. The correlation measure is Pearson's correlation coefficient (13.38) and the optimization decision is determined either parameterically or algorithmically.

The optimization process will calibrate the vision model to a specific set of filters and structures. In this case, the filter is the bi-orthogonal 9/7 tap filter with a 5 level Mallet decomposition structure. For the proposed vision model, a single color channel has 14 parameters to be optimized:

- 6 CSF weights
- 2 scaling constants - k_X and k_Θ
- 2 saturation constants - σ_X and σ_Θ

- 2 exponents - p_X and p_Θ
- 2 gain constants - G_x and G_Θ

If expanded to a full color model (RGB, YC_rC_b, etc.), a total of 42 parameters will need to be optimized (14 per color channel). Realistically, the globally optimal solution to the problem at hand is rather elusive and it is most probable any solution found would be sub-optimal or locally optimal. In spite of this, any solution with a reasonably high correlation ($C_p > 0.9$) will provide a sufficient starting point for the second stage of optimization in Subsection 13.3.9.3.

13.3.9.1 Full parametric optimization

The full parametric optimization is the brute force approach to finding the optimal solution to a function. It searches through all possible permutation of the model parameters[1] to identify the optimal parameter set(s). Full parametric search is computationally intensive and slow. The complexity of a full parametric search increases exponentially with the number of variables in a function that need to be solved. It is particularly effective in cases where the function to be optimized exhibits a smooth monotonic profile (Figure 13.11). However, in real world problems such as vision modeling, this function is unlikely to be smooth or monotonic. Nevertheless, full parametric search remains a useful tool for uncovering the probable locations of the optimal or sub-optimal solutions to a function. For this reason, it is used here to obtain some initial conditions for the algorithmic optimization.

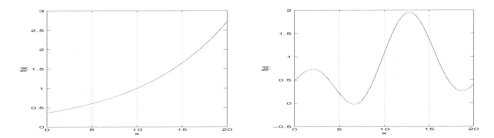

Figure 13.11: Monotonic (*left*) and non-monotonic (*right*) functions.

Parametric searches with fixed intervals covering the entire search space have better chances of uncovering optimal solutions, particularly when these intervals are numerous. However, for a 14 parameter vision model, having more than two intervals using this method would be impractical since the number of intervals, I, dictates the number of sets of parameters, $p = I^{14}$, which have to be evaluated in each iteration of the optimization. A simplistic approach to full parametric optimization is taken here based on

---

[1] within a certain bound or range of parameter values.

a binary search with decimated interval size. Consider a one dimensional problem for the moment. Given an arbitrary parameter, α, bounded by $\alpha_a \leq \alpha \leq \alpha_b$, the parametric optimization adopted here evaluates the cost function, Fc(.) (13.41), at two anchor points, x and y during each iteration. These two points are set to the edges of the parameter's boundary initially. i.e., $x = \alpha_a$ and $y = \alpha_b$. In each iteration, the anchor point which provided the lower correlation is shifted according to the rule:

```
if    Fc(x) < Fc(y)
then  y = (x + y)/2
else  x = (x + y)/2
```

Effectively, the distance between the two anchors halves after each subsequent iteration, thus they converge to a point along the search space. If the cost function is monotonic, the parametric search will eventually produce an optimal solution. However, if the cost functions is non-monotonic, then the end result may not be optimal, but may still reside within the vicinity of an optimal or sub-optimal solution.

With 14 parameters, the number of permutations per iteration of the parametric search amounts to $2^{14} = 16384$. Since the maximum parameter values are unknown, it is only safe to start with a set of large maximum values so as to avoid under estimation. Naturally, this would mean large search interval with lower search resolution, thus requiring more iterations to reach a certain optimal level. This optimal threshold is arbitrarily set at 0.9 on the Pearson's correlation coefficient scale. While it is desirable to have parameters with good correlation as initial starting points for the algorithmic optimization, the correlation need not be overly high since the algorithmic searches will locate the optimal or sub-optimal parameters so long as the initial conditions are in close proximity to the solution. With a large set of permutation, there will be instances where the results within an iteration are similar or the same. In the event of such an occurrence, all these similar/same results will be carried through to the next iteration/phase of optimization. The full parametric optimization was carried out under the cluster computing environment[1] running the **Enfuzion**[2] [Tur00] software.

13.3.9.2 Algorithmic optimization

The algorithmic optimization provides a faster way for locating an optimal/sub-optimal set of parameters of the vision model. The performance in this approach to a degree is dependent on the initial parameter set. A good initial set of parameters will lead to the desired solutions, while poor initial conditions will adversely lead to nowhere. Thus, the safeguard against the latter possibility has been to use of the parameter sets provided by the parametric optimization as the starting points. In addition, other factors which

---

[1]http://hathor.csse.monash.edu.au/
[2]http://www.turbolinux.com/products/enf/

affect algorithmic optimizations are the algorithms themselves and the behavior plus complexity of the vision model to be optimized. In the former case, selecting algorithms appropriate to solving the problem would certainly allow for a more effective model optimization.

Algorithmic optimization of a high dimensional function that exhibits non-smooth characteristics poses a significant challenge with two implications. First, the optimization algorithm must converge to the optimal, or at the very least a sub-optimal, solution to be of any use. Though methods for large scale computation problems have been studied [GMW81], Fletcher [Fle81] pointed out that an effective method for optimizing non-smooth function has yet to be found. He did however, indicate that a variation of the quadratic programming (Lagrange-Newton) method showed some promises. Secondly, given that convergence does occur, one would also need to consider the amount of time required to complete the algorithmic optimization process. The length of which is dependent on the dimensionality of the function, convergence speed and computational complexity of the algorithm.

The algorithms used in this phase of the model optimization are the *Quasi-Newton* algorithm with *BFGS* [DS89] update and the *polytope* "simplex" algorithm which originated from Spendley et al. [SHH62, PH71]. Descriptions of these two and other algorithms can be found in [GMW81, CZ01]. The polytope method has simple implementation and low computational complexity. It is generally more suited for functions with low dimensionality [DS89] and smooth surfaces [GMW81]. The Quasi-Newton method, a modification of Newton's algorithm [DS89], is more complex and slower than polytope. However, it converges to an optimal solution faster than the polytope method. Furthermore, it is equipped to handle functions with large dimensions [GMW81] and non-smooth surfaces [Fle81]. The selection of these two algorithms was limited by the availability of large scale optimization tools in the **Nimrod-O**[1] and **Matlab**[2] software packages. The algorithmic optimizations were carried out with a computer cluster running Nimrod-O and with a half dozen stand-alone workstations running Matlab.

13.3.9.3 Coder optimization

Despite the earlier optimization efforts, the vision model when adapted to the coder requires further manual fine tuning to operate effectively. The advantage of manual optimization is that it provides the subtle subjective adjustments of the parameters which the previous mathematically oriented optimization stage lacked. The adjustment of parameter values are taken at steps of between ± 0.001 and ± 10, depending on how thorough one wishes to be. For the most part, this is a painfully slow exercise without some intuitive guesses.

---

[1] http://www.hathor.csse.monash.edu.au
[2] http://www.mathworks.com

The manual optimization here operates by comparing the visual performances of the perceptual coder operating under two different sets of parameters (with reference to the original images). The parameter set with the overall superior visual performance is chosen as the reference parameter set while a new parameter set is generated based on the reference set. There is no fixed procedure on generating a new parameter set. This process is unfortunately ad hoc and requires some intuition. Due to the time involved in subjective assessment, only four images in Table 13.1, *Lena, Barbara, Goldhill* and *Boat*, were used for the manual optimization. These images were coded at bit rates of 0.125, 0.25 and 0.5 bpp (approximately) for evaluation by one expert subject. Figure 13.12 illustrates the structure for the manual optimization.

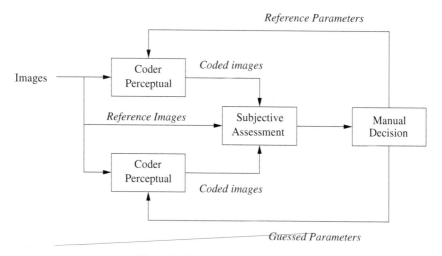

Figure 13.12: Stage two optimization.

13.3.9.4   Remarks

Optimization was problematic with a large number of parameters simply because the search space was far too large. After four iterations, the parametric optimization only revealed general directions of the optimal or sub-optimal parameters. With multiple starting points, the algorithmic searches did produce parameter sets that have high correlations (Table 13.5). However, when adapted to the coder, these parameters did not deliver the expected performance. The end results of the algorithmic search lacked visual sensitivity. Fortunately, the manual optimization in stage two was able to correct this problem.

It is conceivable that the large differences in results between the parametric, algorithmic and manual optimization are caused by the selection of test materials and the way in which subjective results were collected. The later perhaps contributed most to the discrepancy in Table 13.5 where the Pearson's correlation coefficient indicated that both

Table 13.5: Model parameters after various stages of optimization — luminance channel. B_U and B_L are the maximum and minimum attainable values of the model parameters. C_s is the Pearson's correlation coefficient.

Parameters	B_L	B_U	Parametric	Algorithmic	Manual
CSF - LL	0.01	5.00	5.0000	2.6230	1.4800
CSF - 1	0.01	5.00	5.0000	2.6790	1.5500
CSF - 2	0.01	5.00	3.7500	2.6290	1.7700
CSF - 3	0.01	5.00	5.0000	1.4190	1.6800
CSF - 4	0.01	5.00	3.7500	0.1013	1.2900
CSF - 5	0.01	5.00	1.2500	0.0104	0.8050
k_X	0.01	5.00	5.0000	4.4400	1.0880
k_Θ	0.01	5.00	1.2500	4.9380	0.9876
σ_X	1.00	1024	256.00	1.0550	5.5550
σ_Θ	1.00	1024	256.00	22.080	7.6800
p_X	2.10	5.00	2.5000	2.1000	2.5800
p_Θ	2.10	5.00	2.5000	2.1030	2.3950
G_X	0.01	1.00	0.7500	0.7588	0.7588
G_Θ	0.01	1.00	0.2500	0.4834	0.4834
C_p	-	-	0.9170	0.9705	0.8549

the parametric and algorithmic optimization produced better parameter sets with higher correlation than that of the manual optimization with respect to the subjective scores of Tables 13.3 and 13.4. However, visual assessment of images in Figure 13.13 indicates otherwise. While the subjective assessment method described in Section 13.3.3 differs little to that of [YWWC02], which was used to parameterize a similar vision model for an impairment metric, the requirements for vision models in coding and impairment (or quality) measure differ greatly. General impairment measures provide an overall grade for images or videos. Quality measure within the context of a coder requires accurate quantification of sub-images up to the resolution of individual data samples (pixel/coefficients). The subjective test in Section 13.3.3 lacked this accuracy. Thus the results from the parametric and algorithmic optimization differ substantially to the final parameter set depicted in Table 13.5. Another possibility which may have affected the results is the limited range or variety of test images used for optimization.

Irrespective of its shortcomings, the first stage of optimization served its purpose in identifying the approximate optimal/sub-optimal parameter set. This helped in narrowing the search space for the manual optimization and reduced the time needed to find a satisfactory set of model parameters for the perceptual image coder. A final word of note, the parameters from the manual optimization are at best sub-optimal. The optimal parameter set has yet to be determined through this may be somewhat difficult given the fact that finding the optimal solution can only be guaranteed though an exhaustive

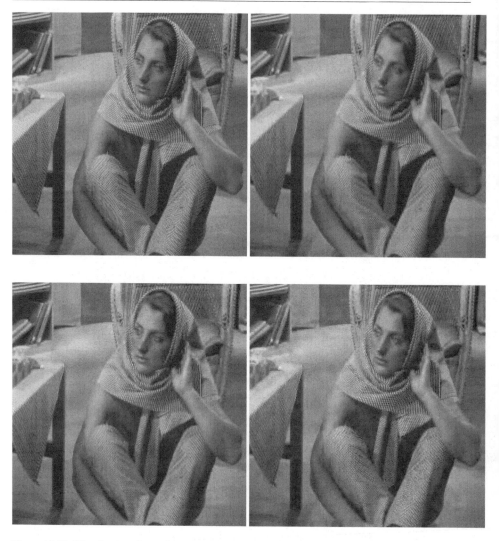

Figure 13.13: Visual comparison of parametric, algorithmic and manual optimizations with the *Barbara* image coded at 0.25 bpp. Clockwise from top left: original, parametric, manual and algorithmic.

parametric search, which is impractical. However, having said that, there is room for further optimization of the model parameters.

13.4 Performance Evaluation

A common approach to evaluating the performances of coders is through some form of fidelity measure. Fidelity metrics provide quality measurements on a quantifiable scale for images which have been processed, either through filtering, compression or by other means. A popular objective fidelity measure is the *Peak Signal-to-Noise Ratio* (PSNR). The subjective equivalence to this is the DSCQS test [ITU00], a variation of which has been described in Section 13.3.3. The performance evaluation for the proposed coder is comparative in nature as opposed to scale oriented quality measure. Thus, the evaluation of the proposed coder calls for comparative assessments with benchmark coders. The current benchmark lossy compression system for images is the EBCOT coder [Tau00, SGE02], which the proposed coder is based upon. Comparisons will be made between the proposed coder and three variations of EBCOT: EBCOT-MSE; EBCOT-CVIS [Tau00] and EBCOT-XMASK [ZDL00]. EBCOT-MSE encodes images with the purpose of minimizing the MSE and hence, it is purely a mathematical approach to coding. Both the EBCOT-CVIS and EBCOT-XMASK employ vision modeling to enhance the perceptual quality of coded images. However, while EBCOT-CVIS adopts a vision model for the sole purpose of predicting - and minimizing - visual distortions, EBCOT-XMASK utilizes vision modeling to enhance visually important features within images (Section 13.1.6).

The coder evaluation process will cover both objective and subjective assessments. While objective evaluations are generally faster and easier to implement than subjective assessments, they do not necessarily provide the best visual assessment. Subjective assessments on the other hand are extremely slow. Nevertheless, they provide a more accurate measurement of image quality, as far as the human vision is concerned. Assessments between the different coding systems are made based upon the quality of images encoded at a target bitrate. Note that due to the block coding structure of the EBCOT coder, the exact bitrate is rarely achieved [ISO00b]. However, the actual encoded bitrate is very close to the target bitrate. For a detailed overview of the performance of these coders, the quality assessments are made at four different bitrates - 0.0625, 0.125. 0.25 and 0.5 *bit per pixel* (bpp).

13.4.1 Assessment Material

The materials used for the coder evaluation consists of a set of eight natural images of dimensions 512×512 pixels. This group of images contains both scenic and close-in type images and covers a variety of content including detail structures. Of the eight images, five were used as part of the vision model optimization process (*Barbara, Boat, Lena, Goldhill* and *Pepper*) covered in Section 13.3. For the subjective evaluations, the dimensions of images were reduced to 512×480 pixels with the top and bottom 16 rows truncated. This reduction in resolution is due to the size of the display device.

13.4.2 Objective Evaluation

The traditional objective metric for measuring signal quality is the *Signal-to-Noise Ratio* (SNR) or its variation, the *Peak Signal-to-Noise Ratio* (PSNR) [Jai89]. Both these quality metrics are derivatives of the MSE, which is written as [JN84, Cla85]:

$$MSE = \frac{1}{N} \sum_i^N |\hat{x}_i - x_i|^2, \tag{13.42}$$

where x and $\hat{x}$ are the original and distorted sample data. The SNR is defined as the log ratio of the source signal to the MSE - between the source and corrupted signals. It is given by,

$$SNR = 10 \log_{10} \frac{\frac{1}{N} \sum_i^N x_i^2}{MSE}, \tag{13.43}$$

where x_i is the sample data. PSNR measures the ratio between the peak signal and the MSE. It is formulated as

$$PSNR = 10 \log_{10} \frac{255^2}{MSE}, \tag{13.44}$$

where the peak signal has a magnitude of 255. Both the SNR and PSNR are popular metrics and enjoy the status as the benchmark tool for measuring the general quality of digital signals/waveforms [JN84] of audible and visual nature. The popularity of these two metrics is driven by their simplicity and perhaps their ability to provide a "reasonable" assessment of signal quality. However, they are unable to accurately quantify the level and type [YW98] of structural errors such as those exhibited in compressed images and videos [PMH96, YWWC02]. Recently, Wang and Bovik [WB02] proposed a new image quality measure coined *Universal Image Quality Index* (UIQI), expressed as

$$UIQI = \frac{\sigma_{xy}}{\sigma_x \sigma_y} \cdot \frac{2\bar{x}\bar{y}}{(\bar{x})^2 + (\bar{y})^2} \cdot \frac{2\sigma_x \sigma_y}{\sigma_x^2 + \sigma_y^2} \tag{13.45}$$

where $\bar{z}$ and σ_z^2 are the mean and (unbiased) variance of data set z given by

$$\bar{z} = \frac{1}{N} \sum_{i=1}^N z_i \tag{13.46}$$

and

$$\sigma_z^2 = \frac{1}{N-1} \sum_{i=1}^N (z_i - \bar{z})^2 \tag{13.47}$$

respectively, with $z \in \{x, y\}$. σ_{xy}, the co-variance is defined as

$$\sigma_{xy} = \frac{1}{N-1} \sum_{i=1}^N (x_i - \bar{x})(y_i - \bar{y}). \tag{13.48}$$

The UIQI metric is in essence a correlation measure with responses within the range of $[-1, 1]$. -1 indicates maximum decorrelation while 1 signifies maximum correlation. This metric has shown good matching responses to subjective opinions for images distorted through filtering operations, additive noise and JPEG [Wal91] compression.

The above mentioned objective metrics were designed without the specific influence of vision modeling. Though it is incorrect to say that these metrics provide no accurate measurements of subjective image quality, they are somewhat less reliable in the critical assessment of the subjective quality of imagery. Alternative vision based objective image and video quality metrics have been proposed to fulfill this end. A study conducted by the Video Quality Experts Group (VQEG) [RLCW00, Roh00] lists some of these video fidelity metrics [BL97, WHMM99, Win99] and reported their performances with respect to subjective data. Similar comparative studies for image fidelity metrics have been reported by Watson [Wat00] and Pappas et al. [PMH96]. At this point, it should be noted that though the image and video metrics differ, in general, one could consider the video fidelity metric as an extension of the image metric as in the case of the Perceptual Distortion Metric (PDM) of Winkler [Win99, Win98]. Hence, video quality metrics can be easily modified to assess image fidelity as well.

The correlation between responses of the vision based quality metrics to those of subjective assessments is generally higher than the correlation between the responses of pure mathematical metrics, of (13.43) and (13.44), and those from subjective tests. However, the implementation of these vision based metrics are rather involved and many require, in addition, calibration to suit the various viewing conditions and display mediums/equipment. Furthermore, Pappas et al. [PMH96] noted that some of these vision based objective metrics lacked consistency in predicting the perceived quality of image. Thus, for the current coder evaluation, vision based objective metrics have been excluded as assessment tools.

13.4.3   Objective Results

In Table 13.6, the objective results based on the PSNR unequivocally indicate the superior performance of the the EBCOT-MSE coder to the other three coders. PSNR also suggests that the performance of the proposed coder is generally superior to EBCOT-CVIS while EBCOT-XMASK has the poorest performance of all. However, according to the UIQI metric from Table 13.7, the performances of the four coders are dependent on image contents and bitrates. The EBCOT-XMASK coder performed the best at higher bitrates (0.25 and 0.5 bpp) and very well for all other bitrates. It had some trouble with the Mandrill image, perhaps due to the high frequency signals contained within that image. EBCOT-MSE rated well on the UIQI scale overall with good performances at the low bitrate (0.0625 bpp). EBCOT-CVIS had good performances which bested EBCOT-MSE on numerous occasions. The proposed coder for the most part had very similar performance rating to EBCOT-CVIS and comparable to that of EBCOT-MSE across the board. Overall, the UIQI metric identifies EBCOT-XMASK as the best performing coder out of the four. The remaining three coders have very similar performance which the UIQI is unable to separate.

Table 13.6: PSNR results. (PC) = Proposed perceptual coder, (MSE) = EBCOT-MSE, (CVIS) = EBCOT-CVIS, (XMASK) = Point-wise extended masking.

Image	Bitrate (bpp)	PSNR (dB)			
		MSE	PC	CVIS	XMASK
Barbara	0.0625	24.00	23.62	23.69	23.43
	0.125	25.97	25.74	24.56	24.25
	0.25	29.04	28.49	27.58	26.20
	0.5	33.06	32.19	32.17	30.21
Boat	0.0625	25.98	25.63	25.49	25.06
	0.125	28.34	28.20	27.90	27.26
	0.25	31.25	31.11	30.69	30.05
	0.5	34.84	34.73	34.35	33.44
Bridge	0.0625	22.25	22.15	22.02	21.57
	0.125	23.47	23.20	23.35	22.68
	0.25	25.02	24.83	24.86	24.07
	0.5	27.38	27.34	27.06	26.57
Goldhill	0.0625	26.82	26.58	26.45	26.20
	0.125	28.65	28.38	28.33	27.82
	0.25	30.71	30.52	30.33	29.76
	0.5	33.38	33.25	33.05	32.19
Lena	0.0625	28.32	27.94	27.84	27.36
	0.125	31.22	30.90	30.79	29.94
	0.25	34.32	33.79	33.80	32.98
	0.5	37.41	36.84	36.86	36.21
Mandrill	0.0625	20.79	20.71	20.55	20.34
	0.125	21.79	21.68	21.54	21.01
	0.25	23.32	23.25	22.84	22.14
	0.5	25.73	25.51	25.10	24.72
Pepper	0.0625	27.83	27.48	27.17	27.23
	0.125	30.91	30.60	30.49	29.92
	0.25	33.64	33.33	33.39	32.70
	0.5	36.02	35.73	35.70	35.04
Plane	0.0625	26.35	25.97	26.00	25.47
	0.125	29.56	29.07	29.23	28.18
	0.25	32.98	32.59	32.68	31.76
	0.5	37.00	36.47	36.65	35.31

Table 13.7: UIQI results. (PC) = Proposed perceptual coder, (MSE) = EBCOT-MSE, (CVIS) = EBCOT-CVIS, (XMASK) = Point-wise extended masking.

Image	Bitrate (bpp)	UIQI			
		MSE	PC	CVIS	XMASK
Barbara	0.0625	0.417	0.425	0.424	0.412
	0.125	0.539	0.533	0.523	0.500
	0.25	0.659	0.651	0.651	0.622
	0.5	0.764	0.760	0.767	0.775
Boat	0.0625	0.380	0.371	0.351	0.366
	0.125	0.475	0.483	0.466	0.484
	0.25	0.572	0.571	0.575	0.596
	0.5	0.667	0.668	0.671	0.681
Bridge	0.0625	0.365	0.355	0.352	0.349
	0.125	0.506	0.492	0.498	0.505
	0.25	0.644	0.639	0.637	0.612
	0.5	0.769	0.768	0.770	0.783
Goldhill	0.0625	0.415	0.406	0.403	0.409
	0.125	0.537	0.536	0.534	0.549
	0.25	0.650	0.644	0.646	0.672
	0.5	0.772	0.768	0.768	0.768
Lena	0.0625	0.463	0.451	0.458	0.463
	0.125	0.567	0.558	0.561	0.571
	0.25	0.652	0.648	0.650	0.667
	0.5	0.726	0.736	0.733	0.740
Mandrill	0.0625	0.289	0.280	0.263	0.248
	0.125	0.412	0.418	0.406	0.387
	0.25	0.547	0.548	0.559	0.525
	0.5	0.684	0.702	0.702	0.707
Pepper	0.0625	0.462	0.451	0.437	0.452
	0.125	0.544	0.535	0.534	0.542
	0.25	0.607	0.608	0.604	0.627
	0.5	0.677	0.686	0.685	0.695
Plane	0.0625	0.355	0.349	0.344	0.362
	0.125	0.446	0.445	0.449	0.491
	0.25	0.551	0.563	0.563	0.597
	0.5	0.645	0.667	0.664	0.699

13.4.4   Subjective Evaluation

The subjective evaluation of images is conducted based on visual assessment of the overall quality between images encoded with two different coding systems. The general technique adopted for the subjective assessment is the *forced choice method* (FCM) [Fri95, KS01], similar to an approach taken in [BB01]. The selection of the FCM over the DSCQS [ITU00] method is due primarily to the inadequate assessment structure of DSCQS. One problem with DSCQS is that it has no provision for a side-by-side assessment of two images encoded by two different coding systems, an important factor in the collation of images. Another is that the five level quality scale used for recording subjective responses is prone to deviations which may affect the results. Therefore, the best option is a direct side-by-side comparative test of images along the forced choice methodology. In the FCM, an assessor is given a set number of choices to choose from (i.e., which image looks better, left or right?), but he or she is forced to choose one and one only. The performance of a coder relative to another is then validated by the percentage of subjects who chose it. Two variations of the FCM are used for assessments of the coders: the dichotomous and trichotomous FCM.

13.4.4.1   Dichotomous FCM

In the dichotomous FCM (D-FCM), the subjects must choose between two coded images which he or she believes has the best overall visual quality. This approach measures the performance differences between two coders. However, for a thorough analysis, it is also important to measure the performance similarities between coding systems so as to identify the common characteristics between coders. Performance similarities lead to uncertainty, a problem which arises in comparative subjective tests. Uncertainty may occur when images presented for comparison are perceptually indistinguishable or have vary similar perceptual quality. In instances such as this, the outcome of the D-FCM is, presumably, randomly selected by the subject. Statistically, if the number of participants for the subject test is sufficiently large, then the results of the test will be split approximately 50% each way, provided that the coded images are indeed perceptually indistinguishable or have the same perceptual quality. However, in cases where the sample population for assessment is small, then random selection may not reflect the similar perceptual quality which may exist between two coded images. Thus, the performance of a particular coding system may be misinterpreted as either superior or inferior to another, where in actual fact, they are equal.

13.4.4.2   Trichotomous FCM

In light of the fact that two different coders may produce images of similar visual quality, it is important that these images be categorized as having the same subjective quality.

For a concise performance evaluation of the proposed coder, there is a need to account for this equality in the subjective assessments. The trichotomous FCM (T-FCM) is introduced to alleviate the problem of uncertainty in comparative subjective tests with a small population of subjects. In the T-FCM, subjects have, in addition to the two choices from the D-FCM, a third choice which quantifies both coded images as having subjectively the same overall quality. Bear in mind that the term "same" is made in the context of subjective opinions. It is perfectly conceivable that two images that are visually different may attract the "same" subjective appeal.

13.4.4.3 Assessment arrangements

The subjective assessment is conducted through a Sun Ultra 160 workstation in a room with minimal illumination. Images are displayed on a 21-inch Sun color monitor with 0.24mm dot pitch at resolution 1280×1024 pixels. The viewing distance is set at three times the image height as opposed to six times [ITU00] with the assumption that images are unlikely to be viewed at long distances. Each test session consists of up to 32 sets of test images. The length of assessment for each image set is dependent on the subject as there is no time restriction involved. However, to reduce the effects of fatigue during tests, the overall duration of each test session will not exceed 30 minutes.

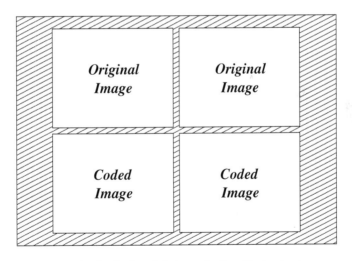

Figure 13.14: Image presentation for the forced choice evaluation. The top two images are original. The bottom two are images derived from two different coders.

The presentation of images for assessment is arranged in the manner as depicted in Figure 13.14. Four images are presented simultaneously in each quadrant of the viewable area of the monitor. Each image has a dimension of 512×480 pixels due to the display resolution. These images are separated by a six pixels deep buffer space.

The buffer spacing helps clarify the definition of image boundaries during testing for the subjects. Pixels in viewable area not occupied by images, which include the buffer space, are considered as background. The background luminance is set to pitch black, a magnitude of zero on a 256 luminance scale. Having a dark background eliminates luminance glare which one might otherwise have with a luminous background. This helps facilitate a proper subjective assessment of details in low luminance regions of images. The top two images in Figure 13.14 are originals and serve as reference for the evaluation of the coded images. The bottom two images are the outputs of two different coding systems to be evaluated. The order (bottom left or bottom right) in which the coded images appear between different presentations is pseudo randomized.

13.4.5 Performance Analysis

Due to the length involved in subjective assessments, some results were collected with less than the desirable number of 15 subjects. The assessment between EBCOT-MSE and the proposed coder in Table 13.8 was conducted with 15 subjects. The results reported in Table 13.9 for EBCOT-CVIS were tabulated from 13 subjects, while the results for EBCOT-XMASK reflected the opinions of seven subjects only. The performance of a coder is measured by the *majority* preference, a term which indicates the support of at least 50% of subjective opinions. A coder with the majority preference is considered to have superior visual performance. Naturally, the greater the majority, the better the performance. In the D-FCM, coder performances are clearly defined given that there are only two choices, hence always a majority. However, for the T-FCM, a majority preference may not materialize since opinions may be split three ways. In such an event, conclusive evidence supporting a coder one way or another may be difficult. However, one could consider the proportion of the distribution of opinions and/or which options received the most votes in deciding coder performances.

13.4.5.1 PC versus EBCOT-MSE

For the most part, the D-FCM results in Table 13.8 favored the proposed coder. Approximately 72% of the D-FCM responses have majority preference for the proposed coder. In the T-FCM tests, this preference dropped to roughly 44%. However, by the same token, preference for the EBCOT-MSE also dropped from 28% to 3%. Opinions in the T-FCM tests indicate that 25% of images tested clearly have the same subjective quality. Another 28% of images lack the majority preference as subjective opinions are split in three directions. Nevertheless, the fact that some of these opinions (6%) are divided suggests that the quality of images may be very similar or indeed the same. A case in point is in the *Lena* image coded at 0.0625 bpp where the result is evenly split. Another is the *Mandrill* image at 0.0625 bpp where both the proposed coder and EBCOT-MSE solicited 40% of the votes each, with the remaining 20% choosing neither. Of the remaining 28%, 13% strongly supported the proposed coder, 6% favored the EBCOT-MSE and

Table 13.8: Subjective results: Proposed Coder (PC) vs. EBCOT-MSE (MSE).

Image	Bitrate	Percentage of scores %				
		Dichotomous		Trichotomous		
	bpp	MSE	PC	MSE	PC	SAME
Barbara	0.0625	27	73	26	73	0
	0.125	0	100	0	100	0
	0.25	7	93	0	87	13
	0.5	40	60	20	47	33
Boat	0.0625	0	100	0	67	33
	0.125	20	80	20	67	13
	0.25	27	73	13	47	40
	0.5	47	53	7	7	86
Bridge	0.0625	33	67	7	27	66
	0.125	33	67	27	60	13
	0.25	20	80	13	74	13
	0.5	47	53	7	7	86
Goldhill	0.0625	33	67	27	53	20
	0.125	60	40	53	33	14
	0.25	60	40	27	13	60
	0.5	67	33	7	0	93
Lena	0.0625	53	47	33	33	34
	0.125	27	73	27	60	13
	0.25	40	60	20	47	33
	0.5	60	40	20	7	73
Mandrill	0.0625	60	40	40	40	20
	0.125	40	60	13	47	40
	0.25	0	100	0	73	27
	0.5	20	80	13	74	13
Pepper	0.0625	27	73	20	53	27
	0.125	40	60	40	27	33
	0.25	20	80	0	67	33
	0.5	67	33	7	0	93
Plane	0.0625	33	67	7	60	33
	0.125	53	47	40	27	33
	0.25	80	20	40	13	47
	0.5	33	67	0	40	60

Table 13.9: Subjective results: Proposed Coder (PC) vs. EBCOT-CIVS (CVIS).

| Image | Bitrate | Percentage of scores % | | | | |
| | | Dichotomous | | Trichotomous | | |
	bpp	CVIS	PC	CVIS	PC	SAME
Barbara	0.0625	23	77	23	69	8
	0.125	15	85	15	85	0
	0.25	8	92	0	85	15
	0.5	85	15	23	8	69
Boat	0.0625	8	92	0	85	15
	0.125	31	69	23	69	8
	0.25	23	77	8	69	23
	0.5	46	54	31	8	61
Bridge	0.0625	31	69	15	62	23
	0.125	46	54	38	46	16
	0.25	15	85	8	69	23
	0.5	23	77	8	46	46
Goldhill	0.0625	23	77	23	69	8
	0.125	15	85	8	77	15
	0.25	31	69	8	54	38
	0.5	38	62	8	31	61
Lena	0.0625	54	46	46	31	23
	0.125	38	62	31	54	15
	0.25	31	69	23	46	31
	0.5	54	46	15	15	70
Mandrill	0.0625	23	77	23	69	8
	0.125	46	54	38	54	8
	0.25	62	38	46	39	15
	0.5	46	54	8	31	61
Pepper	0.0625	31	69	23	69	8
	0.125	15	85	15	54	31
	0.25	38	62	8	69	23
	0.5	77	23	0	0	100
Plane	0.0625	0	100	0	100	0
	0.125	31	69	15	62	23
	0.25	46	54	38	31	31
	0.5	62	38	23	0	77

Table 13.10: Subjective results: Proposed Coder (PC) vs. EBCOT-XMASK (XMASK).

Image	Bitrate	Percentage of scores %				
		Dichotomous		Trichotomous		
	bpp	XMASK	PC	XMASK	PC	SAME
Barbara	0.0625	29	71	29	71	0
	0.125	0	100	0	100	0
	0.25	0	100	0	100	0
	0.5	29	71	29	71	0
Boat	0.0625	0	100	0	100	0
	0.125	0	100	20	100	0
	0.25	29	71	15	57	29
	0.5	43	57	29	42	29
Bridge	0.0625	14	86	0	86	14
	0.125	14	86	14	86	0
	0.25	0	100	0	100	0
	0.5	100	0	57	0	43
Goldhill	0.0625	0	100	0	100	0
	0.125	29	71	29	71	0
	0.25	43	57	43	57	0
	0.5	71	29	0	0	100
Lena	0.0625	0	100	0	86	14
	0.125	0	100	0	100	0
	0.25	71	29	43	14	43
	0.5	29	71	29	14	57
Mandrill	0.0625	14	86	14	86	0
	0.125	14	86	14	86	0
	0.25	57	43	43	43	14
	0.5	57	43	57	43	0
Pepper	0.0625	43	57	29	57	14
	0.125	29	71	15	71	14
	0.25	57	43	29	14	57
	0.5	71	29	43	0	57
Plane	0.0625	0	100	0	100	0
	0.125	0	100	0	100	0
	0.25	0	100	0	100	0
	0.5	29	71	29	42	29

3% took a neutral stance. Overall, for the T-FCM test, the proposed coder performed better than EBCOT-MSE in 57% of the images. In comparison, EBCOT-MSE only out performed the proposed coder in 9% of the images. Performances for the remaining 34% of images for both coders are more or less equal.

13.4.5.2 PC versus EBCOT-CVIS

Results for the D-FCM tests in Table 13.9 showed that subjective preferences for images coded by the proposed coder and EBCOT-CVIS are at 81% and 19% respectively. In the T-FCM assessments, the results indicated clearly - by majority of opinion - that the proposed coder performed better for 59% of images while both coders performed equally well for another 22% of images. The final 19% of images dwell in the contentious region with no majority preferences. Approximately half of these images have opinions leaning towards the perceptual coder while the other half favors the EBCOT-CVIS. With everything considered, in the T-FCM, 66% of images coded with the proposed coder appeared to be better than those coded by EBCOT-CVIS. In contrast, only 9% of images produced by EBCOT-CVIS were visually better than the proposed coder. The remaining 25% of coded images are quantified as visually the same. A point of note is that for the *Bridge* image coded at 0.5 bpp, 46% of subjects believed that the image encoded by the perceptual coder is better than the image encoded by EBCOT-CVIS. However, the same percentage of subjects believe there are no perceived differences in quality between these two images. In this situation, given the ambiguity, it is safer to assume that the visual quality of images from both coders is the same.

13.4.5.3 PC versus EBCOT-XMASK

Subjective responses in Table 13.10 for the D-FCM test has 78% against 22% of coded images in favor of the proposed coder. Again, for the T-FCM test, the results with majority preferences registered a 72% support for the proposed coder. The EBCOT-XMASK coder attracted substantial attention from only 6% of images with another 13% deemed as having same quality. 9% of images fell into contentious category for which 6% are leaning towards the perceptual coder and 3% appears indecisively similar. Overall, the proposed coder came out on top for 78% of coded images while EBCOT-XMASK has 6% with the remaining 16% of coded images having indistinguishable quality. Note that for the T-FCM, the result for the *Lena* image coded at 0.25 bpp is treated as having the "same" quality.

13.4.6 Analysis and Discussion

The subjective results of Tables 13.8, 13.9 and 13.10 only provide performance measure through comparison. Both the PSNR (Table 13.6) and UIQI (Table 13.7) offered quantifiable measurements. The quality rating from PSNR and UIQI can be used to compare the performances of the proposed coder to the benchmark coders. With this one could measure the accuracy of the two objective metrics relative to subjective responses. Results of the objective measures based on PSNR and UIQI painted an image which is somewhat contradictory to that of subjective assessments. When comparing the performance of the proposed coder to the EBCOT-MSE coders on the PSNR scale, there is only a 28% match to the T-FCM results of Table 13.8. This is because all the PSNR results for EBCOT-MSE were better than that of the proposed coder. Surprisingly however, the correlation between PSNR and subjective results for EBCOT-CVIS and EBCOT-XMASK were higher at 47% and 78% match respectively. The high matching responses between PSNR and the T-FCM subjective test for the EBCOT-XMASK is easily explained by the fact that all the PSNR results for EBCOT-XMASK were lower than that of the proposed coder. This bias behavior gives a false impression of the subjective accuracy of PSNR. Similarly, but to a lesser extent, PSNR results between EBCOT-CVIS and the proposed coder are biased towards the latter. It is thus clear that PSNR measurements for images have little or no perceptual bearing in comparative assessments. For the UIQI metric, the match to subjective results was below average with 41% for EBCOT-MSE, 59% for EBCOT-CVIS and 44% for EBCOT–XMASK. However, unlike PSNR, the UIQI results are not entirely one sided. It is able to predict in some instances, two coded images having the same quality. In addition, the UIQI metric is content dependent and therefore appears to have some perceptual orientations. In spite of this, its correlation to subjective responses remains poor, although it may be better than PSNR in assessing the perceived quality of images.

Subjective assessment based on the T-FCM appeared to have cleared up some ambiguity in the D-FCM regarding the quality between images produced by different coders. Tables 13.8 to 13.10 clearly demonstrated the differences in subjective sentiments between both FCMs. For example, a coder found to have superior visual performance in the D-FCM has been interpreted by the T-FCM as having comparable performance. This is especially true for images coded at higher bitrates, as examplify by the pepper image coded at 0.5 bpp and illustrated in Figure 13.15. At high bitrates, the perceptual quality of images from the various coders are exceptionally high, thus making it difficult, if not impossible to identify the coder with the best visual performance. At intermediate bitrates of 0.125 bpp and 0.25 bpp, coder differences are more evident. Within this range of coding rate, the proposed coder performed better than the benchmark coders for most images on both FCM tests. Figures 13.16 and 13.17 show the *Goldhill* and *Barbara* images coded at 0.125 bpp and 0.25 bpp respectively.

In Figure 13.16, images produced by the proposed coder are comparable to those of EBCOT-MSE, though one would find that the former has slightly more texture on the road and sharper details in parts of the background. In comparison with EBCOT-CVIS, the proposed coder has sharper details about the windows. This is in addition to the better background and road texture. With regards to EBCOT-XMASK, the proposed

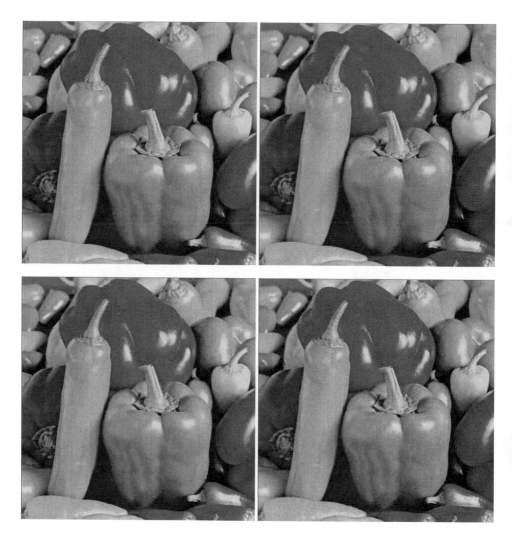

Figure 13.15: *Peppers* image coded at 0.5 bpp. Clockwise from top left: EBCOT-MSE, Proposed Coder, EBCOT-XMASK and EBCOT-CVIS.

coder has less texture on the road and the background trees compared to the EBCOT-XMASK. However, EBCOT-XMASK introduced a substantial amount of blurring into the coded image. For Figure 13.17, the differences between the proposed coder and the benchmark coders are clearer. The proposed coder while maintaining the stripes on the various article of clothing worn by the woman, is able to retain more visually important facial details such as the left eye and the mouth. Substantial blurring can be found in the EBCOT-XMASK image and to a lesser extent, in the EBCOT-MSE image as well. The lost of detail is also apparent in EBCOT-CVIS coupled with some aliasing problems.

Figure 13.16: *Goldhill* image coded at 0.125 bpp. Clockwise from top left: EBCOT-MSE, Proposed Coder, EBCOT-XMASK and EBCOT-CVIS.

At low bitrate (0.0625 bpp), the quality of compressed images, typified in Figure 13.18, is generally quite poor. As a consequence, subjective assessment of low bitrate images is somewhat arbitrary, being more reliant on a subject's taste than anything else. In Figure 13.18, one can be certain that the bottom-right image is the odd one out and perhaps, the least appealing of all, seeing that the whole image has been blurred. The remaining three images are reasonable similar to one another. Despite having questionable image quality, one cannot rule out the practicability of low bitrate coding in light of the constraints placed by the limitations of transmission bandwidths or storage capacity.

Figure 13.17: *Barbara* image coded at 0.25 bpp. Clockwise from top left: EBCOT-MSE, Proposed Coder, EBCOT-XMASK and EBCOT-CVIS.

Figure 13.18: *Bridge* image coded at 0.0625 bpp. Clockwise from top left: EBCOT-MSE, Proposed Coder, EBCOT-XMASK and EBCOT-CVIS.

13.5   Perceptual Lossless Coder

Lossy image coders are separated into two distinct types. The constant bitrate coder attains the optimum picture quality[1] of images at specified bitrates. In contrast, the constant quality coder attains the lowest possible bitrate for specified picture quality levels. While there are different grades for picture quality, ascertaining the quality of images is a highly subjective matter for which there is only one quality level that all observers could consistantly agree on, the level of indistinguishable quality or the Just-Not-Noticeable-Difference (JNND) level. Perceptually Lossless Coding (PLC) is simply constant quality coding at the JNND level. Methods for performing PLC have been well documented [SJ89, HK97, CL95]. However, with past techniques, modeling of the perceptual lossless level has been intertwined with source coding and thereby making portability of PLC model difficult. A possible solution to this is to have modular model, independent of the source coder as in the case of the Visual Pruning function (Figure 13.19).

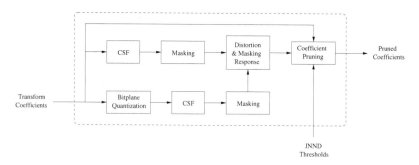

Figure 13.19: Visual pruning function

13.5.1   Coding Structure

The proposed VLGI coder (Figure 13.20) utilizes the EBCOT framework of Taubman [Tau00] implementing the PIDM described in Section 13.2.2. The sensitivity of the vision model requires separate calibration for separate set of transform filters. Therefore, only the bi-orthogonal 9/7 [ABMD92] filter operating in a five level dyadic wavelet decomposition will be employed. This simplification means that both the coder and the vision model will have identical filter sets for their respective multi-resolution decomposition. The perceptual filtering of images is applied through progressive bitplane masking of transform coefficients, from the least significant bit (*lsb*) upwards. For each filtered coefficient, the distortion, D, and the percentage response, R_p, are calculated,

---

[1]Quality in this case is a function of the distortion function used.

Input Image   Compressed Bit Stream

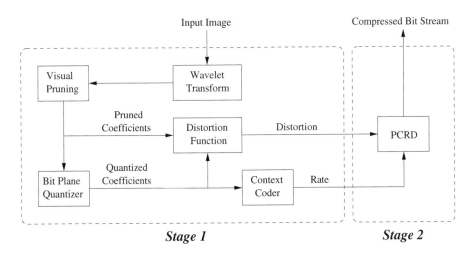

Stage 1   **Stage 2**

Figure 13.20: Generalized structure of the perceptually lossless EBCOT coder. Input images are wavelet transformed (WT), then pruned to remove visually insignificant information. The remaining information is coded in a progressive manner with bitplane quantization and context base adaptive arithmetic coding. The arrangement of the final bitstream is optimized by the *Post Compression Rate-Distortion* (PCRD) function.

then measured against a pre-determined set of holds, T_D and T_P, respectively. D is obtained through (13.30), while R_p is defined as

$$R_p = \frac{\bar{R}_\Theta + \bar{R}_\Upsilon}{R_\Theta + R_\Upsilon}, \tag{13.49}$$

where $\bar{R}$ and R are respectively, the filtered and unfiltered responses taken from (13.20). Θ and Υ signify the orientation and local responses, respectively. If both D and R_p are below their respective thresholds, then the target coefficient is filtered. All transform coefficients are subjected to the filtering operation, except for those in the isotropic lowpass (LL) band. The values of both T_D and T_P are derived from subjective experiments. For each orientation (θ) of each frequency level (l), there is a unique pair of pre-determined thresholds $T_D(\theta, l)$ and $T_P(\theta, l)$, for $\theta = \{\theta_1, \theta_2, \theta_3\}$, $l = \{1, 2, 3, 4, 5\}$.

Optimization is a critical process in ensuring the proper and effective application of the vision model to the VLCI coder. This involves the calibration of model parameters and the determination of the visually lossless threshold levels T_D and T_P. With color vision, this process would require model optimization for each composite color channel Y_D, C_r and C_b. However, to simplify matters here, only the luminance channel (Y_D) will be optimized and used to prune/encode the images in all color channels. The model optimization is conducted in two stages. Stage one collects the subjective data through the Double Stimulus Continuous Quality Scale (DSCQS) [ITU00] testing of 60 images. Stage two then maps the model response to the subjective data with the quasi-

Newton and simplex [SHH62] optimization algorithms. This automated mapping operation produces sub-optimal model parameters which are further refined through manual calibrations and subjective evaluations. For the algorithmic optimization, the Pearson's correlation coefficient [JB87] has been used as the cost function. The single color channel vision model described in Section 13.2.2 has a total of 14 parameters. Optimizing this model is a challenging task considering the fact that the number of permutations is in the order of $P = C^{14}$, with C being the number of possible variations in parameter values. Thus, the solution provided in Table 13.5 reflects, most probably, a set of sub-optimal model parameters. Only an extensive, but lengthy brute force search will provide the optimal solution. Thresholding for the proposed coder is set at the Just-Not-Noticeable-Difference (JNND) level, since the purpose here is to perform perceptually lossless compression. For the threshold tests, one base image (greyscale *Barbara* 512 × 512) has been used to generate 25 distorted images through bitplane filtering. In addition, each distorted image is partitioned into 256 (32 × 32) sub-images, for a total of 6400 sub-images, for the JNND test. Segmenting the test images into smaller portions would provide a better JNND map of images. This would allow the quantification of different JNND levels exhibited by different objects and areas within an image. Comparative forced choice tests [Fri95, KS01] between the distorted and the original sub-images were conducted at a distance equivalent to the image height with one subject. A short viewing distance was chosen in this test with the aim of maintaining a consistent visual performance at varying distances. It is anticipated that if the proposed coder is calibrated to perform visually lossless quality coding at short distances, then it would also perform equally well at longer distances due to the degradation of visual acuity of the HVS. It is also assumed that with the sole subject, any threshold values obtained here are approximated JNND thresholds. Once the JNND map is fully uncovered, the thresholds T_D and T_P can be determined by soliciting the responses (13.23) and (13.49) of the sub-images in the JNND map. i.e., only sub-images at the JNND level will be used to determine the thresholds T_D and T_P. Employing one subject and one base image for the threshold calibration is far from the ideal situation. However, the duration of this calibration process, $d_c = 6400 \times I_b \times S_n \times l$, is dependent on the number of base images, I_b, the number of subjects, S_n, and the average length required to assess each sub-image, l. It is obvious the inclusion of additional subjects and/or base images would increase substantially the JNND threshold calibration process. As a rough indication, assuming a generous assessment length of 5 seconds per sub-image, each subject would require 8.88 continuous hours testing to complete the assessment of all materials related to one base image.

13.5.2   Performance Analysis

Evaluation of the proposed coder is conducted in two parts: subjective evaluation and compression performance. The first assessment measures the visual quality of compressed images in terms of the perceptually lossless (JNND) criterion, while the second

Table 13.11: Compression results for the VLGI, LOCO (LS) and near-lossless LOCO (D2) coders.

Image	Dimensions	VLGI Coder	LOCO (LS)	LOCO (D2)
Actor	512x512	2.990	5.267	3.012
Barbara	512x512	1.986	4.733	2.552
Boat	512x512	1.866	4.250	2.149
Bridge	512x512	3.256	5.500	3.278
Butterfly	512x512	1.500	3.822	1.888
Couple	512x512	2.039	4.262	2.141
Crowd	512x512	1.910	3.914	2.065
Elaine	512x512	2.633	4.898	2.668
Goldhill	512x512	2.424	4.712	2.495
House	512x512	1.904	4.097	2.115
Lake	512x512	2.677	4.983	2.755
Lena	512x512	1.796	4.244	2.097
Mandrill	512x512	3.556	6.036	3.727
Peppers	512x512	2.152	4.489	2.270
F16 Jet	512x512	1.582	3.793	1.830
Sail	512x512	2.810	5.212	2.968
Tank	512x512	2.577	4.808	2.565
Tiffany	512x512	2.309	4.672	2.469
Tulip	512x512	1.841	4.176	2.127
Washington	512x512	2.047	4.129	2.112

measurement focuses on the compression ability of the VLGI coder. Performance of the proposed coder is compared with the LOCO and the near lossless LOCO coders [WSS96]. In this experiment, the near-lossless LOCO encoding has been performed with a maximum error magnitude of 2. Both assessment methods have been carried out with 20 different natural images, covering a variety of image content.

13.5.2.1 Subjective evaluation

For evaluating the visual quality of encoded images, a set double blind test has been conducted with 48 non-expert subjects.[1] A 21-inch SUN monitor with 0.25mm dot pitch running at a screen resolution of 1280×1024 pixels has been used as the primary display device for the test. All tests have been carried out at a viewing distance equivalent to

---

[1] A double-blind test is a control group test where neither the evaluator nor the subject knows which items are controls.

twice the image height in a room with minimal illumination [ITU00]. In this subjective test, each of the 20 source images is represented by three test images: original (A); VLGI coder (B) and near-lossless LOCO (C). Each test session is composed of 20 sets of test images, presented in pairs, side-by-side. The order of presentation within a test session, image 1 to image 20, is randomized. Similarly, the order of presentation within each pair of test images is also randomized. However, since only two images may be presented in each test set, the possible number of variations in pairs, ϑ, is 9, i.e., $\vartheta = \{AA, AB, AC, BA, BB, BC, CA, CB, CC\}$. Hence, while the images are randomized left to right, i.e., AB and BA for example, they are also randomized between images A, B and C. This method of random sampling with replacement leads to pairs such as AA, BB and CC. Though testing identical images may not contribute directly to the assessment of the coders, it does provide some ideas as to the realibility of the subjective test results. For example, the impact of scanning effects and fatigue may be measured from results of identical image tests. Finally, since the purpose of the experiment here is to acertain whether or not, the VLGI coder does indeed operate at the JNND level, there is need to measure directly if images encoded by proposed coder are indistinguishable to their originals. Therefore, all subjective tests have been run as three way tests instead of a two way forced-choice tests. The possible output choices for the three way tests are: *left, right* or *either*. A full description of this subjective testing procedure is given in [WWT+04].

13.5.2.2   Results

Compression performance of the VLGI coder versus the two benchmark coders are given in Table 13.11. The results suggest unanimously that the proposed coder has superior compression power to both the LOCO and the near-lossless LOCO ($d=2$) coders [WSS96]. As far as visual quality is concerned, there are no visible impairments in the images coded by the VLGI coder (Figures 13.21, 13.22, 13.23 and 13.24), irrespective of differences in PSNR values to the near lossless LOCO coder. This is further reinforced by subjective results listed in Tables 13.12 and 13.13.

13.5.2.3   Discussions

As with all experiments, it is important to analyze the data collected not merely for the purpose of evaluating the behavior or performance of a system in question, but also to determine if the results of an experiment are indeed reliable. The testing of identical image pairs (AA, BB and CC) provides a measurement for the scanning effect which to some extent, may be affected by fatigue. The scanning effect is a result of subjects selecting the first image he or she sees. The order of scanning is dependent on individuals which are from either left to right or right to left. From table 13.14, one could surmise that there is a noticeable bias toward the right image with a 4.2% difference over the left. Irrespective of this, the overall variation is marginal and is effectively negated by

Table 13.12: Subjective results — Trichotomous test. Image coded with the VLGI coder (PLC), Original Image (ORG), Indistinguishable/Neutral (NEU).

Image	Selection counts			No. of
	PLC	ORG	NEU	Subjects
Actor	10	5	2	17
Barbara	8	1	1	10
Boats	4	4	2	10
Bridge	8	6	2	16
Butterfly	3	4	6	13
Couple	5	4	2	11
Crowd	2	1	2	5
Elaine	4	2	1	7
Goldhill	4	0	5	9
House	5	3	2	10
Lake	1	3	4	8
Lena	7	6	2	15
Mandrill	6	1	9	16
Peppers	4	4	3	11
F16 Jet	5	7	3	15
Sail	4	1	7	12
Tank	4	4	2	10
Tiffany	2	4	2	8
Tulip	2	1	2	5
Washington	4	1	2	7
Total	92	62	61	215

randomized design of the subjective test.

The results in Table 13.12 clearly favor the VLGI coder in terms of overall subjective preference. However, since the ultimate purpose of this test has been to measure whether or not, the VLGI coder does indeed encode images to visually lossless quality, one should consider the neutral selection and the VLGI selection as favorable outcomes which in this case, amount to 71.2 % of the total selections. The neutral selection is a direct measurement of indifference in image quality while the remaining two choices (VLGI and ORG) can be considered indirect measures through mutual cancellation with a two way split. It is true that while the attraction of 50% more selections for VLGI coded images over the original images may contradict the imperceptible quality of coding, this at the very least suggests that the VLGI coder appears to enhance the subjective quality of images. From Table 13.13, the overall preference for the VLGI coder only accounts for 37% of selections. When combined with neutral scores, this increases to

Table 13.13: Subjective results — Trichotomous test. Image coded with the VLGI coder (PLC), near-lossless LOCO coder (nLC), Indistinguishable/Neutral (NEU).

Image	Selection counts			No. of
	PLC	nLC	NEU	Subjects
Actor	1	5	3	9
Barbara	8	3	1	13
Boats	5	4	3	12
Bridge	6	5	2	13
Butterfly	5	5	2	12
Couple	8	3	2	13
Crowd	4	4	8	16
Elaine	5	8	2	15
Goldhill	5	6	2	13
House	3	3	0	6
Lake	6	3	3	12
Lena	2	7	3	12
Mandrill	1	7	2	10
Peppers	2	4	3	9
F16 Jet	8	2	0	10
Sail	2	3	5	10
Tank	2	4	3	9
Tiffany	3	4	2	9
Tulip	8	2	3	13
Washington	1	8	6	15
Total	85	90	55	230

Table 13.14: Identical image test — E (Either), R (Right) and L (Left).

Image Set	E	R	L	Total
AA	23	42	46	111
BB	31	40	30	101
CC	27	41	34	102
Total	81	123	110	314
%	25.8	39.2	35	100

60.1%. Though preferences for the near-lossless LOCO coder are higher than the VLGI coder, the difference is reasonably small that one could consider their visible quality as equivalent. A comparative assessment between original images and those encoded by

Table 13.15: Subjective results — Trichotomous test. Image coded with the near-lossless LOCO coder (nLC), Original Image (ORG), Indistinguishable/Neutral (NEU).

Image	Selection counts			No. of
	ORG	nLC	NEU	Subjects
Actor	3	1	1	5
Barbara	4	9	3	16
Boats	6	3	3	12
Bridge	2	3	1	6
Butterfly	3	3	1	7
Couple	7	0	3	10
Crowd	3	2	1	6
Elaine	5	3	2	10
Goldhill	3	3	2	8
House	3	9	1	13
Lake	4	4	1	9
Lena	2	5	3	10
Mandrill	2	4	2	8
Peppers	0	1	10	11
F16 Jet	3	2	1	6
Sail	3	5	4	12
Tank	5	3	2	10
Tiffany	5	7	4	16
Tulip	1	3	8	12
Washington	0	6	8	14
Total	64	76	61	201

the near-lossless LOCO coder also suggests, with some strength, an indistinguishable level of encoding by the near-lossless LOCO coder (d=2) (Table 13.15). Although the quality of the near lossless LOCO coder is excellent, and while its compression rate is reasonably good, it has no facility for determining the JNND level. Hence, it is unable to provide visually lossless compression in an automated fashion, at a competitive bitrate.

13.6 Summary

The transform coding of images is a spectral based lossy compression scheme. It consists of three core components which are frequency transformation (or decomposition), quantization and entropy coding. Each of these core components has multiple forms of realization, leading to very efficient coders under the right combinations. A particular class of transform coder which is of interest of late is the hierarchical bit-plane coder popularized by the EZW (Embedded Zerotree Wavelet). The performance of the EZW and its various derivatives are generally superior to that of a traditional block-based coder such as the DCT used in the JPEG standard [Wal91]. Amongst its many virtues, hierarchical bit-plane coding allows for scalability and appears to produce images with superior visual quality corresponding to that of perceptual coders.

Perceptual coding addresses the issue of visual redundancy in image data. Though the reduction of statistical redundancy is important, the removal of perceptual redundancy is considered more relevant since what the HVS cannot see, it would not miss. Perceptual coding comes in two flavors, rate constrained and quality driven. The first approach encodes images to the best visual quality, relative to a specific vision model, for a given bitrate. The second approach encodes images to visually lossless quality with minimum bitrates. In the proposed coder, the former method was adopted with the EBCOT coding structure. EBCOT utilizes a two pass compression system. The first pass generates the rates and distortions associated with the various levels of quantization. The second pass applies the PCRD optimization, with the rate and distortion information accumulated in the first pass, to determine the final compressed bitstream which yields the best R-D trade-offs. In the current proposition, adaptation of vision modeling to the EBCOT coder is realized through the embedding of a vision model into the distortion function. The resulting perceptual image distortion metric then measures perceived errors due to quantization which in turn regulates the PCRD optimizer and generates compressed images with the optimal visual quality (relative to the vision model), for a given bitrate. Unlike the method in [ZDL00], this approach places no additional overheads in the bitstream, thus it is fully standard compliant. Further, it also maintains the scalability property of the coder. As far as vision model is concerned, the model presented in Subsections 13.2.2 and 13.2.3 is more advanced than previous implementation within the EBCOT framework [Tau00, ZDL00].

References

[ABMD92]   M. Antonini, M. Barlaud, P. Mathieu, and I. Daubechies. Image Coding Using Wavelet Transform. *IEEE Trans. on Image Processing*, 1(2):205–220, April 1992.

[AP92]   A. J. Ahumada and H. A. Peterson. Luminance-model-based DCT quantization for color image compression. *in B. E. Rogowitz, Ed., Human Vision, Visual Processing and Digital Display III, Proc. SPIE*, 1666:365–374, 1992.

[BB99]   C. H. Brase and C. P. Brase. *Understandable Statistics: Concepts and Methods.* Boston: Houghton Mifflin Company, 6th ed., 1999.

[BB01]   V. Baroncini and M. Buxton. *Preliminary Results of Assessment of Responses to Digital Cinema Call for Proposals*, July 2001. ISO/IEC JTC 1/SC 29/WG 11 N4241.

Original

nLOCO coder, *error* = 2

PLC

Figure 13.21: *Actor* image. PSNR: *nLOCO* - 45.182 dB; *PLC* - 39.049 dB.

Original   *nLOCO* coder, *error* = 2

PLC

Figure 13.22: *Lena* image. PSNR: *nLOCO* - 45.146 dB; *PLC* - 41.924 dB.

Original   *nLOCO* coder, *error* = 2

PLC

Figure 13.23: *Goldhill* image. PSNR: *nLOCO* - 45.154 dB; *PLC* - 41.094 dB.

Original   *nLOCO* coder, *error* = 2

PLC

Figure 13.24: *Mandrill* image. PSNR: *nLOCO* - 45.124 dB; *PLC* - 36.931 dB.

[BL97]   M. H. Brill and J. Lubin. Methods and Apparatus for Assessing the Visibility of Differenes Between Two Image Sequences, Dec. 1997. US Patent No. 5,694,491.

[Bor57]   E. G. Boring. *A History of Experimental Psychology.* New York: Appleton Century Crofts, Inc., 1957.

[CL95]   C.-H. Chou and Y.-C. Li. A Perceptually Tuned Subband Image Coder Based on the Measure of Just-Noticable-Distortion Profile. *IEEE Trans. on Circuits and Systems for Video Technology*, 5(6):467–476, Dec. 1995.

[Cla85]   R. Clarke. *Transform Coding of Images.* Orlando, FL: Academic Press, 1985.

[Cor90]   T. N. Cornsweet. *Visual Perception.* Orlando, FL: Academic Press, 1990.

[CP84]   W. H. Chen and W. K. Pratt. Scene Adaptive Coder. *IEEE Trans. on Communications*, 32(3):225–232, Mar. 1984.

[CZ01]   E. K. P. Chong and S. H. Żak. *An Introduction to Optimization.* Hoboken, NJ: Wiley, 2nd ed., 2001.

[DAT82]   R. L. De Valois, D. G. Albrecht, and L. G. Thorell. Spatial Frequency Selectivity of Cells in Macaque Visual Cortex. *Vision Research*, 22(5):545–559, 1982.

[DKMM99]   Y. Deng, C. Kenney, M. S. Moore, and B. S. Manjunath. Peer Group Filtering and Perceptual Color Image Quantization. In *Proc. of IEEE Int. Symp. on Circuits and Systems*, 4:21–24, 1999.

[DS89]   J. E. Dennis Jr. and R. B. Schnabel. A View of Unconstrained Optimization. In G. L. Nemhauser, A. H. G. Rinnooy Kan, and M. J. Todd, Eds., *Handbooks in Operations Research and Management Science: Optimization*, 1:1–72. Amsterdam: Elsevier Science Publishers, 1989.

[DSKK58]   R. L. De Valois, C. J. Smith, A. J. Karoly, and S. T. Kitai. Electrical Responses of Primate Visual System. I. Different Layers of Macaque Lateral Geniculate Nucleus. *Journal of Comparative Physiology and Psychology*, 51:622–668, 1958.

[DZLL00]   S. Daly, W. Zeng, J. Li, and S. Lei. Visual Masking in Wavelet Compression for JPEG2000. In *IS&T/SPIE Conf. on Image and Video Communication and Processing*, 3974:66–80, Jan. 2000.

[FB94]   J. M. Foley and G. M. Boynton. A New Model of Human Luminance Pattern Vision Mechanisms: Analysis of the Effects of Pattern Orientation Spatial Phase and Temporal Frequency. In T. A. Lawton, Ed., *Computational Vision Based on Neurobiology, Proceedings of SPIE*, 2054:32–42, 1994.

[FCH97]   W. C. Fong, S. C. Chan, and K. L. Ho. Determination of Frequency Sensitivities for Perceptual Subband Image Coding. *IEE Electronics Letters*, 33(7):581–582, March 1997.

[Fle81]   R. Fletcher. *Practical Methods of Optimization.* Hoboken, NJ: Wiley, 2nd ed., 1981.

[Fol94]   J. M. Foley. Human Luminance Pattern-Vision Mechanisms: Masking Experiments Require a New Model. *Journal of the Optical Society of America A*, 11(6):1710–1719, June 1994.

[Fri95]   L. Friedenberg. *Psychological Testing: Design, Analysis and Use.* New York: Allyn & Bacon, A Simon & Schuster Company, 1995.

[GMW81]   P. E. Gill, W. Murray, and M. H. Wright. *Practical Optimization*. Orlando, FL: Academic Press, 1981.

[Hem97]   S. S. Hemami. Visual Sensitivity Consideration for Subband Coding. In *Proceedings of the 31st Asilomar Conf. on Signals, Systems and Computers*, 1:652–656, 1997.

[HK97]   I. Höntsch and L. Karam. APIC: Adaptive Perceptual Image Coding Based on Subband Decomposition with Locally Adaptive Perceptual Weighting. In *Proc. of IEEE Int. Conf. on Image Processing*, 1:37–40, 1997.

[HK00]   I. Höntsch and L. J. Karam. Locally Adaptive Perceptual Image Coding. *IEEE Trans. on Image Processing*, 9(9):1472–1483, Sept. 2000.

[HK02]   I. Höntsch and L. J. Karam. Adaptive Image Coding with Perceptual Distortion Control. *IEEE Trans. on Image Processing*, 11(3):213–222, Mar 2002.

[HKS97]   I. Höntsch, L. J. Karam, and R. J. Safranek. A Perceptually Tuned Embedded Zerotree Image Coder. In *Proc. of IEEE Int. Conf. on Image Processing*, 1:41–44, 1997.

[HN97]   M. J. Horowitz and D. L. Neuhoff. Image Coding by Matching Pursuit and Perceptual Pruning. In *Proc. of IEEE Int. Conf. on Image Processing*, 3:654–657, 1997.

[ISO00a]   ISO/IEC JTC 1/SC 29. *Information Technology – JPEG 2000 image coding system – Part 2: Extension*, 2000. ISO/IEC 15444-1:2000.

[ISO00b]   ISO/IEC JTC1/SC9/WG1. JPEG2000 Verification Model 8.0 Software, 2000. ISO/IEC JTC1/SC9/WG1 N1819.

[ITU94a]   ITU-R Recommendation BT.814-1. Specifications and Alignment Procedures for Setting of Brightness and Contrast of Displays. Technical report, ITU-R, July 1994.

[ITU94b]   ITU-R Recommendation BT.815-1. Specification of a Signal for Measurement of the Contrast Ratio of Displays. Technical report, ITU-R, July 1994.

[ITU00]   ITU-R Recommendation BT.500-10. Methodology for the Subjective Assessment of the Quality of Television Pictures. Technical report, ITU-R, Mar. 2000.

[Jai89]   A. K. Jain. *Fundamentals of Digital Image Processing*. Englewood Cliffs, NJ: Prentice–Hall, 1989.

[JB87]   R. A. Johnson and G. K. Bhattacharyya. *Statistics: Principles and Methods*. Hoboken, NJ: Wiley, 2nd ed., 1987.

[JN84]   N. S. Jayant and P. Noll. *Digital Coding of Waveforms: Principles and Applications to Speech and Video*. Englewood Cliffs, NJ: Prentice–Hall, 1984.

[KS99]   I. Kopilovic and T. Sziranyi. Non-Linear Scale-Selection for Image Compression Improvement Obtained by Perceptual Distortion Criteria. In *Proc. of IEEE Int. Conf. on Image Analysis and Processsing*, 197–202, 1999.

[KS01]   R. M. Kaplan and D. P. Saccuzzo. *Psychological Testing: Principles, Applications and Issues*. Belmont, CA: Thomson Learning Inc., 5th ed., 2001.

[LF80]   G. E. Legge and J. M. Foley. Contrast Masking in Human Vision. *Journal of the Optical Society of America*, 70(12):1458–1471, Dec. 1980.

[Lim78]   J. Limb. On the Design of Quantisers for DPCM Coders - A Functional Relation-
            ship Between Visibility, Probability and Masking. *IEEE Trans. on Communica-
            tions*, 26:573–578, 1978.

[LJ97]   J. Li and J. S. Jin. Structure-Related Perceptual Weighting: A Way to Improve Em-
            bedded Zerotree Wavelet Image Coding. *IEE Electronics Letters*, 33(15):1305–
            1306, July 1997.

[LK98]   Y.-K. Lai and C.-C. Kuo. Perceptual Image Compression with Wavelet Transform.
            In *Proc. of IEEE Int. Symp. on Circuits and Systems*, 4:29–32, 1998.

[LTK99]   T.-H. Lan, A. H. Tewfik, and C.-H. Kuo. Sigma Filtered Perceptual Image Coding
            at Low Bit Rates. In *Proc. of IEEE Int. Conf. on Image Processing*, 2:371–375,
            1999.

[LW98]   C.-M. Liu and C.-N. Wang. On the Perceptual Interband Correlation for Octave
            Subband Coding. In *Proc. of IEEE Int. Conf. on Image Processing*, 4:166–169,
            1998.

[LZFW00]   K. T. Lo, X. D. Zhang, J. Feng, and D. S. Wang. Univeral Perceptual Weighted
            Zerotree Coding for Image and Video Compression. *IEE Proc. on Vision, Image
            and Signal Processing*, 147(3):261–265, June 2000.

[MH98]   M. Miloslavski and Y.-S. Ho. Zerotree Wavelet Image Coding Base on the Human
            Visual System Model. In *Proc. of IEEE Asia-Pacific Conference on Circuit and
            Systems*, 57–60, 1998.

[Mic64]   A. A. Michelson. *Experimental Determination of the Velocity of Light*. Minneapo-
            lis, MN: Lund Press, 1964.

[ML95]   A. Mazzarri and R. Leonardi. Perceptual Embedded Image Coding using Wavelet
            Transforms. In *Proc. of IEEE Int. Conf. on Image Processing*, 1:586–589, 1995.

[Nad00]   M. Nadenau. *Integration of Human Colour Vision Models into High Quality Image
            Compression*. Ph.D. thesis, EPFL, CH-1015 Lausanne, Switzerland, 2000.

[NLS89]   K. N. Ngan, K. S. Leong, and H. Singh. Adaptive Cosine Transform Coding
            of Images in Perceptual Domain. *IEEE Trans. on Acoustics, Speech and Signal
            Processing*, 37(11):1743–1750, Nov 1989.

[NP77]   A. Netravali and B. Prasada. Adaptive Quantisation of Picture Signals Using Spa-
            tial Masking. *Proceedings of the IEEE*, 65(4):536–548, April 1977.

[NR77]   A. Netravali and C. Rubinstein. Quantisation of Colour Signals. *Proceedings of
            the IEEE*, 65(8):1177–1187, Aug. 1977.

[OS95]   T. P. O'Rourke and R. L. Stevenson. Human Visual System Based Wavelet De-
            composition for Image Compression. *Journal of Visual Communication and Image
            Representation*, 6(2):109–121, June 1995.

[PAW93]   H. A. Peterson, A. J. Ahumada, and A. B. Watson. An Improved Detection Model
            for DCT Coefficient Quantization. *SPIE Proceedings*, 1913:191–201, 1993.

[Pel90]   E. Peli. Contrast in Complex Images. *Journal of the Optical Society of America
            A*, 7(10):2032–2040, Oct. 1990.

[PH71]   J. M. Parkinson and D. Hutchinson. An Investigation into the Efficiency of Vari-
            ants on the Simplex Method. In F. A. Lootsma, Ed., *Numerical Methods for Non-
            Linear Optimization*. Orlando, FL: Academic Press, 1971.

[PMH96]   T. N. Pappas, T. A. Michel, and R. O. Hinds. Supra-Threshold Perceptual Image Coding. In *Proc. of IEEE Int. Conf. on Image Processing*, 1:237–240, 1996.

[RH99]   M. G. Ramos and S. S. Hemami. Activity Selective SPIHT Coding. In *Proc. of SPIE Conf. on Visual Communications and Image Processing*, 3653:315–326, 1999.

[RH00]   M. G. Ramos and S. S. Hemami. Perceptual Quantization for Wavelet-Based Image Coding. In *Proc. of IEEE Int. Conf. on Image Processing*, 1:645–648, 2000.

[RLCW00]   A. M. Rohaly, J. Libert, P. Corriveau, and A. Webster. Final Report from the Video Quality Experts Group on the Validation of Objective Models of Video Quality Assessment. Technical report, VQEG, Mar. 2000.

[Roh00]   A. M. Rohaly et al. Video Quality Experts Group: Current Results and Future Directions. In *Proc. of SPIE 2000 Conference on Visual Communications and Image Processing*, 4067:742–753, June 2000.

[RW96]   R. Rozenholtz and A. B. Watson. Perceptual Adaptive JPEG Coding. In *Proc. of IEEE Int. Conf. on Image Processing*, 1:901–904, 1996.

[SFAH92]   E. P. Simoncelli, W. T. Freeman, E. H. Adelson, and D. J. Heeger. Shiftable Multiscale Transform. *IEEE Trans. on Information Theory*, 38(2):587–607, Mar. 1992.

[SGE02]   D. Santa-Cruz, R. Grosbois, and T. Ebrahimi. JPEG 2000 Performance Evaluation and Assessment. *Signal Processing: Image Communication*, 17(1):113–130, Jan. 2002.

[SHH62]   W. Spendley, G. R. Hext, and F. R. Himsworth. Sequential Application of Simplex Designs in Optimisation and Evolutionary Operation. *Technometrics*, 4(4):441–461, Nov. 1962.

[SJ89]   R. Safranek and J. Johnston. A Perceptually Tuned Sub-band Image Coder with Image Dependent Qunatization and Post-quantization Data Compression. In *Proc. of IEEE Int. Conf. on Acoustics, Speech and Signal Processing*, 1945–1948, 1989.

[Tau00]   D. Taubman. High Performance Scalable Image Compression with EBCOT. *IEEE Trans. on Image Processing*, 7(9):1158–1170, July 2000.

[TH94a]   P. C. Teo and D. J. Heeger. Perceptual Image Distortion. *Proceedings of SPIE*, 2179:127–141, 1994.

[TH94b]   P. C. Teo and D. J. Heeger. Perceptual Image Distortion. In *Proc. of IEEE Int. Conf. on Image Processing*, 2:982–986, Nov. 1994.

[TNH97]   T. D. Tran, T. Q. Nguyen, and Y. H. Hu. A Perceptually-Tuned Block-Transform-Based Progressive Transmission Image Coder. In *Proc. of 31st Asilomar Conf. on Signal, Systems and Computers*, 2:1000–1004, 1997.

[TPN96]   S. H. Tan, K. K. Pang, and K. N. Ngan. Classified Perceptual Coding with Adaptive Quantization. *IEEE Trans. on Circuits and Systems for Video Technology*, 6(4):375–388, Aug 1996.

[Tur00]   Turbolinux. Enfuzion Manual, Mar. 2000. http://www.turbolinux.com.

[usc]   USC-SIPI Image Database. http://sipi.usc.edu/services/database/.

[van96]   C. J. van den Branden Lambrecht. *Testing Digital Video System and Quality Metrics based on Perceptual Models and Architecture*. Ph.D. thesis, EPFL, CH-1015 Lausanne, Switzerland, May 1996.

[Wal91]   G. K. Wallace. The JPEG Still Picture Compression Standard. *Communication of ACM*, 34:30–44, April 1991.

[Wan95]   B. A. Wandell. *Foundations of Vision*. Suderland, MA: Sinauer Associates, Inc, 1995.

[Wat93]   A. B. Watson. DCTune: A Technique for Visual Optimization of DCT Quantization Matrices for Individual Images. *Society for Information Display Digest of Technical papers XXIV*, 946–949, 1993.

[Wat94]   A. B. Watson. Perceptual Optimization of DCT Colour Quantization Matrices. In *Proc. of IEEE Int. Conf. on Image Processing*, 100–104, 1994.

[Wat00]   A. B. Watson. Visual Detection of Spatial Contrast Patterns: Evaluation of Five Simple Models. *Optics Express*, 6(1):12–33, 2000.

[WB01]   Z. Wang and A. C. Bovik. Embedded Foveation Image Coding. *IEEE Trans. on Image Processing*, 10(10):1397–1410, Oct. 2001.

[WB02]   Z. Wang and A. C. Bovik. A Universal Image Quality Index. *IEEE Signal Processing Letters*, 9(3):81–84, Mar. 2002.

[WHMM99]   A. Watson, J. Hu, J. F. McGowan III, and J. B. Mulligan. Design and Performance of a Digital Video Quality Metric. In *Proc. of the SPIE Conference on Human Vision and Electronic Imaging IV*, 3644:168–174, 1999.

[Win98]   S. Winkler. A Perceptual Distortion Metric for Digital Color Images. In *Proc. of IEEE Int. Conf. on Image Processing*, 3:399–403, 1998.

[Win99]   S. Winkler. Perceptual Distortion Metric for Digital Color Video. In *Proc. of the SPIE Conference on Human Vision and Electronic Imaging IV*, 3644:175–184, 1999.

[Win00]   S. Winkler. *Vision Models and Quality Metrics for Image Processing Applications*. Ph.D. thesis, EPFL, CH-1015 Lausanne, Switzerland, December 2000.

[WLB95]   S. J. P. Westen, R. L. Lagendijk, and J. Biemond. Perceptual Image Quality Based on a Multiple Channel HVS Model. In *Proc. of IEEE Int. Conf. on Acoustics, Speech and Signal Processing*, 4:2351–2354, 1995.

[WS97]   A. B. Watson and J. A. Solomon. A Model of Visual Contrast Gain Control and Pattern Masking. *Journal of the Optical Society of America A*, 14(9):2379–2391, 1997.

[WSS96]   M. Weinberger, G. Seroussi, and G. Sapiro. LOCO-I: A Low Complexity, Context-Based, Lossless Image Compression Algorithm. In *Proceedings of IEEE Data Compression Conference*, 140–149, 1996.

[WWT+04]   C. White, H. R. Wu, D. Tan, R. L. Martin, and D. Wu. *Experimental Design for Digital Image Quality Assessment*. Zetland, NSW, Australia: Epsilon Publishing, 2004.

[YW98]   M. Yuen and H. R. Wu. A Survey of Hybrid MC/DPCM/DCT Video Coding Distortions. *Signal Processing*, 70(3):247–278, Nov. 1998.

[YWWC02]   Z. Yu, H. R. Wu, S. Winkler, and T. Chen. Vision-Model-Based Impairment Metric to Evaluate Blocking Artifacts in Digital Video. *Proc. of the IEEE*, 90(1):154–169, Jan. 2002.

[YZF99]   K. H. Yang, W. Zhu, and A. F. Faryar. Perceptual Quantizatin for Predictive Coding in Images. In *Proc. of IEEE Int. Conf. on Image Processing*, 2:381–385, 1999.

[ZDL00]   W. Zeng, S. Daly, and S. Lei. Point-Wise Extended Visual Masking for JPEG-2000 Image Compression. In *Proc. of IEEE Int. Conf. on Image Processing*, 1:657–660, 2000.

[Zen99]   Y. Zeng. Perceptual Segmentation Algorithm and its Application to Image Coding. In *Proc. of IEEE Int. Conf. on Image Processing*, 2:820–824, 1999.

[ZZQ97]   W. Zheng, H. Zhu, and Z. Quan. Lattice Vector Quantization of Wavelet Coefficients with Perceptual Visibility Model and Hybrid Entropy Coding. In *IEEE Int. Symp. on Circuits and Systems*, 1117–1120, 1997.

Chapter 14

Foveated Image and Video Coding

Zhou Wang[†] and Alan C. Bovik[‡]

†   *University of Texas at Arlington, U.S.A.*
‡   *University of Texas at Austin, U.S.A.*

The human visual system (HVS) is highly space-variant in sampling, coding, processing, and understanding of visual information. The visual sensitivity is highest at the point of fixation and decreases dramatically with distance from the point of fixation. By taking advantage of this phenomenon, foveated image and video coding systems achieve increased compression efficiency by removing considerable high-frequency information redundancy from the regions away from the fixation point without significant loss of the reconstructed image or video quality.

This chapter has three major purposes. The first is to introduce the background of the foveation feature of the HVS that motivates the research effort of foveated image processing. The second is to review various foveation techniques that have been used to construct image and video coding systems. The third is to provide in more detail a specific example of such systems, which deliver rate scalable codestreams ordered according to foveation-based perceptual importance, and has a wide range of potential applications, such as video communications over heterogeneous, time-varying, multi-user and interactive networks.

14.1   Foveated Human Vision and Foveated Image Processing

Let us start by looking at the anatomy of the human eye. A simplified structure is illustrated in Figure 14.1. The light that passes through the optics of the eye is projected onto the retina and sampled by the photoreceptors in the retina. The retina has two major types of photoreceptors known as cones and rods. The rods support achromatic vision in low level illuminations and the cone receptors are responsible for daylight

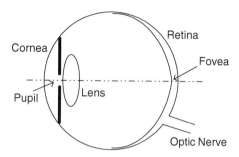

Figure 14.1: Structure of the human eye.

vision. The cones and rods are non-uniformly distributed over the surface of the retina [GB95, Wan95]. The region of highest visual acuity is the fovea, which contains no rods but has the highest concentration of approximately 50,000 cones [Wan95]. Figure 14.2 shows the variation of the densities of photoreceptors with retinal eccentricity, which is defined as the visual angle (in degree) between the fovea and the location of the photoreceptor. The density of the cone cells is highest at zero eccentricity (the fovea) and drops rapidly with increasing eccentricity. The photoreceptors deliver data to the plexiform layers of the retina, which provide both direct and inter-connections from the photoreceptors to the ganglion cells. The distribution of ganglion cells is also highly non-uniform as shown in Figure 14.2. The density of the ganglion cells drops even faster than the density of the cone receptors. The receptive fields of the ganglion cells also vary with eccentricity [GB95, Wan95].

The density distributions of cone receptors and ganglion cells play important roles in determining the ability of our eyes to resolve what we see. When a human observer gazes at a point in a real-world image, a variable resolution image is transmitted through the front visual channel into the information processing units in the human brain. The region around the point of fixation (or foveation point) is projected onto the fovea, sampled with the highest density, and perceived by the observer with the highest contrast sensitivity. The sampling density and the contrast sensitivity decrease dramatically with increasing eccentricity. An example is shown in Figure 14.3, where Figure 14.3(a) is the original *Goldhill* image and Figure 14.3(b) is a foveated version of that image. At certain viewing distance, if attention is focussed at the man at the lower part of the image, then the foveated and the original images are almost indistinguishable.

Despite the highly space-variant sampling and processing features of the HVS, traditional digital image processing and computer vision systems represent images on uniformly sampled rectangular lattices, which have the advantages of simple acquisition, storage, indexing and computation. Nowadays, most digital images and video sequences are stored, processed, transmitted and displayed in rectangular matrix format, in which each entry represents one sampling point. In recent years, there has been growing interest in research work on *foveated image processing* [Sch77, Sch80, Bur88, BS89,

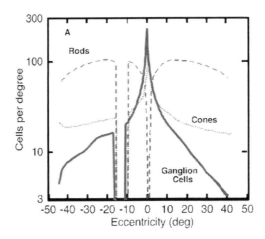

Figure 14.2: Photoreceptor and ganglion cell density versus retinal eccentricity [GB95].

RS90, Wei90, LW90, ZS93, SBC93, BSW93, WOBS94, TEHM96, CAS96, KG96, AG96, RR96, CY97, BW98, GP98, KB98, KGG99, Kin99, ECvdSMZ00, CMY00, Lee00, LB00, DMRC01, LPB01, WB01b, WB01a, WBL01, Wan01, RWB01, GP02, LPB02, Liu02, SLWB02, Sab02, KA02, SEB03, WLB03, LB03, LPKB03], which is targeted at a number of application fields. Significant examples include image quality assessment [WBL01, LPB02], image segmentation [Kin99], stereo 3D scene perception [KB98], volume data visualization [LW90], object tracking [ECvdSMZ00], and image watermarking [KA02]. Nevertheless, the majority of research has been focused on foveated image and video coding, communication and related issues. The major motivation is that considerable high frequency information redundancy exists in the peripheral regions, thus more efficient image compression can be obtained by removing or reducing such information redundancy. As a result, the bandwidth required to transmit the image and video information over communication channels is significantly reduced. Foveation techniques also supply some additional benefits in visual communications. For example, in noisy communication environments, foveation provides a natural way for unequal error-protection of different spatial regions in the image and video streams being transmitted. Such an error-resilient coding scheme has shown to be more robust than protecting all the image regions equally [Lee00, LPKB03]. For another example, in an interactive multi-point communication environment where information about the foveated regions at the terminals of the communication networks is available, higher perceptual quality images can be achieved by applying foveated coding techniques [SLWB02].

Perfect foveation of discretely-sampled images with smoothly varying resolution turns out to be a difficult theoretical as well as implementation problem. In the next section, we review various practical foveation techniques that approximate perfect

(a)   (b)

Figure 14.3: Sample foveated image. (a) original *Goldhill* image; (b) foveated *Glodhill* image.

foveation. Section 14.3 discusses a continuously rate-scalable foveated image and video coding system that has a number of good features in favor of network visual communications.

14.2   Foveation Methods

The foveation approaches proposed in the literature may be roughly classified into three categories: geometric method, filtering-based method, and multiresolution method. These methods are closely related and the third method may be viewed as a combination of the first two.

14.2.1   Geometric Methods

The general idea of the geometric methods is to make use of the foveated retinal sampling geometry. We wish to associate such a highly non-uniform sampling geometry with a spatially-adaptive coordinate transform, which we call the foveation coordinate transform. When the transform is applied to the non-uniform retinal sampling points, uniform sampling density is obtained in the new coordinate system. A typically used solution is the logmap transform [WOBS94] defined as

$$\mathbf{w} = \log(\mathbf{z} + a) \,, \tag{14.1}$$

where a is a constant, and $\mathbf{z}$ and $\mathbf{w}$ are complex numbers representing the positions in the original coordinate and the transformed coordinate, respectively. While the logmap

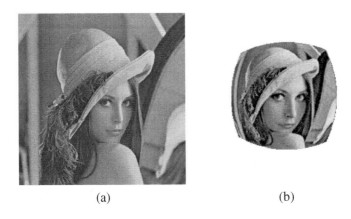

(a)   (b)

Figure 14.4: Application of foveation coordinate transform to images. (a) original image; (b) transformed image.

transform is empirical, it is shown in [Wan01] that precise mathematical solutions of the foveation coordinate transforms may be derived directly from given retinal sampling distributions.

The foveated retinal sampling geometry can be used in different ways. The first method is to apply the foveation coordinate transform directly to a uniform resolution image, thus the underlying image space is mapped onto the new coordinate system as exemplified by Figure 14.4. In the transform domain, the image is treated as a uniform resolution image, and regular uniform-resolution image processing techniques, such as linear and non-linear filtering and compression, are applied. Finally, the inverse coordinate transform is employed to obtain a "foveatedly" processed image. The difficulty with this method is that the image pixels originally located at integer grids are moved to non-integer positions, making it difficult to index them. Interpolation and resampling procedures have to be applied in both the transform and the inverse transform domains. These procedures not only significantly complicate the system, but may also cause further distortions.

The second approach is the superpixel method [WOBS94, KG96, BS89, CAS96, TEHM96], in which local image pixel groups are averaged and mapped into superpixels, whose sizes are determined by the retinal sampling density. Figure 14.5 shows a sophisticated superpixel look-up table given in [WOBS94], which attempts to adhere with the logmap structure. However, the number and variation of superpixel shapes make it inconvenient to manipulate. In [KG96], a more practical superpixel method is used, where all the superpixels have rectangular shapes. In [TEHM96], a multistage superpixel approach is introduced, in which a progressive transmission scheme is implemented by using variable sizes of superpixels in each stage. There are two drawbacks of the superpixel methods. First, the discontinuity across superpixels is often very per-

Logmap lookup tables (S, R) for 60 by 64 TV image.

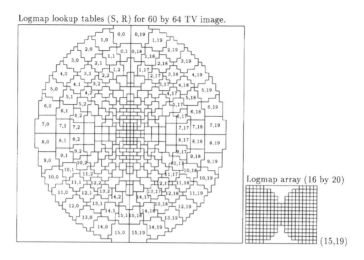

Figure 14.5: Logmap superpixel representation [WOBS94]. The superpixel mask is applied in the pixel coordinates.

ceptually annoying. Blending methods are usually used to reduce the boundary effect, leading to additional computational cost. Second, when the foveation point moves, the superpixel mask has to be recalculated.

In the third method, the foveated retinal geometry is employed to guide the design of a non-uniform subsampling scheme on the uniform resolution image. An example of foveated sampling design is shown in Figure 14.6 [WOBS94]. In [KGG99], uniform grid images are resampled with variable resolution that matches the human retina sampling density. B-Spline interpolation is then used to reconstruct the foveated images. The subsampling idea has also been used to develop foveated sensoring schemes to improve the efficiency of image and video acquisition systems [RS90, WOBS94].

14.2.2   Filtering Based Methods

The sampling theorem states that the highest frequency of a signal that can be represented without aliasing is one-half of the sampling rate. As a result, the bandwidth of a perceived local image signal is limited by local retinal sampling density. In the category of filtering-based foveation methods, foveation is implemented with a shift-variant low-pass filtering process over the image, where the cut-off frequency of the filter is determined by the local retinal sampling density.

Since retinal sampling is spatially varying (smoothly), an ideal implementation of foveation filtering would require using a different low-pass filter at each location in the image. Although such a method delivers very high quality foveated images, it is extremely expensive in terms of computational cost when the local bandwidth is low.

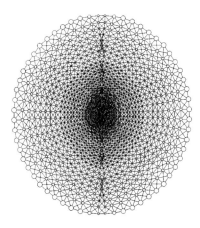

Figure 14.6: Foveated sensor distribution [WOBS94].

The filter bank method provides a flexible trade-off between the accuracy and the cost of the foveation filtering process. As illustrated in Figure 14.7 [Wan01], a bank (finite number) of filters with varying frequency responses are first uniformly applied to the input image, resulting in a set of filtered images. The foveated image is then obtained by merging these filtered images into one image, where the merging process is space-variant according to the foveated retinal sampling density. There are a number of issues associated with the design of such filter banks and merging processes. First, the bank of filters can be either low-pass or band-pass, and thus the merging process should be adjusted accordingly. Second, there are a number of filter design problems. For instance, the filters can be designed either in the spatial or in the frequency domain. For another example, in the design of finite impulse response filters, it is important to consider the trade-offs between transition band size, ripple size, and implementation complexity (e.g., filter length). Usually, small ripple size is desired to avoid significant ringing effect. Third, since both foveation filtering and transform-based (e.g., discrete cosine transform (DCT) or wavelet-based) image compression require transforming the image signal into frequency subbands, they may be combined to reduce implementation and computational complexity. For example, only one-time DCT is used and then both foveation filtering and compression can be implemented by manipulating the DCT coefficients.

In [Lee00, LPB01, LPB02], the filter bank method was employed as a preprocessing step before the standard video compression algorithms such as MPEG and H.26x were applied. Significant compression improvement over uniform resolution coding was obtained because a large amount of visually redundant high frequency information is removed during the foveation filtering processes. Another important advantage of this system is that it is completely compatible with the video coding standards, because no modification on the encoder/decoder of the existing video coding systems is needed, ex-

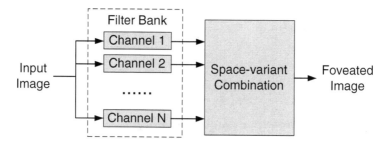

Figure 14.7: Filter bank foveation method.

cept for adding a foveation filtering unit in front of the video encoder. A demonstration of this system is available at [LB00].

The filter bank method was also used to build an eye tracker-driven foveated imaging system at the Laboratory for Image and Video Engineering (LIVE) at the University of Texas at Austin [RWB01]. The system detects the fixation point of the subject in real-time using an eye-tracker. The detected fixation point is then used to promptly foveate the image or video being displayed on a computer monitor or projected on a large screen mounted on the wall. Since all the processes are implemented in real time, the subject feels as if he/she were watching the original image sequence instead of the foveated one, provided the calibration process has been well-performed.

In [SLWB02, SEB03], foveation filtering was implemented in the DCT domain and combined with the quantization processes in standard H.26x and MPEG compression. In [Liu02], such a DCT-domain foveation method is merged into a video transcoding system, which takes compressed video streams as the input and re-encodes them into lower bit rates. Implementing these systems is indeed very challenging because the coding blocks in the current frame need to be predicted from the regions in the previous frame that may cover multiple DCT blocks and have different and varying resolution levels. It needs to be pointed out that although the existing standard video encoders need to be modified to support foveated coding, these systems are still standard compatible in the sense that no change is necessary in order for any standard decoders to correctly decompress the received video streams.

14.2.3   Multiresolution Methods

The multiresolution method can be considered a combination of the geometric and the filtering-based methods, in which the original uniform resolution image is transformed into different scales (where certain geometric operations such as downsampling are involved), and the image processing algorithms are applied separately at each scale (where certain filtering processes are applied).

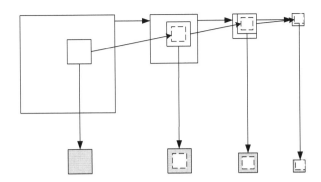

Figure 14.8: Foveated multiresolution pyramid. Adapted from [GP98].

The multiresolution method has advantages over both geometric and filtering-based methods. First, no sophisticated designs for the geometric transforms or superpixels are necessary since scaling can be implemented by simple uniform downsampling. This saves computation as well as storage space, and makes pixel indexing easy. Second, after downsampling, the number of transformed coefficients in each scale is greatly reduced. As a result, the computational cost of the filtering process decreases.

In [Bur88], a multiresolution pyramid method [BA83] is applied to an uniform resolution image and a coarse-to-fine spatially adaptive scheme is then applied to select the useful information for the construction of the foveated image. In [GP98], a very efficient pyramid structure shown in Figure 14.8 is used to foveate images and video. In order to avoid severe discontinuities occurring across stage boundaries in the reconstructed image and video, strong blending postprocessing algorithms were employed. This system can be used for real-time foveated video coding and transmission. A demonstration of the system and its software implementation is available at [GP02].

As a powerful multiresolution analysis tool, the wavelet transform has been extensively used for various image processing tasks in recent years [Mal99]. A well-designed wavelet transform not only delivers a convenient, spatially localized representation of both frequency and orientation information of the image signal, but also allows for perfect reconstruction. These features are important for efficient image compression. In [CY97, CMY00], a non-uniform foveated weighting model in the wavelet transform domain is employed for wavelet foveation. A progressive transmission method was also suggested for foveated image communication, where the ordering of the transmitted information was determined by the foveated weighting model.

In the next section, we will mainly discuss a wavelet-based foveated scalable coding method proposed in [WB01b, Wan01], where the foveated weighting model was developed by joint consideration of multiple HVS factors, including the spatial variance of the contrast sensitivity function, the spatial variance of the local visual cutoff frequency, and the variance of the human visual sensitivity in different wavelet subbands. The ordering of the encoded information not only depends on the foveated weighting model, but also on the magnitudes of the wavelet coefficients. This method was extended for the design of a prototype for scalable foveated video coding [WLB03, Wan01]. The prototype was implemented in a specific application environment, where foveated scalable coding was combined with an automated foveation point selection scheme and an adaptive frame prediction algorithm.

14.3   Scalable Foveated Image and Video Coding

An important recent trend in visual communications is to develop continuously rate scalable coding algorithms (e.g., [Sha93, TZ94, SP96, SD99, CW99, TM01, WB01b, WLB03]), which allow the extraction of coded visual information at continuously varying bit rates from a single compressed bitstream. An example is shown in Figure 14.9, where the original video sequence is encoded with a rate scalable coder and the encoded bitstream is stored frame by frame. During the transmission of the coded data on the network, we can scale, or truncate, the bitstream at any place and send the most important bits of the bitstream. Such a scalable bitstream can provide numerous versions of the compressed video at various data rates and levels of quality. This feature is especially suited for video transmission over heterogeneous, multi-user, time-varying and interactive networks such as the Internet, where variable bandwidth video streams need to be created to meet different user requirements. The traditional solutions, such as layered video (e.g., [AS00]), video transcoding (e.g., [SKZ96]), and simply repeated encoding, require more resources in terms of computation, storage space and/or data management. More importantly, they lack the flexibility to adapt to time-varying network conditions and user requirements, because once the compressed video stream is generated, it becomes inconvenient to change it to an arbitrary data rate. By contrast, with a continuously rate scalable codec, the data rate of the video being delivered can exactly match the available bandwidth on the network.

The central idea of foveated scalable image and video coding is to organize the encoded bitstream to provide best decoded visual information at an arbitrary bit rate in terms of foveated perceptual quality measurement. Foveation-based HVS models play important roles in these systems. In this section, we first describe a wavelet-domain foveated perceptual weighting model, and then explain how this model is used for scalable image and video coding.

14.3.1   Foveated Perceptual Weighting Model

Psychological experiments have been conducted to measure the contrast sensitivity as a function of retinal eccentricity (e.g., [GP98, RG81, BSA81]). In [GP98], a model that

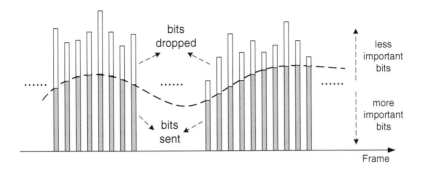

Figure 14.9: Bitstream scaling in rate scalable video communications. Each bar represents the bitstream for one frame in the video sequence. The bits in each frame are ordered according to their importance.

fits the experimental data was given by

$$CT(f,e) = CT_0 \exp\left(\alpha f \frac{e + e_2}{e_2}\right) , \qquad (14.2)$$

where

f:	Spatial frequency (cycles/degree)
e:	Retinal eccentricity (degrees)
CT_0:	Minimal contrast threshold
α:	Spatial frequency decay constant
e_2:	Half-resolution eccentricity constant
CT:	Visible contrast threshold

The best fitting parameters given in [GP98] are $\alpha = 0.106$, $e_2 = 2.3$, and $CT_0 = 1/64$, respectively. The contrast sensitivity is defined as the reciprocal of the contrast threshold: $CS(f,e) = 1/CT(f,e)$.

For a given eccentricity e, equation (14.2) can be used to find its critical frequency or so called cutoff frequency f_c in the sense that any higher frequency component beyond it is imperceivable. f_c can be obtained by setting CT to 1.0 (the maximum possible contrast) and solving for f:

$$f_c(e) = \frac{e_2 \ln\left(\frac{1}{CT_0}\right)}{\alpha(e + e_2)} \left(\frac{cycles}{degree}\right) . \qquad (14.3)$$

To apply these models to digital images, we need to calculate the eccentricity for any given point $\mathbf{c} = (c_1, c_2)^T$ (pixels) in the image. Figure 14.10 illustrates a typical viewing geometry. For simplicity, we assume the observed image is N-pixel wide and

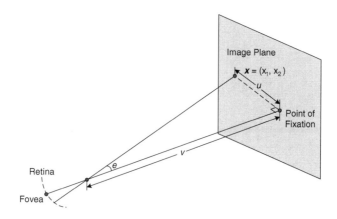

Figure 14.10: A typical viewing geometry. Here, v is the distance to the image measured in image width, and θ is eccentricity measured in degrees.

the line from the fovea to the point of fixation in the image is perpendicular to the image plane. Also assume that the position of the foveation point $\mathbf{c}^f = (c_1^f, c_2^f)^T$ (pixels) and the viewing distance v (measured in image width) from the eye to the image plane are known. The distance from $\mathbf{c}$ to $\mathbf{c}^f$ is given by $d(\mathbf{c}) = \|\mathbf{c} - \mathbf{c}^f\|_2 = [(c_1 - c_1^f)^2 + (c_2 - c_2^f)^2]^{1/2}$ (measured in pixels). The eccentricity is then calculated as

$$e(v, \mathbf{c}) = \tan^{-1}\left(\frac{d(\mathbf{c})}{Nv}\right) . \tag{14.4}$$

With (14.4), we can convert the foveated contrast sensitivity and cutoff frequency models into the image pixel domain. In Figure 14.11, we show the normalized contrast sensitivity as a function of pixel position for $N = 512$ and $v = 3$. The cut-off frequency as a function of pixel position is also given. The contrast sensitivity is normalized so that the highest value is always 1.0 at 0 eccentricity. It can be observed that the cut-off frequency drops quickly with increasing eccentricity and the contrast sensitivity decreases even faster.

In real-world digital images, the maximum perceived resolution is also limited by the display resolution, which is approximately:

$$r \approx \frac{\pi N v}{180} \left(\frac{pixels}{degree}\right) . \tag{14.5}$$

According to the sampling theorem, the highest frequency that can be represented without aliasing by the display, or the display Nyquist frequency, is half of the display resolution: $f_d(v) = r/2$. Combining this with (14.3), we obtain the cutoff frequency for a given location $\mathbf{c}$ by:

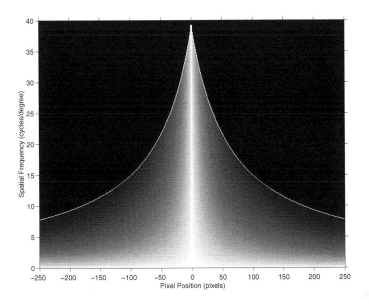

Figure 14.11: Normalized contrast sensitivity for $N = 512$ and $v = 3$. Brightness indicates the strength of contrast sensitivity and the white curves show the cutoff frequency.

$$f_m(v, \mathbf{c}) = min(f_c(e(v, \mathbf{c})), f_d(v)) . \tag{14.6}$$

Finally, we define the foveation-based error sensitivity for given viewing distance v, frequency f and location $\mathbf{c}$ as:

$$S_f(v, f, \mathbf{c}) = \begin{cases} \frac{CS(f, e(v, \mathbf{c}))}{CS(f, 0)} & if \quad f \leq f_m(v, \mathbf{c}) \\ 0 & otherwise \end{cases} . \tag{14.7}$$

S_f is normalized so that the highest value is always 1.0 at 0 eccentricity.

The wavelet coefficients at different subbands and locations supply information of variable perceptual importance to the HVS. In [WYSV97], psychovisual experiments were conducted to measure the visual sensitivity in wavelet decompositions. Noise was added to the wavelet coefficients of a blank image with uniform mid-gray level. After the inverse wavelet transform, the noise threshold in the spatial domain was tested. A model that provided a reasonable fit to the experimental data is [WYSV97]:

$$\log Y = \log a + k(\log f - \log g_\theta f_0)^2 \tag{14.8}$$

where

Y: Visually detectable noise threshold;
θ: Orientation index, representing LL, LH,
                HH, and HL subbands, respectively;
f: Spatial frequency (cycles/degree);
k, f_0, g_θ: Constant parameters.

f is determined by the display resolution r and the wavelet decomposition level λ: $f = r2^{-\lambda}$. The constant parameters in (14.8) are tuned to fit the experimental data. For gray scale models, a is 0.495, k is 0.466, f_0 is 0.401, and g_θ is 1.501, 1, and 0.534 for the LL, LH/HL, and HH subbands, respectively. The error detection thresholds for the wavelet coefficients can be calculated by:

$$T_{\lambda,\theta} = \frac{Y_{\lambda,\theta}}{A_{\lambda,\theta}} = \frac{a10^{k(\log(2^\lambda f_0 g_\theta / r))^2}}{A_{\lambda,\theta}} , \tag{14.9}$$

where $A_{\lambda,\theta}$ is the basis function amplitude given in [WYSV97]. We define the error sensitivity in subband (λ, θ) as $S_w(\lambda, \theta) = 1/T_{\lambda,\theta}$.

For a given wavelet coefficient at position $\mathbf{c} \in \mathbf{B}_{\lambda,\theta}$, where $\mathbf{B}_{\lambda,\theta}$ denotes the set of wavelet coefficient positions residing in subband (λ, θ), its equivalent distance from the foveation point in the spatial domain is given by

$$d_{\lambda,\theta}(\mathbf{c}) = 2^\lambda \left\| \mathbf{c} - \mathbf{c}^f_{\lambda,\theta} \right\|_2 \qquad for \quad \mathbf{c} \in \mathbf{B}_{\lambda,\theta} , \tag{14.10}$$

where $\mathbf{c}^f_{\lambda,\theta}$ is the corresponding foveation point in subband (λ, θ). With the equivalent distance, and also considering (14.7), we have

$$S_f(v, f, \mathbf{c}) = S_f(v, r2^{-\lambda}, d_{\lambda,\theta}(\mathbf{c})) \qquad for \quad \mathbf{c} \in \mathbf{B}_{\lambda,\theta} . \tag{14.11}$$

Considering both $S_w(\lambda, \theta)$ and $S_f(v, f, \mathbf{c})$, a wavelet domain foveation-based visual sensitivity model is achieved:

$$S(v, \mathbf{c}) = [S_w(\lambda, \theta)]^{\beta_1} \cdot \left[S_f(v, r2^{-\lambda}, d_{\lambda,\theta}(\mathbf{c})) \right]^{\beta_2} \quad \mathbf{c} \in \mathbf{B}_{\lambda,\theta} , \tag{14.12}$$

where β_1 and β_2 are parameters used to control the magnitudes of S_w and S_f, respectively.

For a given wavelet coefficient at location $\mathbf{c}$, the final weighting model is obtained by integrating $S(v, \mathbf{c})$ over v:

$$W_w(\mathbf{c}) = \int_{0+}^{\infty} p(v)S(v, \mathbf{c}) \, dv , \tag{14.13}$$

where $p(v)$ is the probability density distribution of the viewing distance v [WB01b]. Figure 14.12 shows the importance weighting mask in the DWT domain. This model can be easily generated for the case of multiple foveation points:

$$W_w(\mathbf{c}) = W_w^j(\mathbf{c}), \quad j \in \underset{i \in \{1,\cdots,K\}}{\arg\min} \left\{ \left\| \mathbf{c} - \mathbf{c}_{i,\lambda,\theta}^f \right\|_2 \right\}, \tag{14.14}$$

where K is the number of foveation points, $\mathbf{c}_{i,\lambda,\theta}^f$ is the position of the i-th foveation point in the subband (λ, θ), and $W_w^i(\mathbf{c})$ is the wavelet-domain foveated weighting model obtained with the i-th foveation point.

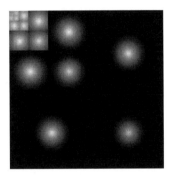

Figure 14.12: Wavelet domain importance weighting mask of a signal foveation point. Brightness (logarithmically enhanced for display purpose) indicates the importance of the wavelet coefficient.

14.3.2 Embedded Foveation Image Coding

The embedded foveation image coding (EFIC) system [WB01b] is shown in Figure 14.13. First, the wavelet transform is applied to the original image. The foveated perceptual weighting mask calculated from given foveation points or regions is then used to weight the wavelet coefficients. Next, we encode the weighted wavelet coefficients using a modified set partitioning in hierarchical trees (SPIHT) encoder, which is adapted from the SPIHT coder proposed in [SP96]. Finally, the output bitstream of the modified SPIHT encoder, together with the foveation parameters, is transmitted to the communication network. At the receiver side, the weighted wavelet coefficients are obtained by applying the modified SPIHT decoding algorithm. The foveated weighting mask is then calculated in exactly the same way as at the encoder side. Finally, the inverse weighting and inverse wavelet transform are applied to obtain the reconstructed image. Between the sender, the communication network and the receiver, it is possible to exchange information about network conditions and user requirements. Such feedback information can be used to control the encoding bit-rate and foveation points. The decoder can also

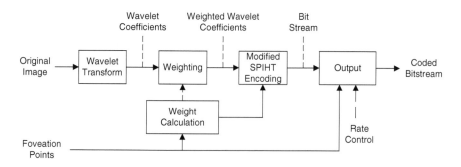

Figure 14.13: EFIC encoding system.

truncate (scale) the received bitstream to obtain any bit rate image below the encoder bit rate.

The modified SPIHT algorithm employed by the EFIC system uses an embedded bit-plane coding scheme. The major purpose is to progressively select and encode the most important remaining bit in the wavelet representation of the image. An important statistical feature of natural images that has been successfully used by the embedded zero tree wavelet (EZW) [Sha93] and SPIHT [SP96] algorithms is that the wavelet coefficients which are less significant have structural similarity across the wavelet subbands in the same spatial orientation. The zerotree structure in EZW and the spatial orientation tree structure in SPIHT capture this structural similarity very effectively. During encoding the wavelet coefficients are scanned multiple times. Each time consists of a sorting pass and a refinement pass. The sorting pass selects the significant coefficients and encodes the spatial orientation tree structure. A coefficient is significant if its magnitude is larger than a threshold value, which decreases by a factor of 2 for each successive sorting pass. The refinement pass outputs one bit for each selected coefficient. An entropy coder can be employed to further compress the output bitstream. In EFIC, the wavelet coefficients being encoded are weighted, which leads to increased dynamic range of the coefficients. This not only increases the number of scans, but also increases the number of bits to encode the large coefficients. The modified SPIHT algorithm employed by EFIC limits the maximum number of bits for each coefficient and scans only the strongly weighted coefficients in the first several scans. Both of these modifications reduce computational complexity and increase the overall coding efficiency.

Figure 14.14 shows the 8 bits/pixel gray scale *Zelda* image encoded with SPIHT and EFIC, where the foveated region is at the center of the image. At a low bit-rate of 0.015625 bits/pixel with compression ratio (CR) equaling 512:1, the mouth, nose, and eye regions are hardly recognizable in the SPIHT coded image, whereas those regions in the EFIC coded image exhibit some useful information. At a medium bit-rate of 0.0625 bits/pixel (CR = 128:1), SPIHT still decodes a quite blurry image, while EFIC gives much more detailed information over the face region. Increasing the bit-rate to as high as 0.25 bits/pixel (CR = 25), the EFIC coded image approaches uniform resolution. The decoded SPIHT and EFIC images both have high quality and are almost indistinguishable. More demonstration images for EFIC can be found at [WB01a].

The EFIC decoding procedure can also be viewed as a progressive foveation filtering process with gradually decreasing foveation depth. The reason may be explained as follows: Note that the spectra of natural image signals statistically follow the power law $1/f^p$ (see [Rud96] for a review). As a result, the low-frequency wavelet coefficients are usually larger than the high-frequency ones, thus generally have better chances to be reached earlier in the embedded bit-plane coding process. Also notice that the foveated weighting process shifts down the bit-plane levels of all the coefficients in the peripheral regions. Therefore, at the same frequency level, the coefficients at the peripheral regions generally occupy lower bit-planes than the coefficients at the region of fixation. If the available bit-rate is limited, then the embedded bit-plane decoding process corresponds to applying a higher-bandwidth low-pass filter to the region of fixation and a lower-bandwidth low-pass filter to the peripheral regions, thereby foveating the image. With the increase of bit-rate, more bits for the high-frequency coefficients in the peripheral regions are received, thus the decoded image becomes less foveated. This is well demonstrated by the EFIC coded images shown in Figure 14.14.

14.3.3   Foveation Scalable Video Coding

The foveated scalable video coding (FSVC) system [WLB03] follows the general method of motion estimation/motion compensation-based video coding. It first divides the input video sequence into groups of pictures (GOPs). Each GOP has one intra-coding frame (I frame) at the beginning and the rest are predictive coding frames (P frames). The diagram of the encoding system is shown in Figure 14.15. The I frames are encoded the same way as in the EFIC algorithm described above. The encoding of P frames is more complicated and is different from other video coding algorithms in that it uses two instead of one version of the previous frames. One is the original previous frame and the other is a feedback decoded version of the previous frame. The final prediction frame is the weighted combination of the two motion compensated prediction frames. The combination is based on the foveated weighting model.

The prototype FSVC system allows one to select multiple foveation points, mainly to facilitate the requirements of large foveation regions and multiple foveated regions of interest. It also reduces the search space of the foveation points by dividing the image space into blocks and limiting the candidate foveation points to the centers of blocks. This strategy not only decreases implementation and computational complexity, but also reduces the number of bits needed to encode the positions of the foveation points. In practice, the best way of foveation point(s) selection is application dependant. The FSVC prototype is very flexible such that different foveation point selection schemes can be applied to a single framework.

Figure 14.14: *Zelda* image compressed with SPIHT and EFIC algorithms. (a) SPIHT compressed image, compression ratio (CR) = 512:1; (b) EFIC compressed image, CR = 512:1; (c) SPIHT compressed image, CR = 128:1; (b) EFIC compressed image, CR = 128:1; (e) SPIHT compressed image, CR = 32:1; (f) EFIC compressed image, CR = 32:1.

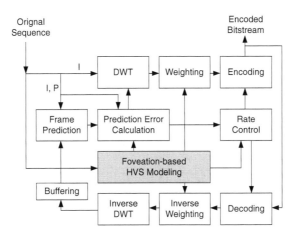

Figure 14.15: FSVC encoding system.

We implemented the FSVC prototype in a specific application environment for video sequences with human faces. A face-foveated video coding algorithm is useful to effectively enhance the visual quality in specific video communication environments such as videoconferencing.

The methods to choose foveation points for I frames and P frames are different. In the I frames, a face detection algorithm similar to that in [WC97] is used, which detects possible face regions by the skin color information [GT99] and uses a binary template matching method to detect human faces in the skin-color regions. A different strategy is used for P frames, where we concentrate on the regions in the current P frame that provide us with new information from its previous frame, in which the prediction errors are usually larger than other regions. The potential problem of this method is that the face regions may lose fixation. To solve this problem, an unequal error thresholding method is used to determine foveation regions in P frames, where a much smaller prediction error threshold value is used to capture the changes occurring in the face regions. In Figure 14.16, we show five consecutive frames in the *Silence* sequence and the corresponding selected foveation points, in which the first frame is an I frame and the rest are P frames.

In fixed-rate motion compensation-based video coding algorithms, a common choice is to use the feedback decoded previous frame as the reference frame for the prediction of the current frame. This choice is infeasible for continuously scalable coding because the decoding bit rate may be different from the encoding bit rate and is unavailable to the encoder. In [SD99], a low base rate is defined and the decoded and motion compensated frame at the base rate is used as the prediction. This solution avoids the significant error propagation problems, but when the decoding bit rate is much higher than the base rate, large prediction errors may occur and the overall coding efficiency may be seriously

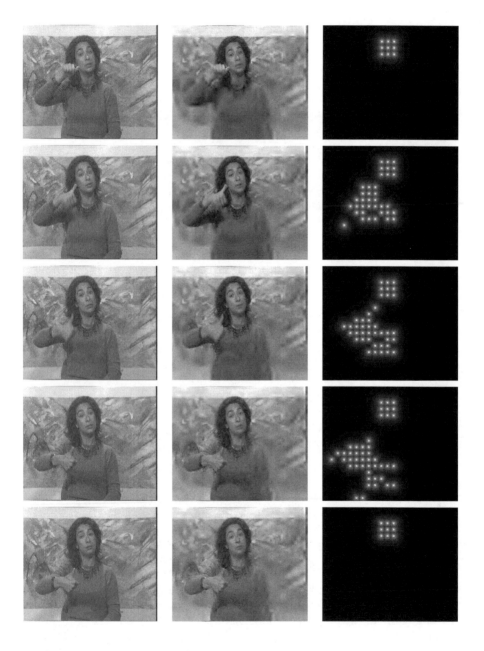

Figure 14.16: Consecutive frames of the *Silence* sequence (left); the FSVC compression results at 200 Kbits/sec (middle); and the selected foveation points (right).

Figure 14.17: Frame 32 of the *Salesman* sequence (a) compressed using FSVC at 200 Kbits/sec (b), 400 Kbits/sec (c), and 800 Kbits/sec (d), respectively [WLB03].

affected. A new solution to this problem is used in the FSVC system, where the original motion compensated frame and the base rate decoded and motion compensated frame are adaptively combined using the foveated weighting model. The idea is to assign more weight to the base rate motion compensated frame for difficult prediction regions, and less weight to the original motion compensated frame for easy prediction regions. By using this method, error propagation becomes a small problem, while at the same time, better frame prediction is achieved, leading to smaller prediction errors and better compression performance.

Figure 14.16 shows the FSVC compression results of the *Silence* sequence. It can be observed that the spatial quality variance in the decoded image sequences is well adapted to the time-varying foveation point selection scheme. Figure 14.17 demonstrates the scalable feature of the FSVC system, which shows the reconstructed 32nd frame of the *Salesman* video sequence decoded at 200, 400 and 800 Kbits/sec, respectively. The reconstructed video sequences are created from the same FSVC-encoded bitstream by truncating the bitstream at different places. Similar to Figure 14.14, the decoded images exhibit decreased foveation depth with increasing bit rate.

14.4 Discussions

This chapter first introduces the background and motivations of foveated image processing, and then reviews the various foveation techniques that are used for the development of image and video coding systems. To give examples on specific implementations of such systems, we described in more detail the EFIC and the FSVC systems, which supply continuously rate-scalable codestreams ordered according to foveation-based perceptual importance. Such systems have a number of potential applications.

One direct application is network image browsing. There are two significant examples. In the first example, prior to using the encoding algorithm, the foveation point(s) are predetermined. The coding system then encodes the image with high bit-rate and high quality. One copy of the encoded bitstream is stored at the server side. When the image is required by a client, the server sends the bitstream to the client progressively. The client can stop the transmission at any time once the reconstructed image quality is satisfactory. In the second example, the foveation point(s) are unknown to the server before transmission. Instead of a fully encoded bitstream, a uniform resolution coarse quality version of the image is precomputed and stored at the server side. The client first sees the coarse version of the image and clicks on the point of interest in that image. The selected point of interest is sent back to the server and activates the scalable foveated encoding algorithm. The encoded bitstream that has a foveation emphasis on the selected point of interest is then transmitted progressively to the client.

Another application is network videoconferencing. Compared with traditional videoconferencing systems, a foveated system can deliver lower data rate video streams since much of the high frequency information redundancy can be removed in the foveated encoding process. Interactive information such as the locations of the mouse, touch screen and eye-tracker can be sent as feedback information to the other side of the network and used to define the foveation points. Face detection and tracking algorithm may also help to find and adjust the foveation points. Furthermore, in a highly heterogeneous network, the available bandwidth can change dramatically between two end users. A fixed bit-rate video stream would either be terminated suddenly (when the available bandwidth drops below the fixed encoding bit-rate) or suffer from the inefficient use of the bandwidth (when the fixed bit-rate is lower than the available bandwidth). By contrast, a rate scalable foveated videoconferencing system can deal with these problems more smoothly and efficiently.

The most commonly used methods for robust visual communications on noisy channels are error resilience coding at the source or channel coders and error concealment processing at the decoders [VZW99]. Scalable foveated image and video streams provide us with the opportunity to do a better job by taking advantage of its optimized ordering of visual information in terms of perceptual importance. It has been shown that significant improvement can be achieved by unequal error protection for scalable foveated image coding and communications [Sab02].

Active networks are a hot research topic in recent years [TSS⁺97]. They allow the customers to send not only static data, but also programs that are executable at the routers or switches within the network. An active network becomes more useful and effective for visual communications if an intelligent scheme is employed to modify the visual contents being delivered in a smart and efficient way. The properties of scalable foveated image/video streams provide a good match to the features of active networks because the bit rate of the video stream can be adjusted according to the network conditions monitored at certain routers/switches inside the network (instead of at the sender side), and the feedback foveation information (points and depth) at the receiver side may also be dealt with at the routers/switches. This may result in quicker responses that benefit real-time communications.

Finally, a common critical issue in all foveated image processing applications is how the foveation points or regions should be determined. Depending on the application, this may be done either interactively or automatically. In the interactive method, an eye tracker is usually used to track the eye movement and send the information back to the foveated imaging system in real time. In most application environments, however, the eye tracker is not available or is inconvenient. A more practical way is to ask the users to indicate fixation points using a mouse or touch screen. Another possibility is to ask the users to indicate the object of interest, and an automatic algorithm is then used to track the user-selected object as the foveated region in the image sequence that follows. Automatic determination of foveation points is itself a difficult but interesting research topic, and is closely related to psychological visual search research (see [Itt00] for a review). In the image processing literature, there also has been previous research towards understanding high level and low level processes in deciding human fixation points automatically (e.g., [KB98, PS00, WLB03, RCB03]). High level processes are usually context dependent and involve a cognitive understanding of the image and video being observed. For example, once a human face is recognized in an image, the face area is likely to become a heavily fixated region. In a multimedia environment, audio signals may also be linked to image objects in the scene and help to determine foveation points [Wan01]. Low level processes determine the points of interest using simple local features of the image [PS00, RCB03]. In [KB98], three-dimensional depth information is also employed to help find foveation points in an active stereo vision system. Although it is argued that it is always difficult to decide foveation points automatically, we believe that it is feasible to establish a statistical model that predicts them in a measurably effective way.

References

[AG96]   T. L. Arnow and W. S. Geisler. Visual detection following retinal damage: Prediction of an inhomogeneous retino-cortical model. *Proc. SPIE: Human Vision and Electronic Imaging*, 2674:119–130, 1996.

[AS00]   S. Aramvith and M.-T. Sun. *Handbook of Image and Video Processing(MPEG-1 and MPEG-2 video standards).* San Diego, CA: Academic Press, May 2000.

[BA83]   P. J. Burt and E. H. Adelson. The Laplacian pyramid as a compact image code. *IEEE Trans. Communications,* 31:532–540, April 1983.

[BS89]   C. Bandera and P. Scott. Foveal machine vision systems. *IEEE Inter. Conf. Systems, Man and Cybernetics,* 569–599, November 1989.

[BSA81]   M. S. Banks, A. B. Sekuler, and S. J. Anderson. Peripheral spatial vision: Limits imposed by optics, photoreceptors, and receptor pooling. *Journal of the Optical Society of America,* 8:1775–1787, 1981.

[BSW93]   A. Basu, A. Sullivan, and K. J. Wiebe. Variable resolution teleconferencing. *IEEE Inter. Conf. Systems, Man, and Cybernetics,* 170–175, October 1993.

[Bur88]   P. J. Burt. Smart sensing within a pyramid vision machine. *Proc. IEEE,* 76:1005–1015, August 1988.

[BW98]   A. Basu and K. J. Wiebe. Videoconferencing using spatially varying sensing with multiple and moving fovea. *IEEE Trans. Systems, Man and Cybernetics,* 28(2):137–148, March 1998.

[CAS96]   P. Camacho, F. Arrebola, and F. Sandoval. Shifted fovea multiresolution geometries. *IEEE Inter. Conf. Image Processing,* 1:307–310, 1996.

[CMY00]   E.-C. Chang, S. Mallat, and C. Yap. Wavelet foveation. *Journal of Applied and Computational Harmonic Analysis,* 9(3):312–335, October 2000.

[CW99]   S.-J. Choi and J. W. Woods. Motion-compensated 3-D subband coding of video. *IEEE Trans. Image Processing,* 8(2):155–167, February 1999.

[CY97]   E.-C. Chang and C. Yap. A wavelet approach to foveating images. *ACM Symposium on Computational Geometry,* 397–399, June 1997.

[DMRC01]   S. Daly, K. Matthews, and J. Ribas-Corbera. As plain as the noise on your face: Adaptive video compression using face detection and visual eccentricity models. *Journal of Electronic Imaging,* 10:30–46, January 2001.

[ECvdSMZ00] R. Etienne-Cummings, J. van der Spiegel, P. Mueller, and M.-Z. Zhang. A foveated silicon retina for two-dimensional tracking. *IEEE Trans. Circuits and Systems II,* 47(6):504–517, June 2000.

[GB95]   W. S. Geisler and M. S. Banks. *Handbook of Optics:Visual performance.* New York: McGraw–Hill, 1995.

[GP98]   W. S. Geisler and J. S. Perry. A real-time foveated multiresolution system for low-bandwidth video communication. *Proc. SPIE: Human Vision and Electronic Imaging,* 3299(2):294–305, July 1998.

[GP02]   W. S. Geisler and J. S. Perry. Space variant imaging. *Center for Perceptual Systems, The University of Texas at Austin,* 2002. http://fi.cvis.psy.utexas.edu/.

[GT99]   C. Garcia and G. Tziritas. Face detection using quantized skin color regions merging and wavelet packet analysis. *IEEE Trans. Multimedia,* 1(3):264–277, September 1999.

[Itt00]   L. Itti. *Models of bottom-up and top-down visual attention.* Ph.D. thesis, California Institute of Technology, Pasadena, CA, 2000.

[KA02]   A. Koz and A. Alatan. Foveated image watermarking. *IEEE Inter. Conf. Image Processing*, 3:661–664, September 2002.

[KB98]   W. N. Klarquist and A. C. Bovik. FOVEA: A foveated vergent active stereo vision system for dynamic three-dimensional scene recovery. *IEEE Trans. Robotics and Automation*, 14(5):755–770, October 1998.

[KG96]   P. Kortum and W. S. Geisler. Implementation of a foveal image coding system for image bandwidth reduction. *Proc. SPIE: Human Vision and Electronic Imaging*, 2657:350–360, 1996.

[KGG99]   T. Kuyel, W. Geisler, and J. Ghosh. Retinally reconstructed images: digital images having a resolution match with the human eyes. *IEEE Trans. Systems, Man and Cybernetics, Part A: Systems and Humans*, 29(2):235–243, March 1999.

[Kin99]   J. M. Kinser. Foveation from pulse images. *IEEE Inter. Conf. Information Intelligence and Systems*, 86–89, 1999.

[LB00]   S. Lee and A. C. Bovik. Foveated video demonstration, 2000. Department of Electrical and Computer Engineering, The University of Texas at Austin. http://live.ece.utexas.edu/research/foveated_video/demo.html.

[LB03]   S. Lee and A. C. Bovik. Fast algorithms for foveated video processing. *IEEE Trans. Circuits and Systems for Video Tech.*, 13(2):148–162, February 2003.

[Lee00]   S. Lee. *Foveated video compression and visual communications over wireless and wireline networks*. Ph.D. thesis, Dept. of ECE, University of Texas, Austin, May 2000.

[Liu02]   S. Liu. *DCT domain video foveation and transcoding for heterogeneous video communication*. Ph.D. thesis, Dept. of ECE, University of Texas, Austin, 2002.

[LPB01]   S. Lee, M. S. Pattichis, and A. C. Bovik. Foveated video compression with optimal rate control. *IEEE Trans. Image Processing*, 10(7):977–992, July 2001.

[LPB02]   S. Lee, M. S. Pattichis, and A. C. Bovik. Foveated video quality assessment. *IEEE Trans. Multimedia*, 4(1):129–132, March 2002.

[LPKB03]   S. Lee, C. Podilchuk, V. Krishnan, and A. C. Bovik. Foveation-based error resilience and unequal error protection over mobile networks. *Journal of VLSI Signal Processing*, 34(1/2):149–1666, May 2003.

[LW90]   M. Levoy and R. Whitaker. Gaze-Directed Volume Rendering. *Computer Graphics*, 24(2):217–223, 1990.

[Mal99]   S. G. Mallat. *A wavelet tour of signal processing*, vol. 31. San Diego, CA: Academic Press, 2nd ed., September 1999.

[PS00]   C. M. Privitera and L. W. Stark. Algorithms for defining visual regions-of-interest: comparison with eye fixations. *IEEE Trans. Pattern Analysis and Machine Intelligence*, 22(9):970–982, September 2000.

[RCB03]   U. Rajashekar, L. K. Cormack, and A. C. Bovik. Image features that draw fixations. *IEEE Inter. Conf. Image Processing*, 3:313–316, September 2003.

[RG81]   J. G. Robson and N. Graham. Probability summation and regional variation in contrast sensitivity across the visual field. *Vision Research*, 21:409–418, 1981.

[RR96]   T. H. Reeves and J. A. Robinson. Adaptive foveation of MPEG video. *ACM Multimedia*, 231–241, 1996.

[RS90]   A. S. Rojer and E. L. Schwartz. Design considerations for a space-variant visual sensor with complex-logarithmic geometry. *IEEE Inter. Conf. Pattern Recognition*, 2:278–285, 1990.

[Rud96]   D. L. Ruderman. The statistics of natural images. *Network: Computation in Neural Systems*, 5:517–548, 1996.

[RWB01]   U. Rajashekar, Z. Wang, and A. C. Bovik. Real-time foveation: An eye tracker-driven imaging system, 2001. Accessible at http://live.ece.utexas.edu /research/realtime_foveation/.

[Sab02]   M. F. Sabir. Unequal error protection for scalable foveated image communication. Master's thesis, Dept. of ECE, University of Texas, Austin, May 2002.

[SBC93]   P. L. Silsbee, A. C. Bovik, and D. Chen. Visual pattern image sequence coding. *IEEE Trans. Circuits and Systems for Video Tech.*, 3(4):291–301, August 1993.

[Sch77]   E. L. Schwartz. Spatial mapping in primate sensory projection: Analytic structure and relevance to perception. *Biological Cybernetics*, 25:181–194, 1977.

[Sch80]   E. L. Schwartz. Computational anatomy and functional architecture of striate cortex: a spatial mapping approach to perceptual coding. *Vision Research*, 20:645–669, 1980.

[SD99]   K. S. Shen and E. J. Delp. Wavelet based rate scalable video compression. *IEEE Trans. Circuits and Systems for Video Technology*, 9(1):109–122, February 1999.

[SEB03]   H. R. Sheikh, B. L. Evans, and A. C. Bovik. Real-time foveation techniques for low bit rate video coding. *Real-time Imaging*, 9(1):27–40, February 2003.

[Sha93]   J. M. Shapiro. Embedded image coding using zerotrees of wavelets coefficients. *IEEE Trans. Signal Processing*, 41:3445–3462, December 1993.

[SKZ96]   H. Sun, W. Kwok, and J. W. Zdepski. Architectures for MPEG compressed bitstream scaling. *IEEE Trans. Circuits and Systems for Video Technology*, 6(2):191–199, April 1996.

[SLWB02]   H. R. Sheikh, S. Liu, Z. Wang, and A. C. Bovik. Foveated multipoint video-conferencing at low bit rates. *IEEE Inter. Conf. Acoust.,Speech, and Signal Processing*, 2:2069–2072, May 2002.

[SP96]   A. Said and W. A. Pearlman. A new, fast, and efficient image codec based on set partitioning in hierarchical trees. *IEEE Trans. Circuits and Systems for Video Technology*, 6(3):243–250, June 1996.

[TEHM96]   N. Tsumura, C. Endo, H. Haneishi, and Y. Miyake. Image compression and decompression based on gazing area. *Proc. SPIE: Human Vision and Electronic Imaging*, 2657:361–367, 1996.

[TM01]   D. S. Taubman and M. W. Marcellin. *JPEG2000: Image Compression Fundamentals, Standards, and Practice.* Norwell, MA: Kluwer Academic Publishers, November 2001.

[TSS+97]   D. L. Tennenhouse, J. M. Smith, W. D. Sincoskie, D. J. Wetherall, and G. J. Mindenk. A survey of active network research. *Proc. IEEE*, 35, January 1997.

[TZ94]   D. Taubman and A. Zakhor. Multirate 3-D subband coding of video. *IEEE Trans. Image Processing*, 3:572–588, September 1994.

[VZW99]   J. D. Villasenor, Y.-Q. Zhang, and J. Wen. Robust video coding algorithms and systems. *Proc. IEEE*, 87:1724–1733, October 1999.

[Wan95]   B. A. Wandell. *Foundations of Vision*. Sunderland, MA: Sinauer Associates, Inc., 1995.

[Wan01]   Z. Wang. *Rate scalable foveated image and video communications*. Ph.D. thesis, Dept. of ECE, The University of Texas at Austin, December 2001.

[WB01a]   Z. Wang and A. C. Bovik. Demo images for 'embedded foveation image coding', 2001. http://www.cns.nyu.edu/˜ zwang/files/research/efic/demo.html.

[WB01b]   Z. Wang and A. C. Bovik. Embedded foveation image coding. *IEEE Trans. Image Processing*, 10(10):1397–1410, October 2001.

[WBL01]   Z. Wang, A. C. Bovik, and L. Lu. Wavelet-based foveated image quality measurement for region of interest image coding. *IEEE Inter. Conf. Image Processing*, 2:89–92, October 2001.

[WC97]   H. Wang and S.-F. Chang. A highly efficient system for automatic face region detection in MPEG video. *IEEE Trans. Circuits and Systems for Video Technology*, 7(4):615–628, 1997.

[Wei90]   C. Weiman. Video compression via a log polar mapping. *Proc. SPIE: Real Time Image Processing II*, 1295:266–277, 1990.

[WLB03]   Z. Wang, L. Lu, and A. C. Bovik. Foveation scalable vidoe coding with automatic fixation selection. *IEEE Trans. Image Processing*, 11(2):243–254, February 2003.

[WOBS94]   R. S. Wallace, P. W. Ong, B. Bederson, and E. L. Schwartz. Space variant image processing. *International Journal of Computer Vision*, 13(1):71–90, 1994.

[WYSV97]   A. B. Watson, G. Y. Yang, J. A. Solomon, and J. Villasenor. Visibility of wavelet quantization noise. *IEEE Trans. Image Processing*, 6(8):1164–1175, 1997.

[ZS93]   Y. Y. Zeevi and E. Shlomot. Nonuniform sampling and antialiasing in image representation. *IEEE Trans. Signal Processing*, 41(3):1223–1236, March 1993.

Chapter 15

Artifact Reduction by Post-Processing in Image Compression

Tao Chen[†][1] and Hong Ren Wu [‡]

†   *Panasonic Hollywood Laboratory, U.S.A.*
‡   *Royal Melbourne Institute of Technology, Australia*

In this chapter, the post-processing techniques developed for reducing artifacts in low bit rate image compression will be discussed. They mainly emphasize the reduction of blocking and ringing artifacts, on which brief discussions will be given first. Then a review of recent developments on deblocking filters will be presented. Specifically, an adaptive DCT domain postfilter will be described in detail. Though the approach was designed to remove blocking artifacts, interestingly it is also observed to be capable of suppressing ringing artifacts in block-based DCT image compression. Following that, recent works on the ringing artifact reduction will be reviewed. The de-ringing techniques tailored for the DCT and wavelet transform based compression will both be discussed.

15.1   Introduction

Due to the rapid evolution and proliferation of telecommunication and computer technologies, a broad range of new applications in visual communications have been made feasible. Almost all the applications, including mobile or PSTN (public switched telephone network) videotelephony, videoconferencing, and video over the Internet, require very efficient data compression techniques. It is to fit a large amount of visual information into a usually narrow bandwidth of communication or transmission channels. Meanwhile, acceptable quality of reconstructed signal has to be preserved. This requirement in turn becomes challenging for most existing image and video compression algorithms and standards. This is because highly perceptible, sometimes objectionable, coding artifacts appear especially at low bit rates.

---

[1]T. Chen was with Sarnoff Corporation, U.S.A.

The perceptual quality of compressed images and video is normally influenced by three factors, i.e., signal source, compression methods, and coding bit rates. Given a coding algorithm, more information contained in the source signal generally requires more bits for representation. When compressed at similar bit rates, images that contain more details in general degrade more than those with fewer detailed features. The coding bit rate is another crucial aspect with regard to the visual quality. It is well known that, in lossy image compression, there is a trade-off between the bit rate and the resultant distortion. The lower the bit rate, the more severe the coding artifacts manifest. In addition, the type of artifacts is dependent on the compression approaches used. For instance, coding schemes based on BDCT (block-based discrete cosine transform) introduce the artifacts which are mainly characterized by blockiness in flat regions and ringing artifacts along object edges [KK95]. In contrast, in images coded by wavelet-based techniques [SCE01], blurring and ringing effects dominate as the most visible artifacts [FC00]. In general, reduction of coding artifacts can result in significant improvement in the overall visual quality of the reconstructed or decoded images.

In general, there are two categories of techniques that are commonly used to reduce coding artifacts. One is to overcome the problem at the encoder side, which is known as the pre-processing scheme. Differently, the other employs post-processing techniques at the decoder side. Pre-processing techniques have been widely used in modern speech and audio coding [J.-92, JJS93]. In image and video coding, for a given encoding algorithm and a bit rate constraint, the quality of reconstructed images can be enhanced by prefiltering approaches [Kun95, OST93]. Prefilters remove unnoticeable details from the source signal so that less information has to be coded. More recently, there have been a variety of efforts made at the encoder end in order to alleviate blocking artifacts suffered in the BDCT based coding at low bit rates. Different image compression approaches have been proposed to avoid blocking artifacts, such as the interleaved block transform [FJ86, PW84], the lapped transform [HBS84, MS82, Mal92], and the combined transform [ZPL93]. However, each of these techniques requires specific coding schemes of its own, such as transform, quantization, and bit allocation. As a result, these approaches can hardly be applied to the commercial coding system products compliant with the existing standards. The JPEG [PM93], H.261/2/3 and MPEG-1/2/4 [J. 96, ISO99b, Gha03] standards employ the recommended block-based DCT algorithm.

On the other hand, post-processing techniques that are employed at the decoder side have received a lot of attention and have been an active research area in recent years. Post-processing is adopted in order to reduce coding artifacts without any increment in the bit rate or any modification in the encoding procedure. It aims at the reduction of artifacts in a way independent of the encoding scheme, and it improves the overall visual quality for a given bit rate. As an equivalence, it increases the compression ratio in terms of a given quality requirement. Furthermore, post-processing techniques maintain the compatibility with current industrial or international standards. Consequently, they possess good potential to be integrated into image and video communication systems, since they are applicable to the JPEG, H.26x and MPEG standards [Gha03].

This chapter covers mainly the discussion on the post-processing methods that have been developed to reduce the blocking and ringing artifacts. To serve this, brief discussions on blocking and ringing artifacts will be presented. Detailed discussion on other coding artifacts is outside the scope of this chapter and can be found in other chapters of this handbook. The rest of this chapter is organized as follows. Coding artifacts introduced in image compression are briefly discussed in Section 15.2, which in turn motivates the formulation of a DCT domain postfiltering approach [CWQ01]. A review of recent developments on deblocking filters and a detailed description of the DCT domain postfilter [CWQ01] will be given in Section 15.3. Section 15.4 will review the recent developments on ringing artifact reduction. Methods designed for DCT and wavelet transform coding will be discussed. Finally, a summary for the chapter is given in Section 15.5.

15.2   Image Compression and Coding Artifacts

Over the past decades, a number of image coding approaches have been proposed [Gha03]. Generally, the main difference among these algorithms lies in the first stage of the coder, i.e., the transformation. The transform coefficients in some schemes are formed in the intensity or spatial domain, while some others are in a domain significantly different from the intensity domain, e.g., the frequency domain. Regardless of the approach used, the main objective of transformation is to reduce redundancies existing among the pixels. However, certain desirable properties such as energy compaction and close match to the characteristics of the human visual system can be achieved simultaneously.

Transform-based compression is by far the most popular choice in image and video coding. Due to its near-optimal decorrelation and energy compaction properties and the availability of fast algorithms and hardware implementations, the discrete cosine transform (DCT) [RY90, Jai89] is the dominant one among various transforms. As a result, the block-based DCT is recommended and used in most of the current image and video compression standards, such as the JPEG [PM93], the MPEG-1/2/4 [J. 96], and the H.261/2/3 [Gha03, ITU98]. In transform based compression, the image is transformed to a domain significantly different from the image intensity (spatial domain) [NH94, Cla85]. This is achieved by transforming blocks of data by a linear transformation. It is well known that, at low bit rates, a major problem associated with the BDCT compression is that the reconstructed images manifest visually objectionable artifacts [YW98]. One of the best known artifacts in low bit rate transform coded images appears as the blocking effect, which is noticeable especially in the form of undesirable conspicuous block boundaries.

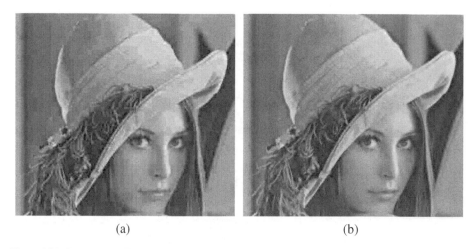

(a)   (b)

Figure 15.1: Reconstructed image *Lena* after compression by (a) JPEG and (b) JPEG-2000, respectively, at 0.22 bpp. (The original image is represented at 8 bpp)

With the continuous expansion of multimedia and Internet applications, the needs and requirements of the technologies used grew and evolved. In March 1997, a new call for contributions was launched for the development of a new standard for the compression of still images, the JPEG-2000 standard [ISO00, ISO99a, TM01, RJ02]. The standardization process has already (since December 2000) produced the International Standard [SCE01], which is based on wavelet transform. At low bit rates, images compressed by JPEG-2000 bear obvious artifacts among continuous areas and around sharp edges as a result of considerable quantization errors of wavelet coefficients. As in other wavelet-based compression methods [Sha93, SP96], the quantization errors in high-frequency subbands generally result in ringing effects, as well as blurring effects near sharp edges.

Figure 15.1 shows an example of the different artifacts observed in the low bit rate coding with JPEG and JPEG-2000 standards. Wavelet-based image coding has demonstrated some advantages over the traditional block-based DCT methods (i.e., JPEG) in terms of visibility and severity of coding artifacts in compressed images. In Figure 15.1a, objectionable blocking artifacts dominate in the reconstructed image. However, the image in Figure 15.1b shows mostly ringing artifacts around sharp edges. In the following two sub-sections, blocking and ringing artifacts will be discussed. Note that the original images discussed in this chapter are quantized and represented at 8 bpp.

15.2.1 Blocking Artifacts

With respect to the visual manifestation, the blocking artifacts are observed as discontinuities between adjacent image blocks. The horizontal and/or vertical straight edges

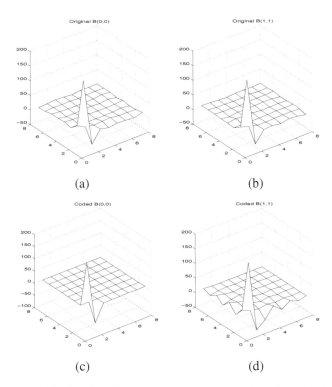

Figure 15.2: Energy distribution of the DCT coefficients of 8×8 blocks in the original and compressed *Lena* image, respectively, at different shifts. Here, $B[0, 0]$ denotes the DCT coefficients of an image block and $B[1, 1]$ are the DCT coefficients of the shifted version of that block with a $[1, 1]$ shift. The description of shifted block DCT coefficients is given in Section 15.3.1.

induced by the block boundary discontinuities, as well as the regular spacing of the blocks, result in the blocking effect being highly visible. It is a common practice to measure and reduce the blocking artifacts by only taking into account the pixels straddling the block boundaries [KK95, Zak92, YGK93]. Therefore, the blocking artifacts are in general described as the discontinuities that appear at the boundaries of adjacent blocks in a reconstructed image [RL84, Tzo83]. It was also termed sometimes as *grid noise* [RG86].

One of the main reasons for the compression of image data as block units is to exploit the high inter-pixel correlation in local areas. Unfortunately, compression of a block as an independent unit does not take into consideration the possibility that the correlation of the pixels may extend beyond the borders of a block into adjacent blocks. This consequently leads to the boundary discontinuities perceived in a reconstructed image, especially at low bit rates.

Figure 15.2 shows an example of the energy distribution of the DCT coefficients of blocks obtained from the original and the compressed *Lena* images, respectively. In each of the images, the DCT coefficients are calculated by transforming two blocks at different (i.e., shifted) locations. In the original image, those energy distributions of DCT coefficients are virtually similar when measured at different shifts. However, in the compressed image, the difference of the DCT coefficients calculated at different shifts can easily be distinguished. Such an observation can be explained as follows. In the original image, one expects by symmetry that the average amplitude of DCT coefficients is approximately similar at different shifts, especially in smooth regions. The experiment clearly demonstrates this intuition (see Figure 15.2a and b). On the other hand, compression, more specifically the quantization, eliminates many of the high frequency DCT coefficients. As a result, this operation also creates discontinuities at the block boundaries that contribute to high frequencies in the transform of shifted image block. However, these discontinuities are not perceived by the block transform at the $[0, 0]$ shift (see Figure 15.2c), since the discontinuities only lie at the block boundaries. However, if the block DCT spectrum is computed at other shifts, the discontinuities contribute to high frequency DCT energy to varying degrees (see Figure 15.2d).

The visual prominence of blocking effect is primarily a consequence of the regularity of the size and spacing of the block-edge discontinuities [YW98]. Each of the 8×8 DCT basis images has a distinctive regular horizontally or vertically oriented pattern which makes it visually conspicuous [WSA94]. The visual characteristics of the DCT basis images may produce coding artifacts when adjacent blocks are taken into consideration. For instance, for basis images that have a zero, or very low, frequency content in either horizontal or vertical direction, the blocking effect may result along the sections of the block boundary which contain limited spatial activity [YW98].

15.2.2 Ringing Artifacts

As discussed above, blocking artifacts are introduced due to the use of short and non-overlapping basis functions, like block based DCT. Overlapping transforms such as wavelet and GenLOT [dNR96] were introduced to reduce or eliminate blocking artifacts in compression, but they have spurious oscillations in the vicinity of major edges at low bit rates. Such a coding artifact is commonly referred to as ringing artifact, which is due to the abrupt truncation of the high frequency coefficients. This effect has also been observed in DCT-based image compression.

Ringing artifacts are caused by lossy quantization of transform coefficients. This cause is similar to that of the Gibbs phenomenon when long basis functions are cut short due to heavy quantization. In Figure 15.3, the original signal is a step function. In the case of truncating its DCT coefficients and keeping only the low frequency coefficients, ringing effect occurs. Truncation of frequency domain coefficients corresponds

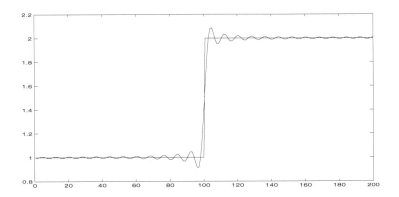

Figure 15.3: An example of ringing effect in 1-D signal.

to the convolution of the time domain sequence with a *sinc* function. The side lobes manifest themselves as oscillations in spatial domain around step discontinuities. Ringing artifacts usually appear as noticeable oscillations (spurious edges) along major (high contrast and large scale) edges in an image. It is mostly prominent in the areas with relatively smooth background. The main issue in dealing with ringing artifacts is that it is difficult to separate ringing artifacts from true edges of texture.

Due to its unique joint space-frequency characteristics, the discrete wavelet transform (DWT) has recently attracted considerable attention for image compression. In DWT, the hierarchical representation allows efficient quantization and coding strategies (e.g., zero-tree quantization [Sha93, SP96]) that exploits both the spatial and frequency characteristics of DWT coefficients. Wavelet-based image coding performs significantly better than the traditional block-based methods in terms of image quality, especially at low bit rates. In the case of large quantization errors of wavelet coefficients in coded images, however, the decoded images still carry obvious artifacts among smooth regions and around sharp edges. Specifically, the quantization errors in high frequency subbands generally result in the ringing effect as well as the blurring effect near sharp edges. Moreover, the errors in both low-frequency and high-frequency subbands cause the contouring effect, graininess, and blotches in smooth regions.

15.3   Reduction of Blocking Artifacts

Since blocking artifacts can be considered as high frequency artifacts, a straightforward solution is to apply lowpass filtering to the regions where artifacts occur. A spatial invariant filtering approach to reduce blocking artifacts in image coding was first proposed in [RL84]. In order to maintain the sharpness of the processed image, filter coefficients must be chosen carefully. However, the lack of consideration of the local image characteristics often causes the loss of details in filtering. In order to tackle this problem, a number of spatially variant filters have been proposed [Kun95, KH95, LRL95, Sau91]. Generally, adaptation uses classification and edge detection in order to categorize pixels into different classes before filtering. Then, different spatial filters, either linear or non-linear, are devised to remove coding artifacts based on the classification information. Classification is essential to adaptive filtering that attempts to exploit local statistics of image regions and the sensitivity of human visual system. More recently, Gao and Kim [GK02] proposed a deblocking algorithm based on the number of connected blocks in a relatively homogeneous area, the magnitude of abrupt changes among neighboring blocks, and the quantization step size used for DCT coefficients.

In these conventional post-processing approaches, the efforts were generally devoted to the spatial domain processing of decoded images on the basis of some *a priori* knowledge about the original image. Commonly, one would assume the intensity smoothness of images or block boundaries in the original images. A general form of postprocessors employed at the decoder side is shown in Figure 15.5a.

Post-processing approaches employed at the decoder side, such as filtering techniques [RL84, RG86], the maximum *a posteriori* probability approach [Ste93], and iterative methods based on the theory of POCS (projections onto convex sets) [Zak92, YGK93] are compatible with the current standardized image and video coding techniques. Hence, these post-processing schemes have the potential to be integrated into image and video communications, as they are in conformity with the JPEG, MPEG-1/2/4, and H.26x standards. In the recent development of low bit rate video coding, deblocking postfilters [ISO99b, ITU97] are also regarded as crucial, since they can suppress the propagation of artifacts in the image sequence and therefore improve the overall visual quality. In the latest video compression standard JVT/H.264/MPEG-4 AVC [P. 03, Joi03, Ric03, IEE03], a deblocking filter has been a normative part of the specification and has been put into the encoding and decoding loops as a loop filter. The location of loop filter in the JVT codec block diagram is shown in Figure 15.4. Regarding this adaptive in-loop deblocking filter [P. 03], the basic idea is that if a relatively large absolute difference between samples near a block edge is measured, it is quite likely a blocking artifact and should therefore be reduced. However, if the magnitude of that difference is so large that it cannot be explained by the coarseness of the quantization used in the encoding, the edge is more likely to reflect the actual behavior of the source picture and should not be smoothed over. The deblocking and deringing filters described in the video coding standards usually require that quantization step size be available in order to derive the filtering strength. In application to still image post-processing, one usually needs to fine tune this parameter for the best performance. This makes it not practically useful in some cases.

In addition to their iterative nature, POCS based methods either excessively blur the image when reducing blocking artifacts (e.g., [Zak92]), or only deal with the pixels straddling the block boundaries that could lead to a new discontinuity between each altered boundary pixel and the pixel immediately adjacent to it within the block

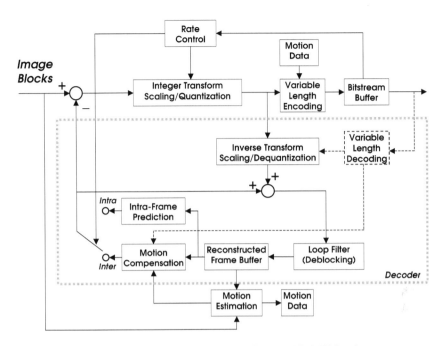

Figure 15.4: Simplified block diagram of the MPEG-4 AVC codec.

(e.g., [YGK93]), which makes such an adjustment not sufficient to reduce blocking artifacts effectively. Recently, Paek, Kim and Lee [PKL98] introduced a POCS based
technique by assuming that the global frequency characteristics in two adjacent blocks
should be similar to the local ones in each block. They employed N-point and $2N$-point
1-D DCTs to describe the local and global characteristics of adjacent blocks, respectively. Then they proposed to reduce blocking artifacts by discarding undesirable high
frequency components of the $2N$-point DCT coefficients prior to the inverse transformation. Figure 15.5b illustrates a framework of postfilters based on the POCS in general
form.

It is well known that blocking artifacts are introduced by the coarse quantization
of transform coefficients in low bit rate coding as well as by the independent quantization applied to each block. From this point of view, alleviating blocking artifacts
in the transform domain would be more efficient than in the spatial domain. Recent
years have seen investigations into transform domain processing to alleviate blocking
artifacts [TTS02]. In [TH93], a post-processing method that adjusts the quantized (and
dequantized) transform coefficients of a decoded image was proposed. The compensation of quantization error patterns obtained through a training process was applied to
edge regions. In smooth areas, the DC component as well as the first-order horizontal and vertical AC coefficients were modified using the DCT coefficients of adjacent
blocks to reduce the discontinuity over the block boundary. Later, a block boundary

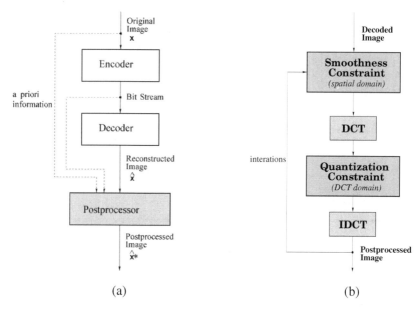

(a)   (b)

Figure 15.5: (a) A post-processor used at the decoder end, and (b) a general structure of POCS based postfilters

discontinuity measure was defined in [JJ98], and the minimization of the measure was realized by compensating the accuracy loss resulting from the quantization process for some selected transform coefficients. It was shown in the experiments that, although the block discontinuity was reduced, the performance with respect to PSNR was found, in general, degraded after post-processing. This is due to the fact that additional ringing effect is introduced in this case. More recently, Nosratinia [Nos99] introduced a post-processing approach based on the restoration of the local stationarity of image signal. Shifted versions of the decoded image were transformed and quantized as done in the encoding process. The dequantized and inverse transformed results were shifted back and then averaged to form the new stationarized estimate. In [LW03], an adaptive approach is proposed to perform blockiness reduction in both the spatial and DCT domains.

In [CWQ01], we proposed an alternative approach to blocking artifacts reduction by postfiltering the block DCT coefficients. The proposed scheme is conceptually similar to a weighted sum filter, however, it is applied in the DCT domain instead. The new postfilter makes use of the transform coefficients of the shifted blocks rather than those of the adjacent blocks in order to obtain a close correlation among the DCT coefficients at the same frequency. The filtering is operated variantly based on the local activity of blocks in order to achieve the artifacts reduction and detail preservation at the same time. In particular, an adaptively weighted low pass filtering technique is applied to image blocks of different activities, which represent the inherent abilities to mask coding artifacts. To characterize the block activity, human visual system sensitivity at different frequencies is considered. For low activity blocks, blocking artifacts are more perceptible and thus the postfiltering of the transform coefficients is applied within a large neighborhood (as will be defined later) to smooth out the artifacts. On the other hand, for blocks of high activity, a small window and a large central weight are used to preserve image details. This is because the eye has difficulty in discerning small intensity variations in portions of an image where strong edges and other abrupt intensity changes occur. Finally, the quantization constraint [Zak92, YGK93] is also applied to the filtered DCT coefficients prior to the reconstruction of the image by the inverse transformation of DCT coefficients. Recently, a fast version of the proposed filter was developed in [LB02]. The computational complexity has been reduced, however, at the cost of slightly degraded performance.

In what follows, we describe in detail the adaptive DCT domain post-filtering approach [CWQ01]. Computational complexity of the method will be discussed. Its implementation and performance evaluation will be presented.

15.3.1 Adaptive Postfiltering of Transform Coefficients

Let $\hat{\mathbf{x}}[\mathbf{c}]$ denote a pixel value in a reconstructed image of size $H \times W$, where $\mathbf{c} = (c_1, c_2)$, and $1 \leq c_1 \leq H$ and $1 \leq c_2 \leq W$. In block-based transform coding, an image $\mathbf{x}[\mathbf{c}]$ for $\mathbf{c} \in \mathcal{C}$ is composed of non-overlapping blocks $b_{m,n}[i,j]$, where $b_{m,n}[i,j]$ represents an $N \times N$ block with the top left pixel being $\mathbf{x}[m,n]$ and the bottom right pixel $\mathbf{x}[m + N - 1, n + N - 1]$. In other words, pixelwise $b_{m,n}[i,j] = \mathbf{x}[m + i, n + j]$, where $i, j = 0, \cdots, N - 1$. Clearly, m and n are integer multiples of N in this case. It should be noted that block $b_{m,n}[i,j]$ is adjacent to $b_{m-N,n}[i,j]$ in the vertical direction, and $b_{m,n}[i,j]$ is adjacent to $b_{m,n-N}[i,j]$ in the horizontal direction. These are shown in Figure 15.6 for $N = 8$, which is specifically used in most of the current image and video compression standards [Gha03].

The block-based 2-D DCT [RY90] can be expressed as follows.

$$B[u,v] = C(u)C(v) \sum_{i=0}^{N-1} \sum_{j=0}^{N-1} b[i,j] \cos \frac{(2i+1)u\pi}{2N} \cos \frac{(2j+1)v\pi}{2N}, \quad (15.1)$$

where $C(k) = \sqrt{\dfrac{1}{N}}$ for $k = 0$ and $C(k) = \sqrt{\dfrac{2}{N}}$ for $k = 1, \cdots, N - 1$. Here, $b[i,j]$ denotes the spatial domain pixel values of an image block, and $B[u,v]$, where $u, v = 0, \cdots, N - 1$, represent its transform coefficients. In other words, $B_{m,n}[u,v]$ represent the DCT coefficients of block $b_{m,n}[i,j]$. Due to the orthonormal property of the DCT, the inverse DCT (IDCT) can then be defined as follows.

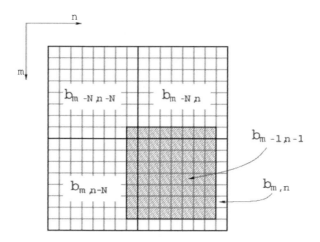

Figure 15.6: An illustration of adjacent and shifted blocks for $N=8$ [18].

$$b[i,j] = \sum_{u=0}^{N-1}\sum_{v=0}^{N-1} B[u,v]C(u)C(v)\cos\frac{(2i+1)u\pi}{2N}\cos\frac{(2j+1)v\pi}{2N}, \qquad (15.2)$$

for $i, j = 0, \cdots, N-1$.

For a specific coefficient, $B_{m,n}[u,v]$, its neighboring transform coefficients are defined as those at the same frequency of the shifted blocks, $B_{m+k,n+l}[u,v]$, where $k, l = -h, -h+1, \cdots, h$, within a window of size $(2h+1) \times (2h+1)$. Note that, in the spatial domain, blocks $b_{m,n}[i,j]$ and $b_{m+k,n+l}[i,j]$ overlap as long as $|k| < N$ and $|l| < N$ (see Figure 15.6). The assumption made in this investigation is that, in the original image, the neighboring transform coefficients at the same frequency are similar or do not vary radically within a local small region. This is obvious in monotone or flat areas as shown in Section 15.2, and thus we can apply a low pass filter to the neighboring transform coefficients in order to smooth out the artifacts. However, in texture or edge areas, this seems in general not to be the case. If we use the similar filtering operation, it would blur image details while reducing blocking artifacts. This is alleviated in the proposed postfiltering technique by an adaptive weighting mechanism. For those blocks containing details, filtering is applied within a small neighborhood. In addition, a large center weight is set in order to preserve the image details by preventing the current DCT coefficient from a significant alteration. A block activity classification is then involved in the proposed post-processing technique and will be described later on.

Prior to the introduction of the weighting strategy, we describe the general postfiltering procedure of the transform coefficients. The weighted average filtering is employed here to postprocess the DCT coefficients. Specifically, it can be expressed as follows.

$$\hat{B}_{m,n}[u,v] = \frac{1}{W} \sum_{k=-h}^{h} \sum_{l=-h}^{h} w_{k,l} B_{m+k,n+l}[u,v], \tag{15.3}$$

where $\hat{B}_{m,n}[u,v]$ denote the filtered coefficients, $w_{k,l}$ are the associated weights for different neighboring coefficients, and

$$W = \sum_{k=-h}^{h} \sum_{l=-h}^{h} w_{k,l} \tag{15.4}$$

is the total weightage of the filter.

The following subsections present how the proposed postfilter works in such a way that it adaptively filters the transform coefficients based on the local signal characteristics to efficiently smooth out blocking artifacts, while preserving image details at the same time.

15.3.1.1   Consideration of masking effect

For a good post-processing method, it is desirable to efficiently reduce blocking artifacts and simultaneously maintain the detailed features. In order to achieve this, we propose to vary the local smoothing correction in accordance with the visibility of artifacts. To measure the visibility of blocking artifacts, the visual masking effect is considered here. Visibility of artifacts is dependent on the local content of a given image. Due to the masking effect of the human visual system (HVS) [JJS93, NH94], artifacts in regions of high activity are less perceptible than those in low activity regions. In order to take into account the masking effect, an activity index is calculated for each image block.

The activity of a block represents its inherent masking capability for blocking artifacts. Generally, the HVS sensitivity function [Nil85, NLS89], $\hat{H}(\omega)$, measures the relative sensitivity of the eye at different frequencies. The inverse of this function, $\hat{H}^{-1}(\omega)$, while used to weigh the AC coefficients, would make a good estimate that approximates the amount of distortion masking [TPN96]. Due to the nonuniformity of HVS transfer function, Ngan, Leong and Singh [NLS89] proposed an HVS sensitivity function as follows:

$$\hat{H}(\omega) = |A(\omega)| \, H(\omega). \tag{15.5}$$

Here, the MTF (modulation transfer function), $H(\omega)$, with a peak at $\omega = 3$ cpd (cycles/degree) is given by

$$H(\omega) = (0.31 + 0.69\omega)\exp(-0.29\omega). \tag{15.6}$$

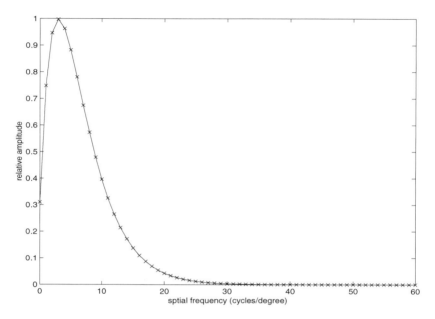

Figure 15.7: Plot of the modulation transfer function, Eq.(15.6).

The plot of this MTF is shown in Figure 15.7, and a comparison of different forms of MTFs can be found in [RY90]. The multiplicative function, $A(\omega)$, introduced by Nill [Nil85] is defined as

$$A(\omega) = \left\{ \frac{1}{4} + \frac{1}{\pi^2} \left[\ln \left(\frac{2\pi\omega}{\sigma} + \sqrt{\frac{4\pi^2\omega^2}{\sigma^2} + 1} \right) \right]^2 \right\}^{\frac{1}{2}} \tag{15.7}$$

where $\sigma = 11.636$ degree^{-1}. In the above equations, the frequency variable ω is in unit of cpd, and it varies corresponding to different 2-D DCT variables u and v. That is, each frequency component $[u, v]$ corresponds to a specific ω. To relate different u and v to the radial frequency ω, the conversion formula proposed in [CR90] is used, which can be described as

$$\omega = \omega_d \times \omega_s, \tag{15.8}$$

where

$$\omega_d = \frac{\sqrt{u^2 + v^2}}{2N} \text{ cycles/pixel}, \qquad u, v = 0, \cdots, N-1, \tag{15.9}$$

and ω_s in pixels/degree is the sampling density that is dependent on the viewing distance. For commonly used 512×512 test images, ω_s is chosen to be 64 pixels/degree, which is divisible by $2N$ as suggested in [CR90] and corresponds approximately to a viewing distance at which 65 pixels subtend 1 degree [RY90, MS74].

15.3.1.2 Block activity

To characterize the activity of an image block in the DCT domain, it can be approximated as the sum of its AC coefficients. Therefore, the activity index of block $b_{m,n}[i,j]$ is represented as

$$A_{m,n} = \frac{1}{B_{m,n}[0,0]} \left[\sum_{u=0}^{N-1} \sum_{v=0}^{N-1} \tilde{B}_{m,n}[u,v]^2 - \tilde{B}_{m,n}[0,0]^2 \right]^{\frac{1}{2}}, \qquad (15.10)$$

where $\tilde{B}_{m,n}[u,v]$ are the perceptually weighted DCT coefficients described in [TPN96] as

$$\tilde{B}_{m,n}[u,v] = \hat{H}^{-1}(\omega) B_{m,n}[u,v]. \qquad (15.11)$$

In Eq.(15.10), the normalization is based on the amplitude of DC coefficient, $B_{m,n}(0,0)$. This is to account for the adaptation of the local background luminance, which is in accordance with the Weber's law [NH94], since the DC coefficient is proportional to the local mean luminance of a block.

With assistance of this activity index, blocks in the image can be classified into either low activity or high activity category. Different classes of blocks are then processed in different ways. In the proposed framework, the block classification is achieved by simply thresholding the activity value of each block. If $A_{m,n} < \tau$, block $b_{m,n}$ is classified as a low activity block; otherwise, it is a high activity block. Threshold τ is determined experimentally as will be shown in the simulations.

15.3.1.3 Adaptive filtering

The proposed filter works in different ways for the blocks of low and high activities, respectively. In particular, for blocks of low activity, where the blocking artifacts appear to be visually more detectable, a 5×5 window (i.e., $h = 2$) is used in the filtering. In this case, the weights used in Equation (15.3) are employed such that $w_{k,l} = 1$ $(k, l = -2, \cdots, 2)$. On the other hand, for high activity blocks, we use a 3×3 filter (i.e., $h = 1$) whose weights are defined as

$$w_{k,l} = \begin{cases} 3 & \text{for } (k,l) = (0,0)) \\ 1 & \text{for } (k,l) \neq (0,0) \end{cases} \qquad (15.12)$$

and $k, l = -1, 0, 1$. The choice of a small neighborhood and a large weight set to the center is to avoid excessive blurring of high activity blocks.

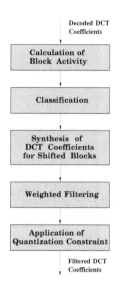

Figure 15.8: Flowchart of the DCT domain postfiltering scheme.

15.3.1.4   Quantization constraint

Prior to the reconstruction by applying the 2-D IDCT [PM93] to the filtered coefficients, the quantization constraint [Zak92, YGK93] is applied to each filtered coefficient. Since the quantization matrix used is known at the decoder, the corresponding original quantization interval $\left[B_{m,n}^{\min}[u,v], B_{m,n}^{\max}[u,v]\right]$ of coefficient $B_{m,n}[u,v]$ can be derived such that

$$\begin{cases} B_{m,n}^{\min}[u,v] = Q(u,v)\left[B_{m,n}[u,v] - \dfrac{1}{2}\right] \\[2mm] B_{m,n}^{\max}[u,v] = Q(u,v)\left[B_{m,n}[u,v] + \dfrac{1}{2}\right] \end{cases} \tag{15.13}$$

where $Q(u,v)$ is the quantization step designed for coefficient $B[u,v]$. The quantization constraint is applied by projecting any filtered DCT coefficient that is outside its quantization interval onto an appropriate value. The value $\hat{B}_{m,n}^{*}[u,v]$ of a filtered DCT coefficient after applying the quantization constraint can be finally given as

$$\hat{B}_{m,n}^{*}[u,v] = \begin{cases} B_{m,n}^{\min}[u,v] & \text{for } \hat{B}_{m,n}[u,v] < B_{m,n}^{\min}[u,v]; \\ \hat{B}_{m,n}[u,v] & \text{for } B_{m,n}^{\min}[u,v] \le \hat{B}_{m,n}[u,v] \le B_{m,n}^{\max}[u,v]; \\ B_{m,n}^{\max}[u,v] & \text{for } \hat{B}_{m,n}[u,v] > B_{m,n}^{\max}[u,v]. \end{cases} \tag{15.14}$$

The image is then reconstructed by applying the block-based 2-D IDCT to the quantization constrained coefficients $\hat{B}_{m,n}^{*}[u,v]$.

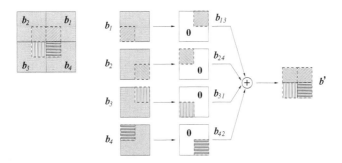

Figure 15.9: An example of block reassembling scheme in the spatial domain.

15.3.2 Implementation

A block diagram of the proposed DCT domain postfiltering is shown in Figure 15.8. Note that the computational complexities of the block classification and the subsequent filtering operations are comparable to those of spatial domain filtering. The extra computational load is introduced by the synthesis of DCT coefficients of shifted blocks, which are used in the filtering. In order to avoid the unnecessary IDCT and DCT pair to obtain the data in the spatial domain and then transform it back to the DCT domain, the DCT coefficients of shifted blocks can be assembled directly from the existing block DCT coefficients [CM95]. In other words, the conversion processes back and forth between the DCT domain and spatial domain can be eliminated. It should be noted that, in the POCS based methods, the DCT and IDCT operation pair is required in an iterative manner (as shown in Figure 15.5b) for the projection onto the quantization constraints. Due to its non-iterative nature, the proposed algorithm is computationally more efficient than the POCS based filtering approaches [Zak92, YGK93]. The latter in general require many iterations until the convergence is attained.

Regarding the DCT coefficient synthesis, let us consider a simple example shown in Figure 15.9. In the spatial domain, we have that

$$b' = b_{13} + b_{24} + b_{31} + b_{42}. \tag{15.15}$$

One can then calculate that [CM95]

$$\text{DCT}(b') = \text{DCT}(b_{13}) + \text{DCT}(b_{24}) + \text{DCT}(b_{31}) + \text{DCT}(b_{42}). \tag{15.16}$$

In this case, each subblock is obtained by using a pair of pre-multiplication and post-multiplication. These operations shift a subblock in the vertical and horizontal directions, respectively. Figure 15.10 illustrates the process of relocating subblock b_{42} by

$$b_{42} = s_{(1)} b_4 s_{(2)}. \tag{15.17}$$

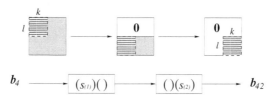

Figure 15.10: An example of subblock relocation operation.

Here, the pre-matrix and post-matrix [CM95] are in the form of

$$s_{(1)} = \begin{bmatrix} 0 & 0 \\ I_l & 0 \end{bmatrix} \quad \text{and} \quad s_{(2)} = \begin{bmatrix} 0 & I_k \\ 0 & 0 \end{bmatrix},$$

respectively, where I_i is an identity matrix of dimension i.

Using the distributive property of the DCT in matrix multiplication [SY00], i.e.,

$$\text{DCT}(XY) = \text{DCT}(X)\text{DCT}(Y), \tag{15.18}$$

we can write, assuming that $B = \text{DCT}(b)$,

$$B_{42} = \text{DCT}(b_{42}) = \text{DCT}(s_{(1)})B_4\text{DCT}(s_{(2)}), \tag{15.19}$$

where B_4, i.e., the DCT of block b_4, is available at the beginning of filtering as shown in Figure 15.8.

In other words, the DCT coefficients of a shifted block can be obtained as [CM95]

$$B_{m-k,n-l} = \sum_{x=0}^{1}\sum_{y=0}^{1} S_{xy(1)} B_{m-xN,n-yN} S_{xy(2)} \quad (0 < k,l < N). \tag{15.20}$$

In the above equation, $S_{xy(1)}$ and $S_{xy(2)}$ are the DCTs of the pre-matrix $s_{xy(1)}$ and post-matrix $s_{xy(2)}$, respectively. These matrices, as defined in [CM95], depend on k and l and can be pre-computed and stored in memory. Given $B_{m-xN,n-yN}$ $(0 \leq x,y \leq 1)$ and $S_{xy(c)}$ $(c = 1,2)$, Eq.(15.20) can be implemented by matrix multiplications. Recently, fast multiplication by a matrix was demonstrated by factorizing it into a product of sparse matrices whose entries are mostly 0, 1, and -1 [MB97]. Inverse motion compensation in the DCT domain and its fast algorithm [MB97, SY00] hopefully facilitate the application of the proposed technique to video processing. For detailed discussions of the DCT domain manipulation, interested readers are referred to [CM95, MB97, SY00] and the references therein.

In the proposed postfilter, the values of k and l are confined such that $-1 \leq k,l \leq 1$ for high activity blocks and $-2 \leq k,l \leq 2$ for low activity blocks as mentioned earlier. Additionally, in images coded at low bit rates, most of the block DCT coefficients at

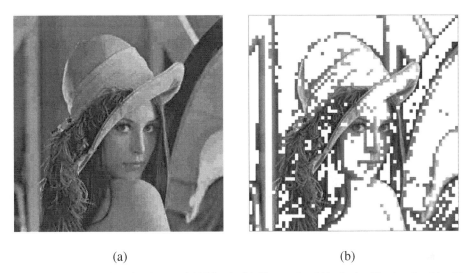

(a)   (b)

Figure 15.11: (a) JPEG coded image (0.219 bpp); (b) The result of block classification for (a) with threshold $\tau = 0.10$, where the blocks labeled in white are classified as low activity blocks [CWQ01].

high frequencies are zero, especially in low activity areas. Therefore, the 5×5 filter activated for low activity blocks may work on the DC coefficient, the most left column AC coefficients and upper top row AC coefficients only. As a result, this reduces the computational complexity by 49/64 for each low activity block. In our simulations, more than 70% of blocks are classified as low activity blocks for the test images used. It is also observed that the degradation in terms of visual quality and PSNR performance is negligible when doing so. Nevertheless, the results presented in the following section are obtained by filtering all the DCT coefficients. Furthermore, hardware implementation with parallel processing of the proposed technique is feasible as the filtering of individual block DCT coefficients is independent of one another.

15.3.3 Simulation Results

In order to gauge the efficacy of the proposed postfiltering technique in reducing blocking artifacts, experiments have been conducted over a broad range of 512×512 test images.[1] The images were compressed by using a base-line JPEG encoder [PM93]. In addition, its performance has been compared with those of other post-processing methods.

---

[1] The test images were downloaded from http://www.cipr.rpi.edu/resource/stills/misc1.html

Table 15.1: Results of performance comparison in terms of the PSNR of different post-processing methods for reducing blocking artifacts in JPEG coded images [18].

Image	Bitrate (bpp)	PSNR (dB)						
		Decoded	H.263	MPEG4	POCS-Z	POCS-Y	POCS-P	Proposed
Airplane	0.241	28.72	29.34	29.57	27.91	29.34	29.07	29.39
Barbara	0.295	25.08	25.31	25.52	25.06	25.52	25.34	25.62
Boats	0.235	28.52	29.07	29.30	28.01	29.16	28.76	29.25
Goldhill	0.230	27.91	28.50	28.53	27.72	28.46	28.12	28.54
Lena	0.219	29.47	30.20	30.43	29.02	30.23	29.88	30.39
Peppers	0.221	29.21	30.02	30.15	28.82	29.85	29.65	29.95
Sailboat	0.286	27.00	27.51	27.68	26.46	27.57	27.26	27.66
Splash	0.178	31.23	32.00	32.29	30.61	31.77	31.75	31.89
Tiffany	0.188	29.59	30.10	30.38	29.36	30.11	29.97	30.15

15.3.3.1   Results of block classification

Figure 15.11 shows an example of the results of block classification employed in the proposed postfiltering scheme. The block activity indexes are calculated and then classified for the JPEG coded image. The blocks labeled in white in Figure 15.11b are classified as low activity blocks, and the blocking artifacts appear to be more discernible in those areas. On the other hand, in high activity areas, the blocking artifacts are less noticeable due to the masking effect by the background, which can be observed from Figure 15.11b. The threshold τ is set to 0.10. According to the extensive simulations conducted, such a choice works satisfactorily in the block classification based on the activity indexes for different test images. In the following experiments, the proposed postfiltering technique is applied with threshold $\tau = 0.10$ for different images coded at various bit rates.

15.3.3.2   Performance evaluation

In this subsection, the performance of the proposed technique is evaluated and compared with those of other postprocessers. The postfilters suggested by the H.263 [ITU98] and the MPEG-4 [ISO99b] standards, and three POCS based methods, i.e., POCS-Z [Zak92], POCS-Y [YGK93], and POCS-P [PKL98], have been simulated for comparison. The filtering strength used in the H.263 postfilter and the quantization parameter used in the MPEG-4 deblocking filter are tuned towards their best PSNR performance for different cases, respectively. It should be noted that this is in fact not feasible in practice where the original image is not available, and thus makes the two deblocking filters less useful practically for still image compression.

The comparative results in PSNR for different test images are provided in Table 15.1. The proposed postfilter exhibits excellent performance as compared to the POCS based

Table 15.2: Comparative results using GBIM for different deblocking approaches. Note that the bit rate of each coded image is the same as that used in Table 15.1 [18].

Image	M_{GBIM}							
	Original	JPEG	H.263	MPEG4	POCS-Z	POCS-Y	POCS-P	Proposed
Airplane	1.1177	4.0615	1.3917	1.4209	1.5470	1.7439	2.0292	1.2864
Barbara	1.0415	3.0789	1.7398	1.1465	1.3447	1.6753	1.7578	1.2404
Boats	1.0737	4.1674	1.9803	1.4785	1.5707	1.4402	2.1040	1.3387
Goldhill	1.0351	4.7412	1.5288	1.5091	1.5285	1.4169	2.2451	1.3979
Lena	1.1009	4.9905	1.5514	1.5931	1.5702	1.4780	2.4191	1.3777
Peppers	1.1368	5.6746	1.7197	1.7831	1.5581	1.8250	2.2613	1.3986
Sailboat	1.1414	3.8388	1.5359	1.4880	1.6358	1.4654	2.1186	1.3321
Splash	1.0435	5.7207	1.8760	1.4846	1.4191	2.3544	1.8109	1.2102
Tiffany	1.0139	6.0632	1.8979	1.5240	1.7656	2.2406	1.9322	1.3784

approaches, in general providing better results for all the test images. The deblocking filters defined in H.263 and MPEG-4 work well for images containing large smooth regions, such as *Peppers* and *Splash*. However, for images having more details, such low-pass filtering brings blurring or insufficient reduction of blocking artifacts in some high activity areas due to the switching of filter modes, depending on the pixel conditions around a block boundary [ISO99b]. This will be shown in the following evaluation of perceived quality.

It is well known that the PSNR is not always a good measure to reflect subjective image and video quality, albeit it is one of the most popular objective metrics employed in image processing. In [WY97], Wu and Yuen introduced a blockiness measurement, called GBIM (generalized block-edge impairment metric), for image and video coding. The GBIM was devised by taking into account the visual significance of blocking artifacts in a given image, based on a formulation of constraint sets applied in the POCS filtering algorithms [YGK93], and was demonstrated to be consistent with subjective quality evaluation. It is interesting to note that, from Table 15.2, the proposed technique also provides better performance than other rivals according to the GBIM evaluation results, where a larger value indicates that blockiness is more prominently associated with the image. This has also been verified with the following examples showing the perceptual quality of post-processed images.

In Figure 15.12, the central portions of respective *Lena* images processed using different postfilters are presented to show the visual quality. In the image processed by the H.263 postfilter (i.e., Figure 15.12c), blocking artifacts are not sufficiently reduced. Although the MPEG-4 deblocking filter reduces the blockiness sufficiently in smooth regions, it blurs the details or retains blocking artifacts perceptible in some high activity areas probably due to the filter model switching. Figure 15.12e is free of blocking artifacts at the expense of excessive blurring caused by a spatial-invariant low pass

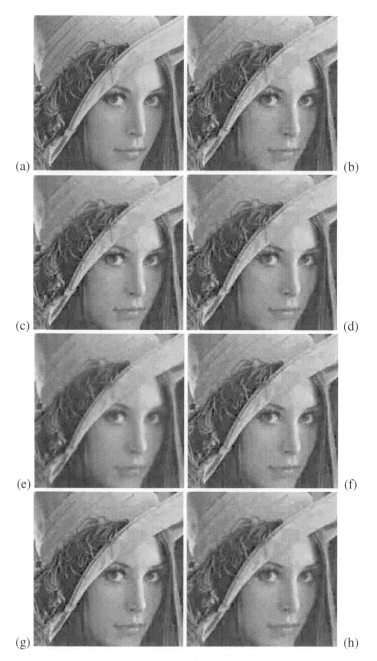

Figure 15.12: Subjective quality comparison of *Lena* compressed at 0.219 bpp and postfiltered by different methods. (a) Original and (b) decoded images; images postprocessed by (c) H.263 post-filter, (d) MPEG-4 deblocking filter, (e) POCS-Z, (f) POCS-Y, (g) POCS-P, (h) the proposed approach.

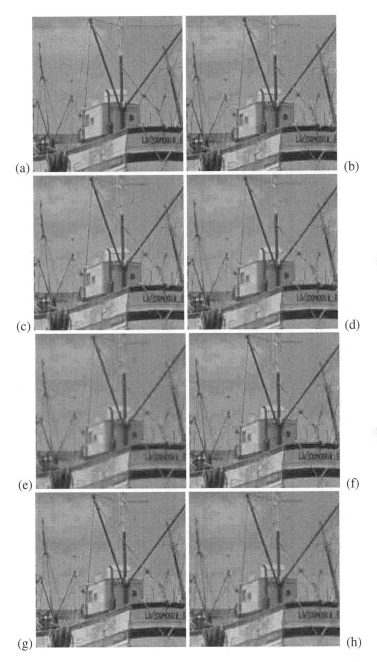

Figure 15.13: Subjective quality comparison of *Boats* compressed at 0.235 bpp and postfiltered by different methods. (a) Original and (b) decoded images; images postprocessed by (c) H.263 post-filter, (d) MPEG-4 deblocking filter, (e) POCS-Z, (f) POCS-Y, (g) POCS-P, (h) the proposed approach.

filtering [Zak92]. Using the POCS based filter proposed in [YGK93], the blocking artifacts are not sufficiently alleviated especially in the flat areas (see Figure 15.12f), since only the values of adjacent pixels on the block boundaries are adjusted. In the image processed using the approach introduced in [PKL98], i.e., Figure 15.12g, the blocking artifacts in high activity regions remain perceptually noticeable. Our proposed postfiltering provides a good performance, achieving favorably both aspects of reducing blocking artifacts and preserving image details. The processed image (see Figure 15.12h) gives subjectively better quality than others. Among those post-processing methods, the best PSNR and GBIM values have been obtained in this case by using the proposed postfilter.

An interesting observation from the extensive experiments conducted is that, in addition to reducing blocking effect, the proposed filtering technique is also capable of suppressing ringing artifacts. The post-processing results of the *Boats* image are shown in Figure 15.13. One can discern a better performance in reducing ringing effect and preserving details provided by the proposed postfilter. This is probably due to the principle that the formulation of the DCT domain filtering is motivated in general to recover the information loss of the coefficients.

15.4   Reduction of Ringing Artifacts

Though different coding methods bring in different artifacts, the ringing artifacts have been observed in both the DCT and wavelet based image compression at low bit rates. The design of post-processing technique for ringing artifact reduction should be tailored for a specific compression method.

For DCT-based image coding, the ringing effect is considered the major artifact around edges. Thus, some methods based on edge-preserving maximum a posteriori (MAP) estimation [Ste93, OS95, LCP97], or adaptive filtering [MSF97, FE96] were proposed to cope with this problem. In the recent video coding standards H.263 [ITU97] and MPEG-4 [ISO99b], deringing filters have been described as part of post-processing similar to deblocking filters at the decoder side. These deringing filters work in the spatial domain. They adaptively change filtering mode or strength based on the detection of edges in the decoded images. However, besides the ringing effect, edges have also been blurred by the low pass filtering effect introduced by the allocation of zero bits to high-frequency coefficients. The previous techniques for reducing the ringing effect cannot eliminate the blurring effects and are unable to reproduce the edge sharpness. Figure 15.14 shows an example of using the MPEG-4 deringing filter. In Figure 15.14a, ringing artifacts are noticed around strong edges. The ringing artifacts have been reduced in Figure 15.14b, however the image is blurred by the deringing filter.

In order to attain sufficient image quality for low bit-rate wavelet-based image coding, post-processing is also an efficient technique for improving compression re-

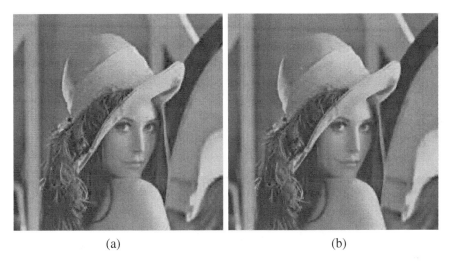

(a)   (b)

Figure 15.14: (a) *Lena* image compressed using MPEG-4 with QP=15 and then processed by MPEG-4 deblocking filter; (b) the image in (a) processed by MPEG-4 deringing filter.

sults. The post-processing techniques for wavelet-based coding have been studied [LCP97, LK97, SK98, SK99] by using optimization functions that share ideas similar to those used in DCT-based post-processing. These methods can achieve certain PSNR gains [LCP97, SK99] without further blurring image details [LK97, SK98]. Nevertheless, these approaches do not deblur edges that were blurred due to the quantization errors or truncation of the high frequency wavelet coefficients. Efforts have been made to improve continuity of smooth regions in the coded images [LCP97], and recently some work has been devoted to the edge recovery, which is worthwhile for low bit-rate wavelet-based image coding. Since edges define the most recognizable features for objects in an image, the distortions around edges are disturbing and annoying to human perception.

In [FC00], edge reconstruction or recovery is studied for low bit-rate wavelet image compression. It can recover distorted edges based on an edge model and a degradation model. However, it does not consider the artifacts in smooth regions. An edge model is developed for coded images by the addition of two new terms to the original edge model, which is able to characterize two artifacts, ringing effects and blurring effects, around lossy edges. The problem of edge reconstruction is formulated as that of image restoration and can be solved by recovering original edge structure and reducing quantization noise. The former involves estimating original model parameters from the coded signal. The latter is implemented by local Gaussian filtering of edges in the coded signal. Furthermore, it introduces a parametric reconstruction model which is able to provide a flexible tradeoff between visual enhancement and fidelity improvement for the reconstructed image.

A method based on maximum-likelihood (ML) estimation is proposed in [S. 01] for ringing artifact reduction. It assumes that an image is a montage of flat surfaces, and the flat surface model is fitted to the observed image by estimating the parameters of the model. Specifically, it uses the flat-surface image model to distinguish true edges from ringing artifacts that have similar magnitudes. The ML method is employed to estimate true pixel value within a sliding window. The k-means algorithm and a hierarchical clustering method were employed to achieve the parameter fitting and estimation. The proposed methods can be simplified towards a non-iterative version.

15.5 Summary

We have discussed in this chapter the post-processing techniques developed for reducing artifacts in low bit rate image compression. The main focus has been on the reduction of blocking and ringing artifacts. The recent developments on deblocking filters have been reviewed. A detailed description of an adaptive DCT domain postfilter has been presented. Its performance in reducing coding artifacts has been evaluated. In addition, recent work on the ringing artifact reduction has been discussed. The de-ringing techniques that were designed for both the DCT and wavelet transform based compression have been reviewed. It is an attractive topic to improve the quality of compressed images since it is equivalent to reducing the bit rate while maintaining the similar quality. On the other hand, it is a challenging task and more effort needs to be made in this field. In recent years, the research on this topic has been closely followed as can be seen from the number of papers published.

References

[Cla85]   R. Clarke. *Transform coding of images*. Orlando, FL: Academic Press, 1985.

[CM95]   S.-F. Chang and D. G. Messerschmitt. Manipulation and compositing of MC-DCT compressed video. *IEEE J. Select. Areas Commun.*, 13(1):1–11, January 1995.

[CR90]   B. Chitprasert and K. R. Rao. Human visual weighted progressive image transmission. *IEEE Trans. Commun.*, 38(7):1040–1044, July 1990.

[CWQ01]   T. Chen, H. R. Wu, and B. Qiu. Adaptive postfiltering of transform coefficients for the reduction of blocking artifacts. *IEEE Trans. Circuits Syst. Video Technol.*, 11:594–602, May 2001.

[dNR96]   R. L. de Queiroz, T. Q. Nguyen, and K. R. Rao. GenLOT: generalized linear-phase lapped orthogonal transform. *IEEE Trans. Signal Processing*, 44:497–507, March 1996.

[FC00]   G. Fan and W. K. Cham. Model-based edge reconstruction for low bit-rate wavelet compressed images. *IEEE Trans. Circuits Syst. Video Technol.*, 10(1):120–132, January 2000.

[FE96]   Z. Fan and R. Eschbach. JPEG decompression with reduced artifacts. In *Proc. SPIE Symp. Image and Video Compression*, 2186:50–55, February 1996.

[FJ86]   P. Farrelle and A. K. Jain. Recursive block-coding — a new approach to transform coding. *IEEE Trans. Commun.*, COM-34:161–179, February 1986.

[Gha03] M. Ghanbari. *Standard Codecs: Image Compression to Advanced Video Coding*. London, UK: IEE, 2003.

[GK02]   W. Gao and Y. Kim. A de-blocking algorithm and a blockiness metric for highly compressed images. *IEEE Trans. Circuits Syst. Video Technol.*, 12(12):1150–1159, December 2002.

[HBS84] B. Hinman, J. Bernstein, and D. Staelin. Short-space Fourior transform image processing. In *Proc. IEEE ICASSP*, 481–484, San Diego, CA, March 1984.

[IEE03] IEEE Trans. Circuits Syst. Video Technol. *Special issue on H.264/MPEG-4 Part 10*, July 2003.

[ISO99a] ISO/IEC. *JPEG 2000 Requirements and Profiles*, March 1999.

[ISO99b] ISO/IEC JTC1/SC29/WG11. MPEG4 Video Verification Model version 15.0 (VM15.0). N3093, December 1999.

[ISO00] ISO/IEC. *JPEG 2000 Part I: Final Draft International Standard (ISO/IEC FDIS15444-1)*, August 2000.

[ITU97] ITU-T Video Coding Experts Group. Video Codec Test Model, Near-term, version 8 (TMN8). Release 0, June 1997.

[ITU98] ITU-T Video Coding Experts Group. Video coding for low bit rate communication. ITU-T recommendation H.263, January 1998.

[J.-92] J.-H. Chen *et al.* A low delay CELP coder for the CCITT 16 kbps speech coding standard. *IEEE J. Select. Areas Commun.*, 10:830–849, June 1992.

[J. 96] J. L. Mitchell *et al. MPEG Video: Compression Standard*. New York, NY: Chapman & Hall, 1996.

[Jai89] A. K. Jain. *Fundamentals of Digital Image Processing*. Englewood Cliffs, NJ: Prentice–Hall, 1989.

[JJ98]   B. Jeon and J. Jeong. Blocking artifacts reduction in image compression with block boundary discontinuity criterion. *IEEE Trans. Circuits Syst. Video Technol.*, 8(3):345–357, June 1998.

[JJS93] N. Jayant, J. Johnston, and R. Safranek. Signal compression based on models of human perception. *Proc. IEEE*, 81(10):1385–1422, October 1993.

[Joi03] Joint Video Team of ISO MPEG and ITU-T VCEG. Draft ITU-T recommendation and final draft international standard of joint video specification (ITU-T Rec. H.264 and ISO/IEC 14496-10 AVC). Approved Output Document of JVT, JVT-G050r1 (ftp://standards.polycom.com), May 2003.

[KH95]   C. J. Kuo and R. J. Hsieh. Adaptive postprocessing for block encoded images. *IEEE Trans. Circuits Syst. Video Technol.*, 5:298–304, August 1995.

[KK95]   S. A. Karunasekera and N. K. Kingsbury. A distortion measure for blocking artifacts on images based on human visual sensitivity. *IEEE Trans. Image Processing*, 4(6):713–724, June 1995.

[Kun95]   A. Kundu. Enhancement of JPEG coded images by adaptive spatial filtering. In *Proc. IEEE ICIP*, 187–190, October 1995.

[LB02]   S. Liu and A. C. Bovik. Efficient DCT-domain blind measurement and reduction of blocking artifacts. *IEEE Trans. Circuits Syst. Video Technol.*, 12(12):1139–1149, December 2002.

[LCP97]   J. Luo, C. Chen, and K. J. Parker. Image enhancement for low bit rate wavelet-based image compression. In *Proc. IEEE ISCAS*, 1081–1084, June 1997.

[LK97]   J. Li and C.-C. J. Kuo. Coding artifact removal with multiscale postprocessing. In *Proc. IEEE ICIP*, 529–532, 1997.

[LRL95]   W. E. Lynch, A. R. Reibman, and B. Liu. Postprocessing transform coded images using edges. In *Proc. IEEE ICASSP*, 2323–2326, May 1995.

[LW03]   Y. Luo and R. K. Ward. Removing the blocking artifacts of block-based DCT compressed images. *IEEE Trans. Image Processing*, 12(7):838–842, July 2003.

[Mal92]   H. S. Malvar. *Signal Processing with Lapped Transforms*. Boston, MA: Artech House, 1992.

[MB97]   N. Merhav and V. Bhaskaran. Fast algorithms for DCT-domain image down-sampling and for inverse motion compensation. *IEEE Trans. Circuits Syst. Video Technol.*, 7(3):468–476, June 1997.

[MS74]   J. L. Mannos and D. J. Sakrison. The effect of a visual fidelity criterion on the encoding of images. *IEEE Trans. Infom. Theory*, IT-20:525–536, July 1974.

[MS82]   H. S. Malvar and D. H. Staelin. The LOT: Transform coding without blocking effects. *IEEE Trans. Acoust., Speech, Signal Processing*, 37:553–559, April 1982.

[MSF97]   J. D. McDonnell, R. N. Shortern, and A. D. Fagan. Postprocessing of transform coded images via histogram-based edge classification. *J. Electron. Imaging*, 6(1):114–124, January 1997.

[NH94]   A. N. Netravali and B. G. Haskell. *Digital Pictures: Representation, Compression, and Standards*. New York, NY: Plenum Press, 2nd ed., 1994.

[Nil85]   N. B. Nill. A visual model weighted cosine transform for image compression and quality assessment. *IEEE Trans. Commun.*, COM-33(6):551–557, June 1985.

[NLS89]   K. N. Ngan, K. S. Leong, and H. Singh. Adaptive cosine transform coding of images in perceptual domain. *IEEE Trans. Acoust., Speech, Singal Processing*, 37(11):1743–1750, November 1989.

[Nos99]   A. Nosratinia. Embedded post-processing for enhancement of compressed images. In *Proc. IEEE Data Compression Conference (DCC)*, 62–71, Snowbird, UT, March 1999.

[OS95]   T. P. O'Rourke and R. L. Stevenson. Improved image decompression for reduced transform coding artifacts. *IEEE Trans. Circuits Syst. Video Technol.*, 5:490–499, December 1995.

[OST93]   M. Ozkan, M. Sezan, and A. Tekalp. Adaptive motion-compensated filtering of noisy image sequences. *IEEE Trans. Circuits Syst. Video Technol.*, 3(4):277–290, August 1993.

[P. 03]   P. List *et al.* Adaptive deblocking filter. *IEEE Trans. Circuits Syst. Video Technol.*, 13(7):614–619, July 2003.

[PKL98]   H. Paek, R.-C. Kim, and S.-U. Lee. On the POCS-based postprocessing technique to reduce the blocking artifacts in transform coded images. *IEEE Trans. Circuits Syst. Video Technol.*, 8(3):358–367, June 1998.

[PM93]   W. B. Pennebaker and J. L. Mitchell. *JPEG Still Image Data Compression Standard.* New York, NY: Van Nostrand Reinhold, 1993.

[PW84]   D. Pearson and M. Whybray. Transform coding of images using interleaved blocks. *Proc. IEE*, 131:466–472, August 1984.

[RG86]   B. Ramamurthi and A. Gersho. Nonlinear space-variant postprocessing of block coded images. *IEEE Trans. Acoust., Speech, Signal Processing*, ASSP-34:1258–1267, October 1986.

[Ric03]   I. E. G. Richardson. *H.264 and MPEG-4 Video Compression: Video Coding for Next Generation Multimedia.* Hoboken, NJ: Wiley, 2003.

[RJ02]   M. Rabbani and R. Joshi. An overview of the JPEG 2000 still image compression standard. *Signal Processing: Image Communication*, 17:3–48, January 2002.

[RL84]   H. Reeve and J. Lim. Reduction of blocking effects in image coding. *Optical Engineering*, 23(1):34–37, January 1984.

[RY90]   K. R. Rao and P. Yip. *Discrete Cosine Transform - Algorithms, Advantages, Applications.* Boston, MA: Academic Press, 1990.

[S. 01]   S. Yang *et al.* Maximum-likelihood parameter estimation for image ringing-artifact removal. *IEEE Trans. Circuits Syst. Video Technol.*, 11(5):963–973, August 2001.

[Sau91]   K. Sauer. Enhancement of low bit-rate coded images using edge detection and estimation. *Computer Vision, Graphics, Image Processing*, 53:52–62, January 1991.

[SCE01]   A. Skodras, C. Christopoulos, and T. Ebrahimi. The JPEG2000 still image compression standard. *IEEE Signal Process. Magazine*, 18(5):36–58, December 2001.

[Sha93]   J. M. Shapiro. Embedded image coding using zerotrees of wavelet coefficients. *IEEE Trans. Signal Processing*, 41(12):3445–3462, December 1993.

[SK98]   M.-Y. Shen and C.-C. J. Kuo. Artifact reduction in low bit-rate wavelet coding with robust nonlinear filtering. In *Proc. IEEE Workshop Multimedia Signal Processing*, 480–485, 1998.

[SK99]   M.-Y. Shen and C.-C. J. Kuo. Real-time compression artifact reduction via robust nonlinear filter. In *Proc. IEEE ICIP*, 565–569, 1999.

[SP96]   A. Said and W. A. Pearlman. A new fast and efficient image codec based on set partitioning in hierarchical trees. *IEEE Trans. Circuits Syst. Video Technol.*, 6(3):243–250, June 1996.

[Ste93]   R. L. Stevenson. Reduction of coding artifacts in transform image coding. In *Proc. IEEE ICASSP*, 5:401–404, Minneapolis, MN, March 1993.

[SY00]   J. Song and B.-L. Yeo. A fast algorithm for DCT-domain inverse motion compensation based on shared information in a macroblock. *IEEE Trans. Circuits Syst. Video Technol.*, 10(5):767–775, August 2000.

[TH93]   C.-N. Tien and H.-M. Hang. Transform-domain postprocessing of DCT-coded images. In *Proc. SPIE Visual Commun. & Image Processing (VCIP)*, 2094:1627–1638, 1993.

[TM01]   D. S. Taubman and M. W. Marcellin. *JPEG2000: Image Compression Fundamentals, Standards, and Practice.* Hingham, MA: Kluwer, 2001.

[TPN96] S. H. Tan, K. K. Pang, and K. N. Ngan. Classified perceptual coding with adaptive quantization. *IEEE Trans. Circuits Syst. Video Technol.*, 6(4):375–388, August 1996.

[TTS02] G. A. Triantafyllidis, D. Tzovaras, and M. G. Strintzis. Blocking artifact detection and reduction in compressed data. *IEEE Trans. Circuits Syst. Video Technol.*, 12(10):877–890, October 2002.

[Tzo83] K.-H. Tzou. Post-filtering of transform-coded images. In *Proc. SPIE Applications of Digital Image Processing XI*, 121–126, April 1983.

[WSA94] A. B. Watson, J. A. Solomon, and A. Ahumada. DCT basis function visibility: Effects of viewing distance and contrast masking. In *Proc. SPIE Human Vision, Visual Processing, and Digital Display*, 2178:99–108, May 1994.

[WY97]   H. R. Wu and M. Yuen. A generalized block-edge impairment metric for video coding. *IEEE Signal Processing Letters*, 4(11):317–320, November 1997.

[YGK93] Y. Yang, N. P. Galastanos, and A. K. Katsaggelos. Regularized reconstruction to reduce blocking artifacts of block discrete cosine transform compressed images. *IEEE Trans. Circuits Syst. Video Technol.*, 3:421–432, December 1993.

[YW98]   M. Yuen and H. R. Wu. A survey of hybrid MC/DPCM/DCT video coding distortions. *Signal Processing* (Special Issue on Image Quality Assessment), 70(3):247–278, November 1998.

[Zak92] A. Zakhor. Iterative procedures for reduction of blocking effects in transform image coding. *IEEE Trans. Circuits Syst. Video Technol.*, 2:91–95, March 1992.

[ZPL93] Y. Zhang, R. Pickholtz, and M. Loew. A new approach to reduce the blocking effect of transform coding. *IEEE Trans. Commun.*, 41:299–302, February 1993.

Chapter 16

Reduction of Color Bleeding in DCT Block-Coded Video

François-Xavier Coudoux and Marc G. Gazalet
University of Valenciennes, France

This chapter discusses a post-filtering technique which reduces the color bleeding artifacts commonly encountered in block-based transform coded pictures and enhances visual quality performance of picture coding systems.

16.1   Introduction

Among the many existing digital image and video compression approaches, transform based methods using the Discrete Cosine Transform (DCT) are by far the most popular [RY90, BK95]. However, these methods are generally lossy, and introduce visually annoying artifacts in the reconstructed sequences [MH98]. Blocking effect is the most well-known one, and numerous pre/post-processing algorithms have been studied in order to ameliorate the blocking artifacts [PL99, AFR01]. Nevertheless, other types of artifacts have been barely considered; this is the case of the so-called color bleeding phenomenon. Color bleeding appears as a smearing of the color between areas of strongly contrasting chrominance, and results from the quantization of the higher order AC coefficients. For chrominance edges of very high contrast, the color artifact corresponding to the ringing effect also occurs. The colors produced as a result of ringing often do not correspond to the color of the surrounding area, as illustrated in Figure 16.1. This kind of distortion is highly annoying for a human observer, and greatly affects the overall image quality. So, reduction of color bleeding among with other artifacts can result in a significant improvement in the overall visual quality of the decoded images.

In this chapter, a post-processing algorithm is described that removes the color bleeding distortion in DCT-coded color images [CGC04]. The chapter is organized as follows. First, we present an overview of existing digital color video formats, and the psychovi-

Figure 16.1: Illustration of color bleeding on one frame from the *Table-Tennis* sequence (see above the racket, as well as along the arm).

sual properties related to color image compression. Then, we give a thorough analysis of the color bleeding phenomenon, as a result of both quantization and decimation of chrominance data. Finally, an adaptive post-filtering algorithm is developed, based on the previous analysis. Simulation results for different color images show the improvement of the reconstructed video, both objectively and subjectively.

16.2 Analysis of the Color Bleeding Phenomenon

Color bleeding occurs when one color in the image bleeds into or overlaps into another color inappropriately. We demonstrate in this section that this distortion occurs because of both decimation and quantization of the chrominance components at the compression stage.

16.2.1 Digital Color Video Formats

Most current digital image compression standards handle colors as three separate components [Wan95]. Among the various existing colorimetric spaces, the YUV color-space is widely used in digital video communications, where Y corresponds to the luminance part of the video signal, and U/V are the so-called chrominance components. Since the human eye is less sensible to colored details, the amount of chrominance information data can be reduced without decreasing the subjective image quality.

Several formats have been developed in recent years, in order to represent digital color video signals. The YUV 4:2:2 video format constitutes the reference for studio

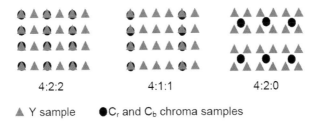

$$4{:}2{:}2 \qquad\qquad 4{:}1{:}1 \qquad\qquad 4{:}2{:}0$$

▲ Y sample   ●C_r and C_b chroma samples

Figure 16.2: Illustration of digital color video formats.

applications: in this format, the U/V components have half the resolution horizontally, when compared to the Y component (Figure 16.2). These are further interpolated at the reconstruction stage, to restore the full-resolution color image data.

In order to further reduce chrominance information data, other digital video formats have been proposed; the U/V color components are decimated once more horizontally (4:1:1), or vertically (4:2:0). These formats are widely used in actual image and video applications, including Digital Video Broadcasting (DVB), Digital Versatile Disk (DVD), or M-JPEG, DV and DVCPRO25 digital video recording formats [MPFL96].

16.2.2   Color Quantization

Color space conversion described above is a first step toward compressing the video. Then, the encoding algorithm itself is applied to each separate color component. It can be divided into three main stages:

- The removal of the data redundancy by means of the Discrete Cosine Transform (DCT), associated with a motion/compensation process in the case of video coding.

- The quantization of the DCT coefficients.

- The entropy encoding of the quantized data, using a variable-length-word encoder.

The quantization step constitutes the lossy part of the encoding process. The frequency coefficients are quantized uniformly by normalization and rounding, using psychovisual weighting functions optimized for the human visual system (HVS). It is well known that the sensitivity of the human eye drops off with increasing spatial frequency [Gle93]. Furthermore, this characteristic is more strongly verified in the two chrominance channels than in the luminance one. As a result, the quantizer step size varies by the coefficient frequency range, and can be tuned for each color component. Since the HVS is less sensitive to the chrominance details, coarse quantization of C_b and

Table 16.1: Example of quantization tables for the luminance (left) and chrominance (right) channels [ISO94, CGC04]. (©2004 IEEE)

16	11	10	16	24	40	51	61		17	18	24	47	99	99	99	99
12	12	14	19	26	58	60	55		18	21	26	66	99	99	99	99
14	13	16	24	40	57	69	56		24	26	56	99	99	99	99	99
14	17	22	29	51	87	80	62		47	66	99	99	99	99	99	99
18	22	37	56	68	109	103	77		99	99	99	99	99	99	99	99
24	35	55	64	81	104	113	92		99	99	99	99	99	99	99	99
49	64	78	87	103	121	120	101		99	99	99	99	99	99	99	99
72	92	95	98	112	100	103	99		99	99	99	99	99	99	99	99

C_r chrominance components can be applied. This is illustrated in Table 16.1, showing typical JPEG default quantization matrices, that have been optimized for the luminance and chrominance components, respectively.

16.2.3   Analysis of Color Bleeding Distortion

We consider in the present section how both decimation and quantization operations are involved in the manifestation of the color bleeding distortion. Let us first evaluate the effects of coarse quantization of the DCT coefficients on the reconstructed color video quality. In Figure 16.1, blocking effect is clearly visible in uniform areas, while ringing and DCT basis image effect mainly appear on edges and textures. It can be seen also that color bleeding is highly present above the racket, and along the player's arm. In order to isolate this phenomenon, we first consider the case where the color image is reconstructed from the coded Y component, while the original chrominance components are used. In this case, most compression artifacts remain despite of the use of C_b and C_r original components; nevertheless, color bleeding disappears (Figure 16.3).

In contrast, if the color image is reconstructed using the Y original part, and the C_b and C_r coded parts, respectively, blocking effect and ringing almost disappear, while color bleeding persists (Figure 16.4). We can therefore reasonably assume that color bleeding is only due to the quantization of the chrominance components of compressed video.

In order to investigate the role of the sub-sampling operation, we consider hereafter in a schematic way the one-dimensional case. Figure 16.5a represents a simple 16×16 color test pattern used in our experiments; it consists in a strong red/yellow vertical transition, located in the left part of the image. The luminance value differs between the red and the yellow parts because, in practice, iso-luminance conditions are extremely rare in natural scenes. Figure 16.5b represents one image row extracted from the C_r

Figure 16.3: One frame from the *Table-Tennis* sequence reconstructed from the Y coded component, and the C_b and C_r original components, respectively.

Figure 16.4: One frame from the *Table-Tennis* sequence: in this case, the image is reconstructed from the Y original component, and the C_b and C_r coded components, respectively.

component. Up to there, no sub-sampling is applied, so that the vector's length is equal to 16 pixels; the right half part is uniform, while a sharp transition is located in the left half part of the 1-D vector, corresponding to the presence of the color edge.

If the chrominance components are not sub-sampled, each of the 8-pixel half-vectors is transformed independently by means of DCT, and quantized. After decoding, coding errors are limited to the corresponding 8-pixel half-vector. In particular, spurious components due to ringing along the color edge are restricted to the left half-part, and do not bleed in the right one. However, in practice, chrominance data are sub-sampled prior to encoding (Figure 16.5c). After decimation (here, by a factor of 2), a single 8-pixel vector is assumed to represent the C_r color content of the whole 16-pixel length block. This 8-pixel vector is transformed by means of DCT, and then quantized. Due to quantization, the original balance of the DCT coefficients is corrupted, and distortions are introduced in the whole reconstructed signal (Figure 16.5d). Finally, the C_r chrominance component is interpolated in order to reconstruct the 16-pixel vector. Unfortunately,

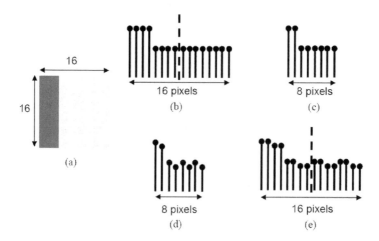

Figure 16.5: Schematic illustration of sub-sampling effect on the manifestation of color bleeding [CGC04]. (©2004 IEEE)

this process propagates errors to the right side of the signal, so that color bleeds locally beyond the image area where it should physically be restricted. This is clearly visible in Figure 16.5e where spurious chrominance variations have been artificially introduced in the right part of the 16-pixel decoded vector.

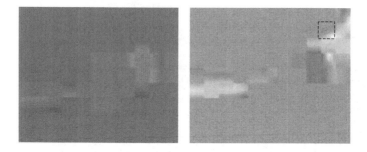

Figure 16.6: Enlarged parts of the C_b (left) and C_r (right) components of the coded frame shown in Figure 16.1; note that red color bleeds in the left part of the macro-block.

This analysis is easily extended to bi-dimensional signals. Figure 16.6 illustrates the C_b and C_r components of the coded frame presented in Figure 16.1. A 16×16 pixels macro-block, made from four adjacent blocks of 8×8 pixels each is selected (represented by dashed square lines). This macro-block contains a strong color edge, corresponding to the player's red shirt. Clearly, the original content of chrominance blocks has been widely modified: in particular, spurious red color components smear in the up-left part of the macro-block, leading to color bleeding distortion. In the case the same color image is compressed without chrominance decimation, no more color bleeding is present: this confirms that color bleeding is related to both quantization and decimation processes, and occurs when strong color edges are present in the macro-block to be coded. In the following section, an adaptive post-processor for color bleeding removal is proposed, in order to improve the visual quality of displayed color video.

16.3 Description of the Post-Processor

The post-processor is described in the case of 4:2:0 signals; the color bleeding defect does not exist in 4:4:4, and is present in 4:2:2 and 4:1:1 formats. For these two last digital video formats, the algorithm proposed here can be adapted. In the case of 4:2:0 format, a macro-block consists of four blocks of Y, one block of C_r and one block of C_b.

The de-color bleeding algorithm can be divided into two main steps. In the first step, the chrominance blocks subject to color bleeding distortion are detected. These macro-blocks are then processed, in order to reduce color bleeding artifacts. The flow diagram of the complete post-processing algorithm is shown in Figure 16.7; the different parts of the algorithm will be described hereafter in a progressive manner.

We have seen that color bleeding occurs when one color in the image bleeds into another color inappropriately. This is strongly related to the presence of chrominance edges in the coded blocks of an image. In the case of natural images, chrominance edges are mostly accompanied by luminance edges, while a luminance edge may often appear without any corresponding chrominance variation [NH95]. It would be interesting to compute edge detection in luminance and chrominance components separately; the manifestation of color bleeding would be associated with the presence of chrominance edges, with no corresponding edge in the luminance part. But results from edge detection are generally poor, due to coarse quantization of the chrominance components. Therefore, a robust detector has been proposed, based on the combination of two different criteria for luminance and chrominance components:

- We first compute the variance of each chrominance block, as any significant variance in C_r or C_b components usually corresponds to regions of higher detail in the luminance component. Because of 4:2:0 sub-sampling, the variances are computed on 4×4 chrominance blocks.
- In order to locate the position of edge pixels within the color image, while avoiding false detections due to spurious color values, horizontal and vertical gradient operators are applied to the luminance component. A binary edge map is then obtained by appropriate threshold. The threshold value is determined automatically based on the RMS estimate in the derivative image. In order to remove isolated edge pixels and refine the edge map, morphological filtering is finally applied.

The detection of the chrominance edge blocks subject to color bleeding is done by comparison of the variance of the chrominance components to a threshold T. The value

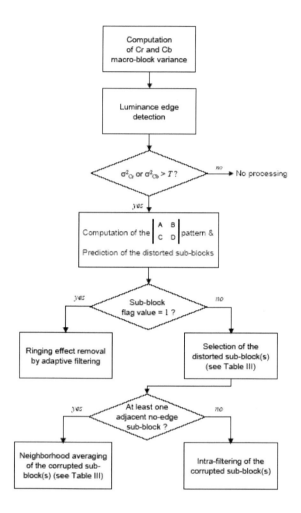

Figure 16.7: Flow diagram of the proposed post-processor [CGC04]. (©2004 IEEE)

proposed in [Web96] has been used in order to identify color detailed blocks. No post filtering is applied if both C_r and C_b variances are less than T. In the opposite case, the so-called chrominance edge block is processed as described hereafter[1].

Let us consider each 8×8 C_r or C_b chrominance edge block separately. This chrominance block is composed of four adjacent 4×4 blocks, noted A, B, C, D (Figure 16.8). For each of these sub-blocks, a flag is set to 1 if: (i) it contains a strong chrominance edge, and (ii) at least one edge pixel has been detected in the corresponding luminance

---

[1]Note that chrominance edge blocks detected in the previous step, which do not contain luminance edge pixels, correspond typically to blocks distorted by color bleeding phenomenon.

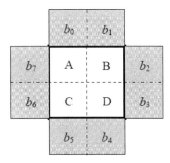

Figure 16.8: Adaptive signal filtering with respect to the image edge block map [CGC04]. (©2004 IEEE)

block. The flag is set to 0 elsewhere. This leads to a complete set of sixteen [A B C D] combination patterns, as described in Table 16.2. Each of these patterns is directly related to the position and local orientation of color edges in the image block.

Post-processing is then proposed, which is based on the computed [A B C D] pattern. For each detected macro-block, we predict in which blocks the color is expected to bleed. Indeed, it is shown that, in the case of an edge block, coarse quantization of DCT coefficients leads to spurious variations in the direction parallel to the edge orientation [vdBL96]. In order to prevent extensive computational cost, edge orientation could be approximated here from the local configuration of the 4×4 sub-blocks with flag value 1 into the [A B C D] pattern. For example, [A B C D] = [0 0 1 1] is typically associated with the presence of a horizontal edge in the bottom part of the corresponding chrominance blocks.

We then deduce the 4×4 chrominance sub-blocks to be filtered that are, roughly speaking, at the opposite location in the given macro-block. Considering the previous example, the chrominance blocks should be a careful balance of vertical frequency DCT coefficients, in order to reconstruct the horizontal edge accurately. After quantization, this balance is disrupted, and some of the DCT eigenimages corresponding to vertical frequency content become predominant in the reconstructed macro-block. This leads to spurious chrominance values that appear mainly as a color horizontal bar in the top of the chrominance blocks. So, in our example, A and B sub-blocks are predicted to be subject to color bleeding, and therefore have to be filtered. This approach can be extended to other combination patterns; Table 16.2 indicates which 4×4 chrominance sub-blocks are to be filtered for the different configurations.

Finally, we consider the filtering process. It consists of restoring the corrupted color data in the image, based on the flag value previously computed for the 4×4 chrominance sub-blocks.

If the flag value equals 1, the corresponding chrominance blocks should overlap the

Table 16.2: Adaptive filtering in consideration of the chrominance edge block map: in the left four columns, flag value 0 means a no edge 4×4 chrominance block, and flag value 1 means a 4×4 edge chrominance block. In the two last columns, a dash means that no filtering is applied [CGC04]. (ⓒ2004 IEEE)

CONFIGURATION OF THE CURRENT 8x8 CHROMINANCE BLOCK				4x4 CHROMINANCE BLOCK TO BE FILTERED	4x4 NEIGHBORING BLOCKS TO BE ACCOUNTED FOR IN THE AVERAGING PROCESS
A	B	C	D		
0	0	0	0	–	–
0	0	0	1	A	b_0, b_7
0	0	1	0	B	b_1, b_2
0	0	1	1	A and B	b_0, b_1
0	1	0	0	C	b_5, b_6
0	1	0	1	A and C	b_6, b_7
0	1	1	0	–	–
0	1	1	1	A	b_0, b_7
1	0	0	0	D	b_3, b_4
1	0	0	1	–	–
1	0	1	0	B and D	b_2, b_3
1	0	1	1	B	b_1, b_2
1	1	0	0	C and D	b_4, b_5
1	1	0	1	C	b_5, b_6
1	1	1	0	D	b_3, b_4
1	1	1	1	–	–

boundary between different color areas: in this case, color bleeding manifests itself as ringing along strong color edges. Ringing effect is then removed by smoothing the pixels located on both sides of the color edge, while keeping the edge pixel values untouched. This filtering operation is applied to the Y, C_r and C_b components, which are involved altogether in the manifestation of ringing distortion. As a result, color edges are free of spurious oscillations.

If the flag value equals 0, the sub-blocks corrupted by color bleeding have to be corrected. Two different cases are considered, based on the local properties of the color image. The first case is the most common one, and typically occurs when the 4×4 distorted chrominance block belongs to a large uniform color area, with homogeneous color properties. Then, at least one of the 4×4 adjacent chrominance blocks $[b_0 \quad b_7]$

Table 16.3: Improvement of color fidelity for the de-color-bleeding post-processor.

Color frame from sequence	C_r component – ΔPSNR (dB)	C_b component – ΔPSNR (dB)
Table-tennis #001	+0.32	+0.10
Foreman #225	+0.41	+0.43

(see Figure 16.8), issued from the surrounding macro-blocks, is expected to belong to the same color area, and so to have very similar chrominance pixel values. Thus, the proposed algorithm consists of restoring the chrominance values in the distorted area from the mean value of the 4×4 adjacent no-edge blocks, selected from the 4-connected neighborhoods. Of course, only the adjacent sub-blocks not belonging to a chrominance edge block are taken into account in the averaging process.

A specific case arises when three out of the four sub-blocks are edge ones: indeed, the corresponding image area is expected to be highly detailed, and the averaging process is applied only if all neighboring blocks are free of edges, in order to avoid misfiltering. To conclude, Table 16.2 summarizes for the different [A B C D] configurations, the neighboring blocks involved in the averaging process. We can note that, in some cases, no preferred direction can be deduced in a simple way: then, no filtering is applied.

The second case arises when none of the adjacent blocks is free of edges, for example in textures or detailed areas of a color image. Then, a simple average of the corrupted values inside the processed block is used in order to soften the chrominance mismatch.

16.4 Experimental Results — Concluding Remarks

Simulations have been performed on a large color image and video database, in order to validate the adaptive post-processor. Both subjective and objective evaluations of the algorithm have been used. The performances were mainly evaluated by visual judgment, for there is no effective measure available at the present. Nevertheless, the peak signal-to-noise ratio (PSNR) was also used for the objective evaluation of color fidelity, although it is not well correlated with the properties of human color vision. In particular, Table 16.3 presents the values of the improvement in PSNR due to post-processing, for the C_r and C_b macro-blocks corrupted by color bleeding. According to these results, the PSNR values of the chrominance components are increased after post-processing, and so the objective quality of color images is improved.

Finally, we present some color images in order to illustrate the visual improvement due to post-processing. Figure 16.9 shows the results of applying color bleeding removal to the color image shown in Figure 16.1. It is very noticeable that color defects have been efficiently reduced by adaptive neighborhood averaging. Note that the smearing of the color red has been efficiently cleaned up above the racket, as well as along the player's arm, and the ringing phenomenon has been removed, while preserving color edge sharpness. This leads to a great improvement in the visual appearance of the displayed color image.

Figure 16.9: Results of applying the de-color-bleeding post-processing to Figure 16.1.

Figure 16.10 shows another example of a color image subject to color bleeding. This image from the well-known *Foreman* sequence has been coded by means of DCT, using the quantization tables given in Table 16.1. Color bleeding is particularly noticeable on the foreman's chin (see the pink blob), as well as along the edges of the shoulder (middle left) and the crane (middle right). The right part of the figure illustrates the results of applying the post-filtering algorithm: clearly, the chrominance mismatches are greatly reduced, and the visual quality of reconstructed video is consequently better.

This chapter provides a thorough analysis of the color bleeding artifacts caused by standard digital image and video coding algorithms. A new post-filtering algorithm for the reduction of color bleeding artifacts is then proposed, that efficiently improves color fidelity. This algorithm is applied to the Y, C_r, C_b components of compressed video, and so can be implemented directly in classical decoders, without changing of colorimetric space. It can be integrated in a more general digital video post-processor addressing various image/video coding artifacts to improve rate-distortion optimization performance of the codecs.

Figure 16.10: One decoded frame from the Foreman sequence before (left) and after (right) post-processing.

References

[AFR01]   A. Al-Fahoum and M. Reza. Combined Edge Crispiness and Statistical Differencing for Deblocking JPEG Compressed Images. *IEEE Transactions on Image Processing*, 10(9):1288–1298, September 2001.

[BK95]   V. Bhaskaran and K. Konstantinides. *Image and Video Compression Standards.* Norwell, MA: Kluwer Academic Publishers, 1995.

[CGC04]   F.-X. Coudoux, M. Gazalet, and P. Corlay. An adaptive Post-processing Technique for the Reduction of Color Bleeding in DCT Coded Images. *IEEE Transactions on Circuits and Systems for Video Technology*, 14(1):114–121, January 2004.

[Gle93]   W. Glenn. Digital Image Compression Based on Visual Perception and Scene Properties. *SMPTE Journal*, 102(5):392–397, 1993.

[ISO94]   ISO/IEC JTC1 110918-1. ITU-T Rec. T.8.1. Information Technology–Digital Compression and Coding of Continuous-tone Still Images: Requirements and Guidelines. Technical report, ISO/IEC, 1994.

[MH98]   M.Yuen and H.R.Wu. A Survey of Hybrid MC/DPCM/DCT Video Coding Distortions. *Signal Processing*, 70(3):247–278, November 1998.

[MPFL96]   J. Mitchell, W. Pennebaker, C. Fogg, and D. LeGall. *MPEG Video Compression Standard.* New York: Chapman and Hall, 1996.

[NH95]   A. Netravali and B. Haskell. *Digital Pictures: Representation and Compression, Second edition.* New York and London: Plenum Press, 1995.

[PL99]   H. Park and Y. Lee. A Postprocessing Method for Reducing Quantization Effects in Low Bit-Rate Moving Picture Coding. *IEEE Transactions on Circuits and Systems for Video Technology*, 9(1):161–171, Febuary 1999.

[RY90]   K. R. Rao and P. Yip. *Discrete Cosine Transform — Algorithms, Advantages, Applications.* Boston, MA: Academic Press, 1990.

[vdBL96] C. van den Branden Lambrecht. *Perceptual Models and Architectures for Video Coding Applications*. Ph.D. thesis, Swiss Federal Institute of Technology, Lausanne, Switzerland, 1996.

[Wan95] B. A. Wandell. *Foundations of Vision*. Sunderland, MA: Sinauer Associates, Inc, 1995.

[Web96] J. Webb. Postprocessing to Reduce Blocking Artifacts for Low Bit-rate Video Coding Using Chrominance Information. In *Proceedings of IEEE International Conference on Image Processing*, 2:9–12, September 1996.

Chapter 17

Error Resilience for Video Coding Service

Jian Zhang
National Information Communication Technology Australia (NICTA)

17.1   Introduction to Error Resilient Coding Techniques

Error resilience is an important issue when coded video data is transmitted over wired and wireless networks. Error can be introduced by network congestion, mis-routing and channel noise. These transmission errors can result in bit errors being introduced into the transmitted data, or packets of data being completely lost. Consequently, the quality of the decoded video is degraded significantly. This chapter describes techniques for minimizing this degradation.

The provision of error resilience consists of a number of parts, each of which are discussed briefly below:

- **Error Detection** Even a single error in a video bitstream can have a large effect on video quality, especially if it causes the decoder to loose synchronization with the arriving bitstream. In order to minimize this effect, it is desirable to detect quickly that an error has occurred and immediately take steps to prevent catastrophic degradation in video quality.

- **Resynchronization** When synchronism is lost with the arriving bitstream, the decoder will normally go looking for a unique resynchronization codeword within the arriving bitstream. The codeword is unique in the sense that it cannot occur other than at a resynchronization point within a non-errored bitstream. The resynchronization word is followed by sufficient data to allow the decoder to continue the decoding process from that point.

- **Data Recovery** Techniques exist (such as two-way variable length codeword decoding [Ad-96]) which allow some of the data between the point where decoding synchronism is lost to the point where it is regained to be utilized.

- **Concealment** Finally, the impact of any errors that has occurred as a result of errors (and in particular lost data) can be concealed using correctly decoded information from the current or previous frames.

In this Chapter, error resilient coding methods compatible with MPEG-2 are described.

17.2 Error Resilient Coding Methods Compatible with MPEG-2

MPEG-2 [ISO93, ISO95] divides error resilient coding methods into three categories:

- Temporal localization, which aims to minimize the propagation of errors with time.

- Spatial localization, which minimizes the effect of errors within a frame, and

- Concealment processes, which hide the effect of errors once they have occurred within frames.

17.2.1 Temporal Localization

One significant effect of cell loss and bit error within a video coded stream (called a decoded sequence) is that of the error propagation. By reviewing the processed frame order of the MPEG-2 coding scheme [ISO93, M. 93a] for example, the errors that occur in Intra-frame (I-frame) could be propagated into next P or B-frames since the P and B-frames will be predicted by the I-frame. Therefore, the initial idea of this method tries stopping these error propagations from picture to picture in the temporal sequence, by providing early re-synchronization of pictures that are coded differentially. Techniques include:

- Cyclic intra-coded pictures: If extra intra-coded I-frames are sent to replace B or P-frames in the frame order such as MPEG-2 frame order as shown in figure 17.1, the error propagation can be stopped with the cost of extra bit amount expense in the bit stream. This will cause the lower coding efficiency than the normal coding, and also bring higher possibility of cell loss and erroneous bits within the same frame than the normal coding due to the sudden large amount of bits added into bit stream.

- Cyclic intra-coded slices: Instead of using intra-coded frames to erase the errors in the screen, which could be the high cost of disadvantage penalty, the extra intra-coded slices can be used to periodically refresh the frame from the top to the bottom over the number of P-frames as shown in Figure 17.2. This will reduce the error propagation and not add lots of bits in the same frame suddenly. The disadvantage is that the partially updating of the screen in a frame period will produce the "window screen wipe" [ISO93] effect noticeably.

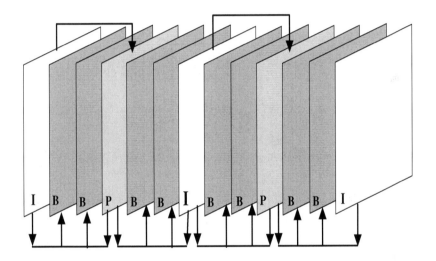

Figure 17.1: Cyclic intra-coded picture to protect error propagation

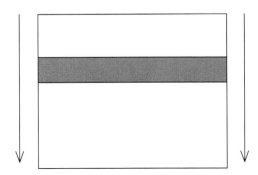

Figure 17.2: Cyclic intra-coded slices to protect error propagation

17.2.2   Spatial Localization

Spatial localization encompasses those methods aimed at minimizing the extent to which errors propagate within a picture, by providing early re-synchronization of the elements in the bitstream that are coded differentially between MBs. There are three methods:

Small slice if a slice in a picture consists of one row of macroblocks (e.g., 44 MB/slice in a CCIR Rec. 601 Frame [Rec90]), the resynchronization point will be the next available slice header and the damaged area in a picture could be up to one row of MBs after any cell loss. If this resynchronization scheme is used for video services in a wireless ATM network, which may sometimes have a cell loss probability of 10^{-2} or more, it will not be possible to achieve an acceptable decoded picture quality. The small slice scheme can reduce the damage to decoded pictures after cell loss by reducing the slice size. Figures 17.3 and 17.4 illustrate a good case and a bad case of small slice resynchronization.

There is a significant variation in the number of bits required to code a slice in a picture, depending on the coding mode, picture activity, etc. The good case can protect the data from the second half of a row of macroblocks from being discarded. On the other hand, two slice headers may be packed into one ATM cell [M. 96]. In this case, once the cell is lost, the simple small slice method cannot provide any advantage. Obviously, coding efficiency is reduced due to the extra overhead for the additional slice headers. Therefore, offering multiple resynchronization points in the same cell serves no purpose. To achieve high error resilient performance using spatial localization, the following technique can be used.

Adaptive Slice Scheme The small slice scheme does not take account of the packing structure. Therefore, an improved technique called adaptive slices will now be introduced.

The adaptive slices change the slice size based on the spacing of ATM cells as shown in Figure 17.5. The encoder must trace the coding and packetization processes to place the slice start code at the first opportunity in each ATM cell.

17.2.3   Concealment

The purpose of concealment is to minimize the impact of an error once it occurs by taking account of spatial correlation and temporal redundancy in the video sequence.

- Temporal Predictive Concealment (Block Replacement): In areas of the picture that do not change very much with time, it is effective to conceal the effect of cell loss by temporal replacement, i.e. using the corresponding information from the previous frame. Figure 17.6 shows a typical example.

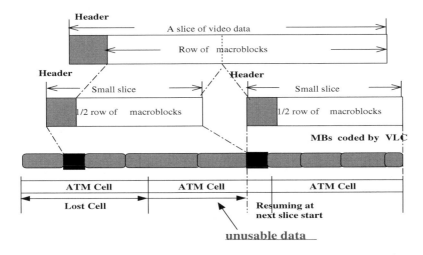

Figure 17.3: Simple small slices able to minimize error effect

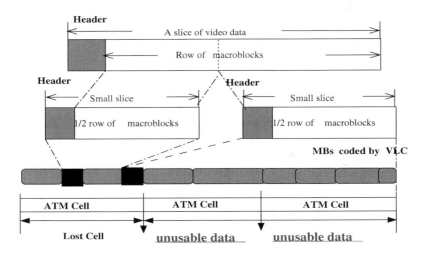

Figure 17.4: Simple small slices unable to minimize error effect

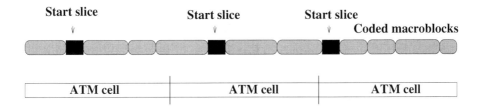

Figure 17.5: Adaptive slice

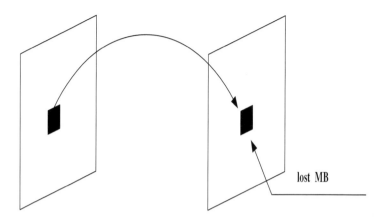

Figure 17.6: Macroblock replacement

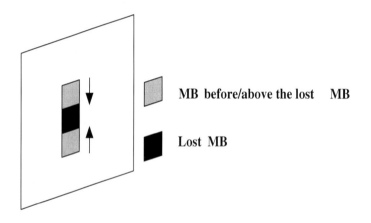

Figure 17.7: Spatial interpolation

- Spatial Interpolation: The above approach is not effective in high motion areas. In this situation, spatial interpolation tends to be more effective. One approach is to get the neighboring blocks (i.e., top, bottom, left and right) and then use interpolation to estimate the lost data. Figure 17.7 gives an example.

- Motion Compensated Concealment: It combines both temporal replacement and motion estimation and can be used to improve the effectiveness of concealment (see Figure 17.8). The technique works by exploiting the fact that there is generally high correlation among nearby motion vectors in a picture. Motion vectors for macroblocks above or below a macroblock lost due to cell loss can be used to predict the motion vector of the lost macroblock. This motion vector is then used to find a block in a previously decoded picture which (it is hoped) will provide

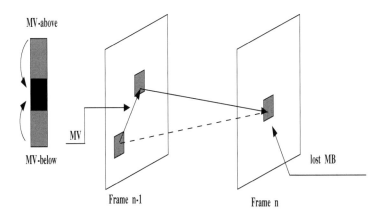

Figure 17.8: Motion compensated error concealment

a good estimate of the lost information. Providing that the estimation of the lost motion vector is accurate, this technique can significantly improve error concealment in moving areas of the picture. However, it is not able to conceal error for a lost macroblock which is surrounded by intra-coded marcoblocks. This is because when intra-coding is employed, no motion vectors are transmitted with the intra-coded information. To avoid this, the MPEG-2 standard allows the encoding process to be optionally extended to include motion vectors for intra-coded macroblocks. Of course, the motion vector and the coded information for a macroblock should be transmitted separately (e.g. in different transmission packets) so that the motion vector is still available in the event that the other macroblock data is lost. This means that an intra-coded macroblock will contain the motion vector which should be used to conceal an error in the macroblock immediately below it. Of course, motion vectors for intra-coded macroblocks are only used for error concealment.

17.2.4 Scalability

It is well known that the use of scalability can enhance the error robustness of a video service. In this application, the benefit of scalability relies on being able to somehow divide information in the video bitstream into two separate bitstreams. One bitstream (the base layer) is labelled as the most important information; the other (the enhancement layer) is labelled as the remaining information. The two bitstreams are then transmitted in such a way that all cell loss is concentrated in the enhancement layer and the base layer experiences no loss. This approach is feasible in networks where the network can control which cells are discarded, such as in ATM networks where loss is generally caused by congestion. It may, however, be impractical in other networks, such as wireless networks, where loss is caused by radio propagation conditions. The use of scalability for enhancing error resilience is described in [R. 96] scalable approaches are examined here.

Data partitioning Within a video bitstream, not all information is of the same importance. Motion vectors, for example, have more impact on decoded video quality than DCT coefficients, and low frequency DCT coefficients are more important than high frequency coefficients. Data partitioning permits a video bitstream to be divided into two separate bitstreams (see Figure 17.9). The lower layer bitstream contains the more important information, while the upper layer bitstream contains less important information. Data partitioning is achieved simply by partitioning the bitstream: it requires no additional signal processing compared to a standard encoder or decoder. If transmission can be arranged such that no loss occurs in the lower layer bitstream, i.e. all loss is concentrated in the enhancement layer bitstream, improved quality can be obtained. Data partitioning has the advantage that it introduces very small additional overhead, and hence its performance in the absence of cell loss is essentially indistinguishable from the non-layered case. The disadvantage of data partitioning is that considerable drift occurs if only the base layer is available to a decoder, as would be the case at high cell loss probabilities.

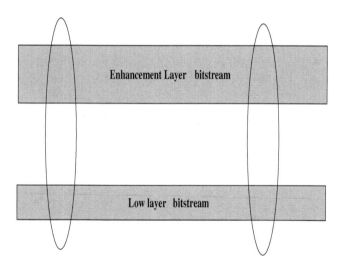

Figure 17.9: Illustration of data partitioning of the two-layer bitstreams

SNR Scalability The base layer of an SNR scalable encoder is the same as a single layer encoder. The upper layer bitstream is derived by taking the residual error after quantizing and re-quantizing this information with a finer quantizer, thus allowing a higher quality reconstruction. This high quality reconstruction is then placed in the picture store for use in motion compensation prediction (MCP) for subsequent pictures (see Figure 17.10). SNR scalability provides high efficiency, i.e. the overhead compared

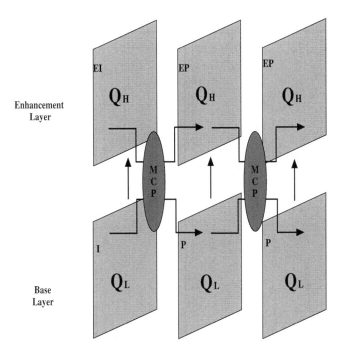

Figure 17.10: Illustration of SNR scalability

to a single layer service is small. As with data partitioning, however, SNR scalability may be subject to drift in the lower layer [R. 96]. The reason for this is that the high quality reconstructed picture is placed in the picture store rather than the lower quality reconstruction that is available to single layer decoders.

Spatial Scalability This scheme overcomes the problem of drift in the lower layer by providing a separate picture store for each layer. Hence, the lower layer picture store sees no data that will be transmitted in the enhancement layer bitstream (see Figure 17.11). In the upper layer, the motion compensated prediction can be derived from either the reconstructed current picture in the base layer or a previous picture in the enhancement layer. If the latter is used, cell loss can cause drift in the upper layer.

Spatial Scalability without Temporal Prediction This scheme is identical to the previous one, except that the prediction in the upper layer is always derived from the base layer (see Figure 17.12). This removes the possibility of drift in the upper layer when cell loss occurs. Once again, there is no drift in the lower layer. The performance in the upper layer is slightly degraded by the restriction placed on the prediction decision.

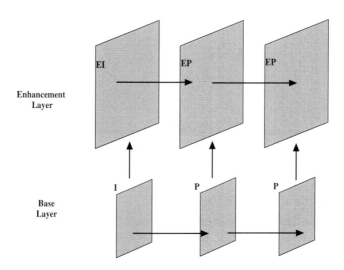

Figure 17.11: Illustration of spatial scalability without temporal predication: MPC can be derived from the reconstructed picture in the base layer only.

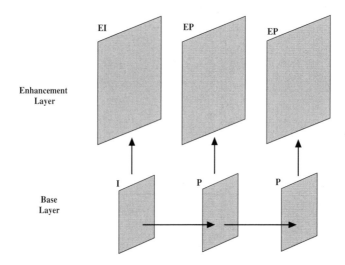

Figure 17.12: Illustration of spatial scalability without temporal predication

The performance of these schemes both at different cell loss rates and for different allocations of bit rates between base and enhancement layers is of interest. A fair comparison can be achieved between different allocations of bits to the two layers by choosing an overall cell loss rate, and conducting an experiment where all of these cells are lost in the enhancement layer. This means that if the base layer is allocated (90%) of the total bit rate and the enhancement layer (10%), and the overall cell loss probability is (2%), then the base layer experiences no loss and the enhancement layer has (20%) loss.

The performance of scalable coders under low loss conditions is limited by the performance of the scalable algorithm itself. The performances of the four techniques outlined above were found to be very similar in [R. 96], with SNR scalability being better than other methods by less than 1 dB PSNR in most cases.

Under conditions of high loss, it is the performance of the lower layer in the absence of any upper layer information that limits quality.

17.3   Methods for Concealment of Cell Loss

As discussed at the beginning of this chapter, the provision of error resilience can be divided into four parts. These are: error detection, resynchronization, data recovery and concealment. While all of these parts are important. it is probably true to say that the use of good error concealment techniques will lead to the greatest improvement in subjective quality.

As described in section 17.2.3, the concealment techniques suggested in the MPEG-2 standard can be summarized as follows:

17.3.1   Spatial Concealment

The simplest form of error concealment is to estimate the lost information by interpolating from surrounding macroblocks. This approach works well for areas of the picture where there is little spatial activity (i.e. in flat areas) but is far less successful in areas where there is significant spatial detail. Consider, for example, the picture from the sequence *Flower Garden* shown in Figure 17.13. Data lost as a result of cell loss is represented by black blocks in the image. In Figure 17.14, a simple interpolation is used to conceal the lost data. In the case of the loss in the low detail sky, the concealment is quite effective. In the flower area, however, the concealment is much less effective. A further measurement of the usefulness of this approach was calculated by performing experiments in which cell loss events are applied to a coded bitstream and then their impact is concealed using simple spatial interpolation. A normal slice structure (with one slice start code per row of macroblocks) was used to provide resynchronization after cell loss.

17.3.2   Temporal Concealment

Another approach which can be used for concealment is to replace any lost information with information from a previously decoded picture. Two approaches are possible and

Figure 17.13: Picture from the sequence *Flower Garden* illustrating the positions of cell loss

Figure 17.14: Effect of concealment using spatial interpolation

these are discussed in the following sections.

Macroblock Replacement The simplest approach is to replace lost macroblocks with co-located macroblocks from a previously decoded picture. This is called macroblock replacement. Clearly, this approach will work best in areas of the picture where there is little or no motion and will have least success in areas of high motion. Figure 17.15 shows a picture from the *Bus* sequence in which a fast panning camera keeps the bus relatively stationary in the center of the picture while the background appears to move with high motion. Cells have been lost both in the fast moving background and in the relatively stationary bus. Data lost as a result of cell loss is represented by black blocks in the image. Figure 17.16 shows the results of using block replacement at 30 frame/second and with IBBPBBPBBPBB... temporal coding structure. Concealment is effective in the relatively stationary bus but much less effective in the fast moving background.

Comparing these two techniques (spatial interpolation and block replacement), it is clear that the performances based on these two techniques are comparable. In MPEG-2 prediction, a picture can be predicted from a reference picture which is temporally displaced by several picture times. For example, the first P picture in a group of pictures may be three picture times from the I picture from which it is predicted. The larger the gap between the reference picture and the picture in which cell loss has occurred, the poorer will be the performance of a block replacement strategy since it will be more likely that motion will have taken place between the two pictures.

Motion Compensated Concealment Cell loss concealment could be improved if information about the motion of any lost macroblock was available at the decoder. Of course, this is not generally the case since if a macroblock is lost as a result of a cell loss event then the motion vector or motion vectors associated with that macroblock are also lost. However, it is well known that there is a high correlation between the motion vectors of spatially adjacent macroblocks. It is therefore often possible to accurately estimate the missing motion vectors using information available in surrounding correctly received macroblocks.

One simple way to achieve this, called AboveMV, is to use the motion vector of the macroblock directly above the lost macroblock (assuming that it has not also been lost) to estimate the motion vector which would have arrived with the lost macroblock. This predicted motion vector can then be used to find the macroblock from the reference picture which would have been identified in the normal motion compensated prediction process carried out at the decoder had the macroblock not been lost. Providing that the correct motion vector is determined, the impact of the cell loss is then reduced to the error introduced by the loss of the quantized DCT coefficients associated with the lost macroblock.

Of course, motion compensated concealment using the motion vector associated with the macroblock above the lost macroblock is impossible in I pictures or where the

Figure 17.15: Picture from the sequence *Bus* illustrating the positions of cell loss

Figure 17.16: Effect of concealment using macroblock replacement

macroblock above the lost macroblock is an intra-coded macroblock. This is because the macroblocks surrounding a lost macroblock do not contain motion vectors and so cannot be used to estimate the motion vector associated with the lost macroblock. However, I pictures are of particular importance since all subsequent pictures within a group of pictures depend on the I picture (either directly or indirectly) for prediction. Further, since I pictures do not use temporal prediction, they require more bits to encode than other types of pictures. As a result, the likelihood of cell loss in I pictures is increased. Recognizing this problem, the MPEG-2 standard allows I macroblocks to have motion vectors associated with them. These motion vectors are not used in the coding process but are available for use in motion compensated concealment. It clearly makes little sense to transmit the motion vectors with the I macroblock that they are supposed to conceal since if the macroblocks are lost then so too are the motion vectors. Consequently, these motion vectors (called concealment motion vectors) are transmitted with the data for the macroblocks immediately above the I macroblocks. They are designed to conceal only. Our experiments have shown that using concealment motion vectors is crucial to successful motion compensated concealment.

Since motion compensated concealment attempts to take account of motion within the video sequence, it allows for effective concealment of both the moving and the non-moving areas of the picture. Thus, overall performance should be enhanced.

Figure 17.17 shows the improvement in decoded picture quality compared to spatial interpolation for picture of the *Flower Garden* sequence. Figure 17.18 shows the improvement in decoded picture quality compared to macroblock replacement for picture of the *Bus* sequence. In generating these results, the same cell loss patterns were applied to identical coded bitstreams. The only difference was in the concealment technique employed. The improvement in performance compared to spatial interpolation and block replacement is quite obvious when Figures 17.17 and 17.18 are compared with Figures 17.14 and 17.16 respectively.

Crucial to the success of motion compensated concealment is the accuracy of estimation of the motion vector of the lost macroblocks. The exact method for performing this task is a decoder function and so is not a matter for standardization. One could, for example, use the average, median or mode of several motion vectors surrounding the lost macroblock in the estimation process. Unfortunately, making an incorrect choice can, in many circumstances, have a highly deleterious impact on decoded service quality.

17.3.3   The Boundary Matching Algorithm (BMA)

This algorithm in [W. 93] attempts to exploit a fact that the best matching block will be highly correlated to the pixels which immediately surround it in the current frame. Based on this concept, the algorithm takes the lines of pixels above, below and to the left of the lost macroblock and uses them to surround candidate blocks from a previously decoded

Figure 17.17: Decoded *Flower Garden* image when motion compensated concealment using the motion vector in the macroblock above the lost macroblock is employed

Figure 17.18: Decoded *Bus* image when motion compensated concealment using the motion vector in the macroblock above the lost macroblock is employed

frame. It then calculates the total squared difference between these three lines and the corresponding three lines on the edge of a 16x16 block of data within a previous decoded frame. This is illustrated in Figure 17.19. The BMA estimates the lost motion vector as the one in which the squared difference between the surrounding lines (from the current decoded frame) and the block (from the previous decoded frame) is a minimum. Referring to Figure 17.19, this means that the total squared difference calculated by summing the following three square differences is minimized:

- squared difference between the pixels above the block and the pixels on the top line of the block (i.e. region A in Figure 17.19)

- the squared difference between the pixels to the left of the block and the pixels on the left edge of the block (i.e. region B in Figure 17.19)

- the squared difference between the pixels below the block and the pixels on the bottom line of the block (i.e. region C in Figure 17.19)

The search method employed to estimate the lost motion vector can be a full search over some area in the previous frame. Alternatively, the search process can be greatly speeded up if only a small number of candidate motion vectors are considered. These may include:

- the motion vector for the same macroblock in the previous frame

- the motion vectors associated with available neighboring macroblocks

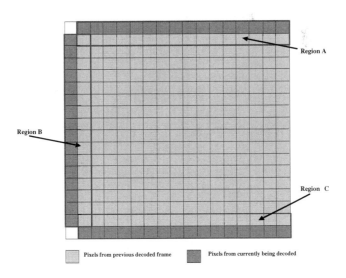

Figure 17.19: Matching technique employed in boundary matching algorithm [W. 93]

- the median of the motion vectors of available neighboring macroblocks

- the average motion vector of the available neighboring macroblocks

- the zero motion vector

As described, the algorithm has two forms. When motion compensation is employed, there are two types of data which need to be transmitted in a macroblock, namely, the motion vector and the coded displaced picture difference. Initially, only the loss of the motion vector was considered with the coded displaced picture difference assumed to be received correctly. In addition, the case where both the motion vector and the coded displaced picture difference were lost was also considered. Only this latter case is relevant in our study since in an MPEG-2 video bitstream it is certain that if the motion vector is lost then the coded DCT coefficients in that macroblock will also be lost since resynchronization cannot occur until the next macroblock at the earliest (and then only if the next macroblock is proceded by a slice header).

This technique has some significant limitations. In the first place, using only the three boundary lines to match the entire 16×16 macroblock is not sufficient in many cases. In addition, it is often the case that all three of these lines are not available for matching. For example, when cell loss occurs several macroblocks of data are lost. This means that when a macroblock is lost then it will be common for the macroblock to its left to be lost as well thus removing one of the three matching criteria. If the macroblock above or below is also lost then the performance of the technique is further weakened. The results shown in Tables 17.1–17.4 allow us to conclude that the performance of the BMA algorithm is similar to the AboveMV scheme.

17.3.4   Decoder Motion Vector Estimation (DMVE)

It is now appropriate to introduce a new algorithm which, like BMA, aims to accurately estimate the motion vectors of any lost macroblocks using correctly received information at the decoder. While the BMA uses spatial correlation to estimate the motion vector, the Decoder Motion Vector Estimation (DMVE) algorithm primarily exploits temporal correlation between correctly received pixels in the current frame and pixels in the previously decoded prediction frame in the estimation process. As explained below, this is achieved by carrying out a process similar to the motion estimation process performed at the encoder to determine the missing motion vector.

When cell loss occurs, several lines (two to eight) of information around any lost macroblocks are taken. This includes available information in the macroblock above the lost macroblock (even if this macroblock is itself a concealed macroblock), the macroblock below the lost macroblock (if received correctly) and the macroblock to the left of the lost macroblock (even if this macroblock is itself a concealed macroblock). In

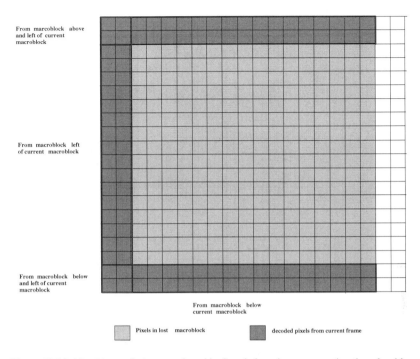

Figure 17.20: Matching technique employed in decoded motion vector estimation algorithm

addition, pixels from the above-left macroblock (even if this macroblock is itself a concealed macroblock) and below-left macroblock (if received correctly) are used to complete the encirclement of the lost macroblock. If it is assumed that only two surrounding lines are used and that macroblocks above, below and right of the lost macroblock are available then the lines used are as shown in Figure 17.20.

The algorithm then performs a full search within the previous frame for the best match to the available lines of decoded pixels from the current frame. The macroblock of data which is surrounded by the lines which best matches the lines from the current frame is assumed to be the best match to the lost macroblock. In the experiments performed for this work, a search area of (-16, +16) pixels is used both horizontally and vertically. However, this is a decoder option and can be used to trade performance against computational complexity. Motion estimation is performed to half pixel accuracy since it was found during the experiments that the filtering associated with half pixel accuracy prediction tends to smooth any blocking artifacts which might otherwise be apparent between concealed macroblocks and normally decoded macroblocks within a frame. Hence, to speed up processing, a search to single pixel accuracy was first performed and then the best match within (-0.5, + 0.5) pixels surrounding that point was chosen for concealment.

Let us now consider the advantages of this method when compared to a simpler implementation of the motion compensated concealment scheme proposed within the MPEG-2 standard. In this simpler approach, it is assumed that the motion vector associated with the macroblock directly above the lost macroblock should be used for motion compensated concealment. Due to the strong spatial correlation among typical motion vector fields, this is often a reasonable assumption. However, significant subjectively annoying artifacts appear when this assumption proves to be incorrect. It was found that this could be a particular problem at the edges of low complexity regions within the frame. For example, in the *Flower Garden* sequence, it was discovered that macroblocks from the roof of a house in the scene background could be relocated into the flat blue sky with obvious subjective impairment resulting. Using several lines of pixels which surround the entire lost macroblock as the basis of lost motion vector estimation ensures that the chosen concealment information is more likely to be a good subjective match.

17.3.5 Extension of DMVE algorithm

1. Optional Candidate Search While the DMVE algorithm is a good technique for determining lost motion vectors, it suffers from the limitation that it has a high computational overhead associated with the requirement for a large number of searches. A fast search method was also investigated, called Optional Candidate Search (OCS), in which the same candidate motion vectors are taken as those used in the BMA described earlier namely:

- the motion vector for the same macroblock in the previous frame

- the motion vectors associated with available neighboring macroblocks

- the median of the motion vectors of available neighboring macroblocks

- the average motion vector of the available neighboring macroblocks

- the zero motion vector

Each candidate motion vector is assumed to point at a potential best matched macroblock in the previous frame. The eight lines surrounding the lost macroblock are then used as usual for DMVE, but the search is performed only at these relatively few motion vector positions. After less computation cycles than full search, these motion vectors are determined. This allows a balance to be achieved between computational complexity (i.e. full search using the surrounding 8-lines) and search speed.

2. DMVE and Bi-directional Frames In the case of macroblocks in B frames, both backward and forward predictions are possible. The DMVE algorithm can be simply extended to perform full search in both the backward and the forward reference frames.

3. Two Line Search Using only two lines surrounding the lost macroblock as opposed to eight lines as the basis of the motion vector search is also investigated. This significantly reduces computational complexity.

17.4 Experimental Procedure

Experiments have been carried out in which replacement of lost macroblocks with the macroblock in the same position in the reference frame as well as the other concealment techniques described in this chapter were used in the video decoding process. Two well-known video test sequences used during the development of MPEG-2 (*Flower Garden* and *Bus*) were used for our experiments. Each video sequence was coded using MPEG-2 compatible software, with an output bit rate of 4 Mbit/s. The resulting bitstream was corrupted by random cell loss as would occur during transmission across a wide range of transmission channels. In all the results reported below, we consider one slice per row of macroblocks (i.e., 44 macroblocks per slice). Four groups of results are shown based on the cell loss probabilities at 10^{-4}, 10^{-3}, 10^{-2} and 0.05. For each error condition, each of the following techniques to conceal the effect of cell loss were studied:

- simple frame replacement (i.e. replace the lost macroblock with the corresponding macroblock in the prediction frame)

- above motion vector scheme (i.e. motion compensated concealment using the motion vector of the macroblock directly above the lost macroblock)

- the BMA algorithm

- the standard DMVE algorithm (8 line search)

- the DMVE algorithm with Optional Candidate Search (8 line search)

- the DMVE algorithm with Bi-directional Prediction (8 line search)

- the DMVE algorithm (2 line search)

So as to achieve statistically significant results, each cell loss experiment was repeated with 25 different cell loss patterns each based on a different random number seed. It should be pointed out that since the coded bitstream in each case is identical and the position of lost cells for a given seed is also identical, any difference in decoded service quality is only a function of the performance of the concealment algorithm.

Random independent cell loss rather than bursty cell loss (as used during the development of the MPEG-2 standard) is chosen, since this represents a more severe test as described in [J. 99, J. 97]. When any cell loss event occurs, the remainder of the received bitstream is lost until the next resynchronization point (e.g., a slice header) is received. If several cells are lost consecutively, it is still quite possible that resynchronization will be achieved at the same point as if a single cell loss had occurred. It is therefore cell loss events, rather than total cells lost, which are most important when studying the impact of cell loss. Random cell loss represents the worst case number of cell loss events.

The experiments simulated in this work have also shown that the decoded service quality is significantly reduced if either

- any I frame or

- a P frame early in a Group of Pictures [ISO95]

is lost due to the loss of the picture header information as a result of a cell loss event. For example, the loss of a single I frame can result in a PSNR drop of up to 1 dB in the decoded quality of a sequence. It therefore seems worthwhile to send this important header information more than once. Fortunately, the MPEG-2 Systems Layers support this redundant information transfer. It was noted that for all of the error concealment schemes studied, the performance of all schemes improved, when this approach was implemented.

Before presenting the results obtained, two more things need to be pointed out:

- for the BMA technique, both full search and optional search are studied. The results in both cases are almost identical and so only the results for optional search are presented.

- as well as considering the quality of the decoded errored bitstream (measured using PSNR), it has been also attempted to quantify the computational complexity of each method. This has been done by measuring the decoding time required for each technique and comparing it to the decoding time for a non-errored sequence. Thus a CPU time of 2.0 indicates that the decoding time for the errored sequence was twice the time required to decode the non-errored sequence.

17.5   Experimental Results

The experimental results obtained when cell loss was introduced are shown in Tables 17.1− 17.4 for the *Flower Garden* and the *Bus* sequences, it is also very important to view Figures 17.21 and 17.22 which shows the decoded PSNR versus cell loss probability for the various concealment strategies.

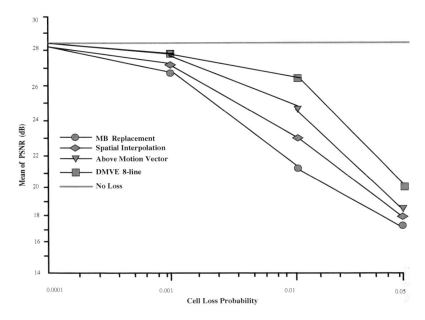

Figure 17.21: Decoded PSNR versus cell loss probability for various concealment strategies (Sequence: *Flower Garden* coded at 4 Mbit/s)

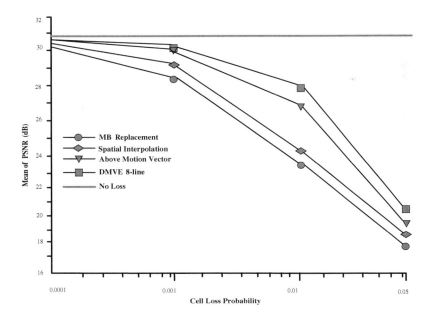

Figure 17.22: Decoded PSNR versus cell loss probability for various concealment strategies (Sequence: Bus coded at 4 Mbit/s)

At low cell loss rates (10^{-4}), all schemes perform well with even simple frame replacement achieving a decoded service quality of only 0.2 dB worse that the non-errored case. The computational complexity (as measured by the CPU time required to decode the sequence) is comparable in all cases and changed little from the non-errored case

At medium cell loss rates (10^{-3}), using only simple frame replacement results in a 1.5 dB loss in decoded service quality compared to the non-errored case. Using motion compensated concealment based on the motion vector for the macroblock directly above the lost macroblock or the BMA recovers around 1.0 dB gain of this loss with motion compensated concealment outperforming BMA marginally (around 0.1 dB) in both of the sequences studied. Using the more sophisticated DMVE algorithm achieves only a further 0.1 dB gain when compared to using motion compensated concealment based on the motion vector in the macroblock directly above the lost macroblock. However, again computational complexity is not an important issue since all schemes have a negligible impact on the overall decoding time. For this reason, the DMVE approach does seem appropriate.

At high cell loss rates (10^{-2}), using only simple frame replacement results in a very significant 6.5 dB to 7.0 dB loss in decoded service quality compared to the non-errored case. Using motion compensated concealment based on the motion vector in the macroblock directly above the lost macroblock recovers a little over 3.0 dB of this loss with the BMA performing slightly better in the case of the *Flower Garden* sequence and somewhat worse in the case of the Bus sequence. Using the DMVE algorithm, further improvements are achieved ranging from a 0.6 dB improvement in the case of OCS to in excess of 1.0 dB when an eight line search is utilized together with bi-directional search in the case of B frames. When this last technique is used, decoded service quality is reduced by only 2.0 dB compared to the non-errored case even though the cell loss rate is 1%.

In this case, computational complexity becomes more important. The results show that DMVE when using the fast OCS approach has computational complexity comparable to the simpler approaches and thus its benefits can be gained with little overhead. The more computationally complex forms of DMVE do result in further improvements in decoded service quality but at a computational overhead of a factor of two (DMVE with 2 line search), five (DMVE with 8 line search) or six (DMVE with 8 line search and bi-directional search in B frames). The implementer can therefore do a simple performance versus computational complexity trade-off to determine which technique to choose.

Finally, at very high cell loss rates (0.05), the drop in decoded service quality is a very substantial drop (i.e., 12 dB). Using motion compensated concealment based on the motion vector in the macroblock directly above the lost macroblock or the BMA recovers around 1.5 dB of this loss with the various forms of DMVE achieving a further gain of around 1.5 dB. Computational complexity rapidly rises as more complex forms of DMVE are employed. Even OCS introduces a 60-70% increase in computational requirements which is comparable to BMA but considerably in excess of the simpler motion compensated concealment based on the motion vector in the macroblock directly above the lost macroblock.

17.6 Conclusions

It is well known that basing subjective quality decisions purely on PSNR is fraught with danger. For this reason, all decoded sequences were subjectively viewed with viewers agreeing that the subjective improvement achieved by DMVE was even more than might have been expected from the PSNR values achieved. In an attempt to demonstrate these improvements, Figures $17.23 - 17.37$ show frames from the two sequences used in author's experiments and compare decoded quality at CLP = 0.05 and 10^{-2} respectively, when

- motion compensated concealment based on the motion vector in the macroblock directly above the lost macroblock is used, and

- when DMVE with 8 line search and bi-directional search for B frames is employed.

In summary, It has been shown that simple application of the error concealment techniques defined within the MPEG-2 standard can improve the quality of the decoded video significantly when cell loss occurs. However, decoded service quality can be improved still further in high to very high cell loss environments by the use of the DMVE algorithm to estimate the best motion vector to use for concealment. This approach can be implemented in a decoder which is completely compliant with the MPEG-2 video standard. Further, this technique can be combined with other error resilient coding techniques, such as temporal localization, spatial localization or macroblock-resynchronization [M. 93b] to achieve still further increases in decoded service quality.

Table 17.1: Comparison of PSNR values (in dB), cell loss probability $= 10^{-4}$

Measurements for *Flower Garden* sequence					
Mean of total damaged MBs = 92, PSNR with no loss = 28.38 dB					
Technique	Single Slice Header per Row				
	Mean	Max	Min	Std. Dev	CPU Time
MB Replacement	28.16	28.30	27.88	0.13	1.0
Above Motion Vector	28.34	28.36	28.31	0.01	1.0
BMA	28.32	28.35	28.26	0.03	1.0
DMVE-OCS (8 line)	28.34	28.36	28.32	0.01	1.0
DMVE (2 line)	28.34	28.36	28.31	0.01	1.0
DMVE (8 line)	28.34	28.36	28.32	0.01	1.0
DMVE-Bi-Dir (8 line)	28.35	28.36	28.32	0.01	1.0
Measurements for Bus sequence					
Mean of total damaged MBs = 106, PSNR with no loss = 30.71 dB					
Technique	Single Slice Header per Row				
	Mean	Max	Min	Std. Dev	CPU Time
MB Replacement	30.49	30.70	29.93	0.22	1.0
Above Motion Vector	30.67	30.71	30.57	0.04	1.0
BMA	30.64	30.70	30.51	0.06	1.0
DMVE-OCS (8 line)	30.68	30.71	30.60	0.03	1.0
DMVE (2 line)	30.67	30.70	30.57	0.04	1.0
DMVE (8 line)	30.68	30.71	30.61	0.03	1.0
DMVE-Bi-Dir (8 line)	30.69	30.71	30.61	0.03	1.0

Table 17.2: Comparison of PSNR values (in dB), cell loss probability $= 10^{-3}$

Measurements for *Flower Garden* sequence					
Mean of total damaged MBs = 864, PSNR with no loss = 28.38 dB					
Technique	Single Slice Header per Row				
	Mean	Max	Min	Std. Dev	CPU Time
MB Replacement	26.90	27.59	26.25	0.39	1.0
Above Motion Vector	28.11	28.26	27.77	0.13	1.0
BMA	28.03	28.21	27.69	0.13	1.0
DMVE-OCS (8 line)	28.13	28.25	27.96	0.09	1.0
DMVE (2 line)	28.16	28.26	27.91	0.10	1.0
DMVE (8 line)	28.17	28.26	27.98	0.08	1.0
DMVE-Bi-Dir (8 line)	28.18	28.27	27.99	0.07	1.0
Measurements for Bus sequence					
Mean of total damaged MBs = 865, PSNR with no loss = 30.71 dB					
Technique	Single Slice Header per Row				
	Mean	Max	Min	Std. Dev	CPU Time
MB Replacement	29.24	29.72	28.80	0.32	1.0
Above Motion Vector	30.37	30.52	30.18	0.11	1.0
BMA	30.21	30.43	30.00	0.15	1.0
DMVE-OCS (8 line)	30.42	30.55	30.25	0.10	1.0
DMVE (2 line)	30.43	30.54	30.15	0.11	1.0
DMVE (8 line)	30.44	30.55	30.25	0.09	1.0
DMVE-Bi-Dir (8 line)	30.46	30.56	30.26	0.09	1.0

Table 17.3: Comparison of PSNR values (in dB), cell loss probability = 10^{-2}

Measurements for *Flower Garden* sequence					
Mean of total damaged MBs = 8373, PSNR with no loss = 28.38 dB					
Technique	Single Slice Header per Row				
	Mean	Max	Min	Std. Dev	CPU Time
MB Replacement	21.78	22.26	21.31	0.26	1.0
Above Motion Vector	24.90	25.51	24.23	0.43	1.0
BMA	25.02	25.64	24.40	0.38	1.0
DMVE-OCS (8 line)	25.83	26.56	25.08	0.39	1.0
DMVE (2 line)	26.14	26.69	25.45	0.30	1.7
DMVE (8 line)	26.21	26.72	25.80	0.27	4.3
DMVE-Bi-Dir (8 line)	26.26	26.76	25.84	0.27	5.2
Measurements for Bus sequence					
Mean of total damaged MBs = 8606, PSNR with no loss = 30.71 dB					
Technique	Single Slice Header per Row				
	Mean	Max	Min	Std. Dev	CPU Time
MB Replacement	23.59	24.19	22.83	0.34	1.0
Above Motion Vector	26.81	27.52	26.21	0.36	1.0
BMA	26.28	27.05	25.65	0.34	1.0
DMVE-OCS (8 line)	27.63	27.83	26.97	0.25	1.0
DMVE (2 line)	27.45	28.09	26.94	0.44	1.9
DMVE (8 line)	27.84	28.22	27.04	0.29	4.8
DMVE-Bi-Dir (8 line)	27.92	28.29	27.06	0.31	5.7

Table 17.4: Comparison of PSNR values (in dB), cell loss probability = 0.05

Measurements for *Flower Garden* Sequence					
Mean of total damaged MBs = 36374, PSNR with no loss = 28.38 dB					
Technique	Single Slice Header per Row				
	Mean	Max	Min	Std. Dev	CPU Time
MB Replacement	17.45	17.55	17.36	0.05	1.0
Above Motion Vector	18.59	18.81	18.46	0.11	1.0
BMA	18.84	18.97	18.61	0.09	1.5
DMVE-OCS (8 line)	19.23	19.40	19.04	0.10	1.6
DMVE (2 line)	19.68	19.95	19.38	0.16	3.9
DMVE (8 line)	20.13	20.55	19.71	0.27	14.1
DMVE-Bi-Dir (8 line)	20.21	20.60	19.84	0.22	17.8
Measurements for Bus Sequence					
Mean of total damaged MBs = 38100, PSNR with no loss = 30.71 dB					
Technique	Single Slice Header per Row				
	Mean	Max	Min	Std. Dev	CPU Time
MB Replacement	17.75	17.95	17.59	0.11	1.0
Above Motion Vector	19.45	19.61	19.28	0.12	1.0
BMA	19.59	19.89	19.36	0.15	1.6
DMVE-OCS (8 line)	20.12	20.53	19.70	0.24	1.7
DMVE (2 line)	20.04	20.73	19.32	0.39	4.3
DMVE (8 line)	20.30	20.84	19.72	0.34	15.0
DMVE-Bi-Dir (8 line)	20.38	20.89	19.75	0.40	18.8

Figure 17.23: Sampled picture (No.18) from original sequence

Figure 17.24: Sampled picture (No.18) decoded by Above-MVs Tech. (CLP=0.05)

Figure 17.25: Sampled picture (No.18) decoded by 8-around-line with bi-directional search (CLP=0.05)

Figure 17.26: Sampled picture from (No.57) from original sequence

Figure 17.27: Sampled picture (No.57) decoded by Above-MVs Tech. (CLP=0.05)

Figure 17.28: Sampled picture (No.57) decoded by 8-around-line with bi-directional search (CLP=0.05)

Figure 17.29: Sampled picture from (No.60) from original sequence

Figure 17.30: Sampled picture (No.60) decoded by Above-MVs Tech. (CLP=0.05)

Figure 17.31: Sampled picture (No.60) decoded by 8-around-line with bi-directional search (CLP=0.05)

Figure 17.32: Sampled picture from (No.18) decoded by Above-MVs Tech. (CLP=0.01)

Figure 17.33: Sampled picture (No.18) decoded by 8-around-line with bi-directional search (CLP=0.01)

Figure 17.34: Sampled picture (No.57) decoded by Above-MVs Tech. (CLP=0.01)

Figure 17.35: Sampled picture (No.57) decoded by 8-around-line with bi-directional search (CLP=0.01)

Figure 17.36: Sampled picture (No.60) decoded by Above-MVs Tech. (CLP=0.01)

Figure 17.37: Sampled picture (No.57) decoded by 8-around-line with bi-directional search (CLP=0.01)

References

[Ad-96]   Ad-hoc group on core experiments on error resilience aspects in MPEG 4 video. Description of Error Resilient Core Experiments. *ISO/IEC JTC1/SC29/WG11 MPEG-96*, August 1996.

[ISO93]   ISO/IEC JTC1/SC29/WG11. MPEG 2 Test Model 4. *ISO/IEC*, 1993.

[ISO95]   ISO/IEC JTC1/SC29/WG11. Generic coding of moving pictures and associated auido information: Video. *ISO/IEC International Standard 13818-2*, 1995.

[J. 97]   J. Zhang, M. R. Frater, J. F. Arnold and T. M. Percival. MPEG-2 Video Service for Wireless ATM Networks. *IEEE Journal on Selected Areas in Communications*, 15(1):119–128, January 1997.

[J. 99]   J. Zhang. *Error Resilience for Video Coding Service over Packet-based Networks.* Ph.D. thesis, University of New South Wales, 1999.

[M. 93a] M. Ghanbari. Standard codecs: Image Compression to Advanced Video Coding. *IEE, UK*, 1993.

[M. 93b] M. Ghanbari and V. Seferidis. Cell-loss concealment in ATM Video Codecs. *IEEE Trans. on Circuits & Systems for Video Technology*, 3:238–247, 1993.

[M. 96]   M. D. Prycker. *Asynchronous Transfer Mode Solution for Broadband ISDN.* Upper Saddle River, NJ: Prentice Hall, 3rd ed., 1996.

[R. 96]   R. Aravind, M. R. Civanlar and A. R. Reibman. Packet Loss Resilience of MPEG-2 Scalable Video Coding Algorithms. *IEEE Trans. on Circuits & Systems for Video Technology*, 6:426–435, October 1996.

[Rec90]   Recommendations and Report of the CCIR Recommendation of 601. Encoding Parameters of Digital Television of Studios: Broadcasting Service (Television) Rec. 601-2. *CCIR*, 1990.

[W. 93]   W. M. Lam, A. R. Reibman and B. Liu. Recovery of lost or erroneously received motion vectors. *Proc. ICASSP*, 5:417–420, 1993.

Chapter 18

Critical Issues and Challenges

H. R. Wu[†] and K.R. Rao[‡]
†   *Royal Melbourne Institute of Technology, Australia*
‡   *University of Texas at Arlington, USA*

Looking to the future research in the field of digital picture processing, coding and communications, this book addresses issues in three areas, including the VHS based picture quality assessment and metrics, perceptual picture coding and compression, and post processing (restoration and concealment) algorithms and techniques for picture reconstruction with improved visual quality. This concluding chapter attempts to highlight a number of issues and challenges which have been raised by contributors of this handbook in previous chapters, as well as by others [Wat93b, vdBL98, FvdBL02, Wuv00, Cav02]. Since these issues are all related to human visual perception, inevitably they are leading back to fundamental questions and, maybe, new challenges to vision research and/or how to transfer vision science, in the context of this handbook, to imaging and visual communication systems engineering.

18.1   Picture Coding Structures

To appreciate or to understand the bases on which digital picture coding framework and structures have been selected and adopted in current state-of-the-art standards and implementations or products, one has to review, to revisit and to examine a series of historical events, along with theoretical premises and performance criteria adopted, in picture coding and compression research. One will need to understand and to compare them with the mathematical framework and structures proposed in the field of vision research, which have been successfully used for analysis, interpretation, understanding and modeling of the human visual system and human vision. On one hand, one must understand that selection criteria for what deems to be an effective mathematical framework are entirely independent and need not be related in the above two fields, i.e., picture coding and vision science and research. On the other hand, one should now have realized that in order to achieve even higher picture coding performance in terms of the bit rate and perceived picture quality, a joint optimization in coder design is mandatory in reduction of both statistical and psychovisual redundancies, for a large number of picture processing and visual communications applications where information lossy compression is acceptable or necessary due to cost-performance considerations. It does appear that more attention, effort and enthusiasm has been devoted to how much more a picture can be compressed than how much better picture quality it could be coded at. For instance, there has been significant emphasis on very low bit rate video coding pushing the bit rate to 4 kbps [GJR96], while there has been little done to the more pronounced temporal granular noise or stationary area temporal fluctuation artifacts around the edges and in texture areas, jerky motion and often blurring in HDTV pictures. It is reminiscent of historical events around the 1980s when Johnstone published his work in the field of perceptual audio coding [Joh88]. A number of critical issues and challenges raised over a decade ago [JJS93, Wat93b] remain today as they did then.

From time to time, someone would ponder whether or not all these made sense. Torres and Kunt pointed out that even the original digital video data were acquired using practical sampling rates which defied sampling theorem and bore little relevance to the 4-D (i.e., 3 spatial and 1 temporal dimensions) analog input signal [TK96]. Tan and Wu explored whether picture compression should be considered as a two stage process, consisting of reorganization of the data set and identifying the best compression strategy (e.g., transform) for the compression [TW99a, TW99b]. The purpose of data reorganization is to introduce additional dimensions achieving a higher degree of correlation among pixels or data points in order to facilitate the use of multidimensional transforms where higher compression can be obtained. An intuitive example is that of reorganizing or folding the 1-D video trace as a result of the PCM process into a 3-D array of data, therefore the name of *data folding*. There are immediate observations, i.e., most of efforts so far in picture compression have been on the second task. Unless the folding has certain structure or regularity, the overhead used to describe how the data are reorganized or folded may jeopardize any potential compression gain by higher dimensional compression techniques.

There are various efforts to improve picture coding performance, which follow categorically, at least, four routes, i.e., pushing the boundaries of existing waveform and entropy coding techniques, incorporating various HVS aspects into the waveform coding techniques, pursuing mathematical alternatives for waveform coding (e.g., matching pursuit [NZ02a, NZ02b], projection on to convex sets [SR87], fractal transform [Cla95], the finite ridgelet transform [DV03], the dual-tree complex wavelet transform [KR03], etc.), searching new coding framework and structures propelled by vision science and research [Wat93b, TK96][1]. Steady and assured incremental improvements and pro-

---

[1] In [Wat93b], Watson groups them into two approaches, i.e., the "conservative" approach which exploits aspects of human vision in existing coding framework via perceptual distortion criteria or the use of contrast sensitivity in quantizer designs, and the "radical" approach which mimics processes by the visual brain.

gresses have been made to traditional waveform coding and associated techniques as seen over the years [LSW03, Gir03, HDZG03, FG03, Ebr05, KTR05]. Nonetheless, wide spread interests in digital picture coding research and standardization activities have increasingly pushed the "first generation" (i.e., waveform coding) techniques to the wall. Gradually and yet surely, it has become clear that the human perception and visual system hold a vital key to further performance improvement of digital picture coding in terms of rate-distortion optimization, albeit the "distortion" here will no longer be straight forward pixel-based differences such as the mean squared error. Not all of these efforts have led to a convincing theoretical breakthrough or better performance (e.g., see the critique by Clarke on fractal based image coding [Cla95]) than the classic waveform coding framework.

In search of new picture compression techniques, many have been led to question if the waveform based coding framework which dominated picture coding up to the mid 1980s should not be challenged [TK96]. Historical events and theoretical premises that led to this dominance are best documented and summarized by Jayant and Noll [JN84], and Clarke [Cla85], which may help to place the current research activities and issues in perspective. Here are a listing of assumptions which are worthy revisiting or re-examining.

18.1.1 Performance Criteria

What constitutes an efficient picture coding technique has a lot to do with certain assumptions which one has made, including the mind setting, available knowledge and technology, applications environment and performance criteria. For example, video signal was naturally considered as a 1-D continuous time signal [Goo51, Hua65, IL96] and as such the DPCM (Differential Pulse Code Modulation) [Cut52] as a signal compression technique was (and still is) more efficient than the PCM (Pulse Code Modulation) [Goo51, Hua65]. The performance criterion used was statistical (or source data) redundancy reduction as measured by the entropy [JN84]. After realizing redundancies not only exist between adjacent samples or pixels along the same scanning line but also along other dimensions, fast developments in picture coding led to techniques exploring picture data redundancies along the other spatial (i.e., vertical) axis and the temporal axis, which were quickly overtaken by multi-dimensional transform and hybrid coding techniques [Cla85].

Once the performance criteria were set in place, such as the self-information and the entropy, the MSE (Mean Square Error), the signal-to-zonal sampling-error ratios in terms of the MSE, the SNR (Signal-to-Noise Ratio) or the peak SNR (PSNR) [JN84], the decorrelation efficiency and the energy packing efficiency [Cla85], majority of the efforts over the past few decades were to find and to determine who got the most efficient technique (prediction models, transforms, algorithms) or a hybrid of techniques (lossy and lossless) for picture compression in terms of these performance criteria. Although introductions to almost all picture coding and image processing books cover the ground of two compression principles, i.e., reduction of statistical redundancies in source data (or sample coding), spatiotemporal and across different scales, and reduction of psychovisual redundancies (or irrelevancies), the decision in selecting the winner was a lot easier, and therefore, made using the above criteria which have little to do with how successful the winner measured up in psychovisual redundancy reduction and, in particular, whether it achieved the best performance in terms of the rate distortion theory where an appropriate perceptual (or visual) distortion measure was used. When the psychovisual redundancy and the HVS were taken into account, it was their simplest forms or aspects such as the CSF and luminance adaptation that were grafted onto a coding framework or mathematical structure which had been selected based on the aforementioned statistical redundancy reduction measures and performance criteria [Cla85, Wat93a, TWY04]. This has been so not without a reason. As Eckert and Bradley once explained, "Our knowledge of perceptual factors is patchy, with well accepted models for visual factors such as contrast sensitivity functions and luminance adaptation, and incomplete models for perceptual factors such as contrast masking and error summation" [EB98].

In selection of the best transform for image compression, for example, the decorrelation efficiency and the energy packing efficiency have been used [Cla85]. Coder performance evaluation has followed a rate distortion minimization process where the MSE is commonly used [JN84, Ber71]. The decorrelation efficiency measures the ability of a transform to convert a set of highly corrected data to a relatively independent data set (therefore, reducing statistical redundancy in the data set), and the energy packing efficiency assesses the ability of a transform to compact energy contents of the given data set to as few transformed data points as possible (thus achieving compression when fewer data points that contain the signal energy in the transform domain are encoded with the minimum error, which is, again, measured by the MSE usually). For a given image that can be represented by a first-order Markov model, the Karkunen-Loève transform (KLT) deems to be the optimal transform since it achieves 100% decorrelation efficiency, while the Discrete Cosine Transform or the DCT has a better energy packing efficiency than the KLT even though its decorrelation efficiency is less than 100% [Cla85]. Taking into account that the DCT is data independent and aided by fast algorithms and implementations, the dominance of the DCT in most of the current international picture coding standards seems only logical and natural [RH96]. It has all made perfect sense until one further examines the aforementioned performance criteria on which the "best" transform is based.

The energy packing efficiency performance criterion works on minimization of the error introduced by using a reduced number of transform coefficients to represent the original data set. The mean square error is commonly used because of mathematical tractability for the minimization process. An immediate observation from a mathematical point of view is that both a few large errors and a large number of small errors in the compressed data set may result in the same MSE, and the MSE places more emphasis on a few errors with large magnitudes for the square operation. This has led to other error measurements, such as the Mean Absolute Difference or the MAD [RY90]. From a traditional signal processing theory stand point, transforms lead to signal decomposition with coefficients representing different frequency components with varying importance, and some are more important than others in reconstruction or synthesis of the original signal. In the context of picture compression, the response of the HVS, which is described by its contrast sensitivity function, varies with spatial frequencies of the image [Wan95]. The HVS also responds to different types of errors differently as shown in Figure 12.2 demonstrating further deficiency of the MSE as a performance measure for picture compression. Weighted MSE criteria, which worked reasonably well for 1-D audio signal processing and compression, were investigated [JN84]. Modified Karhunen-Loève transform using weighted MSE as optimization criterion was attempted. It did not, however, extend too well to picture coding and met only with limited success.

In the search for an alternative mathematical framework for picture analysis and compression, spatial and spatial-frequency localization is closely examined as a performance criterion. A major finding in joint spatial and sptial-frequency representations as stated by the *uncertainty principle* is that we cannot achieve unlimited high resolution simultaneously in both domains [Dau85]. Reed provided an intuitive explanation to the concept of spatial and spatial-frequency localization and its implications in transform coding of images where the CSF and visual masking were applied [Ree93]. He explained that while the DCT has the advantage of having singular functions (or "line spectum" [OS89]) in spatial frequency domain (i.e., DCT coefficients) which aids the quantization of high spatial frequency DCT coefficients using the CSF without *leakage* to lower spatial-frequencies, its basis functions are not localized resulting in the effects of a quantization error in any DCT coefficient to propagate without attenuation throughout the data block, which are the cause of blurring, blocking and ringing artifacts prominent in block-based DCT coded images (see Chapter 3). In terms of this new performance criterion, the DWT (Discrete Wavelet Transform) has been found to have a much superior spatial and spatial-frequency localization property [VK95]. The DWT based image coding algorithms have demonstrated significant performance gains over the traditional DCT image coding [Sha93, SP96, Tau00] which has been adopted in the JPEG2000 image coding standard [BCM00]. However, being a critically-sampled complete transform, the DWT coding of digital pictures is not without its own misgivings, i.e., when quantization is applied in the transform domain to achieve compression, perfect reconstruction is no longer possible, introducing visible coding artifacts such as blurring, pattern aliasing and ringing artifacts for still images [KR03] and "speckles" or temporal fluctuation noise for video sequences [AC96].

In the selection of a mathematical framework for vision modeling, the criteria are somewhat different. The two main criteria as stated by Watson [Wat87] are that the transform must resemble functions of the primate visual cortex (or V1 area [Wan95]) and that it must be computationally convenient. The transform is required to be of or to approximate a 2-D Gabor shape (i.e., function) which resembles the receptive fields of the human primary visual cortex, to have selectable spatial frequency bandwidth about one octave and selectable orientations with bandwidth about 40°, to be scale invariant and of a pyramidal structure. Various multi-channel vision models are implemented using the Gabor filter-bank [Dau88], the cortex decomposition [Wat87], steerable filters [SF95, SFAH92], and wavelets [ABMD92]. An example of steerable pyramid decomposition, consisting of an isotropic low-pass band, four band-pass bands with four orientations each centered around 0°, 45°, 90° and 135°, and an isotropic high-pass band, is given by Figure 8.3 in Chapter 8. Daly, Feng and Speigle pointed out in [DFS02] that of the above implementations, "wavelets are nearly always implemented using Cartesian separability, which gives the coarsest orientation segregation (in fact, the orientations near 45 and 135 (i.e., -45) degrees are combined in a single mixed-orientation band." To design a good vision model, it is desirable that the mathematical framework or transforms are spatial frequency and orientation selectable, localized in both scale and orientation, aliasing-free or shift-invariant, which are usually over-complete [SF95].

The point which must be made is that the mathematical framework and structures for picture coding thus far have been determined mostly based on their mathematical properties and performance in statistical redundancy reduction. Many well-known coding artifacts, such as blurring, blocking and ringing artifacts in block DCT coded pictures, and blurring, ringing and aliasing artifacts in DWT coded pictures, are deeply rooted in how the coding framework and structures are selected. Another interesting point worth noting is that it has long been realized that even aided with the most powerful prediction model and transform domain techniques devised based on statistical redundancy reduction performance criteria, information lossless techniques can only achieve modest compression [Wat93b], e.g., a theoretical compression gain of a little higher than 3:1 shown for the 8-bit grey-scale "*Lena*" image with an entropy of 2.55 bpp in the 2-D Gabor transform domain [Dau88]. This is in sharp contrast to the compression ratios (a typical 10:1 to 16:1 or higher) which are commonly used in digital imaging applications. In other words, the performance criteria that are used to select the mathematical framework for picture coding have little to do with a further compression gain of a factor of 3 to 5 or higher in image coding applications using lossy compression techniques.

Since Watson coined the concept of "perceptually lossless" image coding in 1989 [Wat89], there have been few perceptually (or visually) lossless coders reported [LYC97, WTB97, WTW03] and much less were deployed in applications [SFW$^+$00] even though up to 48% compression gain has been demonstrated for perceptually lossless picture coding where medical grade displays are used in a standard radiological reporting room, compared with the state-of-the-art information lossless coding techniques [WTB$^+$04]. In addition to a number of well known disadvantages against mathematical structures proposed based on vision modeling criteria, such as lacking perfect reconstruction properties, being over-complete, and possessing higher computational complexity [van96], lacking quantified psychovisual criteria seems to be a major challenge or obstacle to further progress in a wide introduction of vision model based framework and structures for picture compression applications.

The choice between a mathematically based coding framework and a psychovisually (or vision model) based one seems to have a lot to do with a mind-setting associated with certain assumptions that are made when considering complete versus over-complete transforms for picture coding. This is further discussed in the next subsection.

18.1.2 Complete vs. Over-Complete Transforms

When considering the criteria for selection of transform techniques for compression, it seemed only natural that a complete (or critically sampled) transform should be favored over an over-complete transform. After all, the aim was about compression and why introduce even more redundancy. The concept of introducing wavelet packets and an over-complete decomposition, such as matching pursuit [NZV94] seemed counter intuitive. The message on importance of matching basis functions to signal structures, (one might have a better chance with an over-complete set than a complete set), seems to have fallen on deaf ears [NZ97]. The view seems rather entrenched that the complete transforms are for picture compression, the over-complete for image analysis and interpretation or modeling of the HVS. The inabilities of the DCT basis functions (see Chapter 3) and the wavelets in matching signal structures and features at very low and considerably high bit rates, which have been known too well, are not going to change that, so it seems. A recent work by Kingsbury and Reeves on applications of the Q-shift dual-tree complex wavelet transform (DT CWT) [Kin01] to digital image compression exemplifies some of the key issues associated with the choice between complete and over-complete transforms [KR03].

Critically-sampled (or complete) transforms, such as the discrete wavelet transform (DWT) are usually preferred to over-complete transforms for picture compression applications. For example, the DWT has the following advantages, such as good energy packing, perfect reconstruction with short support filters, no redundancy and low computation complexity. However, it lacks shift invariance (i.e., small shifts in the input signal result in major variations in the distribution of energy among DWT coefficients at different scales), and suffers from poor directional selectivity for diagonal features [Kin01]. The over-complete DT CWT addresses these issues associated with the DWT with limited redundancy and low extra computational costs. Basis functions of the 2-D Q-shift complex wavelets are shown in Figure 18.1 in comparison with those of 2-D DWT filters. The DT CWT produces six directionally selective subbands which are better equipped to capture non-horizontal or -vertical features of digital pictures.

Shift invariance is an important property for transforms to have in image compression as well as image analysis and synthesis applications. The decimation or down

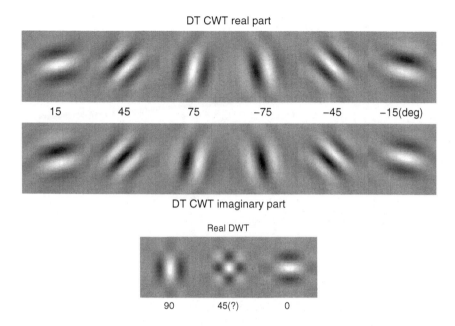

Figure 18.1: Basis functions of 2-D Q-shift complex wavelets (top), and of 2-D real wavelet filters (bottom), all illustrated at level 4 of the transforms. The complex wavelets provide 6 directionally selective filters, while real wavelets provide 3 filters, only two of which have a dominant direction [KR03]. ©SPIE.

sampling processes within the DWT cause aliasing, making the DWT highly shift dependent. Figure 18.2 demonstrates practical advantages of the shift invariance property of the DT CWT in image reconstruction in comparison with the DWT where some of the wavelet subbands are scaled differently from others [Kin01].

An image compression example is given in [KR03] using the *"Lena"* image and the DT CWT at a bit rate of around 0.2 bpp compared with the DWT (Figure 18.3). While the PSNR shows that the DT CWT has a very modest improvement (less than 1 dB) over the DWT, its crucial visual improvement which is representative over a whole range of bit rates is the absence of aliasing artifacts that have plagued the DWT image compression techniques. This can be seen most noticeably along the rim of the straw hat, the frame of the mirror and the strand of hair on the back of the shoulder. Edges and fine textures seem to be better preserved by the DT CWT than the DWT as well.

In the current implementation and preliminary results reported in [KR03], ringing artifacts are more pronounced in DT CWT coded images. It should be noted that the distortion measure used in the DT CWT coding optimization was the Root Mean Square (RMS) error, which does tend to result in Gibb's-type ringing in any bandlimited compression process. The DT CWT with its shift invariance and directional selective properties may serve as an alternative mathematical framework for picture compression.

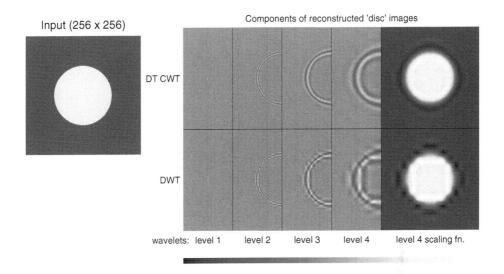

Figure 18.2: Wavelet and scaling function components at levels 1 to 4 of an image of a light circular disc on a dark background, using the 2-D DT CWT (upper row) and 2-D DWT (lower row). Only half of each wavelet image is shown in order to save space. Courtesy of N. G. Kingsbury.

To achieve further performance improvements for digital picture coding in a perceptual quality and bit rate optimization sense, selection of a better coding framework and structure may hold a vital key. In addition to vision modeling inspired transforms such as the Gabor transform [Dau88], the cortex transform [Wat87] and the steerable pyramid decomposition [SFAH92], a number of alternative (complete and over-complete) mathematical frameworks and structures have been made available such as directional wavelet transforms [Vet01, DV03, DVV02, DV01, DV02] and the DT CWT [Kin01], offering solutions to eliminating some hard to solve problems associated with classical complete transforms such as the DCT and the DWT (which have dominated picture coding applications) from their roots. A key issue is what performance criteria should be used in selecting the most suitable framework for the next generation picture compression technology.

18.1.3 Decisions Decisions

When going forward and searching for a better mathematical framework or coding structure for digital picture compression, it all comes down to how much one is prepared to expand or decompose a given picture in an alternative representation (or representations) in order to achieve maximum picture compression. It is one thing to acknowledge that the time honoured philosophy for thousands of years, "What needs be shrunk must first

Non-redundant DWT
0.1994 bit/pel (33.47 dB)

4:1 Overcomplete DT CWT
0.1992 bit/pel (34.12 dB)

Figure 18.3: Comparison between DT CWT and DWT ©SPIE [KR03].

be purposely expanded" by Laotzu [Che81, VK95], does ring true, but it is quite another to practice it. The question is not about whether complete decomposition or over-complete decomposition should or will be considered for use in next generation picture coders. It is rather about the premises on which the selection should be based. Performance criteria for reduction of statistical redundancies and the MSE based rate-distortion optimization theory have led to numerous coding techniques and the current remarkable achievements in picture compression. Many of the popular techniques which are based on complete or critically-sampled transforms or decompositions, such as the DCT and the DWT as well as their variants, have been made to incorporate aspects of the HVS, with varying degrees of success. However, as shown in Chapter 13, while more sophisticated vision model based distortion metrics or pruning techniques may be grafted onto these complete transforms to fend off to a certain extent some fundamental deficiencies, e.g., aliasing artifacts in the DWT coded images (see Figure 13.13), they eventually will cave in under tighter bit rate constraints (see Figure 13.17). This is in a sharp contrast with the over-complete DT CWT coding where aliasing artifacts were eliminated at bit rates ranging from 0.1 to 0.8 bpp [KR03], with ringing artifacts being a major problem.

In the context of image scaling, Daly, Feng and Speigle have a detailed discussion in [DFS02] on the three way tradeoffs among aliasing, blurring and ringing artifacts. The discussion is of particular interest and relevance because in critically-sampled transforms used for picture coding, such as the DWT, sub- or down-sampling is carried out in the analysis or decomposition or encoding process and up-sampling is done in synthesis or reconstruction or decoding process. They thoroughly interrogated the interaction of these distortions involved in applications associated with image scaling. It helps to explain the performance differential discussed previously between the DWT and the DT CWT in picture coding, and it is briefly presented here.

Assuming that the given picture has the Nyquist frequency of 0.5 cycles per pixel (cpp for short), it will result in aliasing if, e.g., sub-sampled by a factor of 2 since the new frequency support in the subsampled image is halved and thus its Nyquist frequency is 0.25 cpp, if expressed in the units of source image pixel frequency. The portion of the spectrum of the original picture above 0.25 cpp will be folded back below 0.25 cpp as aliased signal [OS89]. It is worth noting that in critically-sampled subband coding and wavelet transforms, perfect reconstruction is obtained only by meeting ideal conditions where aliasing effects are cancelled, which is not the case where quantization of transform coefficients is applied to achieve compression [VK95, Cla95]. Low-pass or band limiting or anti-aliasing filters are generally used to minimize or to reduce the aliasing effects, which suppress frequency components above 0.25 cpp at the cost of introducing blurring artifacts since the low-pass filter smoothes images by removing high spatial, high frequency variations. This side effect of anti-aliasing filter can be improved by using a low-pass filter with sharper roll off at its cut-off frequency which makes Gibbs phenomenon visually more prominent in the pixel domain [OS89, DFS02]. They demonstrated the effects of the above tradeoffs using an example with a *disembodied edge* (Gaussian windowed edge) .

Figure 18.4a shows the original disembodied edge at vertical orientation with its 2-D Fourier magnitude spectrum (Fourier transform or FT for short) shown in Figure 18.4b. The same edge at 25° angle from the vertical axis is shown in Figure 18.4c with its FT in Figure 18.4d. Figures 18.4f and h show the frame of the cortex channels overlaid on top of FTs. The center square of Figure 18.5b shows the support in the sub-sampled domain by a factor of 2, and within it is shown the half-amplitude heights of the cortex filters. When the edge is orthogonal, the aliases will land exactly on top of the source spectra and appear as a light doubling of the edge as can be seen in Figure 18.5a. However, if the edge is oblique, the aliases do not overlap the baseband spectra as shown in Figure 18.5d, resulting in the aliasing artifacts commonly known as "jaggies," which are easily seen in Figure 18.5c. When a low-pass filtering is used to eliminate the aliasing effects in a sub-sampled edge as shown in Figure 18.5f, it produces a blurred edge as a result. It is also interesting to note that when an approximately ideal low-pass anti-aliasing filter is used, the resultant reconstructed sub-sampled edge is free from both aliasing and blurring artifacts, but with a hint of ringing artifacts as shown in Figure 19 of [DFS02]. In [DFS02], the authors summarized a number of well-known observations that with edges, aliasing and blurring are easiest to see whereas ringing is less visible. It is also better to allow some aliasing, since the amount of blurring required to remove aliasing entirely is too easily visible. For textures, masking makes all distortions harder to see, and blurring artifact is the visible dominant error. How to achieve the best balance between these tradeoffs is a challenge in selection and design of future picture coding frameworks.

It is plausible to suggest that joint performance criteria be used in selecting future picture coding frameworks. An obvious challenge there is to devise quantified performance criteria and measures for evaluation and automated parameterization of vision models. Relationships and links need to be further investigated between the complete decompositions arrived at by decorrelation and statistical redundancy reduction criteria and over-complete decompositions obtained through vision modeling, in order to search for minimum redundancy decompositions which suit vision modeling. In making decisions on which coding framework is to be used, one is faced with the same old question, i.e., how much one prepares to expand or decompose a given picture for us to achieve maximum picture compression while retaining the best visual picture quality.

18.2   Vision Modeling Issues

A view which is expressed by the contributors of this handbook as well as in published work is that better understanding of the human visual system and better vision models are crucial and are required in digital video image quality assessment and perceptual

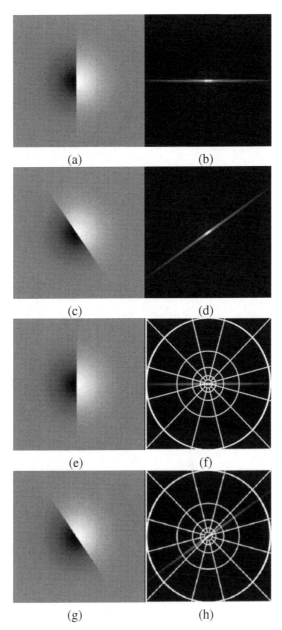

Figure 18.4: Examples of 2-D aliasing and blurring tradeoffs: (a) Original vertical disembodied edge, (b) the 2-D Fourier magnitude spectrum of the disembodied edge, (c) Original edge at 25° from the vertical axis, and (d) the 2-D Fourier spectrum of the edge, (e) Original vertical edge, (f) the 2-D Fourier spectrum of the edge showing Cortex channels, (g) Original edge at 25° from the vertical axis, and (h) the 2-D Fourier spectrum of the edge showing Cortex channels. Courtesy of S. Daly.

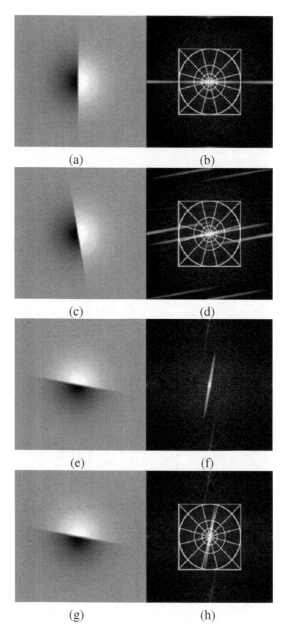

Figure 18.5: Examples of 2-D aliasing and blurring tradeoffs: (a) Sub-sampled edge at vertical orientation without anti-aliasing filtering, (b) the 2-D Fourier magnitude spectrum of the edge showing Cortex channels, (c) Sub-sampled edge at non-orthogonal orientation without anti-aliasing filtering, (d) the 2-D Fourier spectrum of the edge showing Cortex channels, (e) Sub-sampled edge at non-orthogonal orientation with anti-aliasing filtering, (f) the 2-D Fourier spectrum of the edge, (g) Sub-sampled edge at non-orthogonal orientation with anti-aliasing filtering, and (h) the 2-D Fourier spectrum of the edge showing Cortex channels. Courtesy of S. Daly.

picture coder design, which themselves require collaborative work by researchers, at least, in fields of human vision, color science and imaging systems engineering. A number of issues are highlighted here regarding vision modeling which reflect different perspectives and may be of interest for further investigation or pursuit.

Montag and Fairchild reiterated in Chapter 2 the importance of simple stimuli used in visual/psychophysical experiments in order to build up our knowledge of visual perception and vision models based on first principles [DFS02, Dal02]. Natural images or videos may also be required in modeling of more complex visual phenomena, as suggested by Winkler in Chapter 5. A combined approach to applications of vision models to picture coding and quality assessment may require further investigation where vision modeling framework is built on psychophysical experiments based on first principles, and parameterization and optimization of the model for an application are facilitated by application specific visual data and curve fitting techniques [YWWC02].

The vision models currently used in picture quality assessment and coding are based on a number of assumptions and premises. They have led to quite sophisticated, relatively accurate and computationally efficient mathematical models, which are useful for engineering applications. However, the current multiresolution models assume independence among the different spatial frequency and orientation channels, while there is evidence showing nonlinear interactions among the channels as pointed out in Chapter 2.

Even though most visual information processing and communications applications are operating in the suprathreshold range, the vision models that are commonly used in these applications are based on threshold vision experiments. As pointed out in Chapter 2, the effects at threshold are qualitatively different from those at suprathreshold levels and models of detection and discrimination may not be applicable to applications above the threshold. The lack of refined mathematical suprathreshold vision models has seen the application of threshold vision models used, regardless, in picture quality assessment and coding applications [VQE00, Cav02, van96, EB98, Wat93a, TWY04]. When using threshold vision models for applications in suprathreshold range, one would at least like to ascertain modeling inaccuracy quantitatively (see Chapter 5).

There are obvious deficiencies of linear transform to represent HVS, e.g., subband decomposition with equal bandwidth with no 45 and 135 degrees orientation do not represent vision well [EB98]. As well, one cannot expect to use a vision model (e.g., the complete discrete wavelet transform) which has inherited aliasing effects when conditions for perfect reconstruction cannot be met to measure accurately or to assess critically aliasing distortion performance of a picture coder etc. (see Chapters 5 and 13). The tradeoffs among aliasing, blurring and ringing effects need to be balanced in multiresolution representation models where image scaling is involved [DFS02].

Another obvious deficiency in vision modeling is that most of the current work has been focused on the part from the eyes to the primary visual cortex or area V1 [Wan95]. Accurate modeling of higher level vision is required to facilitate further research and development in a number of areas in the field of the HVS based picture quality metrics and perceptual coding techniques such as second generation coding techniques [TK96], foveated picture coders, the structure-similarity based image quality assessment methods and regions of interest based picture quality assessment techniques discussed in this handbook.

The complex visual processing in the retina of the human visual system as described in Chapter 2 results in a remarkable 127:1 compression [THD84]. Better understanding and modeling of this process in the HVS may provide a rich source for human vision inspired perceptual coder design.

18.3   Spatio-Temporal Masking in Video Coding

Temporal adaptation to spatial detail vision was studied by Seyler and Budrikis from 1959 [SB59] to 1965 [SB65]. The magic 160ms to 780ms average recovery time after a scene change was reported during which the detail perception ability of the human visual system was significantly reduced. Temporal response of the HVS has been modeled as a sustained channel by a low-pass filter and a transient channel by a band-pass filter [van96, HS92]. As discussed in Chapter 5, temporal masking may occur before as well as after a temporal discontinuity in intensity. It has been well understood that spatial acuity is reduced as retinal velocity increases [Gir88], and up to 70% reduction in spatial resolution is possible [EB93], even though there is no loss of spatial acuity in the direction orthogonal to motion.

It is interesting to note, however, that all these findings have not quite propagated to video coder design in a more concerted effort for some reason. By and large, an ad hoc or a rule of thumb approach has been commonly adopted in practical video coding systems and products where 2 to 3 frames after a scene change are coded using low bit rates (i.e., coarse quantization is used). Temporal granular noise or fluctuations have also been a problem with video coders based on different techniques (see Chapter 3). While perseverance has been paid off handsomely by introduction of novel motion compensation in terms of conventional rate-distortion optimization, e.g., a coding gain up to 2.7 dB and a bit-rate saving up to 30% have been reported using variable block size and multiple-frame motion compensation [FWG02], a similar level of efforts has been rarely seen and demonstrated to better utilize joint spatio-temporal masking of the HVS in video coding, saving a few [Wat98, YLL+03]. A systematic spatio-temporal masking model or technique is needed either for enhancing the performance of current video coding systems and products or in design of future perceptual video coders. Since, nowadays, post-processing or filtering can be considered as an integral part of a picture decoding system to achieve improved perceptual quality, this spatio-temporal masking model is crucial to the improvement of overall codec system performance as well as the design of high performance post-filtering techniques in minimizing spatial and temporal coding artifacts.

18.4 Picture Quality Assessment

The major focus of this book, in terms of picture quality assessment, is on quantitative or objective measurements and metrics, which is based on the HVS to be exact. This is by no means to overlook the importance and challenges related to subjective testing methods as amply highlighted in Chapter 4. Success and validation of vision models and performance evaluation of HVS based quantitative picture impairment or quality metrics rely heavily on accuracy and validity of subjective test data. More efficient and effective subjective testing methods are required to address issues related to unreliable subjective test data caused by large variance or deviation or contextual effects. Limited investigations on alternative subjective testing methods in order to minimize the variance and contextual effects raise more questions than provide solutions [WYQ02]. Alternative quality scales and different subjective assessment methods are required for video and images which are different from standard TV applications as discussed in Chapter 6. In the rest of this section, however, a number of observations and challenges are summarized regarding HVS based quantitative picture quality assessment.

18.4.1 Picture Quality Metrics Design Approaches

There are different approaches to the design of quantitative picture quality metrics. In terms of the basis of error or distortion formulation, quality or distortion metrics/measures can be categorized as the direct signal waveform (i.e., pixel) based and the HVS based. The typical examples of the former include the MSE, the MAE and the PSNR and other statistics based measures. The latter can be further classified as vision model based or feature extraction based as discussed in Part II. In terms of principal hypothesis as to what affects the process of picture quality assessment by the HVS, structural similarity based or error sensitive based approaches have been formed (see Chapter 6 and Chapter 7).

While it is now widely known that direct pixel-based metrics, such as the MSE and the PSNR, do not correlated well with the perceived picture quality, it has not discouraged investigations into modified pixel-based metrics by incorporating HVS aspects into different weighting schemes or combining them with segmentation and feature extraction techniques (e.g., calculating PSNR values only in the vicinities of block or object boundaries to measure blocking or ringing artifacts). Vision model based metrics have steadily consolidated their theoretical foundation and gradually established their superiority in picture quality and distortion assessment accuracy and versatility over the past decade or two. Nonetheless, they face numerous challenges and obstacles, including improvement of vision model accuracy, vision model parametrization and optimization, high computational complexity, to name a few. As pointed out previously, picture quality assessment is statistical in nature and therefore places the limits on the assessment accuracy that is practically achievable. It is a serious challenge to answer the question whether the large computational overhead of the vision model based metrics is justifiable to obtain limited gains over the pixel-based metrics as reflected in the most recent VQEG reports and activities. In contrast, as discussed in Chapter 5, feature extraction based metrics are computationally more efficient, although they are application specific and not as versatile. An obvious observation here is that the degree of difficulty and the turn around cycle for refinement of vision model based metrics (e.g., framework, structure, parameterization and optimization) are usually much larger and longer, respectively, compared with feature extraction based metrics. Theoretically, performance boundaries of vision model based and feature extraction based metrics require thorough investigation.

Contrast sensitivity and luminance adaptation (Weber's law) were the most commonly used aspects of the HVS in early perceptual quality measures [EB98]. In [Kle93], Klein argued that "image quality is multidimensional and more than one number is needed for its specification." At the question time of Picture Coding Symposium 1996, Musmann[1] asked once again the question: "Is one number enough for picture quality assessment?" In Chapter 6, Miyahara reiterates that at least three distortion factors are to be considered in picture quality or impairment measures, i.e., the "amount of errors," the "location of errors" and the "structure of errors," which are associated with different levels of human vision. Wang, Bovik and Sheikh provide an in-depth discussion in Chapter 7 on the structural similarity based metric which complements the widely used error sensitive based metrics. The structural similarity based metrics rely on our better understanding of higher level of human vision. It points to further investigation into a picture quality/distortion assessment framework which balances impacts of the structural similarity, error sensitivity and error location on perceptual picture quality for a range of applications.

18.4.2 Alternative Assessment Methods and Issues

As discussed in Chapter 4, value judgement and dissimilarity assessment are two methods commonly used for picture quality or impairment evaluations. For human observers, the former task is hard and tends to result in large variances in the measurements and the latter is easy with smaller variances [ITU97, Cea99, WYQ02].

In discussion of no-reference picture quality metrics, Caviedes and Oberti define the concept of "virtual reference" in Chapter 10. This virtual reference does depend on the information or the relevant knowledge on what constitutes a normal and natural looking picture, which an observer uses and pre-stores in his/her memory [F. 05]. While by and large, most reported picture quality or impairment metrics use original pictures as the

---

[1]Dr.-Ing H. G. Musmann is a professor at the University of Hanover, Germany and the author of "Advances in picture coding," *Proc. of. the. IEEE*, 73:523-548, Apr. 1985.

reference, assume that they are noise free, and therefore the assessed pictures will have lower quality ratings than their originals, image processing and enhancement techniques will require picture quality metrics to assess both picture quality gain (or enhancement) and quality loss (or distortion). There are obvious issues regarding criteria and processes that lead to the virtual reference and their impacts on quality assessment outcomes. It is an interesting issue for a multidimensional approach to use different quality and impairment attributes leading to overall image quality ratings.

It is worthwhile noting that various impairment metrics reported in the literature were based on certain assumptions related to the types of coding artifacts or coding techniques, while Budrikis stated that based on the rate distortion theory "ideally, the measure should not be restricted at all as to type of error that it would evaluate," i.e., impairment or quality metric should not be coding artifacts or coder dependent [Bud72].

18.4.3   More Challenges in Picture Quality Assessment

There is little doubt that more research and investigations are required to improve and to enhance the current picture assessment techniques for existing digital picture applications, e.g., modeling of color components [Zha04] and temporal effects for metrics using JND models (see Chapter 9), and no-reference video quality assessment using motion information [F. 05]. At the same time, quantitative metrics are developed for quality and impairment evaluations of new applications and products.

A recent report introduces modifications to Visual Difference Predictor (VDP) [Dal93] for high dynamic range (HDR) images and videos as viewed on advanced HDR display devices where the assumption of global adaptation to luminance in conventional pictures does not hold [MDMS05]. New models and processes are required for monochrome and color HDR picture quality evaluations.

In stereoscopic 3-D TV applications, it has been found that of two video sequences projected for two eyes, one of the sequences can be of considerably lower quality if the other is of fairly good quality [STM+00]. Design and development of an overall picture quality metric for stereoscopic 3-D TV will require not only accurate quality measurements of two video sequences, but also the balance of contributions from and impacts of the quality measures from the two sequences, predicting subjective test data.

As discussed in Chapter 5, it has been known for sometime now, that perceptual picture quality ratings given by human observers are affected by other factors, such as perceived quality of accompanying sound tracks. Integration of perceptual quality indexes of pictures and sound for applications, such as home entertainment and multimedia applications is still a challenging topic.

In diagnostic and medical imaging, various picture coding, processing and enhancement processes are available to assist with efficient and effective diagnosis, archive and transportation of medical images of different modalities for telemedicine and healthcare systems. Picture quality assessment and metrics for different medical imaging applications will be another challenging area of future work.

Along with the gradual and quiet transition from bit-rate driven digital picture storage and transportation services to picture quality driven services, it is inevitable that we face the obvious and fundamental theoretical challenge to establish the quantitative lower bound of perceptually lossless picture compression, or to define quantitatively the somewhat illusive, to date, concept of *psychovisual redundancy*. If nothing else, development of perceptual picture quality/distortion metrics based on accurate vision modeling for this purpose seems to be well justified, albeit with hefty computational burden and tremendous difficulties compared with many other engineering approaches.

18.5   Challenges in Perceptual Coder Design

The definition of perceptual picture coding and decoding system as covered by this book encompasses a much broader concept than what the conventional will normally allow. Chapter 13 presents an account of representative perceptual coding approaches and techniques in the public domain to date. In a symmetric encoding and decoding system, the decoder consists of the inverse operations of those in the encoder in an inverse order. The asymmetric encoding and decoding structures have been around for some time and are very common in standard picture codec algorithms and systems, in the form of pre-processing for noise reduction, bandlimiting for an increase of pixel correlations and loop-filtering to maximize compression performance, as well as having significantly higher computational complexity at the encoder than that at the decoder. This book also considers asymmetric systems to include post-filtering, reconstruction, error correction and concealment, and perceptual dithering as part of a decoding system in order to achieve superior picture coding performance in the rate and perceptual distortion optimization sense. The perceptual picture codec system design therefore requires optimization of the entire encoding and decoding process from picture acquisition to display. For many who have been working on perceptual coder designs, a difficult issue is display dependency, e.g., a perceptually lossless picture coder design for an 8-bit display may not be able to achieve perceptually lossless performance on a 10-bit display [WTW03, WTB+04].

This section discusses what is possible in perceptually lossy and perceptually/visually lossless picture coding.

18.5.1   Incorporating HVS in Existing Coders

There has been a rich pool of waveform based picture coding techniques and algorithms which have been nearly perfected over the past decades [JN84, Cla85, RY90,

VK95, Cla95, RH96]. The research and development momentum so built will further push these techniques and algorithms to their theoretical limits [LSW03]. Incorporating HVS into existing waveform based coding structures is a natural course of action in order to improve the perceptual picture quality for a given bit rate budget [Wat93a, YLL+03, Tau00, DZLL00, TWY04]. In addition to simple contrast sensitivity and luminance adaptation of the HVS which are used in adaptive quantization matrix and bit allocation scheme designs, more sophisticated spatial frequency masking and vision models have been introduced for adaptive quantization algorithms [Wat93a, YLL+03], perceptual pruning [TW03a, TW03b, WTW03] and rate-perceptual distortion optimizations [Tau00, DZLL00, TWY04], as far as underlining mathematical coding structures will allow.

There are a number of theoretical and practical issues as to what can be incorporated in a given coding structure, where and how to optimize the compression and perceived picture quality performance. A theoretical link of what HVS aspects are incorporated by a selected mathematical framework (e.g., a transform or a decomposition) of a picture coder to a full vision model, or lack of it, needs to be established.

For example, Figure 18.6 demonstrates the visual performance of the perceptual coder proposed in [TTW04] where a more sophisticated multi-channel vision model is used in a perceptual distortion measure for JPEG2000 bit-stream compliant coding, compared with JPEG2000 coding with the CVIS [Tau00] and with the MSE, respectively, as the distortion measure in the rate-distortion optimization. This perceptual coder is an extension of the work reported in [TWY04] (also see Chapter 13) to digital YCbCr color space. At 0.125bpp or a compression ratio of 192:1, coding artifacts, such as blurring, aliasing and ringing, are obvious. Perceived picture quality improvement over the other two coders as achieved by the perceptual coder (as shown in Figure 18.6d) is easily identifiable in complete reconstruction of the eyes, reductions of blurring and aliasing at cost of an increase in ringing around the corner of the mouth. A point that has to be made here is that this vision model based coding framework has offered a means of balancing different picture coding artifacts to minimize overall perceived distortions by human viewers. Since the critically sampled wavelet transform decomposition is used as the underpinning coding framework, the perceptual coder is not able to eliminate the aliasing artifacts completely, which provides a sharp contrast to coding results produced by the over-complete DT CWT as shown in Figure 18.3 where aliasing artifacts are absent. It highlights once again the limitations of this approach.

18.5.2 HVS Inspired Coders

Picture coders based on complete modeling of human vision have not been developed due to our incomplete knowledge of the human visual system, whether or not it is necessary for picture compression applications. Obvious redundancy of over-complete decomposition transforms and filters used for vision modeling as discussed previously and

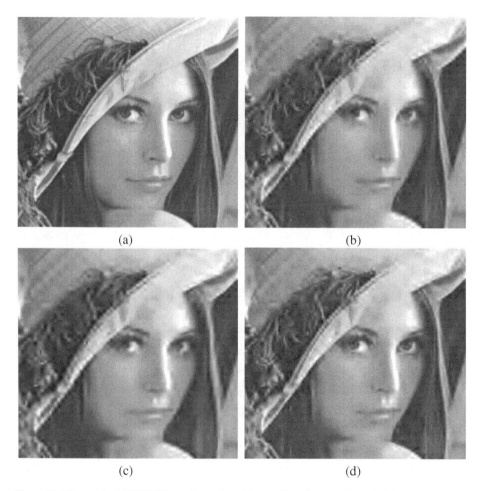

(a)   (b)

(c)   (d)

Figure 18.6: Examples of JPEG2000 coding using different distortion metrics at 0.125bpp: (a) Original *Lena* image, (b) the MSE, (c) CVIS, and (d) a vision model based perceptual metrics [TTW04]. ©IEEE.

in various chapters seems enough a reason for them to be ruled out for consideration as a serious contender for picture compression. This, however, does not prevent introductions of plenty of HVS inspired picture coders over the years, such as model- or object-based coding techniques [TK96], and foveated coders as discussed in Chapter 14. Neither incomplete knowledge of human vision nor engineering/technological difficulties could deter further research and investigations along this direction. From discussions and examples presented in subsections 18.1.2 and 18.5.1, HVS inspired coders such as vision model based coders do offer a more concerted approach to minimization or elimination or balancing of perceptual picture coding artifacts. Significant computational overhead existing in more sophisticated vision model based picture coders will remain as an obstacle in practical applications, notwithstanding that an ever increasing computation power will be made available in future years. Various simplified vision model based techniques have been and will be proposed. A critical issue is to establish theoretical and practical limitations of picture coders where various omissions and simplification of vision model are introduced.

It is expected that as more advanced vision models become available which are more accurate and incorporate high level vision, research and development of HVS inspired coders will flourish.

18.5.3   Perceptually Lossless Coding

Watson defined that "When a compressed image cannot be distinguished visually from the original, the compression is said to be 'perceptually lossless' [Wat89]." The picture coder that achieves this performance, therefore, is called a *perceptually lossless coder* (or PLC).

The theoretical significance and practical importance of perceptual lossless compression cannot be underestimated. It is directly linked to the establishment of picture compression lower bound where both statistical and psychovisual redundancies are completely removed, and of quantitative definition of psychovisual redundancy, if at all possible. While it is much harder to solicit an unambiguous response as to which coder or coding strategy performs perceptually better, say, for a given bit rate budget at supra-threshold level (i.e., where different degrees of different coding artifacts are noticeable), it is relatively easier to ascertain if viewers can see a difference between the original and the compressed image (e.g., a simple "yes" or "no" answer in a double blind test). Although the bit rate generated by a given PLC will most likely be much higher than the theoretical lower bound or practically achievable, improvement of perceptual model and parameterizations can be followed through where an even lower bit rate at just-not-noticeable level may be achieved. This may serve, hopefully, as a better starting point for a perceptual picture coder that can offer superior rate and visual quality performance in suprathreshold applications, which are picture quality driven, rather than bit rate driven.

As discussed in Chapter 13 as well as previous sections, there are a number of tough issues in design and evaluation of PLCs. The majority, if not all, of PLCs reported in the literature to date are based on a certain mathematical framework for picture coding that imposes various limitations to what aspects or mechanisms of the HVS it may incorporate. These limitations in turn will force the PLC to operate at a much higher bit rate to achieve perceptually lossless performance than what will be the theoretical lower bound. Given a mathematical framework, e.g., the DWT decomposition, for picture coding, there is a question of how to select DWT filters which suit a particular application [LYC97]. There is an issue regarding where in the encoding process to incorporate HVS components and, therefore, the performance tradeoffs. For instance, there are at least three approaches to including vision model in a DWT (discrete wavelet transform) based picture coder (see Chapter 13), i.e., a visual pruning component after the DWT, a vision model based perceptual distortion metric using the DWT in rate-distortion optimization, a vision model based perceptual distortion metric using different visual decompositions such as the steerable pyramid filter or the cortex transform. When a vision model based approach is used in PLC design, model parameterization and optimization are considerable challenges. Reliability of the subjective data set is crucial to obtain an appropriate set of parameters for the model as well as what performance criterion should be used for parameterization and optimization processes. Performance of a perceptually lossless coder is also display dependent. There are issues associated with how to evaluate and to ascertain that the designed coder is indeed perceptually lossless, including how to select test pictures (i.e., images or video sequences), how many test pictures are sufficient, how many observers, selection of performance criteria and analysis methods.

Obviously, research in perceptually lossless picture coding is closely related to research in the field of JND computational models as discussed in Chapter 9.

18.6 Codec System Design Optimization

By including three chapters on post-picture-decoding-processing and error concealment or restoration in this book, we hope to emphasize that when searching for future generation picture coding framework, the overall codec system design optimization should include pre-filtering/processing, post-processing and reconstruction techniques in a similar way to transform coding, predictive coding, loop-filtering, motion compensation, entropy coding, etc. in the current picture coding standards [RH96, ISO93a, ISO93b, ITU96, BCM00, ITU03], as a part of the equation. When perceptual coding systems are examined, display environment may also be considered as appropriate. It is understood that techniques such as pre-filtering and post-filtering are purposely left out of the existing standards to encourage innovation and ingenuity. However, in selection and design of the next generation picture codec systems, an approach to overall encoding and decoding system optimization ought be considered in order to achieve higher performance in the rate and perceptual distortion minimization sense.

18.7 Summary

In re-examining the course of digital picture coding and compression research, there is an obvious observation and there exists a continuing challenge. The observation is that although the goal of picture compression is to achieve rate-distortion minimization or rate-quality optimization, most picture coders thus far are rate driven and rarely quality driven. While the historical reasons for rate driven coder design are no longer valid, such as constant bit-rate transmissions, picture quality driven coder design rarely forms a main stream of research and development in the field. For starters, one would have to be sure of how to measure quantitatively perceptual picture quality first to begin considering the design of a quality driven picture coder. As William Thomson Kelvin (Physicist, 1824-1907) once said, "When you can measure what you are speaking about and express it in numbers, you know something about it." This once again highlights the importance of perceptual picture quality metrics in the field. The statement, which was made by van den Branden Lambrecht in 1998, "Today we know more about what are NOT good measures of image quality than we do about what are good measures." [van98] still rings true.

The challenge is that facing glorious achievements in rate driven coder design, one would continue questioning him- or herself and to be questioned whether there is still a place for quality driven coder and perceptual coder design. After all, traditional rate driven coders have been refined to perform almost as well as perceptual coders [TW03a, WTW03] (also see Figures 18.6b and d). The jury is still out on what and how much quality driven or perceptual coders may achieve that the existing rate driven coders have not or cannot. It is hoped that contributions by the authors included in this volume may highlight relevant aspirations on this issue and serve to provide alternative approaches to the challenge.

As is well known in digital picture coding and compression arena, there are two fundamental principles which are frequently practiced, i.e., reduction of statistical redundancies and reduction of psychovisual redundancies. While Claude E. Shannon laid the foundation of modern information theory and marked the lower bound for information lossless compression by quantitatively defining the concept of entropy [Sha48], the rate-distortion optimization theory and techniques are most frequently used in information lossy picture compression, albeit the picture distortions are still commonly represented by direct pixel difference based measures, such as the MSE, which are known not to correlate to visual quality well. An obvious theoretical and practical challenge is to define quantitatively the concept of psychovisual redundancy and to establish theoretical lower bound for perceptually lossless coding.

References

[ABMD92] M. Antonini, M. Barlaud, P. Mathieu, and I. Daubechies. Image coding using wavelet transform. *IEEE trans, image processing*, 1(2):205–219, April 1992.

[AC96] R. Aravind and T. Chen. On comparing mpeg-2 intraframe coding with a wavelet-based codec. In M. Biggar, Ed., *Proceedings of Picture Coding Symposium 1996*, Melbourne, Australia, 1996.

[BCM00] M. Boliek, C. Christopoulos, and E. Majani. JPEG 2000 Part 1 Final Committee Draft Version 1.0. Technical report, ISO/IEC JTC1/SC9/WG1 N1646, March 2000.

[Ber71] T. Berger. *Rate-Distortion Theory*. Englewood Cliffs, NJ: Prentice Hall, 1971.

[Bud72]   Z. L. Budrikis. Visual fidelity criterion and modeling. *Proceedings of the IEEE*, 60(7):771–779, July 1972.

[Cav02]   Special session on objective video quality metrics. In *Proceedings of IEEE International Conference on Image Processing 2002*, Rochester, New York, September 2002. IEEE.

[Cea99]   P. Corriveau and et al. All subjective scales are not created equal: The effects of context on different scales. *Signal Processing*, 77:1–9, 1999.

[Che81]   M.-J. Cheng with translation by T. C. Gibbs. *Lao-Tzu: My words are very easy to understand*. Richmond, California: North Atlantic Books, 1981. Chinese ed., Taipei, Taiwan: Chung Hwa Book Company, Ltd., 1971.

[Cla85]   R. J. Clarke. *Transform Coding of Images*. San Diego, CA: Academic Press, 1985.

[Cla95]   R. J. Clarke. *Digital Compression of Still Images and Video*. San Diego, CA: Academic Press, 1995.

[Cut52]   C. C. Cutler. *Differential Quantization of Communication Signals*: U.S. Patent, No.2,605,361, July 1952.

[Dal93]   S. Daly. The visible differences predictor: An algorithm for the assessment of image fidelity. In A. B. Watson, Ed., *Digital Images and Human Vision*, 179–206. Cambridge, MA: The MIT Press, 1993.

[Dal02]   S. Daly. Using spatio-temporal-chromatic visual models for perceived quality optimization. In *International Congress of Imaging Science (ICIS '02)*, Tokyo, Japan, May 2002. Invited focal talk.

[Dau85]   J. G. Daugman. Uncertainty relation for resolution in space, spatial frequency, and orientation optimized by two-dimensional visual cortical filters. *Journal of the Optical Society of America. A*, 2(7):1160–1169, 1985.

[Dau88]   J. Daugman. Complete discrete 2-d Gabor transforms by neural networks for image analysis and compression. *IEEE Transactions on Acoustics, Speech, and Signal Processing*, 36(7):1169–1179, July 1988.

[DFS02]   S. Daly, X. Feng, and J. Speigle. A few practical applications that require some of the more advanced features of current visual models. In *SPIE/IS&T Electronic Imaging Conference*, 4662:70–83, San Jose, CA, January 2002. SPIE.

[DV01]   M. N. Do and M. Vetterli. Pyramidal directional filter banks and curvelets. In *International Conference on Image Processing*, 3:158–161, Thessaloniki, Greece, October 2001.

[DV02]   M. Do and M. Vetterli. Contourlets: A directional mulitresolution image representation. In *Proc. International Conference on Image Processing*, 1:357–360, September 2002.

[DV03]   M. Do and M. Vetterli. The finite ridgelet transform for image representation. *IEEE Transactions on Image Processing*, 12:16–28, January 2003.

[DVV02]   P. L. Dragotti, M. Vetterli, and V. Velisavljevic. Directional wavelets and wavelet footprints for compression and denoising. In *Proc. International Workshop of Advanced Methods for Multimedia Signal Processing (IWDC)*, Capri, Italy, September 2002.

[DZLL00]   S. Daly, W. Zeng, J. Li, and S. Lei. Visual Masking in Wavelet Compression for JPEG2000. In *IS&T/SPIE Conf. on Image and Video Communication and Processing*, 3974, Jan. 2000.

[EB93]   M. P. Eckert and G. Buchsbaum. The significance of eye moverments and image acceleration for coding television image sequences. In A. B. Watson, Ed., *Digital Iimages and Human Vision*, 89–98. Cambridge, MA: The MIT Press, 1993.

[EB98]   M. P. Eckert and A. P. Bradley. Perceptual quality metrics applied to still image compression. *Singal Processing*, 70:177–200, 1998.

[Ebr05]   T. Ebrahimi, Ed. Proceedings of Workshop on Future Directions in Video Compression. Busan, Korea, April 2005. MPEG/JVT.

[F. 05]   F. Yang and S. Wan and Y. Chang, and H. R. Wu. A Novel Objective No-Reference Metric for Digital Video Quality Assessment. *IEEE Signal Processing Letters*, 2005. (to appear).

[FG03]   M. Flierl and B. Girod. Generalized B pictures and the draft H.264/AVC video-compression standard. *IEEE Transactions on Circuits and Systems for Video Technology*, 13(7):587– 597, July 2003.

[FvdBL02]   J. E. Farrell and C. van den Branden Lambrecht, editors. Special issue on translating human vision research into engineering technology. *Proceedings of the IEEE*, 90(1):3–169, January 2002.

[FWG02]   M. Flierl, T. Wiegand, and B. Girod. Rate-constrained multihypothesis prediction for motion-compensated video compression. *IEEE Transactions on Circuits and Systems for Video Technolgoy*, 12(11):957–969, November 2002.

[Gir88]   B. Girod. Eye movements and coding of video sequences. In T. R. Hsing, Ed., *SPIE Proceedings, Visual Communications and Image Processing*, 1001:398–405. Cambridge, MA: SPIE, November 1988.

[Gir03]   B. Girod. Video coding for compression and beyond. In *Proceedings of IEEE International Conference on Image Processing*, IEEE Press. Barcelona, Spain, September 2003.

[GJR96]   N. García, F. Jaureguizar, and J. R. Ronda. Pixel-based video compression schemes. In L. Torres and M. Kunt, Eds., *Video Coding-The Second Generation Approach*, 31–78. Dordrecht, The Netherlands: Kluwer Academic Publishers, 1996.

[Goo51]   W. M. Goodall. Television by pulse code modulation. *Bell System Technical Journal*, 33–49, January 1951.

[HDZG03]   C. He, J. Dong, Y. F. Zheng, and Z. Gao. Optimal 3-d coefficient tree structure for 3-d wavelet video coding. *IEEE Trans. Circuits Syst. Video Techn.*, 13(10):961–972, Oct. 2003.

[HS92]   R. F. Hess and R. J. Snowden. Temporal properties of human visual filters: Number, shapes and spatial covariation. *Vison Research*, 32(1):47–59, 1992.

[Hua65]   T. S. Huang. Pcm picture transmission. *IEEE Spectrum*, 2:57–63, December 1965.

[IL96]   A. F. Inglis and A. C. Luther. *Video Engineering*. New York, NY: McGraw–Hill, 2 ed., 1996.

[ISO93a]   ISO/IEC. ISO/IEC 11172 Information Technology: Coding of Moving Pictures and Associated Audio for Digital Storage Media at up to about 1.5 Mbit/s. Technical report, ISO/IEC, 1993.

[ISO93b]   ISO/IEC JTC1/SC29/WG11. CD 13818, Generic Coding of Moving Pictures and Associated Audio. Technical report, ISO/IEC, November 1993.

[ITU96]   ITU-T. ITU-T Recommendation H.263, Video Coding for Low Bit Rate Communication. Technical report, ITU-T, 1996.

[ITU97]   ITU-R. *Investigation of Contextual Effects*. ITU-R, Canada, France, Germany, Switzerland, March 1997. ITU-R Document 11E/34-E, 24.

[ITU03]   ITU-T. ITU-T Recommendation H.264, Advanced Video Coding for Generic Audiovisual Services. Technical report, ITU-T, 2003.

[JJS93]   N. Jayant, J. Johnston, and R. Safranek. Signal compression based on models of human perception. *Proceedings of the IEEE*, 81(10):1385–1422, October 1993.

[JN84]   N. S. Jayant and P. Noll. *Digital Coding of Waveforms— Principles and Applications to Speech and Video*. Englewood Cliffs, NJ: Prentice Hall, 1984.

[Joh88]   J. D. Johnston. Transform coding of audio singals using perceptual noise criteria. *IEEE Journal on Selected Areas in Communications*, 6(2):314–323, February 1988.

[Kin01]   N. G. Kingsbury. Complex wavelets for shift invariant analysis and filtering of signals. *Journal of Applied and Computational Harmonic Analysis*, 10(3):234–253, May 2001.

[Kle93]   S. A. Klein. Image quality and image comression: A psychophysicist's viewpoint. In A. B. Watson, Ed., *Digital Iimages and Human Vision*, 73–88. Cambridge, MA: The MIT Press, 1993.

[KR03]   N. G. Kingsbury and T. H. Reeves. Iterative image coding with overcomplete complex wavelet transforms. In *Proc. Conf. on Visual Communications and Image Processing*, SPIE Press. Lugano, Switzerland, July 2003. paper 1160.

[KTR05]   S.-K. Kwon, A. Tamhankar, and K. R. Rao. Overview of H.264/MPEG-4 Part 10. *J. VCIR*, 2005. (to appear).

[LSW03]   A. Luthra, G. Sullivan, and T. Wiegand. Special issue on the h.264/avc video coding standard. *IEEE Trans. Circuits Syst. Video Techn.*, 13(7):557–725, July 2003.

[LYC97]   N. Lin, T. Yu, and A. K. Chan. Perceptually lossless wavelet-based compression for medical images. *Proceedings SPIE, Medical Imaging 1997: Image Display*, 3031:763 – 770, 1997.

[MDMS05]   R. Mantiuk, S. Daly, K. Myszkowski, and H. Seidel. Predicting visible differences in high-dynamic-range images: Model and its calibration. In *SPIE Electronic Imaging Conference: Human Vision and Electronic Imaging*, 5666:204–214. SPIE, 2005.

[NZ97]   R. Neff and A. Zakhor. Very low bit-rate video coding based on matching pursuits. *IEEE Transactions on Circuits and Systems for Video Technology*, 7(1):158–171, July 1997.

[NZ02a]   R. Neff and A. Zakhor. Matching pursuit video coding I. dictionary approximation. *IEEE Trans. Circuits Syst. Video Techn.*, 12(1):13–26, January 2002.

[NZ02b]   R. Neff and A. Zakhor. Matching pursuit video coding II. operational models for rate and distortion. *IEEE Trans. Circuits Syst. Video Techn.*, 12(1):27–39, January 2002.

[NZV94]   R. Neff, A. Zakhor, and M. Vetterli. Very low bitrate video coding using matching pursuits. In *Proceedings SPIE Conference on Visual Communications and Image*, 2308:47–60, Chicago, IL, September 1994.

[OS89]   A. V. Oppenheim and R. W. Schafer. *Discrete-Time Signal Processing*. Englewood Cliffs, NJ: Prentice–Hall, 1989.

[Ree93]   T. R. Reed. Local frequency representations for image sequence processing and coding. In A. B. Watson, Ed., *Digital Iimages and Human Vision*, 3–12. Cambridge, MA: The MIT Press, 1993.

[RH96]   K. R. Rao and J. J. Hwang. *Techniques and Standards for Image, Video and Audio Coding*. Upper Saddle River, NJ: Prentice Hall, 1996.

[RY90]   K. R. Rao and P. Yip. *Discrete Cosine Transform—Algorithms, advantages, applications*. San Diego, CA: Academic Press, 1990.

[SB59]   A. J. Seyler and Z. L. Budrikis. Measurements of temporal adaptation to spatial detail vision. *Nature*, 184:1215–1217, October 1959.

[SB65]   A. J. Seyler and Z. L. Budrikis. Detail perception after scene changes in television image presentations. *IEEE Trans. on Information Theory*, IT-11(1):31–43, January 1965.

[SF95]   E. Simoncelli and W. Freeman. The steerable pyramid: A flexible architecture for multi-scale derivative computation. In *Proceedings of 2nd IEEE Intl. Conf. On Image processing*, III:444–447, Washington, DC, October 1995.

[SFAH92]   E. P. Simoncelli, W. T. Freeman, E. H. Adelson, and D. J. Heeger. Shiftable Multiscale Transform. *IEEE Trans. on Information Theory*, 38(2):587–607, 1992.

[SFW+00]   R. M. Slone, D. H. Foos, B. R. Whiting, E. Muka, D. A. Rubin, T. K. Pilgram, K. S. Kohm, S. S. Young, P. Ho, and D. D. Hendrickson. Assessment of visually lossless irreversible image compression: Comparison of three methods by using an image-comparison workstation. *Radiology-Computer Applications*, 215(2):543–553, 2000.

[Sha48]   C. E. Shannon. A mathematical theory of communication. *Bell System Technical Journal*, 27:379–623, 1948.

[Sha93]   J. M. Shapiro. Embedded image coding using zerotrees of wavelet coefficients. *IEEE Trans. on Signal Processing*, 41(12):3445–3462, 1993.

[SP96]   A. Said and W. A. Pearlman. A new fast and efficient image codec based on set partitioning in hierarchical trees. *IEEE Trans. on Circuits and Systems for Video Technology*, 6:243–250, June 1996.

[SR87]   P. Santago and S. A. Rajala. Using convex set techniques for combined pixel and frequency domain coding of time-varying images. *IEEE Journal on Selected Areas in Communications*, SAC-5(7):1127–1139, August 1987.

[STM+00]   L. B. Stelmach, W. J. Tam, D. V. Meegan, A. Vincent, and P. Corriveau. Human perception of mismatched stereoscopic 3D inputs. In *Proc. ICIP*, Vancouver, BC, Canada, 2000.

[Tau00]   D. Taubman. High performance scalable image compression with EBCOT. *IEEE Trans. Image Proc.*, 9:1158–1170, July 2000.

[THD84]   K. Tzou, R. Hsing, and J. G. Dunham. Applications of physiological human visual system model to image compression. In *SPIE: Applications of Digital Image Processing VII*, 504:419–424. SPIE, 1984.

[TK96]   L. Torres and M. Kunt. *Video Coding-The Second Generation Approach*. Dordrecht, The Netherlands: Kluwer Academic Publishers, 1996.

[TTW04]   C.-S. Tan, D. M. Tan, and H. R. Wu. Perceptual coding of digital colour images based on a vision model. In *Proceedings of IEEE International Symposium on Circuits and Systems*, 441–444, IEEE Press. Vancouver, Canada, May 2004. (An invited paper).

[TW99a]   D. M. Tan and H. R. Wu. Multi-dimensional discrete cosine transform for image compression. In *Proceedings of the 6th IEEE International Conference on Electronics, Communications and Systems (ICECS'99)*, T1E2.1–T1E2.5, September 1999.

[TW99b]   D. M. Tan and H. R. Wu. Quantisation design for multi-dimensional dct for image coding. In *Proceedings of 1999 IEEE Region Ten Conference (TENCON99)*, 1:198–201, Cheju, Korea, September 1999. IEEE Region Ten.

[TW03a]   D. Tan and H. Wu. Perceptual lossless coding of digital monochrome images. In *Proceedings of 2003 International Symposium on Intelligent Signal Processing and Communication Systems (ISPACS 2003)*, IEEE. Awaji Island, Japan, December 2003.

[TW03b]   D. M. Tan and H. Wu. Adaptation of visually lossless colour coding to jpeg2000 in the composite colour space. In *Proceedings of The Fourth International Conference on Information, Communications & Signal Processing and Fourth Pacific-Rim Conference on Multimedia (ICICS-PCM 2003)*, 1B2.4.1–7, Singapore, December 2003.

[TWY04]   D. M. Tan, H. R. Wu, and Z. Yu. Perceptual coding of digital monochrome images. *IEEE Signal Processing Letters*, 11(2):239–242, Febuary 2004.

[van96]   C. J. van den Branden Lambrecht. *Testing Digital Video System and Quality Metrics based on Perceptual Models and Architecture*. Ph.D. thesis, EPFL, CH-1015 Lausanne, Switzerland, May 1996.

[van98]   C. van den Branden Lambrecht. Editorial. *Singal Processing*, 70:153–154, 1998.

[vdBL98] C. van den Branden Lambrecht, editor. Special issue on image and video quality metrics. *Signal Processing*, 70(3):153–294, November 1998.

[Vet01]   M. Vetterli. Wavelets, approximation, and compression. *IEEE Signal Processing Magazine*, 18(5):59–73, September 2001.

[VK95]   M. Vetterli and J. Kovacevic. *Wavelets and Subband Coding*. Englewood Cliffs, NJ: Prentice Hall, 1995.

[VQE00]   VQEG. *Final Report from the Video Quality Experts Group on the Validation of Objective Models of Video Quality Assessment*. VQEG, March 2000. available from *ftp.its.bldrdoc.gov*.

[Wan95]   B. A. Wandell. *Foundations of Vision*. Sunderland, MA: Sinauer Associates, Inc, 1995.

[Wat87]   A. B. Watson. The cortex transfrom: Rapid computation of simulated neural images. *Computer Vision, Graphics and Image Processing*, 39:311–327, 1987.

[Wat89]   A. B. Watson. Receptive fields and visual representations. In *SPIE Proceedings*, 1077:190–197. SPIE, 1989.

[Wat93a] A. B. Watson. DCTune: A technique for visual optimization of dct quantization matrices for individual images. *Society for Information Display Digest of Technical papers XXIV*, 946–949, 1993.

[Wat93b]   A. B. Watson, Ed. *Digital Images and Human Vision.* Cambridge, MA: The MIT Press, 1993.

[Wat98]   A. B. Watson. Toward a perceptual video quality metric. In *Proc. SPIE*, 3299:139–147, San Jose, CA, January 26–29, 1998.

[WTB97]   A. B. Watson, M. Taylor, and R. Borthwick. DCTune perceptual optimization of compressed dental X-rays. In Y. Kim, Ed., *Proceedings SPIE, Medical Imaging*, 3031:358–371, 1997.

[WTB+04]   D. Wu, D. M. Tan, M. Baird, J. DeCampo, C. White, and H. R. Wu. Perceptually lossless medical image coding. In *Proceedings of 2004 International Symposium on Intelligent Multimedia, Video & Speech Processing*, Hong Kong, October 2004.

[WTW03]   D. Wu, D. M. Tan, and H. R. Wu. Visually lossless adaptive compression of medical images. In *Proceedings of The Fourrth International Conference on Information, Communications & Signal Processing and Fourth Pacific Rim Conference on Mulitmedia (ICICS-PCM 2003)*, Singapore, December 2003.

[Wuv00]   Special session on testing and quality metrics for digital video services. In K. N. Ngan, T. Sikora, and M.-T. Sun, Eds., *SPIE Proceedings, Visual Communications and Image Processing 2000*, 4067:741–809, SPIE. Perth, Australia, June 2000.

[WYQ02]   H. R. Wu, Z. Yu, and B. Qiu. Multiple reference impairment scale subjective assessment method for digital video. In *Proceedings of 14th International Conference on Digital Signal Processing*, 1:185–189, IEEE/IEE/EURASIP. Santorini, Greece, July 2002.

[YLL+03]   X. Yang, W. Lin, Z. Lu, E. P. Ong, and S. Yao. Perceptually adaptive hybrid video encoding based on just-noticeable-distortion profile. In *SPIE Proceedings of VCIP*, 1448–1459, 2003.

[YWWC02]   Z. Yu, H. R. Wu, S. Winkler, and T. Chen. Vision model based impairment metric to evaluate blocking artifacts in digital video. *Proc. IEEE*, 90(1):154–169, January 2002.

[Zha04]   X. Zhang. Just-noticeable distortion estimation for images. Masters thesis, School of Electrical and Electronic Engineering, Nanyang Technological University, Singapore, 2004.

Appendix A

VQM Performance Metrics

Philip Corriveau
Intel Media and Acoustics Perception Lab, U.S.A.

This appendix discusses metrics used by the VQEG (Video Quality Experts Group) for performance evaluation of objective or quantitative video quality measurements and metrics. All definitions in statistical analysis used by the VQEG are in line with generally described statistical processes.

Performance of the objective models was evaluated with respect to three aspects of their ability to estimate subjective assessment of video quality, namely,

- Prediction accuracy - the ability to predict the subjective quality ratings with low error;
- Prediction monotonicity - the degree to which the model's predictions agree with the relative magnitudes of subjective quality ratings; and
- Prediction consistency - the degree to which the model maintains prediction accuracy over the range of video test sequences, i.e., that its response is robust with respect to a variety of video impairments.

These attributes were evaluated through seven performance metrics specified in the objective test plan, and are discussed below.

The outputs by the objective video quality model (i.e., the Video Quality Rating, VQR) should be correlated with the viewer Difference Mean Opinion Scores (DMOSs) in a predictable and repeatable fashion. The relationship between the predicted VQR and the DMOS need not be linear, as subjective testing can have nonlinear quality rating compression at the extremes of the test range. It is not the linearity of the relationship that is critical, but the stability of the relationship and a data set's error-variance from the relationship that determines predictive usefulness. To remove any nonlinearity due to the subjective rating process and to facilitate comparison of the models in a common analysis space, the relationship between each model's predictions and the subjective ratings was estimated using a nonlinear regression between the model's set of VQRs and the corresponding DMOSs.

The nonlinear regression was fitted to the [DMOS,VQR] data set and restricted to be monotonic over the range of VQRs. The following logistic function was used:

$$DMOS_p = \frac{b_1}{(1 + exp(-b_2 \times (VQR - b_3)))}$$

fitted to the data [DMOS,VQR]. The nonlinear regression function was used to transform the set of VQR values to a set of predicted MOS values, $DMOS_p$, which were then compared with the actual DMOS values from the subjective tests. Once the nonlinear transformation was applied, the objective model's prediction performance was then evaluated by computing various metrics on the actual sets of subjectively measured DMOS and the predicted $DMOS_p$.

The VQEG test plan mandated six metrics of the correspondence between a video quality metric (VQM) and the subjective data (DMOS). In addition, it requires checks of the quality of the subjective data. The Test Plan does not mandate statistical tests of the difference between different VQMs' fit to DMOS.

A.1   Metrics Relating to Model Prediction Accuracy

Metric 1: The Pearson linear correlation coefficient between $DMOS_p$ and the DMOS.

The correlation coefficient is a quantity that gives the quality of a least squares fitting to the original data [Act66, Edw76, GS93, KK62a, PFTV92, WR67]. To define the correlation coefficient, first consider the sum of squared values ss_{xx}, ss_{xy} and ss_{yy} of a set of N data points (x_i, y_i) about their respective means, μ_x and μ_y, respectively,

$$
\begin{aligned}
ss_{xx} &\equiv \sum (x_i - \mu_x)^2 & \text{(A.1)}\\
&= \sum x^2 - 2\mu_x \sum x + \sum \mu_x^2 \\
&= \sum x^2 - 2N\mu_x^2 + N\mu_x^2 \\
&= \sum x^2 - N\mu_x^2 & \text{(A.2)}
\end{aligned}
$$

$$
\begin{aligned}
ss_{yy} &\equiv \sum (y_i - \mu_y)^2 & \text{(A.3)}\\
&= \sum y^2 - 2\mu_y \sum y + \sum \mu_y^2 \\
&= \sum y^2 - 2N\mu_y^2 + N\mu_y^2 \\
&= \sum y^2 - N\mu_y^2 & \text{(A.4)}
\end{aligned}
$$

$$ss_{xy} \equiv \sum (x_i - \mu_x)(y_i - \mu_y) \tag{A.5}$$

$$= \sum (x_i y_i - \mu_x y_i - x_i \mu_y + \mu_x \mu_y)$$

$$= \sum xy - N\mu_x \mu_y - N\mu_x \mu_y + N\mu_x \mu_y$$

$$= \sum xy - N\mu_x \mu_y. \tag{A.6}$$

For the linear least squares fitting, the coefficient b in

$$y = a + bx \tag{A.7}$$

is given by

$$b = \frac{N \sum xy - \sum x \sum y}{N \sum x^2 - (\sum x)^2} = \frac{ss_{xy}}{ss_{xx}}, \tag{A.8}$$

and the coefficient b' in

$$x = a' + b'y \tag{A.9}$$

is given by

$$b' = \frac{N \sum xy - \sum x \sum y}{N \sum y^2 - (\sum y)^2}. \tag{A.10}$$

The correlation coefficient r (sometimes also denoted R) is defined by

$$r \equiv \sqrt{bb'} = \frac{N \sum xy - \sum x \sum y}{\sqrt{[N \sum x^2 - (\sum x)^2][N \sum y^2 - (\sum y)^2]}}, \tag{A.11}$$

which can be written simply as

$$r^2 = \frac{ss_{xy}^2}{ss_{xx} ss_{yy}}. \tag{A.12}$$

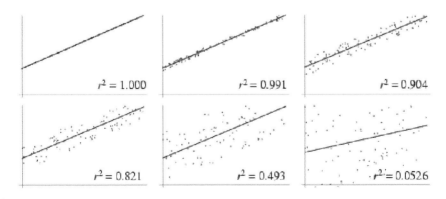

Figure A.1: The correlation coefficient r.

The correlation coefficient is also known as the *product-moment coefficient of correlation* or *Pearson's correlation*. The correlation coefficients for linear fits to increasingly noisy data are shown in Figure A.1. The correlation coefficient has an important physical interpretation. To see this, define

$$A \equiv [\sum x^2 - N\mu_x^2]^{-1} \tag{A.13}$$

and denote the "expected" value for y_i as $\hat{y}_i$, i.e.,

$$
\begin{aligned}
\hat{y}_i &= a + bx_i = \mu_y - b\mu_x + bx_i = \mu_y + b(x_i - \mu_x) \\
&= A(\mu_y \sum x^2 - \mu_x \sum xy + x_i \sum xy - N\mu_x\mu_y x_i) \\
&= A[\mu_y \sum x^2 + (x_i - \mu_x) \sum xy - N\mu_x\mu_y x_i]
\end{aligned} \tag{A.14}
$$

Sums of $\hat{y}_i$, $\hat{y}_i^2$ and $y_i\hat{y}_i$ are, respectively,

$$
\begin{aligned}
\sum \hat{y}_i &= A(N\mu_y \sum x^2 - N^2\mu_x^2\mu_y) \tag{A.15} \\
\sum \hat{y}_i^2 &= A^2[N\mu_y^2(\sum x^2)^2 - N^2\mu_x^2\mu_y^2(\sum x^2) \\
&\quad - 2N\mu_x\mu_y(\sum xy)(\sum x^2) + 2N^2\mu_x^3\mu_y(\sum xy) \\
&\quad + (\sum x^2)(\sum xy)^2 - N\mu_x^2(\sum xy)] \tag{A.16} \\
\sum y_i\hat{y}_i &= A\sum [y_i\mu_y \sum x^2 + y_i(x_i - \mu_x) \sum xy - N\mu_x\mu_y x_i y_i] \\
&= A[N\mu_y^2 \sum x^2 + (\sum xy)^2 - N\mu_x\mu_y \sum xy - N\mu_x\mu_y(\sum xy)] \\
&= A[N\mu_y^2 \sum x^2 + (\sum xy)^2 - 2N\mu_x\mu_y \sum xy] \tag{A.17}
\end{aligned}
$$

The sum of squared residuals (SSR) is then

$$
\begin{aligned}
SSR &\equiv \sum(\hat{y}_i - \mu_y)^2 = \sum(\hat{y}_i^2 - 2\mu_y\hat{y}_i + \mu_y^2) \\
&= A^2(\sum xy - N\mu_x\mu_y)^2(\sum x^2 - N\mu_x^2) = \frac{(\sum xy - N\mu_x\mu_y)^2}{\sum x^2 - N\mu_x^2} \\
&= b\, ss_{xy} = \frac{ss_{xy}^2}{ss_{xx}} = ss_{yy}r^2 = b^2 ss_{xx}, \tag{A.18}
\end{aligned}
$$

and the sum of squared errors is

$$
\begin{aligned}
SSE &\equiv \sum(y_i - \hat{y}_i)^2 = \sum(y_i - \mu_y + b\mu_x - bx_i)^2 \\
&= \sum[y_i - \mu_y - b(x_i - \mu_x)]^2 \\
&= \sum(y_i - \mu_y)^2 + b^2\sum(x_i - \mu_x)^2 - 2b\sum(x_i - \mu_x)(y_i - \mu_y) \\
&= ss_{yy} + b^2 ss_{xx} - 2b ss_{xy}.
\end{aligned}
\tag{A.19}
$$

Note that

$$
b = \frac{ss_{xy}}{ss_{xx}},
\tag{A.20}
$$

$$
r^2 = \frac{ss_{xy}^2}{ss_{xx} ss_{yy}},
\tag{A.21}
$$

therefore,

$$
\begin{aligned}
SSE &= ss_{yy} + \frac{ss_{xy}^2}{ss_{xx}^2} ss_{xx} - 2\frac{ss_{xy}}{ss_{xx}} ss_{xy} \\
&= ss_{yy} - \frac{ss_{xy}^2}{ss_{xx}} \\
&= ss_{yy}\left(1 - \frac{ss_{xy}^2}{ss_{xx} ss_{yy}}\right) \\
&= ss_{yy}(1 - r^2),
\end{aligned}
\tag{A.22}
$$

and

$$
SSE + SSR = ss_{yy}(1 - r^2) + ss_{yy}r^2 = ss_{yy}.
\tag{A.23}
$$

The square of the correlation coefficient, r^2, is therefore given by

$$
r^2 \equiv \frac{SSR}{ss_{yy}} = \frac{ss_{xy}^2}{ss_{xx} ss_{yy}} = \frac{(\sum xy - N\mu_x\mu_y)^2}{(\sum x^2 - N\mu_x^2)(\sum y^2 - N\mu_y^2)}.
\tag{A.24}
$$

In other words, r^2 is the proportion of ss_{yy} which is accounted for by the regression.

If there is complete correlation, the lines obtained by solving for the best-fit for (a, b) and (a', b') coincide (since all data points lie on them). As a result, solving (A.9) for y and equating the result to (A.7) result in

$$
y = -\frac{a'}{b'} + \frac{x}{b'} = a + bx.
\tag{A.25}
$$

Therefore, $a = -\frac{a'}{b'}$ and $b = \frac{1}{b'}$, leading to

$$r^2 = bb' = 1. \tag{A.26}$$

The correlation coefficient is independent of both origin and scale, i.e.,

$$r(u, v) = r(x, y), \tag{A.27}$$

where

$$u \equiv \frac{x - x_0}{h}, \tag{A.28}$$

$$v \equiv \frac{y - y_0}{h}, \tag{A.29}$$

x_0 and y_0 represent a pair of initial values for x and y, and h is a scaling factor.

A.2   Metrics Relating to Prediction Monotonicity of a Model

Metric 2: Spearman rank order correlation coefficient between $DMOS_p$ and the DMOS.

VQR performance was assessed in VQEG tests by correlating subjective scores and corresponding VQR predicted scores after the subjective data were averaged over subjects yielding 64 means for the 64 HRC-SRC combinations. The Spearman correlation and the Pearson correlation and all other statistics were calculated across all 64 HRC-SRC data simultaneously. In particular, these correlations were not calculated separately for individual SRCs (Sources) or for individual HRCs (Hypothetical Reference Circuits). The algorithms for calculating correlations in the SAS statistical package used by the VQEG conform to standard textbook definitions [HC95, LD98, PFTV92, Weic].

A nonparametric (distribution-free) rank statistic was proposed by Spearman in 1904 as a measure of the strength of the associations between two variables [LD98]. The Spearman rank correlation coefficient can be used to give an *R-estimate*, and is a measure of monotonic association that is used when the distribution of the data make Pearson's *correlation coefficient* undesirable or misleading.

The Spearman rank correlation coefficient is defined by

$$r_S \equiv 1 - 6 \sum_{i=1}^{N} \frac{D^2}{N(N^2 - 1)}, \tag{A.30}$$

where N is the number of pairs of values, D is the difference in statistical rank of corresponding variables, and is an approximation to the exact *correlation coefficient*

$$r \equiv \frac{\sum xy}{\sqrt{\sum x^2 \sum y^2}} \tag{A.31}$$

computed from the original data. Because it uses ranks, the Spearman rank correlation coefficient is much easier to compute.

The *variance*, *kurtosis*, and higher-order *moments* are

$$\sigma^2 = \frac{1}{N-1} \tag{A.32}$$

$$\gamma_2 = -\frac{114}{25N} - \frac{6}{5N^2} - \cdots \tag{A.33}$$

$$\gamma_3 = \gamma_5 = \cdots = 0. \tag{A.34}$$

A.3   Metrics Relating to Prediction Consistency

Metric 3: Outlier Ratio

Outlier ratio is defined as the ratio between the number of "outlier points", N_O, and the total number of data points, N, i.e.,

$$Outlier \ Ratio = \frac{N_O}{N}$$

where an outlier is a point for which the prediction error, $Qerror[i]$, for $1 \le i \le N$:

$$|Qerror[i]| > 2 \times DMOS\_Standard\_Error[i]$$

Twice the DMOS Standard Error (i.e., standard deviation) was used in the VQEG tests as the threshold for defining an outlier point.

Metrics 4, 5, 6: These metrics were evaluated based on the method described in [T1.TR.PP.72-2001] ("Methodological Framework for Specifying Accuracy and Cross-Calibration of Video Quality Metrics"). Examples of accepted calculations are shown as follows:

Metric 4: RMS Error

The *root-mean-square (RMS)* of a variate X, also known as the quadratic mean, is the square root of the mean squared value of x [Weib, KK62b]:

$$R(x) \equiv \sqrt{\langle x^2 \rangle} \tag{A.35}$$

$$= \begin{cases} \sqrt{\frac{\sum_{i=1}^{n} x_i^2}{n}}, & \text{for a discrete distribution;} \\ \sqrt{\frac{\int P(x)x^2 dx}{\int P(x)dx}}, & \text{for a continuous distribution.} \end{cases} \tag{A.36}$$

The root-mean-square is the special case, M_2, of the power mean [Bul03].

Hoehn and Niven [HN85] showed that

$$R(a_1 + c, a_2 + c, \ldots, a_n + c) < c + R(a_1, a_2, \ldots, a_n) \tag{A.37}$$

for any positive constant c.

Scientists often use the term root-mean-square as a synonym for standard deviation when they refer to the square root of the mean squared deviation of a signal from a given baseline or fit.

Metric 5: Resolving Power

Resolving power is defined as the delta VQM value (i.e., ΔVQM) above which the conditional subjective-score distributions have means that are statistically different from one another (typically at the 0.95 significance level). Such an "error bar" measure is required in order for video service operators to judge the significance of VQM fluctuations.

Metric 6: Classification Errors

A *classification error* is made when the subjective test and the VQM lead to different conclusions on a pair of situations (i.e., video-source/distortion combinations). There are three different types of classification errors which can arise when using a VQM, including *false tie, false differentiation*, and *false ranking*. The "false tie" error is the least offensive error and occurs when the subjective test says that two situations are different, while the VQM says that they are the same. A "false differentiation" error is usually more offensive and occurs when the subjective test says that two situations are the same, but the VQM says that they are different. The "false ranking" error would generally be the most offensive error and occurs when the subjective test says that situation i is better than situation j, but the VQM says just the opposite.

A.4 MATLAB Source Code

The following Matlab code was provided by S. Wolf of NTIA/ITS[1]

Below is a MATLAB subroutine called vqm_accuracy.m. This version scales the subjective data to [0,1], applies a polynomial fit of the objective to the scaled subjective data, calculates all the metrics, and plots the VQM frequencies of "False Tie," "False Differentiation," "False Ranking," and "Correct Decision." It is sufficient to have Version 5.3.1 of MATLAB (1999) with the Statistics and Optimization toolboxes which are available separately. Software can also be developed which does not use any of the toolboxes. The present code is intended as an illustrative example, and does not include all possible options and fitting functions [ITU04, ATI03, Bri04].

Usage: At the matlab prompt ("≫"), for VQM r0 type:
    ≫load r0.dat
    ≫vqm_accuracy(r0,-1,0,100,2)
For VQM r2, type:
    ≫load r2.dat
    ≫vqm_accuracy(r2,1,0,100,2)

Here, r0.dat and r2.dat are text files that contain a subset of the VQEG 525-line data. Each line in this file corresponds to a situation, and comprises an SRC number, an HRC number, VQM score, number of viewings, mean subjective score, and subjective-score variance. Once the r0 and r2 dat files are loaded, either form of vqm_accuracy may be run again. In the first calling argument of vqm_accuracy, r0 corresponds to the PSNR model in [T1.TR.74-2001], and r2 corresponds to the PQR model in [T1.TR.75.2001]. The second argument is 1 if the objective metric indicates worse image quality when it is larger, else the argument is -1. The third and fourth arguments are the nominal best and worst ratings on the native subjective scale. The final argument is the order of the polynomial to which the VQM is fitted.

Source Code:

```
function vqm_accuracy (data_in, vqm_sign, best, worst, order)
% MATLAB function vqm_accuracy (data_in, vqm_sign, best, worst, order)
%
% Each row of the input data matrix data_in must be organized as
% [src_id  hrc_id  vqm  num_view  mos  variance], where
%
%     src_id is the scene number
%     hrc_id is the hypothetical reference circuit number
%     vqm is the video quality metric score for this src_id x hrc_id
%     num_view is the number of viewers who rated this src_id x hrc_id
```

[1]National Telecommunications and Information Administration/Institute for Telecommunication Sciences.

```
%     mos is the mean opinion score of this src_id x hrc_id
%     variance is the variance of this src_id x hrc_id
%
%     The total number of src x hrc combinations is size(data_in,1).
%
% vqm_sign = 1 or -1 and gives the direction of vqm with respect to
%              the common subjective scale.  For instance, since "0" is
%              no impairment and "1" is maximum impairment on the common
%              scale, vqm_sign would be -1 for PSNR since higher values
%              of PSNR imply better quality (i.e., this is opposite to
%              the common subjective scale).
%
% mos and variance will be linarly scaled such that
%     best is scaled to zero (i.e., the best subjective rating)
%     worst is scaled to one (i.e., the worst subjective rating)
%
% order is the order of the polynomial fit used to map the objective
% data to the scaled subjective data (e.g., order = 1 is a linear fit).
%


% Number of src x hrc combinations
num_comb = size(data_in,1);

% Pick off the vectors we will use from data_in
vqm = data_in(:,3);
num_view = data_in(:,4);
mos = data_in(:,5);
variance = data_in(:,6);

% Scale the subjective data for [0,1]
mos = (mos-best)./(worst-best);
variance = variance./((worst-best)^2);

% Use long format for more decimal places in printouts
format('long');

% Fit the objective data to the scaled subjective data.
% Following code implements monotonic polynomial fitting using
% optimization toolbox routine lsqlin.
%
% Create x and dx arrays.  For the dx slope array (holds the
% derivatives of mos with respect to vqm), the vqm_sign specifies
% the direction of the slope that must not change over the vqm range.
```

```matlab
x = ones(num_comb,1);
dx = zeros(num_comb,1);
for col = 1:order
    x = [x vqm.^col];
    dx = [dx col*vqm.^(col-1)];
end
% The lsqlin routine uses <= inequalities.  Thus, if vqm_sign is -1
% (negative slope), we are correct but if vqm_sign is +1 (positive
% slope), we must multiple each side by -1.
if (vqm_sign == 1)
    dx = -1*dx;
end
fit = lsqlin(x,mos,dx,zeros(num_comb,1));
fit = flipud(fit)' % organize this fit the same as what is output
                   % by polyfit


% vqm fitted to mos
vqm_hat = polyval(fit,vqm);


% Perform the vqm RMSE calculation using vqm_hat.
vqm_rmse = (sum((vqm_hat-mos).^2)/(num_comb-(order+1)))^0.5

% Perform the vqm resolution measurement on both vqm and vqm_hat.
vqm_pairs = repmat(vqm,1,num_comb)-repmat(vqm',num_comb,1);
vqm_hat_pairs=repmat(vqm_hat,1,num_comb)-repmat(vqm_hat',num_comb,1);
mos_pairs = repmat(mos,1,num_comb)-repmat(mos',num_comb,1);
stand_err_diff = sqrt(repmat(variance./num_view,1,num_comb)+ ...
    repmat((variance./num_view)',num_comb,1));
z_pairs = mos_pairs./stand_err_diff;

% Include everything above the diagonal.
delta_vqm = [];
delta_vqm_hat = [];
z = [];
for col = 2:num_comb
    delta_vqm = [delta_vqm; vqm_pairs(1:col-1,col)];
    delta_vqm_hat = [delta_vqm_hat; vqm_hat_pairs(1:col-1,col)];
    z = [z; z_pairs(1:col-1,col)];
end

% Switch on z and delta_vqm for negative delta_vqm
z_vqm = z;
```

```
negs_vqm = find(delta_vqm < 0);
delta_vqm(negs_vqm) = -delta_vqm(negs_vqm);
z_vqm(negs_vqm) = -z_vqm(negs_vqm);

z_vqm_hat = z;
negs_vqm_hat = find(delta_vqm_hat <0);
delta_vqm_hat(negs_vqm_hat) = -delta_vqm_hat(negs_vqm_hat);
z_vqm_hat(negs_vqm_hat) = -z_vqm_hat(negs_vqm_hat);

% Plot scatter plot of z_vqm versus delta_vqm in figure 1.
% Plot scatter plot of z_vqm_hat versus delta_vqm_hat in figure 2.
figure(1) plot(delta_vqm,z_vqm,'.','markersize',1)
set(gca,'LineWidth',1) set(gca,'FontName','Arial')
set(gca,'fontsize',12) xlabel('Delta VQM') ylabel('Subjective Z
Score') grid on print -dpng figure1

figure(2) plot(delta_vqm_hat,z_vqm_hat,'.','markersize',1)
set(gca,'LineWidth',1) set(gca,'FontName','Arial')
set(gca,'fontsize',12) xlabel('Delta VQM Hat') ylabel('Subjective Z
Score') grid on print -dpng figure2

% Plot average confidence that vqm(2) is worse than vqm(1) in figure 3.
% Plot average confidence that vqm_hat(2) is worse than vqm_hat(1) in
% figure 4.  These are the resolving power plots.
%
% One control parameter for delta_vqm resolution plot; number of vqm
% bins equally spaced from min(delta_vqm) to max(delta_vqm).
% Sliding neighbood filter with 50% overlap means that there will
% actually be vqm_bins*2-1 points on the delta_vqm resolution plot.
cdf_z_vqm = .5+erf(z_vqm/sqrt(2))/2;
cdf_z_vqm_hat = .5+erf(z_vqm_hat/sqrt(2))/2;
vqm_bins = 10; % How many bins to divide full vqm range for local
               % averaging
vqm_low = min(delta_vqm); % lower limit on delta_vqm
vqm_high = max(delta_vqm); % upper limit on delta_vqm
vqm_step = (vqm_high-vqm_low)/vqm_bins; % size of delta_vqm bins

vqm_hat_low = min(delta_vqm_hat);
vqm_hat_high = max(delta_vqm_hat);
vqm_hat_step = (vqm_hat_high-vqm_hat_low)/vqm_bins;

% lower, upper, and center bin locations
low_limits = [vqm_low:vqm_step/2:vqm_high-vqm_step];
```

```
high_limits = [vqm_low+vqm_step:vqm_step/2:vqm_high];
centers = [vqm_low+vqm_step/2:vqm_step/2:vqm_high-vqm_step/2];

hat_low_limits=[vqm_hat_low:vqm_hat_step/2:vqm_hat_high-vqm_hat_step];
hat_high_limits=[vqm_hat_low+vqm_hat_step:vqm_hat_step/2:vqm_hat_high];
hat_centers = [vqm_hat_low+vqm_hat_step/2:vqm_hat_step/2: ...
        vqm_hat_high-vqm_hat_step/2];

mean_cdf_z_vqm = zeros(1,2*vqm_bins-1);
mean_cdf_z_vqm_hat = zeros(1,2*vqm_bins-1);
for i=1:2*vqm_bins-1
    in_bin = find(low_limits(i)<=delta_vqm&delta_vqm<high_limits(i));
    hat_in_bin = find(hat_low_limits(i) <= delta_vqm_hat & ...
        delta_vqm_hat < hat_high_limits(i));
    mean_cdf_z_vqm(i) = mean(cdf_z_vqm(in_bin));
    mean_cdf_z_vqm_hat(i) = mean(cdf_z_vqm_hat(hat_in_bin));
end

% The x-axis is vqm(2)-vqm(1).  For figure 3 (the vqm plot), if
% vqm_sign is 1, then the Y-axis is the average confidence that vqm(2)
% is worse than vqm(1).  On the other hand, if vqm_sign is -1, then
% the Y-axis is the average confidence that vqm(1) is worse than
% vqm(2). Figure 4 is the plot for vqm_hat, and since it always has
% the same sign as mos, the Y-axis is always the average confidence
% that vqm_hat(2) is worse than vqm_hat(1).
if (vqm_sign == 1)
    figure(3)
    % VQM resolving power
    plot(centers,mean_cdf_z_vqm)
    grid
    set(gca,'LineWidth',1)
    set(gca,'FontName','Arial')
    set(gca,'fontsize',11)
    xlabel('VQM(2)-VQM(1)')
    ylabel('Average Confidence VQM(2) is worse than VQM(1)')
    print -dpng figure3
else
    figure(3)
    % VQM resolving power
    plot(centers,1-mean_cdf_z_vqm)
    grid
    set(gca,'LineWidth',1)
    set(gca,'FontName','Arial')
```

```
        set(gca,'fontsize',11)
        xlabel('VQM(2)-VQM(1)')
        ylabel('Average Confidence VQM(1) is worse than VQM(2)')
        print -dpng figure3
end

figure(4)
% VQM Hat resolving power.
plot(hat_centers,mean_cdf_z_vqm_hat) grid set(gca,'LineWidth',1)
set(gca,'FontName','Arial') set(gca,'fontsize',11) xlabel('VQM
Hat(2) - VQM Hat(1)') ylabel('Average Confidence VQM Hat(2) is worse
than VQM Hat(1)') print -dpng figure4

% This portion of the code calculates and plots the relative
% frequencies of three types of classification errors.  A
% classification error is made when the subjective test and
% the VQM lead to different conclusions on a pair of data points.
%
% Background:  For any subjective test, one must set a threshold
% that will determine when two results are statistically equivalent,
% and when they are statistically distinguishable.  Then for each
% pair of data points (A,B), the subjective test can yield one of
% three possible outcomes: (1) A better than B, (2) A same as B,
% and (3) A worse than B.
%
% If we define a similar threshold for VQM values, we have the same
% situation.  For each pair of data points, VQM can yield one of
% three possible outcomes: (1) A better than B, (2) A same as B,
% and (3) A worse than B. Since each pair of data points undergoes
% three-way classification by the subjective test and three-way
% classification by the VQM, there are nine possible outcomes.  For
% three of these outcomes, the subjective test and the VQM agree.
% If we take the subjective test to be correct by definition, and
% the VQM to be under test, then we say that for these three
% outcomes, the VQM is correct.  In two other cases the VQM has
% committed the "false-tie" error (subjective test says A better
% than B, or A worse than B, but VQM says A same as B).  In two other
% cases the VQM has committed the "false differentiation" error
% (subjective test says A same as B, but VQM says A better than
% B, or A worse than B.)  Finally, there are two cases where the
% VQM has performed a false ranking (subjective test says A better
% than B, or A worse than B, but VQM says the opposite.)  Thus, all
% nine outcomes are accounted for.  Note that a three by three grid
```

```
% in (delta_vqm, subjective Z score) space describing the above
% could be drawn.
%
% In the code below, the threshold used for the subjective test is
% subj_th. The threshold used for the delta VQM is vqm_th and this
% is left as a free parameter.  The code plots the frequency of
% occurrence for the three different kinds of errors and for no error
% vs. vqm_th.  An optimal value of vqm_th might be one that maximizes
% the frequency of occurrence of no error, or one that minimizes a
% cost-weighted sum of the errors.  Note that in general, it is
% likely that false ties will be the least offensive error, false
% differentiations will be more offensive, and false rankings will be
% the worst sort of error.
%
% For more details, see S. Voran, "Techniques for Comparing Objective
% and Subjective Speech Quality Tests," Proceedings of the Speech
% Quality Assessment Workshop, Bochum, Germany, November 1994.
%
% Note: The nine outcomes and the three by three grid in (delta_vqm,
% subjective Z score) space is the most natural way to describe this
% analysis.  This assumes bipolar values for delta_vqm.  But the code
% has already taken the absolute value of delta_vqm (and replaced Z
% with -Z for all points with negative values of delta_vqm). This does
% not change the math, but the more natural description of the
% situation is now 6 outcomes and a 2 by 3 grid.  Two correct outcomes
% (A better than B and A worse than B) have been folded on top of each
% other.  There are still two false tie outcomes, but only one false
% differentiation outcome and one false ranking outcome.

% Figure 5 is the plot for vqm and figure 6 is the plot for vqm_hat.
subj_th = 1.6;  % 95 percent confidence
num_th = 50;  % number of delta_vqm thresholds to examine
vqm_th_list = [vqm_low:(vqm_high-vqm_low)/num_th:vqm_high];
vqm_hat_th_list = [vqm_hat_low:(vqm_hat_high-vqm_hat_low)/num_th: ...
        vqm_hat_high];
rel_freqs = zeros(vqm_bins+1,4);
rel_hat_freqs = zeros(vqm_bins+1,4);
for i = 1:num_th+1
    vqm_th = vqm_th_list(i);
    vqm_hat_th = vqm_hat_th_list(i);
    % Number of data points in the false tie region
    rel_freqs(i,1) = length(find((delta_vqm < vqm_th) & ...
        (subj_th <= abs(z_vqm))));
```

```
    rel_hat_freqs(i,1) = length(find((delta_vqm_hat < vqm_hat_th) & ...
        (subj_th <= abs(z_vqm_hat))));
    % Number of data points in the false differentiation region
    rel_freqs(i,2) = length(find((vqm_th <= delta_vqm) & ...
        (abs(z_vqm) < subj_th)));
    rel_hat_freqs(i,2) = length(find((vqm_hat_th<=delta_vqm_hat) & ...
        (abs(z_vqm_hat) < subj_th)));
    % Number of data points in the false ranking region
    if (vqm_sign == 1)
        rel_freqs(i,3) = length(find((vqm_th <= delta_vqm) & ...
            (z_vqm <= -subj_th)));
    else
        rel_freqs(i,3) = length(find((vqm_th <= delta_vqm) & ...
            (z_vqm >= subj_th)));
    end
    rel_hat_freqs(i,3) = length(find((vqm_hat_th<=delta_vqm_hat) & ...
        (z_vqm_hat <= -subj_th)));
end
% Normalize counts by total number of points to get relative
% frequencies
rel_freqs = rel_freqs/length(z_vqm);
rel_hat_freqs = rel_hat_freqs/length(z_vqm_hat);
% Calculate relative frequency of correctness
rel_freqs(:,4) = (1-sum(rel_freqs(:,1:3)'))';
rel_hat_freqs(:,4) = (1-sum(rel_hat_freqs(:,1:3)'))';

% Figure 5 is plot for vqm and figure 6 is plot for vqm_hat.
figure(5)
% VQM Subjective Classification Errors
plot(vqm_th_list,rel_freqs(:,1),'m-.',vqm_th_list,rel_freqs(:,2),...
    'r:', vqm_th_list,rel_freqs(:,3),'k-',vqm_th_list,...
    rel_freqs(:,4),'b--');
grid\\
set(gca,'LineWidth',1) set(gca,'FontName','Arial')
set(gca,'fontsize',12) xlabel('Delta VQM Significance Threshold')
ylabel('Relative Frequencies') legend('False Tie','False
Differentiation','False Ranking',...
    'Correct Decision')
print -dpng figure5

figure(6)
% VQM Hat Subjective Classification Errors
plot(vqm_hat_th_list,rel_hat_freqs(:,1),'m-.', ...
```

```
          vqm_hat_th_list,rel_hat_freqs(:,2),'r:', ...
          vqm_hat_th_list,rel_hat_freqs(:,3),'k-', ...
          vqm_hat_th_list,rel_hat_freqs(:,4),'b--');
grid set(gca,'LineWidth',1) set(gca,'FontName','Arial')
set(gca,'fontsize',12) xlabel('Delta VQM Hat Significance
Threshold') ylabel('Relative Frequencies') legend('False Tie','False
Differentiation','False Ranking',...
          'Correct Decision')
print -dpng figure6
```

A.5   Supplementary Analyses

Analysis of variance (ANOVA) has been added to those mandated by the VQEG Test Plan.

1. An ANOVA of the subjective rating data alone shows the amount of noise in the data, and whether the HRCs and SRCs had an effect on the subjective responses (as they should).

2. Each SRC can be characterized by the amount of variance in subjective judgment across HRCs, which measures an SRC's ability to discriminate among HRCs. (The famous Mobile and Calendar discriminates among HRCs.)

3. An "optimal model" of the subjective data can be defined to provide a quantitative upper limit on the fit that any objective model could achieve with the given subjective data. The optimal model defines what a "good fit" is.

ANOVA [Mil97, Weia]

"Analysis of Variance" or ANOVA is a statistical test for heterogeneity of means by analysis of group variances. ANOVA is implemented as $ANOVA[data]$ in the Mathematica package Statistics'ANOVA' (available starting with Version 4.2 and which can be loaded using the command: $<<$Statistics').

To apply the test, assume random sampling of a variate Y with equal variances, independent errors, and a normal distribution. Let N be the number of replicates (sets of identical observations) within each of K factor levels (treatment groups), and y_{ij} be the jth observation within factor level i. Also assume that the ANOVA is "balanced" by restricting N to be the same for each factor level.

Table A.1: SS and MS computations.

category	^{o} freedom	SS	mean squared	F-ratio
model	$K - 1$	SSA	$MSA \equiv \frac{SSA}{K-1}$	$\frac{MSA}{MSE}$
error	$K(N - 1)$	SSE	$MSE \equiv \frac{SSE}{K(N-1)}$	
total	$KN - 1$	SST	$MST \equiv \frac{SST}{KN-1}$	

The total, treatment, and error sums of squares are defined, respectively, as

$$SST \equiv \sum_{i=1}^{K}\sum_{j=1}^{N}(y_{ij} - \mu_{\mu_y})^2 \tag{A.38}$$

$$= \sum_{i=1}^{K}\sum_{j=1}^{N} y_{ij}^2 - \frac{(\sum_{i=1}^{K}\sum_{j=1}^{N} y_{ij})^2}{KN} \tag{A.39}$$

$$SSA \equiv \frac{1}{N}\sum_{i=1}^{K}\left(\sum_{j=1}^{N} y_{ij}\right)^2 - \frac{1}{KN}\left(\sum_{i=1}^{K}\sum_{j=1}^{N} y_{ij}\right)^2 \tag{A.40}$$

$$SSE \equiv \sum_{i=1}^{K}\sum_{j=1}^{N}(y_{ij} - \mu_{y_i})^2 \tag{A.41}$$

$$= SST - SSA, \tag{A.42}$$

where μ_{y_i} is the mean of observations within factor level i, and μ_{μ_y} is the "group" mean (i.e., the mean of means). Compute the entries in Table A.1, obtaining the P-value corresponding to the calculated F-ratio of the mean squared values

$$F = \frac{MSA}{MSE}. \tag{A.43}$$

If the P-value is less than .05, reject the null hypothesis that all means are the same for the different groups.

Comparing residual variances from ANOVAs of the VQMs is an alternative to comparing correlations of VQMs with the subjective data that may yield finer discriminations among the VQMs. A supplementary metric (**Metric 7**) was added to the analyses. Although this metric was not mandated by the VQEG test plan, it was included because it was deemed to be a more informative measure of the prediction accuracy of a model. The metric is an F-test of the residual error of a model versus the residual error of an "optimal model." The metric is explained in more detail in Section 4.6 of the Full Reference VQEG report. A general description is provided by [NIS] and given as follows.

An F-test is used to test if the standard deviations of two populations are equal. This test can be a two-tailed test or a one-tailed test. The two-tailed version tests against the alternative that the standard deviations are not equal. The one-tailed version tests only in one direction, i.e., the standard deviation from the first population is either greater than or less than (but not both) the second population standard deviation. The choice is determined by the problem. For example, if we are testing a new process, we may only be interested in knowing if the new process is less variable than the old process.

Definition: The F hypothesis test is defined as

$$H_0: \quad \sigma_1 = \sigma_2$$
$$H_a: \quad \sigma_1 < \sigma_2 \qquad \text{for a lower one-tailed test}$$
$$\sigma_1 > \sigma_2 \qquad \text{for an upper one-tailed test}$$
$$\sigma_1 \neq \sigma_2 \qquad \text{for a two-tailed test}$$

Test Statistic: $F = \frac{s_1^2}{s_2^2}$ where s_1^2 and s_2^2 are the sample variances. The more this ratio deviates from 1, the stronger is the evidence for unequal population variances.

Significance Level: α

Critical Region: The hypothesis that the two standard deviations are equal is rejected if

$$F > F_{(\alpha, N1-1, N2-1)} \qquad \text{for an upper one-tailed test}$$
$$F < F_{(1-\alpha, N1-1, N2-1)} \qquad \text{for a lower one-tailed test}$$
$$F < F_{(1-\alpha/2, N1-1, N2-1)} \qquad \text{for a two-tailed test}$$
$$\text{or}$$
$$F > F_{(\alpha/2, N1-1, N2-1)}$$

where $F > F_{(\alpha, k-1, N-k)}$ is the critical value of the F distribution with v_1 and v_2 degrees of freedom and a significance level of α.

In the above formulae for the critical regions, the convention is followed [NIS] that F_α is the upper critical value from the F distribution and $F_{1-\alpha}$ is the lower critical value from the F distribution. Note that this is the opposite of the designation used by some texts and software programs. In particular, Dataplot uses the opposite convention.

References

[Act66] F. S. Acton. *Analysis of Straight-Line Data*. New York: Dover, 1966.

[ATI03] ATIS T1.TR.72-2003. Methodological Framework for Specifying Accuracy and Cross-Calibration of Video Quality Metrics. Technical report, Alliance for Telecommunications Industry Solutions (ATIS), Dec. 2003.

[Bri04] M. Brill et al. Accuracy and Cross-Calibration of Video Quality Metrics: New Methods from ATIS/T1A1. *Signal Processing: Image Communication*, 19:101–107, Feb. 2004.

[Bul03] P. S. Bullen. The Power Means. Ch. 3. In *Handbook of Means and Their Inequalities.* Dordrecht, Netherlands: Kluwer, 2003.

[Edw76] A. L. Edwards. The Correlation Coefficient. Ch. 4. In *An Introduction to Linear Regression and Correlation*, 33–46. San Francisco, CA: W. H. Freeman, 1976.

[GS93] L. Gonick and W. Smith. Regression. Ch. 11. In *The Cartoon Guide to Statistics*, 187–210. New York: Harper Perennial, 1993.

[HC95] R. V. Hogg and A. T. Craig. *Introduction to Mathematical Statistics.* New York: MacMillan, 5 ed., 1995.

[HN85] L. Hoehn and I. Niven. Averages on the Move. *Mathematics Magazine*, 58:151–156, 1985.

[ITU04] ITU-T Recommendation J.149. Methodological Framework for Specifying Accuracy and Cross-Calibration of Video Quality Metrics (VQM). Technical report, ITU-Telecommunication Standardization Sector, Mar. 2004.

[KK62a] J. F. Kenney and E. S. Keeping. Linear Regression and Correlation. Ch. 15. In *Mathematics of Statistics, Pt. 1*, 252–285. Princeton, NJ: Van Nostrand, 3 ed., 1962.

[KK62b] J. F. Kenney and E. S. Keeping. Root Mean Square. §4.15. In *Mathematics of Statistics, Pt. 1*, 59–60. Princeton, NJ: Van Nostrand, 3 ed., 1962.

[LD98] E. L. Lehmann and H. J. M. D'Abrera. *Nonparametrics: Statistical Methods Based on Ranks.* Englewood Cliffs, NJ: Prentice–Hall, revised ed., 1998.

[Mil97] R. G. Miller. *Beyond ANOVA: Basics of Applied Statistics.* Boca Raton, FL: Chapman & Hall, 1997.

[NIS] NIST/SEMATECH. e-Handbook of Statistical Methods. NIST/SEMATECH homepage, Accessible at *http://www.itl.nist.gov/div898/handbook/eda/section3/eda359.htm.*

[PFTV92] W. H. Press, B. P. Flannery, S. A. Teukolsky, and W. T. Vetterling. Linear Correlation. §14.5. In *Numerical Recipes in FORTRAN: The Art of Scientific Computing.* Cambridge, England: Cambridge University Press, 2 ed., 1992.

[Weia] E. W. Weisstein. ANOVA. MathWorld homepage, Accessible at *http://mathworld.wolfram.com/ANOVA.html.*

[Weib] E. W. Weisstein. Root-mean-square. MathWorld homepage, *Accessible at http://mathworld.wolfram.com/Root-Mean-Square.html.*

[Weic] E. W. Weisstein. Spearman rank correlation coefficient. MathWorld homepage, Accessible at *http://mathworld.wolfram.com/SpearmanRankCorrelationCoefficient.html.*

[WR67] E. T. Whittaker and G. Robinson. The Coefficient of Correlation for Frequency Distributions Which Are Not Normal. §166. In *The Calculus of Observations: A Treatise on Numerical Mathematics*, 334–336. New York: Dover, 4 ed., 1967.

Index